Dynamics of Visual Motion Processing

Uwe J. Ilg · Guillaume S. Masson
Editors

Dynamics of Visual Motion Processing

Neuronal, Behavioral, and Computational Approaches

Springer

Editors
Uwe J. Ilg
Department of Cognitive Neurology
Hertie-Institute of Clinical Brain Research
University of Tuebingen
Otfried-Mueller Str 27
Tuebingen 72076, Germany
Uwe.Ilg@uni-tuebingen.de

Guillaume S. Masson
Team Dynamics of Visual Perception
and Action
Institut de Neurosciences Cognitives
de la Méditerranée
CNRS & Université de la Méditerranée
31 Chemin Joseph Aiguier
13402 Marseille, France
guillaume.masson@incm.cnrs-mrs.fr

ISBN 978-1-4899-8995-6 ISBN 978-1-4419-0781-3 (eBook)
DOI 10.1007/978-1-4419-0781-3
Springer New York Dordrecht Heidelberg London

Cover illustration: Dynamical extraction and diffusion 2D motion using a particle filter. Blue and red color illustrate the direction of local motion at early and late time steps. See Perrinet & Masson, CoSyne Annual Meeting 2009. Courtesy of Laurent U. Perrinet.

Printed on acid-free paper

Springer is part of Springer Science+Business Media (www.springer.com)

Preface

Biological motion is an essential piece of sensory information for living organism and therefore motion processing units, from simple elementary motion detectors to dedicated motion sensitive cortical areas, have been identified over a broad spectrum of animals. Biological visual motion systems are among the ones having been the most scrutinized at many different levels from microcircuits to perception (see Born and Bradley 2005; Bartels et al. 2008; Callaway 2005; Sincich and Horton 2005; Demb 2007; Britten 2008; Bradley and Goyal 2008; Kourtzi et al. 2008; Orban 2008 for recent reviews). In parallel, since the early work of Reichardt (1961), theoretical approaches of motion detection have been always tightly linked with experimental work so that nowadays, most experiments are conducted within rather well-defined theoretical frameworks (e.g. Carandini et al. 2005). Visual motion has thus become a representative of system neurosciences where similar approaches can be applied across very different levels of brain organization. In particular, neuronal activity at both single-cell and population levels can be accurately linked to simple action system driven by visual motion such as tracking eye movements (Lisberger et al. 1987) as well as motion perception (Parker and Newsome 1998). This integrative approach is rooted on decades of psychophysics that have explored human motion perception (Nakayama 1985; Vaina 1998; Snowden and Freeman 2004; Lu and Sperling 2001). Visual psychophysics provides all of us with a large class of calibrated motion stimuli that can be used to dissect out the different aspects of motion integration and segmentation as needed to accurately measure the velocity of an object that is, the direction and speed of its movement. We decided to open this book with a review paper describing what are the different classes of visual stimuli and what aspects of biological motion processing each of them can unveil. Focusing of low-level motion, Lorenceau presents in great details the different elements of this artillery and how they can be used at both behavioral and neurophysiological levels. By doing so, he set the stage over which most of the work presented inside this book will take place. As for the other chapters, corresponding movies can be found in the DVD joined to the book. However, Lorenceau also stresses out that a motion perception most often involves a tight link between form and motion cues. Such form–motion interactions will be illustrated by other contributions, further demonstrating that biological motion processing escapes the strict modular approach and call for a more integrative view as needed to understand the root of the problem: how to measure the motion of an object, usually represented as a visual surface, singled out from its complex environment. The following chapters will survey how this can be performed at cellular and network levels, with either static or moving eyes.

Dynamics of Neural Mechanisms

Surprisingly, several key aspects of motion perception have not been emphasized over the years. First, although a few psychophysical studies had pointed out that perceived motion undergoes complex time course when human subjects are presented with ambiguous motion signals (e.g. Yo and Wilson 1992; Castet et al. 1993; Lorenceau et al. 1993), it is only very recently that temporal dynamics of motion processing has received attention from physiologists. Before the pioneering work of Pack and colleagues, neurons were classified between those who solve the aperture problem and those who do not. This selection was based on the steady-state properties of their direction selectivity tuning when presented with bars, gratings, or plaid patterns (Movshon et al. 1985). Pack and Born (2001) presented MT neurons with sets of tilted bars, the neuronal counterpart of the elongated moving bars used in psychophysical experiments, and analyzed the time course of direction selectivity of the single-unit responses. They found that such a basic response property of MT neuron is indeed not static. Instead, early part of their tuning showed interactions between direction and orientation while ~100 ms after response onset, optimal direction became independent of lines orientation. Several studies, largely summarized here in the chapters by Pack et al. and Smith et al., have looked at the dynamics of direction selectivity in macaque area MT in response to various 2D motions such as plaid patterns, barber poles, or lines. Although there is a common agreement on the similar timing characteristics across motion stimuli (see chapter by Smith et al.), the origin of such neuronal dynamics is still highly controversial, carrying on a long debate about which, and how, local features are extracted from the image flow. Born and coworkers favor an explanation based on feature-extraction mechanisms such as end-stopped cells found in area V1 (Hubel and Wiesel 1968; Pack et al. 2003). On the other hand, Smith and coworkers argue for a filter-based approach where global motion is computed by merging excitatory and inhibitory inputs from different spatio-temporal channels (see Rust et al. 2006). Within these two frameworks, the dynamics can be seen either as a result of a delayed feature-extraction mechanisms, as the by-product of the different signal strength between channels or by the time course of contextual modulation such as implemented by center-surround interactions or recurrent circuits. The book offers the opportunity for these different views to be presented back to back.

Motion information is extracted locally, but there is many evidence that the brain pools information to solve the aperture problem, to improve signal-to-noise ratio or to normalize inputs across the image, to take a few examples of motion integration. Since all these different aspects involved the diffusion of information between neighboring neurons, there is an urgent need to explore the neural dynamics at population level.

Frégnac and coworkers introduce the concepts and techniques used to investigate the relationships between fundamental properties of individual neurons such as orientation- and direction-selective cells in primary visual cortex and the dynamics of their surrounding network. They point out that descriptive tuning functions in fact reflect the very large diversity of inputs that a single neuron would receive through feed-forward, lateral, and recurrent connectivity. This message is to keep in mind in the design of detailed biophysical models at both cellular and network levels. It remains coherent with the current view that direction selectivity emerges from the convergence of many different feed-forward inputs (both excitatory and inhibitory) covering a very broad range of the spatiotemporal spectrum in Fourier space (see Rust et al. 2006; Lennie and Movshon 2005). However, the evidence gathered by intracellular recordings that responses dynamics of V1 neurons reflect non-iso-oriented inputs (Monier et al. 2003) from distant part of the cortex (Bringuier et al. 1999) urge us to take into account the role of intra- and intercortical connections. The fact that they all have different timing shall help us in constraining dynamical models of motion integration.

Linking population dynamics and integrative properties of individual neurons will be most certainly a future challenge in sensory neuroscience. Visual motion once again shall offer an excellent approach. Jancke, Chavane, and Grinvald provide one very attractive insight into this perspective. Using different and complementary techniques such as voltage-sensitive dye optical imaging and population reconstruction from extracellular recordings, they propose a fresh look at how motion information is represented. In particular, their approach stresses one point often ignored in most electrophysiological, and psychophysical, studies. Motion is primarily a displacement in the visual space and therefore a moving object will elicit a traveling wave along the cortical representation of its trajectory. Moreover, linear and nonlinear interactions along such cortical trajectories can be identified in cat area V1 (Jancke et al. 2004). Most certainly, future work will be able to relate such population dynamics to single-unit activity within direct projection areas such as V2 or MT as well as with perceptual performance in primates (Chen et al. 2006).

Overall, looking at the temporal dynamics of contextual biological motion processing, as well as for other elementary aspects of image features extractions such as orientation, texture, and illusory contours has reinvigorated the investigations on the underpinning neural mechanisms. The results gathered might turned out to be important to decipher which theoretical approach is more closely related to cortical computation. They might also force us to finally take into account the different connectivity rules, operating at different spatial and temporal scales, which are important to compute global object motion.

Visual Motion and Eye Movements

Measuring speed and direction of a moving object is an essential step for many sensorimotor transformations, in particular when controlling eye movements. The impact of low-to-high level motion processing onto the dynamics of oculomotor behavior is reviewed in several chapters. Sheliga and Miles summarize their seminal work in elucidating the basic properties of motion detection in the context of triggering reflexive eye movements at ultrashort latency. Their work illustrates how much can be learned about the spatial and temporal characteristics of the earliest, preattentive stage of local motion extraction when using very accurate behavioral probes. From this, as well as from the work of other groups, it becomes possible to sketch a detailed model of early, and fast, motion processing that incorporates many aspects investigated previously at psychophysical and physiological levels: how is motion information extracted by luminance-based motion detectors, how are their activity normalized across directions and so on and so forth. More than simply confirming what was learned from other approaches, the experiments conducted on ocular following responses unveil functional consequences of such linear and non-linear processing such as automatic motion segmentation and integration (see Miles et al. 2004; Masson 2004). If tracking eye movements are primarily driven by luminance-based local motion detection, this so-called first-order motion mechanisms might not be the only one contributing to a nearly perfect pursuit performance under a wide range of conditions. Other types of motion information can be extracted under constant luminance conditions, either at preattentive or at attentive stages. System view of the primate motion system postulates the existence of three different motion systems, called first order, second order, and third order (see Lu and Sperling 2001 for a review). The exact contribution of second- and third-order motion information to perceptual performance is still a matter of debate and it is unclear where and how they are computed by the primate brain. Chapter by Ilg and Churan reviews the existing evidence, supporting the idea that second-order motion is indeed extracted within posterior parietal areas. The authors point out, however, that investigating second-order motion, as well as pattern motion, had defeated the simplistic view that global motion is computed once for all in area MT and therefore that area MT must be seen as the key, if not unique area responsible for motion perception in both human and nonhuman primates (see Ilg 2008 for a review).

Once motion is locally extracted, several processing steps are still necessary to reconstruct speed and direction of the object to be pursued. Chapters presenting new results on motion integration, obtained at both psychophysical and physiological levels, have introduced the idea that the integration stage presents complex dynamics. This approach is further extended in the chapter by Masson and colleagues, showing such dynamics of motion integration can have a major impact on how the brain controls action. Taking advantage of the fast visuomotor transformation underlying that pursuit eye movements as well as their smooth acceleration, the oculomotor system can trigger tracking responses based only on the coarse estimate of motion direction that arises from the feed-forward motion pathway but then gradually correct the pursuit direction by taking into account features motion extracted at a finer scale. Thus, time course is closely related to the temporal dynamics of motion integration that we have discussed above. In return, this work stresses the fact that eye movements are an exquisite tool to probe the properties of early motion processing stages, since initial eye acceleration reflects visual velocity signals encoded at the level of macaque areas MT and MST (Krauzlis 2004; Masson 2004).

However, it is well known since the early 1970s that pursuit responses depend on both visual and nonvisual signals, the later being related to eye velocity memory (Yasui and Young 1975; Miles and Fuller 1975). Moreover, the perceived direction of oriented after-images presented during on-going pursuit is always biased toward the axis normal to the orientation of the flashed bars (Goltz et al. 2003). This intriguing result suggests first that the aperture problem contaminates egocentric motion and second that more is yet to learn about motion integration during eye movements (Murakami 2004). Indeed, motion integration tasks such as introduced by Lorenceau offer a great deal to investigate the link between perception and action, as well as the dependency or the lack of dependency of early visual stages upon cognitive factors such as prediction or anticipation. Masson and colleagues report results arguing for a mere independence between low-level vision and higher cognitive processing such as engaged in anticipating future motion events or predicting target trajectory. They suggest that low-level motion integration and spatial reconstruction of target motion acts more or less independently, as illustrated by the difference observed between neuronal responses in either areas MT or MSTl/FEF when using complex line drawing stimuli avoiding the center of the receptive field (Ilg and Thier 2003). These latter experiments suggest that pursuit-related neurons in the lateral part of macaque area MST (also called visual-tracking neurons) integrate both visual and nonvisual information (see Ilg 2008 for a review). Whether these neurons compute the motion-in-space of a pursued target (Newsome et al. 1988) or reflect the existence of a more abstract representation of the inferred motion already emerging at the level of area MT (Assad and Maunsell 1995; Schlack and Albright 2007) is still a matter of debate. Recording activities of MSTl neurons during tracking of different line-drawing objects is one piece of evidence. Furthermore, looking at the dynamics of direction selectivity using tilted bars that are transiently occluded (see Assad and Maunsell 1995 for a similar paradigm although with a simple spot) might also largely contribute to a better understanding about what and how information is represented at various stages along the motion pathway. Clearly, more investigations are needed about the dynamical interactions between posterior parietal and prefrontal cortical areas for motion integration in the context of pursuit eye movements, as well as perception (see Pasternak and Greenlee 2005). However, once again, these studies point out how using simple motion stimuli such as designed for psychophysics can highlight the mechanisms of sensorimotor transformation when the biological motion stage is not collapsed into a simple black box extracting retinal velocity in some unspecified way. Obviously, there is need for models of oculomotor behavior with a more complex front-end dynamics.

In the aforementioned chapters, motion is seen as the source of information for driving perception or simple actions such as tracking responses. Although active vision has been a very productive field of research trying to understand how visual information is actively extracted by means of our eye movements, much more attention has been paid to saccadic eye movements rather than smooth pursuit in this context (Findlay and Gilchrist 2003). Tracking an object aims at stabilizing its image onto the retina, but a mere consequence of the eyeball rotation is a steady continuous retinal sweep of the background image. Dozens of studies have been conducted to understand how such background motion can be either eliminated to perceive a stable world during tracking or on the contrary taken into account to compute object motion in a spatial frame of reference (see Abadi and Kulikowski 2008). Hafed and Krauzlis take a different approach, trying to demonstrate that smooth eye movements can be useful to resolve perceptual ambiguities. This approach is rooted on the motion stimuli, and psychophysical paradigms described by Lorenceau but offer a fresh view of the fascinating problem of perception–action coupling. Their experimental work, summarized in Chap. 9, shows that partially occluded objects can be perceived coherently thanks to the pattern of eye movements produced by human subjects. This seminal study opens the door to a closer examination to the interaction between perception and action using both well-defined behavior and calibrated tasks where retinal flows can be matched between pursuit and fixation condition.

Visual motion processing is not only related to the execution of pursuit eye movements. Both saccadic and pursuit eye movements introduce major changes in the retinal images. However, how motion perception and eye movements are coupled with respect to saccades has been a matter of intense debates over the last decades. One acute example is a phenomenon called "saccadic suppression" (see Ross et al. 2001). That visual perception is largely reduced during saccades is a well-documented phenomenon that everyone can experience everyday. Indeed, psychophysical studies have convincingly demonstrated that intrasaccadic detection thresholds are strongly deteriorated at the time of a saccade (e.g. Volkmann 1986; Burr et al. 1994). Several recent physiological studies have demonstrated that some, but not all direction-selective cells in macaque area MT are consistently inhibited during saccade. On the contrary, some cells also show a strong postsaccadic rebound of activity that could be correlated to the postsaccadic enhancement originally reported by Miles and colleagues when recording ocular following responses (Miles et al. 1986). In Chap. 8, Mike Ibbotson summarizes these studies and relates these saccade-related changes in activity at the level of area MT with the changes in perceptual performance described earlier in human subjects. However, the use of the term "suppression" has led to the stronger, but wrong, assumption that vision is prevented during saccades. Textbooks and nonspecialist review articles have even further cartooned this saying that the entire visual system, not only visual perception, is turned off during saccadic flight. The chapter by Castet offers a very helpful re-examination of the different perceptual changes that occur before, during, and after a saccade. He points out the difficulty in interpreting a wide diversity of perceptual phenomena within the unique, stringent hypothesis of an active (i.e. extraretinal) suppression or modulation of visual inflow at its earliest stages (Ross et al. 2001). One goal of this book was to publish back-to-back articles offering different, sometimes even opposite, standpoints onto a specific aspect of motion processing. The chapters by Ibbotson on one hand and Castet on the other hand give such an opportunity and remind us that solving controversies in neuroscience often needs first to (re)clarify key concepts as often popular ideas drift far away from the conclusions that were drawn from the original experimental results.

Modeling Visual Motion: From Natural Scenes Statistics to Motion Recognition

Listing the existing computational models of visual motion would probably take a couple of pages. Computer as well as biological vision researches have produced a huge number of models, based on many different theoretical approaches such as linear filtering, probabilistic inference, or dynamical systems. Several recent books are available from the library shelves (see Blake 1998; Paragios et al. 2005; Stocker 2004; Watanabe 1998 for a few recent examples) to explore these different aspects. There is, however, clearly the need for a more theoretical approach unifying all these computational efforts. Herein, we have preferred to highlight some key aspects of visual motion information processing. First, Dong summarizes the statistical approach trying to understand what is the critical information in sequences of natural images. Relating the window of visibility, and its neuronal counterpart defined as a set of optimal filters, to the statistics of still natural images has been an intensive area of research over the last two decades. The same approach is now conducted using movies of the image flow experienced by an observer moving in complex, natural environments. From these, Dong demonstrates that spatial–temporal contrast sensitivity of human observers is tuned to extract the most pertinent and reliable motion information that is mainly low temporal frequencies.

A second aspect of motion processing is integration, which involves diffusion of information over neighboring parts of the images to reconstruct the global motion of the object of interest and single it out from its surround. Grossberg summarizes the work conducted by his group in implementing dynamical models of motion segmentation and integration. His solution relies on a strong interplay between modules extracting either form (i.e. features) or motion. Diffusion of information is done by means of recurrent connectivity between areas working at different spatial scales. Once again, this class of model reminds us that motion pathways are highly recurrent and that we absolutely need to better understand how feed-forward and feedback flows of information interplay to solve problems such as motion binding. The model reviewed here sums up a decade of progressive improvement of the class of models developed by himself and his group. Clearly, this approach highlights the interest of computational principles that can be implemented by set of differential equations. The cost is then to overview the detailed connectivity rules corresponding to the actual cortical mechanisms. But we clearly need such a more generic approach, complementary to the more detailed, but also more focused, models proposed by others. Lastly, Grossberg introduces one new aspect of the dynamical approach. The brain takes decision about the incoming stimulus speed or direction. It model succeeds in simulating the time course of such a decision, as seen in parietal cortices (e.g. Britten et al. 1992; Shadlen and Newsome 2001; Hanks et al. 2006; Huk and Shadlen 2005) but also question on what information processing such decision is taken. This links to a rapidly growing field of research about sensory decisions along the motion pathways. Recent reviews about this topic can be found elsewhere (Shadlen 2002; Rorie and Newsome 2005; Gold and Shadlen 2007).

Motion is a useful source of information not only for controlling our basic actions but also to solve highest cognitive tasks such as face recognition (see Roark et al. 2003) or biological action recognition (see Blake and Shiffrar 2007). Understanding how biological motion is analyzed by dedicated brain loci within the superior temporal sulcus (STS) for instance has been the focus of a vast literature. However, biological motion stimuli carry information not only the type of action being executed but also more fine-grained, cognitive cues that are used for social interactions. Giese and coworkers detailed their recent modeling work asking how human emotions can be recognized for sequences of point-light walkers. Here again, a key point is to be able to extract remarkable features such as joint-angle trajectories using sparse feature learning. This approach not only defines a compact visual representation for complex information but depart from more classical models assuming that visual recognition involves the activation of motor representations. Instead, this model demonstrates that human subjects can nearly optimally use the visual information extracted from joint trajectories.

Features, trajectories, dynamic motion integration: these terms have been found in nearly all chapters of this book. By highlighting a few recent approaches, the contributors have shown how much an integrative approach can be useful to understand how the brain computes global motion of an object, being a simple line or a full body. Some of these issues still remain controversial and we want to thank the different contributors to have accepted that chapters with different views are presented back to back. We hope that our colleagues and their students will consider this book for what it was originally proposed: an incentive to bridge approaches across levels and models, using tasks and stimuli as an Ariadne's thread.

Marseille, France — Guillaume S. Masson
Tuebingen, Germany — Uwe J. Ilg

References

Abadi RV, Kulikowski JJ (2008) Perceptual stability: going with the flow. Perception 37(9):1461–1463

Assad JA, Maunsell JHR (1995) Neuronal correlates of inferred motion in primate posterior parietal cortex. Nature 373:518–521

Bartels A, Logothetis NK, Moutoussis K (2008) fMRI and its interpretations: an illustration on directional selectivity in area V5/MT. Trends Neurosci 31(9):444–453

Blake A (1998) Active contours: the application of techniques from graphics, vision, control theory and statistics to visual tracking of shapes in motion. Springer, Berlin

Blake R, Shiffrar, M (2007) Perception of human motion. Annu Rev Psychol 58:47–73

Bradley DC, Goyal MS (2008) Velocity computation in the primate visual system. Nat Rev Neurosci 9(9):686–695

Britten KH (2008) Mechanisms of self-motion perception. Annu Rev Neurosci 31:389–410

Britten KH, Shadlen MN, Newsome WT, Movshon JA (1992) The analysis of visual motion: a comparison of neuronal and psychophysical performance. J Neurosci 12(12):4745–4765

Born RT, Bradley DC (2005) Structure and function of visual area MT. Annu Rev Neurosci 28:157–189

Bringuier V, Chavane F, Glaeser L, Frégnac Y (1999) Horizontal propagation of visual activity in the synaptic integration field of area 17 neurons. Science 283(5402):695–699
Britten KH (2008) Mechanisms of self-motion perception. Annu Rev Neurosci 31:389–410
Burr DC, Morrone MC, Ross J (1994) Selective suppression of the magnocellular visaul pathway during saccadic eye movements. Nature 371:511–513
Callaway EM (2005) Neural substrates within primary visual cortex for interactions between parallel visual pathways. Prog Brain Res 149:59–64
Carandini M, Demb JB, Mante V, Tolhurst DJ, Dan Y, Olshausen BA, Gallant JL, Rust NC (2005) Do we know what the early visual system does? J Neurosci 25(46):10577–10597
Castet E, Lorenceau J, Shiffrar M, Bonnet C (1993) Perceived speed of moving lines depends on orientation, length, speed and luminance. Vision Res 33(14):1921–1936
Chen Y, Geilser WS, Seidemann E (2006) Optimal decoding of correlated neural population responses in the primate visual cortex. Nat Neurosci 9(11):1412–1420
Demb JB (2007) Cellular mechanisms for direction selectivity in the retina. Neuron 55(2): 179–286
Dobbins A, Zucker SW, Cynader MS (1987) Endstopped neurons in the visual cortex as a sub strate for calculating curvature. Nature 329:438–441
Findlay JM, Gilchrist ID (2003) Active vision. The psychology of looking and seeing. Oxford University Press, Oxford
Gold JI, Shadlen MN (2007) The neural basis of decision making. Annu Rev Neurosci 30:535–574
Goltz HC, DeSouza JF, Menon RS, Tweed DB, Vilis T (2003) Interactions of retinal image and eye velocity in motion perception. Neuron 39(3):569–579
Hanks TD, Ditterich J, Shadlen MN (2006) Microstimulation of macaque area LIP affects decision making in a motion discrimination task. Nat Neurosci 9(5):682–689
Hubel DH, Wiesel TN (1968) Receptive fields and functional architecture of monkey striate cortex. J Physiol (Lond) 195(1):215–243
Huk AC, Shadlen MN (2005) Neural activity in macaque parietal cortex reflects temporal integration of visual motion signals during perceptual decision making. J Neurosci 25(45): 10420–10436
Ilg UJ (2008) The role of areas MT and MST in coding of visual motion underlying the execution of smooth pursuit. Vision Res 48(20):2062–2069
Ilg UJ, Thier P (2003) Visual tracking neurons in primate area MST are activated by smooth-pursuit eye movements of an "imaginary" target. J Neurophysiol 90(3):1489–1502
Jancke D, Chavane F, Naaman S, Grinvald A (2004) Imaging cortical correlates of illusion in early visual cortex. Nature 428(6981):423–426
Kourtzi Z, Krekelberg B, Van Wezel RJ (2008) Linking form and motion in the primate brain. Trends Cogn Sci 12(6):230–236
Krauzlis RJ (2004) Recasting the smooth pursuit eye movement system. J Neurophysiol 91(2):591–603
Lennie P, Movshon, JA (2005) Coding of color and form in the geniculostriate visual pathway. J Opt Soc Am A 22(10):2013–2033
Lisberger SG, Morris EJ, Tychsen L (1987) Visual motion processing and sensory-motor integration for smooth pursuit eye movements. Annu Rev Neurosci 10:97–129
Lorenceau J, Shiffrar M, Wells N, Castet E (1993) Different motion sensitive units are involved in recovering the direction of moving lines. Vision Res 33(9):1207–1217
Lu ZL, Sperling G (2001) Three-systems theory of human visual motion perception: review and update. J Opt Soc Am A 18(9):2331–2370
Masson GS (2004) From 1D to 2D via 3D: dynamics of surface motion segmentation for ocular tracking in primates. J Physiol (Paris) 1–3:35–52
Miles FA, Fuller JH (1975) Visual tracking and the primate flocculus. Science 189:1000–1002
Miles FA, Kawano K, Optican LM (1986) Short-latency ocular following responses of monkey. I. Dependence on temporospatial properties of visual input. J Neurophysiol 56(5):1321–1354

Miles FA, Busettini C, Masson GS, Yang D-Y (2004) Short-latency eye movements: evidence for parallel processing of optic flow. In: Vaina L, Beardsley SA, Rushton S (eds) Optic flow and beyond. Kluwer, New York, pp 70–103
Monier C, Chavane F, Baudot P, Graham LJ, Frégnac Y (2003) Orientation and direction selectivity of synaptic inputs in visual cortex neurons: a diversity of combinations produces spike tuning. Neuron 37(4):663–680
Murakami I (2004) The aperture problem in egocentric motion. Trends Neurosci 27(4):174–177
Nakayama K (1985) Biological image motion processing: a review. Vision Res 25(5):625–660
Newsome WT, Wurtz RH, Komatsu H (1988) Relation of cortical areas MT and PST to pursuit eye movements. II. Differentiation of retinal from extraretinal inputs. J Neurophysiol 60:604–620
Orban GS (2008) Higher-order visual processing in macaque extrastriate cortex. Physiol Rev 88(1):59–89
Pack CC, Born RT (2001) Temporal dynamics of a neural solution to the aperture problem in visual area MT of macaque brain. Nature 409:1040–1042
Pack CC, Livingstone MS, Duffy KR, Born RT (2003) End-stopping and the aperture problem: two-dimensional motion signals in macaque V1. Neuron 39(4):671–680
Paragios N, Chen Y, Faugeras O (2005) Handbook of mathematical models in computer vision. Springer, Berlin
Parker AJ, Newsome WT (1998) Sense and the single neuron: probing the physiology of perception. Annu Rev Neurosci 21:227–277
Pasternak T, Greenlee MW (2005) Working memory in primate sensory systems. Nat Rev Neurosci 6(2):97–107
Reichardt W (1961). Autocorrelation, a principle for evaluation of sensory information by the central nervous system. In: Rosenblith WA (ed) Sensory communication (p. 303). Wiley, New York, pp 303–317
Roark DA, Barrett SE, Spence MJ, Abdi T, O'Toole AJ (2003) Psychological and neural perspectives on the role of motion in face recognition. Behav Cogn Neurosci Rev 2(1):15–46
Rorie AE, Newsome WT (2005) A general mechanism for decision-making in the human brain? Trends Neurosci 9(2):363–375
Ross J, Morrone MC, Goldberg ME, Burr DC (2001) Changes in visual perception at the time of saccades. Trends Neurosci 24(2):113–121
Rust NC, Movshon JA (2005) In praise of artifice. Nature Neuroscience 8(12):1647–1650
Rust NC, Mante V, Simoncelli EP, Movshon JA (2006) How MT cells analyze the motion of visual patterns. Nature Neuroscience 9(11):1421–1431
Schlack A, Albright TD (2007) Remembering visual motion: neural correlates of associative plasticity and motion recall in cortical area MT. Neuron 53:881–890
Shadlen MN (2002) Pursuing commitments. Nat Neurosci 5(9):819–821
Shadlen MN, Newsome WT (2001) Neural basis of a perceptual decision in the parietal cortex (areas LIP) of the rhesus monkey. J Neurophysiol 86(4):1916–1936
Sincich LC, Horton JC (2005) The circuitry of V1 and V2: integration of color, form and motion. Annu Rev Neurosci 28:303–326
Snowden RJ, Freeman TC (2004) The visual perception of motion. Curr Biol 14(9):R828–R831
Vaina LM (1998) Complex motion perception and its deficits. Curr Opin Neurobiol 8(4):494–502
Volkmann FC (1986) Human visual suppression. Vision Res 26(9):1401–1416
Watanabe T (1998) High-level motion processing: computational, neurobiological and psychophysical perspective. MIT Press, Cambridge, MA
Yasui S, Young LR (1975) Perceived visual motion as effective visual stimulus for pursuit eye movement system. Science 190:906–908
Yo C, Wilson HR (1992) Perceived direction of moving two-dimensional patterns depends on duration, contrast and eccentricity. Vision Res 32(1):135–147

Contents

Part I Low-level Cortical Dynamic Motion Processing

1 **From Moving Contours to Object Motion: Functional Networks for Visual Form/Motion Processing** 3
Jean Lorenceau

2 **Temporal Dynamics of Motion Integration** 37
Richard T. Born, James M. G. Tsui, and Christopher C. Pack

3 **Dynamics of Pattern Motion Computation** 55
Matthew A. Smith, Najib Majaj, and J. Anthony Movshon

4 **Multiscale Functional Imaging in V1 and Cortical Correlates of Apparent Motion** 73
Yves Fregnac, Pierre Baudot, Fréderic Chavane, Jean Lorenceau, Olivier Marre, Cyril Monier, Marc Pananceau, Pedro V. Carelli, and Gerard Sadoc

5 **Stimulus Localization by Neuronal Populations in Early Visual Cortex: Linking Functional Architecture to Perception** 95
Dirk Jancke, Fréderic Chavane and Amiram Grinvald

6 **Second-order Motion Stimuli: A New Handle to Visual Motion Processing** 117
Uwe J. Ilg and Jan Churan

Part II Active Vision, Pursuit and Motion Perception

7 **Motion Detection for Reflexive Tracking** 141
Frederick A. Miles and Boris M. Sheliga

8 When the Brain Meets the Eye: Tracking Object Motion 161
Guillaume S. Masson, Anna Montagnini, and Uwe J. Ilg

9 Interactions Between Perception and Smooth Pursuit Eye Movements 189
Ziad M. Hafed and Richard J. Krauzlis

10 Perception of Intra-saccadic Motion 213
Eric Castet

11 Intrasaccadic Motion: Neural Evidence for Saccadic Suppression and Postsaccadic Enhancement 239
Michael R. Ibbotson

Part III Modeling Dynamic Processing

12 Maximizing Causal Information of Natural Scenes in Motion 261
Dawei W. Dong

13 Neural Model of Motion Integration, Segmentation, and Propabilistic Decision-Making 283
Stephen Grossberg

14 Features in the Recognition of Emotions from Dynamic Bodily Expression 313
Claire L. Roether, Lars Omlor, and Martin A. Giese

Index 341

Contributors

Pierre Baudot
Institut des Systèmes Complexes, CNRS UMR7656, 57/59 rue Lhomond, 75005 Paris, France
baudot@math.jussieu.fr

Richard T. Born
Department of Neurobiology, Harvard Medical School, 220 Longwood Avenue, Boston, MA 02115, USA
rborn@hms.harvard.edu

Predo V. Carelli
Unité de Neurosciences Intégratives et Computationnelles (UNIC), CNRS UPR 2191, 1 Avenue de la Terrasse Bat 32/33, 91198 Gif-sur-Yvette, France
carelli@unic.cnrs-gif.fr

Eric Castet
Dynamics of Visual Perception and Action, Institut de Neurosciences Cognitives de la Méditerranée, CNRS & Université de la Méditerranée, 31 Chemin Joseph Aiguier, 13402 Marseille, France
eric.castet@incm.cnrs-mrs.fr

Fréderic Chavane
Dynamics of Visual Perception and Action, Institut de Neurosciences Cognitives de la Méditerranée, CNRS & Université de la Méditerranée, 31 Chemin Joseph Aiguier, Marseille 13402, France
frederic.chavane@incm.cnrs-mrs.fr

Jan Churan
Department of Neurology and Neurosurgery, Montréal Neurological Institute, McGill University, 3801 University Street, Montréal, QC, Canada H3A2B4

Dawei W. Dong
Center for Complex Systems and Brain Sciences, Florida Atlantic University, 777 Glades Road, Boca Raton, FL 33431, USA
dawei@ccs.fau.edu

Yves Fregnac
Unité de Neurosciences Intégratives et Computationnelles (UNIC), CNRS UPR 2191, 1 Avenue de la Terrasse Bat 32/33, 91198 Gif-sur-Yvette, France
fregnac@unic.cnrs-gif.fr

Martin A. Giese
Section for Computational Sensomotorics, Department of Cognitive Neurology, Hertie Institute for Clinical Brain Research & Center for Integrative Nueroscience, Frondsbergstr. 23, 72074 Tuebingen, Germany
martin.giese@uni-tuebingen.de

Stephen Grossberg
Cognitive & Neural Systems, Boston University, 677 Beacon Street, Boston, MA 02215, USA
steve@cns.bu.edu

Amiram Grinvald
Department of Neurobiology, Weizmann Institute of Science, Rehovot 76100, Israel
Amiram.Grinvald@weizmann.ac.il

Ziad M. Hafed
The Salk Institute for Biological Studies, 10010 North Torrey Pines Road, La Jolla, CA 92037, USA
zhafed@salk.edu

Michael Ibbotson
Visual Sciences, Research School of Biological Sciences, Australian National University, Canberra, Australia
Michael.Ibbotson@anu.edu.au

Uwe J. Ilg
Department of Cognitive Neurology, Hertie-Institute of Clinical Brain Research, University of Tuebingen, Otfried-Mueller Str 27, Tuebingen 72076, Germany
Uwe.Ilg@uni-tuebingen.de

Dirk Jancke
Department of Neurobiology, ND 7/72, Ruhr-University Bochum, Bochum 44780, Germany
jancke@neurobiologie.rub.de

Richard J. Krauzlis
The Salk Institute for Biological Studies, 10010 North Torrey Pines Road,
La Jolla, CA 92037, USA
rich@salk.edu

Jean Lorenceau
Equipe Cogimage, UPMC Univ Paris 06, CNRS UMR 7225,
Inserm UMR_S 975, CRICM 47 boulevard de l'Hôpital, Paris, F-75013, France
jean.lorenceau@chups.jussieu.fr

Najib Majaj
McGovern Institute for Brain Research, Massachusetts Institute of Technology,
in Cambridge, 46-6161, 77 Massachusetts Avenue, MA 02139, USA
majaj@mit.edu

Olivier Marre
Unité de Neurosciences Intégratives et Computationnelles (UNIC), CNRS UPR
2191, 1 Avenue de la Terrasse Bat 32/33, 91198 Gif-sur-Yvette, France
marre@unic.cnrs-gif.fr

Guillaume S. Masson
Dynamics of Visual Perception and Action, Institut de Neurosciences Cognitives
de la Méditerranée, CNRS & Université de la Méditerranée, 31 Chemin Joseph
Aiguier, 13402 Marseille, France
guillaume.masson@incm.cnrs-mrs.fr

Anna Montagnini
Dynamics of Visual Perception and Action, Institut de Neurosciences Cognitives
de la Méditerranée, CNRS & Université de la Méditerranée, 31 Chemin Joseph
Aiguier, 13402 Marseille, France
Anna.Montagnini@incm.cnrs-mrs.fr

Frederick A. Miles
Laboratory of Sensorimotor Research, National Eye Institute/NIH,
Bldg 49 Rm 2A50, Bethesda, MD 20892, USA
fam@lsr.nei.nih.gov

Cyril Monier
Unité de Neurosciences Intégratives et Computationnelles (UNIC), CNRS UPR
2191, 1 Avenue de la Terrasse Bat 32/33, 91198 Gif-sur-Yvette, France
monier@unic.cnrs-gif.fr

J. Anthony Movshon
Center for Neural Science, New York University, 4 Washington Place, Room 809,
New York, NY 10003, USA
movshon@nyu.edu

Lars Omlor
Section for Computational Sensomotorics, Department of Cognitive Neurology, Hertie Institute for Clinical Brain Research & Center for Integrative Nueroscience, Frondsbergstr. 23, 72074 Tuebingen, Germany
lars.omlor@medizin.uni-tuebingen.de

Christopher C. Pack
Montréal Neurological Institute, McGill University, 3801 University Street, Montréal, QC, Canada H3A2B4
christopher.pack@mcgill.ca

Marc Pananceau
Unité de Neurosciences Intégratives et Computationnelles (UNIC), CNRS UPR 2191, 1 Avenue de la Terrasse Bat 32/33, 91198 Gif-sur-Yvette, France
pananceau@unic.cnrs-gif.fr

Claire L. Roether
Section for Computational Sensomotorics, Department of Cognitive Neurology, Hertie Institute for Clinical Brain Research & Center for Integrative Nueroscience, Frondsbergstr. 23, 72074 Tuebingen, Germany
claire.roether@medizin.uni-tuebingen.de

Gérard Sadoc
Unité de Neurosciences Intégratives et Computationnelles (UNIC), CNRS UPR 2191, 1 Avenue de la Terrasse Bat 32/33, 91198 Gif-sur-Yvette, France
sadoc@unic.cnrs-gif.fr

Boris M. Sheliga
Laboratory of Sensorimotor Research, National Eye Institute/NIH, Bldg 49, Rm 2A50, Bethesda, MD 20892, USA
bms@lsr.nei.nih.gov

Matthew A. Smith
Center for Neural Basis of Cognition, University of Pittsburgh, 4400 Fifth Avenue, Mellon Institute, Room 115, Pittsburgh, PA 15213, USA
masmith@cnbc.cmu.edu

James M.G. Tsui
Montréal Neurological Institute, McGill University, 3801 University Street, Montréal, QC H3A2B4, Canada
james.tsui@mail.mcgill.ca

Part I
Low-Level Cortical Dynamic Motion Processing

Chapter 1
From Moving Contours to Object Motion: Functional Networks for Visual Form/Motion Processing

Jean Lorenceau

Abstract Recovering visual object motion, an essential function for living organisms to survive, remains a matter of experimental work aiming at understanding how the eye–brain system overcomes ambiguities and uncertainties, some intimately related to the sampling of the retinal image by neurons with spatially restricted receptive fields. Over the years, perceptual and electrophysiological recordings during active vision of a variety of motion patterns, together with modeling efforts, have partially uncovered the dynamics of the functional cortical networks underlying motion integration, segmentation and selection. In the following chapter, I shall review a subset of the large amount of available experimental data, and attempt to offer a comprehensive view of the building up of the unitary perception of moving forms.

1.1 Introduction

An oriented slit of moving light, a microelectrode and an amplifier! Such were Hubel and Wiesel's scalpel used during the 1960s (1959–1968) to uncover the properties of the visual brain of cat and monkey. A very simple visual stimulus indeed which, coupled with electrophysiological techniques, nevertheless allowed the analysis of many fundamental aspects of the functional architecture of primary visual cortex in mammals: distribution of orientation and direction selective neurons in layers, columns and hyper columns, discovery of simple, complex and hyper complex cells, distribution of ocular dominance bands, retinotopic organization of striate visual areas, etc.

J. Lorenceau (✉)
Equipe Cogimage, UPMC Univ Paris 06, CNRS UMR 7225, Inserm UMR_S 975, CRICM 47 boulevard de l'Hôpital, Paris, F-75013, France
e-mail: jean.lorenceau@chups.jussieu.fr

U.J. Ilg and G.S. Masson (eds.), *Dynamics of Visual Motion Processing: Neuronal, Behavioral, and Computational Approaches*,
DOI 10.1007/978-1-4419-0781-3_1, © Springer Science+Business Media, LLC 2010

Equipped with the elementary brick of information processing – the oriented receptive fields – the house of vision was ready to be built up and the Nobel price was in view. However, recording isolated neurons with a microelectrode might, for a while, have been the tree hiding the forest. If an oriented slit of moving light optimally gets a neuron to fire spikes, how many neighboring neurons also fire in response to that stimulus? What is the size and functional role of the neuronal population presumably recruited by this simple stimulus? Is there more than redundancy? An indirect answer to this question is inverse engineering: what are the requirements for recovering the direction of motion of a slit of moving light, e.g. a moving contour?

Fennema and Thompson (1979), Horn and Schunk (1981) and Hildreth (1984) raised the question and uncovered intrinsic difficulties in answering it, as many problems paved the way, like the "correspondence" and "aperture" problems, also identified on experimental grounds by Henry and Bishop (1971)[1]. Imagine two frames of a movie describing the motion of a 1D homogeneous contour (Fig. 1.1a): what part of the contour in the first frame should be associated to its counterpart in the second frame? The shortest path, corresponding to the motion vector orthogonal to the contour orientation seems the obvious answer, but may not correspond to the distal direction of motion. Applying the shortest path rule between two successive frames – a priority towards low speed – might leave parts of the contour unpaired thus facing a "correspondence" problem. Recovering the direction of an infinite 1D contour soon appeared as an ill posed problem, as an infinity of directions of motion are compatible with a single "local" measurement – e.g. through a biological or artificial motion sensor with a spatially restricted field of "view" (Fig. 1.1b).

In order to overcome this "aperture" problem, a solution is to combine at least two measurements from two 1D contours at different orientations. Amongst the large family of possible motion vectors associated to each contour motion, only one is compatible with both and may therefore correspond to the searched solution (Adelson and Movshon 1982). According to this scheme, motion processing would require two processing stages: the first one would extract local – ambiguous – directions and these measurements would be combined at a second stage. Numerous models (Nowlan and Sejnowski 1995; Liden and Pack 1999; Simoncelli and Heeger 1998; Wilson and Kim 1994) rely on this idea: the small receptive fields of V1 cells would first calculate motion energy locally (Adelson and Bergen 1985), followed by the integration of these local responses at a larger spatial scale at a second stage, which has been associated to area MT on experimental grounds

[1] "Although aware that units may be direction selective, Hubel and Wiesel have not emphasized this property and it is not considered in any of their tables. In this connection, however, it is interesting to note that, for an extended edge or slit much longer than the dimensions of the receptive field there are only two possible directions of movement, namely the two at right angles to the orientation. This is simply a geometrical necessity. Although orientation necessarily determines the direction of stimulus movement, which of the two possible directions will be effective is independent of the orientation". Bishop et al. (1971). See also Henry and Bishop (1971).

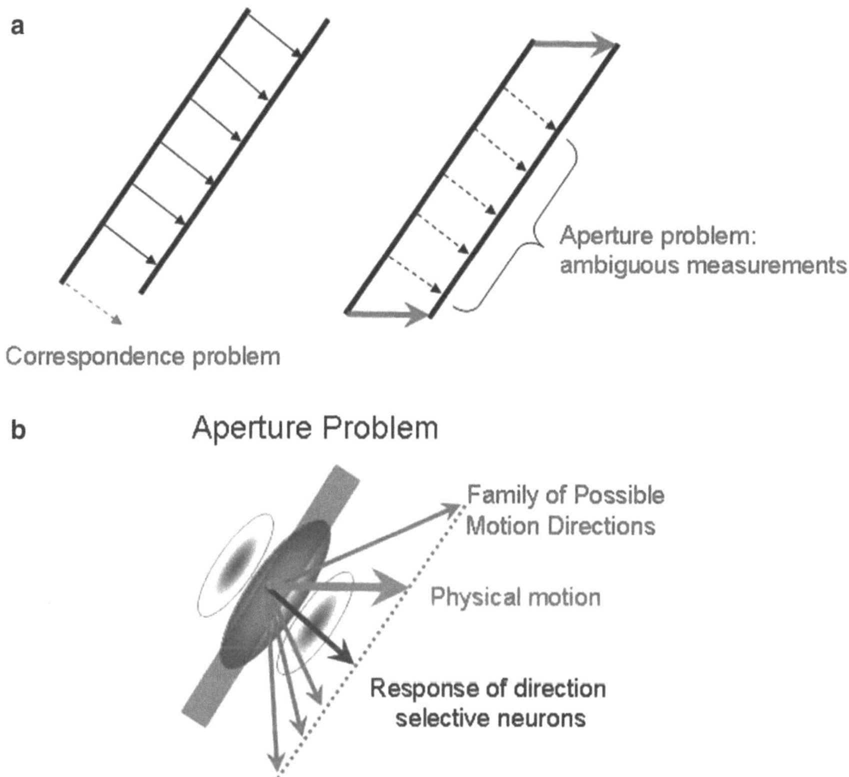

Fig. 1.1 Aperture and correspondence problem. *Top*: Illustration of the correspondence problem. Two frames of a bar moving horizontally are shown. A correspondence established over time using the shortest path – i.e. the lowest speed – leaves parts of the contour unpaired. . *Bottom*: A straight contour crossing the receptive field of a single direction selective neuron elicits the same response for a large family of physical motions. The cell responds only to the motion vector orthogonal to the cell preferred orientation

(Movshon et al. 1986; but see Majaj et al. 2007). Do, however, these two contours belong to a single translating shape or object, conditions required to justify the combination of both measurements, or do they belong to two different shapes or objects, in which case combining these motion measurements would distort the physical stimulus and yield a false solution? Answering the question clearly requires additional constraints for this calculation to be functionally relevant, a point addressed later on.

Another way of solving the "aperture problem" is to use the motion energy available at 2D singularities such as the line-endings of a finite unitary contour. These singularities can be seen as local geometrical features in visual space but are also characterized by their spatial frequency spectrum. As a matter of fact, these singularities of limited spatial extent have a wide energy spectrum in the Fourier plane with a large distribution of orientations and spatial frequencies with different power and phase.

As visual neurons behave as spatial frequency filters (Campbell and Robson 1968; De Valois 1979), singularities provide a rich set of possible motion measurements to spatial frequency and orientation tuned sensors, whose combination can signal the veridical direction of motion, at least for translations in the fronto-parallel plane. In addition or alternately, these local features can be matched or tracked from one position to the next, offering a "simple" solution to the correspondence problem through a feature matching process.

The influence of line-ends or terminators on motion interpretation was first analyzed by Wallach (1935; but also see Silverman and Nakayama 1988; Shimojo et al. 1989) who found that the perceived direction of a moving contour was strongly determined by the apparent direction of line-ends motion, whether these were real line-ends intrinsically belonging to the contour itself or spurious line-ends extrinsically defined by occluders. One question remains: what happens to the measurements of motion performed by direction selective neurons stimulated by the inner part of a moving contour? Consider the following alternatives:

1. Each motion signal from a single neuron is an independent labeled line on which the brain relies to infer the distribution of movements in the outside world. Under this assumption, a single moving contour would appear to break into the different directions of motion signaled by different direction selective neurons. This would not favor the survival of organisms endowed with such an apparatus!
2. Ambiguous motion signals that may not carry reliable information about the physical direction of contour motion are ignored or discarded. Only motion signals corresponding to line-endings are taken into consideration. Under this assumption, what would be the neuronal response that substantiates the contour unity? In addition, discriminating a translation from an expansion would be difficult if each line-end was processed independently.
3. All neurons have the same status regarding the encoding of stimulus direction that is; each response to a moving bar is considered an equal "vote" in favor of a particular direction of motion. Under this assumption, the resulting direction, processed through some kind of averaging of neuronal responses, would not necessarily correspond to the physical motion.

How then is it ever possible to recover the direction of a contour moving in the outside world? One possibility is to weigh these different votes according to some criterion, such as their reliability or salience (Perrone 1990). But again, what homunculus decides that this particular "vote" has less or more weight than the other one, especially if the "voters" are neurons whose receptive fields have similar spatio-temporal structure and function, like the simple and complex direction selective cells discovered by Hubel and Wiesel and thoroughly studied since?

Hildreth (1984) proposed an alternative according to which the ambiguous responses of neurons confronted with the aperture problem would be constrained so as to match the reliable measurements at 2D singularities. She offered a "smoothness constraint" rule – whereby information from measurements at singularities "propagates" along a contour – and elaborated a computational model that recovers the velocity of curved contours. However, the neural implementation of the mechanisms underlying the propagation process along contours still remains an open issue. Others (Nowlan and Sejnowski 1995) developed computational models that implement a selective integration through a weighting process in which the reliability of a measure results from an estimation procedure. However, it remains unclear how this estimation might be implemented in the brain.

Thus, although the seminal work of Hubel and Wiesel helped us to understand what the trees are, we still need to understand what is the forest, which, in modern terms, is captured by the concept of "functional assembly," still to be constrained by experimental data to fully characterize what constitute the "unitary representation" of a moving contour and the activity within a neuronal assembly that provides a "signature" of this unity.

More generally, central questions that should be answered to understand how biological organisms recover the velocity – speed and direction – of objects are the following:

1. When should individual neuronal responses be "linked" into functional assemblies?
2. What is the functional anatomy that underlies this linking, or binding, process?
3. Are the mechanisms identical throughout the visual system, or are there specific solutions at, or between, different processing stages?
4. What are the rules to select, and mechanisms used to select weight and combine the responses of direction selective neurons?

In the following, I briefly review experimental data that suggest a possible neuronal mechanism to smoothing, analyze the dynamics of contour processing and its contrast dependency, address the issue of motion integration, segmentation and selection across moving contours and describe how form constraints are involved in these processes. In the end, I'll attempt to ascribe a functional network to these different processes.

1.2 Propagating Waves Through Contour Neurons: Dynamics Within Association Fields

Neighboring positions in the visual field are analyzed by neighboring neurons in the primary visual cortex, acting as a parallel distributed spatio-temporal processor. However, distant neurons with non-overlapping receptive fields but tuned to similar orientations aligned in the visual field do not process incoming information independently. Instead, these neurons may form a "perceptual association field" linking local orientations into an extended contour. Reminiscent of the Gestalt rule of good continuity and closure, its characteristics were experimentally specified by Field et al. (1993) and Polat and Sagi (1993), although with different paradigms. The particular structure of association fields fits well with the architectony of long-range connections running horizontally in primary visual cortex over long distances (up to 8 mm, Gilbert and Wiesel 1989; Sincich and Blasdel 2001). Moreover electrophysiological responses to contextual stimuli (Kapadia et al. 1995, 2000; Bringuier et al. 1999) suggest that horizontal connectivity is functionally relevant for contour processing (Seriès et al. 2003 for a review).

In addition, optical imaging (Jancke et al. 2004) and intracellular recordings (Bringuier et al. 1999) bring support to the idea that lateral interactions through long-range horizontal connections propagate across the cortex with propagation speeds ranging between 0.1 and 0.5 m/s, which corresponds to speeds around 50–100 °/s in visual space. Recent work in psychophysics, modeling and intracellular recordings further suggest that these slow dynamics can influence the perception of motion (Georges et al. 2002; Seriès et al. 2002; Lorenceau et al. 2002; Alais and Lorenceau 2002; Cass and Alais 2006; Frégnac et al., this volume).

This is for instance the case with the Ternus display (Fig. 1.2) in which, the perception of group or element motion can be seen in a two frames movie, depending upon, amongst many other parameters, the time delay between frames. Alais and Lorenceau (2002) observed that for a given delay, group motion is seen more frequently when the Ternus elements are collinear and aligned as compared to non-oriented or non-aligned elements. This finding indicates that "links" between

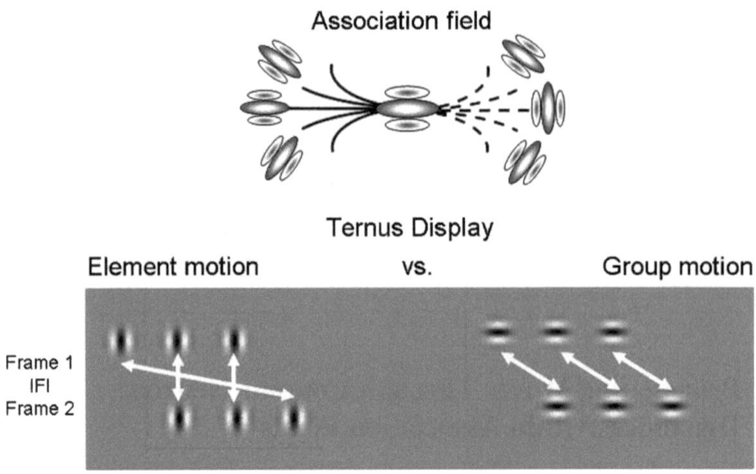

Fig. 1.2 Illustration of the "association field" depicting the spatial configurations that can (*left*) or cannot (*right*) be detected in an array of randomly oriented contour elements. This perceptual "association field" is presumably implemented in the network of long-range horizontal connections running horizontally within V1 (Gilbert and Wiesel 1995; Sincich and Blasdel 2001). In this figure, schematic oriented receptive fields interact through facilitatory long-range horizontal connections when the gestalt criterion of good continuity is met (*black lines*). When it is not (*dashed lines*), these long-range connections may be absent, ineffective or suppressive, a point that is still debated. *Bottom*: Illustration of the Ternus display of Alais and Lorenceau (2002) consisting in three oriented elements presented successively in a two frames movie. When the oriented elements are aligned and collinear (*right*), group motion is seen more often than when they are not (*left*). In this case element motion is seen more often. It is proposed that these different percepts of group and element motion reflect the links established between collinear and aligned element trough long-range associations

elements defining a pseudo continuous contour have been established and that strengthen the association between elements then considered a "whole."

A possible explanation is that horizontal connections provide a mean to bind individual neuronal responses into a functional assembly signaling a unitary contour moving as an ensemble in a single direction. This mechanism would have the advantage of being highly flexible, such that a functional assembly would easily adapt, within limits, to contours of varying length and curvature. An open issue is whether and how the association field characterized with static stimuli is used in motion processing. In this regard, it should be noted that eye movements of different kinds constantly shift the image on the retina such that different neurons, forming different assemblies, are recruited, even with static images. Thus, a continuous updating of the links to the incoming stimulus is required for "static" images as well as for moving stimuli, raising the possibility that association fields are relevant in motion processing as well.

1.3 Dynamics of Contour Motion Perception

Up to now, the need for combining motion measurements across space and time to recover a contour direction stems from theoretical considerations related to the initial sampling of the retinal image by cells with restricted receptive fields. If true, the computation of a global solution – e.g. Hildreth's smoothing process – may not be instantaneous and could take time. The finding that indeed the perception of a moving contour smoothly develops and unfolds over a period of time in a measurable way (Yo and Wilson 1992; Lorenceau et al. 1993) brings support to the idea that recovering the direction of moving contours involves an integration process endowed with a slow dynamics. In psychophysical experiments, Lorenceau et al. (1993) found that an oblique contour moving along a horizontal axis first appears to move in a direction orthogonal to contour orientation which smoothly shifts over tens of milliseconds towards the real contour direction (Fig. 1.3a, see Movie 1). This perceptual dynamics was found to depend on contour length and contrast such that a biased direction was seen for a longer time with longer contours and lower contrasts. In the framework described above, the effect of contour length is expected as it can be accounted for by the recruitment of a larger population of cells facing the aperture problem relative to those processing line-ends, thereby contributing to a strong bias toward an orthogonal perceived direction (Fig. 1.3b), that takes time to overcome.

The larger bias observed at low contrasts remains a matter of debate, although there is an agreement to consider that the sensitivity to the direction of 2D singularities the grating's or contour's line-ends is in cause. As mentioned above, these singularities are characterized by a broad spatial frequency and orientation spectrum. Decreasing contrast may therefore bring some frequencies close or below detection threshold in which case cells tuned to spatial frequencies and orientations with weak energy would respond poorly and with long latencies, thus degrading the global directional response or slowing down its recovery (Majaj et al. 2002). A model based on a parallel neuronal filtering through V1 receptive fields, followed by response pooling by MT neurons could thus account for the contrast effect.

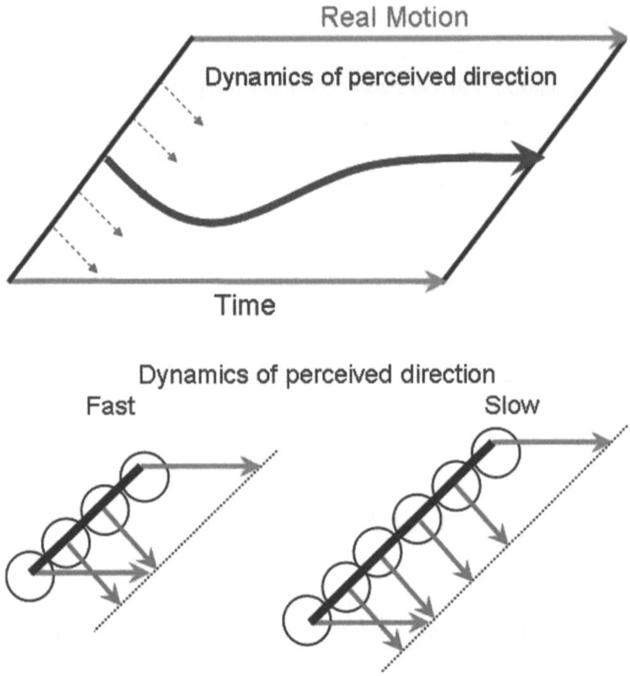

Fig. 1.3 Perceptual dynamics of an oblique bar moving horizontally. The perceived direction at motion onset is orthogonal the segment orientation and then smoothly tunes with the physical motion. The dynamics of the directional shift depends on the contour length (*bottom*), presumably because of the imbalance between the number of cells recruited by the inner part of the contour and its line-ends. The dependence of the dynamics on contrast may reflect a lower sensitivity to line-ends. (Lorenceau et al. 1993; see Movie 1). These perceived shifts are well correlated to the dynamics of MT cell response (Pack and Born 2001) are found in ocular following itself in and pursuit eye movements (Masson 2004)

A second possibility is that these singularities are analyzed by neurons with center-surround organization, often referred to as hyper complex or end-stopped cells (Hubel and Wielsel 1968) whose structure and function make them well suited for the processing of line-endings' motion or local curvature (Dobkins et al. 1987; Pack et al. 2003a, b; Lalanne and Lorenceau 2006). Sceniak et al. (1999) recorded such neurons in macaque V1 and observed that the end-stopping behavior found at a high contrast is decreased at low contrast, such that their capability to process line-ends' motion is degraded. This pattern of response could explain the longer integration time found at low contrast. Interestingly, this type of neurons mostly lies in the superficial layer of V1 where response latencies are longer than in other intermediate layers (Maunsell and Gibson 1992), suggesting that their contribution to motion computation is delayed relative to the simple direction selective neurons of layer 4.

In an attempt to decipher between these explanations (although different mechanisms could be simultaneously at work), Lalanne and Lorenceau (2006) used a Barber pole stimulus – an oblique drifting grating seen as moving in the direction of the line-ends present at the grating's borders. A localized adaptation paradigm was used in order to selectively decrease the sensitivity of the putative neurons underlying line-ends processing. Decreasing the contribution of these neurons to the global motion computation should increase the directional biases toward orthogonal motion thus allowing to isolate the spatial location and structure of the adapting stimulus that entails the largest biases. To get insights into the neuronal substrate at work, high contrast adapters were positioned in different locations at the border or within the grating and their effects on the subsequent grating's perceived direction measured. The results show that the largest directional biases are produced by adapters located within the grating itself and not at the line-endings positions. Although this "remote" effect of adaptation may seem surprising at first sight, it is compatible with a model in which the difference in response of two simple cells gives rise to end-stopping (Dobkins et al. 1987), but at odd with the idea that line-ends' direction is recovered by the parallel filtering of V1 receptive fields at line-ends positions (e.g. Löffler and Orbach 1999).

Neuronal counterparts of the perceptual dynamics underlying the recovery of moving contours described above have been found in macaque MT (Pack and Born 2001; Majaj et al. 2002; Born et al. this issue). In addition, ocular following was also found to manifest similar dynamical directional biases during its early phase with pursuit being deviated towards the normal to contour orientation (Masson et al. 2000; Barthélemy et al. 2008; see Chap. 8).

Altogether these psychophysical, behavioral and electrophysiological results indicate that recovering the motion of the simple moving bar used by Hubel and Wiesel in the sixties is a complex time consuming process that involves a large population of neurons distributed across visual cortex and endowed with different functional characteristics. As complex objects are generally composed of a number of contours at different orientations, understanding how biological systems overcome the aperture problem when processing objects' motion should take these findings into account.

1.4 Integration, Segmentation and Selection of Contour Motions

As stated above, the combination of responses to multiple component motions offers a way to overcome the aperture problem so as to recover object motion (e.g. Fennema and Thompson 1979; Adelson and Movshon 1982). In order to assess the underlying perceptual processes several classes of stimuli have been used to:

1. Measure the global perceived velocity and to determine the computational rules involved in motion integration
2. Evaluate the conditions under which component motions can, or cannot, be bound into a whole
3. Identify the neural substrate and physiological mechanisms that implement these perceptual processes

The numerous kinds of stimuli used to explore these issues can be broadly divided in three classes: Plaids, Random Dot Kinematograms (RDKs) and "aperture" stimuli. Before trying to offer a synthetic view of the results, let us spend some time discussing the appearance and relative significance of these different stimuli (Fig. 1.4).

Made of two extended overlapping gratings at different orientations, drifting plaids can be seen as a single moving surface or as two sliding transparent surfaces, depending on their coherency. As plaids are well defined in the Fourier plane by their

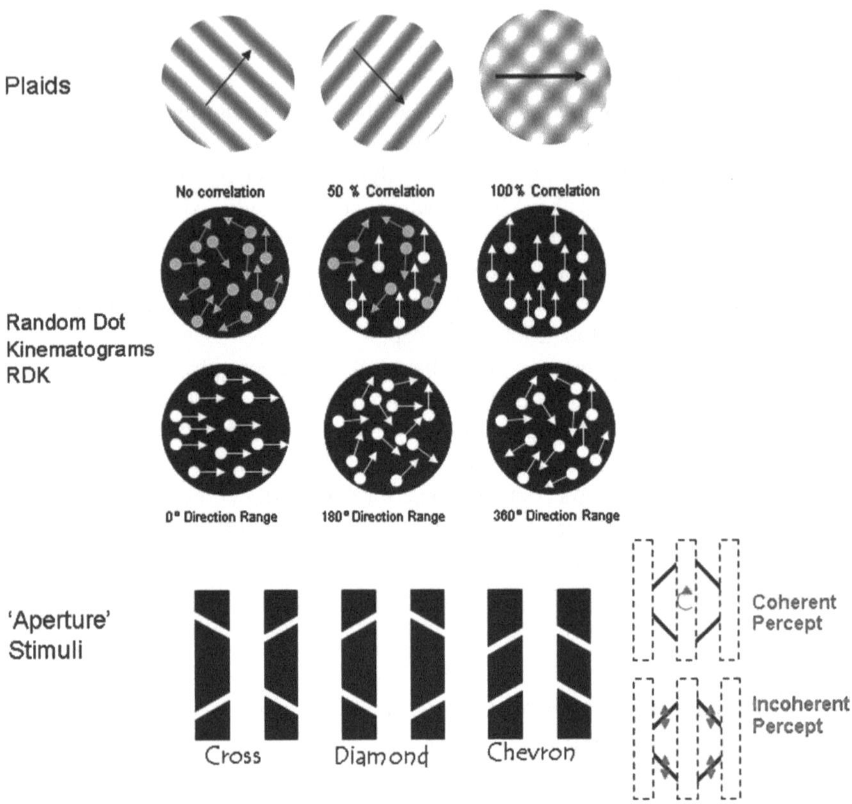

Fig. 1.4 Different stimuli used to probe contour integration. *Top*: Plaid patterns made of two superimposed gratings. Changes of relative orientation, contrast, speed, spatial frequency have been used to determine the conditions of perceived coherence, the perceived direction and speed and the nature of the underlying combination rule. *Middle*: Two types of random dot kinematograms (RDKs). In one, the percentage of coherently moving dot is used to assess motion sensitivity. In the second, dot directions are chosen amongst a distribution of direction varying in width to characterize directional – and or speed – integration. *Bottom*: "Aperture" stimuli where a moving geometrical figure is partially visible behind aperture or masks. Each figure segment appears to move up and down. Recovering figure motion requires the spatio-temporal integration of segment motions. Changing figure contrast or shape, aperture visibility or luminance, duration, eccentricity deeply influences perceived rigidity and coherence and may impair the ability to recover object motion

component spatial and temporal frequencies, they proved useful to study how the output of different spatio-temporal frequency channels are combined and to investigate the combination rule underlying the perceived direction and speed of plaid patterns (e.g. Adelson and Movshon 1982; Movshon et al. 1986; Welch 1989; Gorea and Lorenceau 1990; Yo and Wilson 1992; Stoner and Albright 1992, 1998; Stone et al. 1990; Van der Berg and Noest 1993; Delicato and Derrington 2005; Bowns 1996, 2006). However, with rare exceptions only plaids made of two overlapping gratings have been used in these studies, limiting the generality of the findings. In addition, the gratings' intersections that carry relevant information at a small spatial scale raised questions concerned with the nature of the process at work (see below).

Similar issues have been addressed with random dot kinematograms (RDKs) in which dots randomly distributed across space move in different directions (Marshak and Sekuler 1979; Watamaniuk et al. 1989; Watamaniuk and Sekuler 1992). A variety of RDKs have been used in studies of motion integration. This variety is related to the way each dot is moving, allowing the assessment of several characteristics of motion processing. For instance, a RDK can be made of a percentage of dots moving in a given direction embedded in a cloud of incoherently moving dots. Measures of motion coherence thresholds, corresponding to the percentage of coherently moving dots yielding a directional percept, are routinely used in electrophysiological recordings to assess both perceptual and neuronal sensitivities in behaving – and possibly lesioned – monkeys (e.g. Britten and Newsome 1989; Newsome and Paré 1988) or in patients with brain damage (Vaina 1989; Vaina et al. 2005). Perceptual data show that ultimately, a single dot moving consistently in a single direction can be detected in a large cloud of incoherently moving dots (Watamaniuk et al. 1995). Other versions of RDKs have been used, either with dots moving for a limited life time, thus allowing the measure of the temporal integration of motion mechanisms, or dots moving along a random walk thereby allowing the measure of the directional bandwidth of the integration process. One critical outcome of these studies is that global motion coherence depends upon the salience and reliability of each dot motion. For instance, if two sub ensembles of dots move rigidly in two different directions – i.e. if the relationships between dots remain constant over space and time – transparency dominates over coherence. In addition perceptual repulsion between close directions is observed, suggesting inhibitory interactions between direction selective neurons (Marshak and Sekuler 1979; see also Qian et al. 1994), a finding consistent with the center-surround antagonism of MT receptive fields (Allman et al. 1985; but see Huang et al. 2007). If each dot follows a random walk, changing direction from frame to frame within limits, the cloud of dots appears to move globally in the averaged direction, even with wide distributions of directions (Watamaniuk et al. 1989; Lorenceau 1996). Additional studies show that not only direction, but also speed can be used to segregate motion into different transparent depth planes, in accordance with the layout of speed distributions during locomotion in a rich and complex environment (Watamaniuk and Duchon 1992; Masson et al. 1999).

Non-overlapping moving contours or drifting gratings distributed across space have also been used (Shimojo et al. 1989; Anstis 1990; Mingolla et al. 1992;

Lorenceau and Shiffrar 1992, 1999; Lorenceau 1998; Rubin and Hochstein 1993; McDermott et al. 2001; McDermott and Adelson, 2004). In several studies, these "aperture stimuli" consist in geometrical shapes partially hidden by masks that conceal their vertices, such that recovering the global motion requires the integration of component motion across space and time (Fig. 1.4 bottom). One advantage of this class of stimuli, in addition to their "ecological validity," is the large space of parameters that can be studied and the lack of confounding factors such as the intersections existing in plaids. The parameters controlling the different possible interpretations and the coherency of these stimuli have been thoroughly investigated (review in Lorenceau and Shiffrar 1999; see below), providing insights into the mechanisms involved in form/motion binding.

1.5 What "Combination Rule" for Motion Integration?

It is out of the scope of the present article to thoroughly review the abundant literature concerned with the modeling of the motion combination rule: predictions of the "Intersection of Constraints" (IOC, Adelson and Movshon 1982; Lorenceau 1998), "Vector Averaging" (Kim and Wilson 1989), "Feature Based" (Gorea and Lorenceau J. 1991; Bowns 1996), or Bayesian rules (Weiss and Adelson 2000) which have been tested experimentally and the debate still develops/goes on (e.g. Delicato and Derrington 2005; Bowns and Alais 2006). In parallel to psychophysics a number of computational models, with varying degrees of biological plausibility, have been proposed (Koechlin et al. 1999; Nowlan and Sejnoswki 1995; Liden and Pack 1999; Grossberg et al. 2001; Rust et al. 2006). One difficulty in accurately modeling perceptual data might come from the fact that perceived coherence and perceived speed and direction are not measured simultaneously, although they could interact. As a matter of fact, one may perceive a "global" direction with stimuli of low or high coherence or rigidity. Disentangling the different models often requires specific combinations of oriented gratings – known as Type II plaids – for which the models' predictions disagree. However, the perceptual coherency of type II plaids, understood herein as the degree of rigidity, sliding or transparency, may be equivocal and bistable. Does the same "combination rule" apply similarly during these different perceptual states? One possibility is that perceived coherence and perceived direction are interdependent because several combination rules are implemented, each being used according to the task and stimulus at hand (Bowns and Alais 2006; see also Jazayeri and Movshon 2007). Examples from everyday life suggest it might be the case: the flight of a large ensemble of birds or the falling of the snow might give rise to a perception of motion in a single global direction as is the case with random dot kinematograms. However, not every bird or snowflake really moves in that direction. Segmenting a particular element and perceiving its particular direction remains possible (Bulakowski et al. 2007), although it may be biased by the surrounding context (Duncker 1929). By contrast, a car or a plane appears to move rigidly and coherently in a single direction that needs to be accurately recovered and thus segmented from other cars or planes. Perceptually dissociating the "local direction" of object's parts and accessing a "local measurement of motion" is very difficult, despite the fact that some neurons facing the aperture problem do "signal" different contour directions. The differences between these examples might lie in the binding "strength," reflected in part in the perceived motion rigidity and coherency, although this latter subjective notion remains difficult to fully characterize.

1.6 Bound or Not Bound?

The second issue regarding the combination of multiple component motions is the definition of the space of stimulus parameters that yields either motion integration or segmentation into independent motions. In other words, do human observers always combine component motions into a whole, or are there specific constraints controlling the integration process that need to be characterized?

The principle of common fate promoted by the Gestalt school states that "what moves in the same direction with the same speed is grouped together." Although a simple and powerful concept, common fate is loosely defined, especially when taking the aperture problem – and complex 3D motions – into account, and must be revised. The need for combining component motions in order to recover object's motion indicates that common fate in not directly available to perception and requires sophisticated mechanisms to extract the common direction and speed of moving objects.

Plaids allowed the exploration of four main variables that could influence the combination process: relative orientation/direction, spatial frequency, speed – or temporal frequency – color and contrast. For the four former variables, limits in the possibility of combining drifting gratings into a coherent whole were found. Small relative gratings angles, very different spatial and temporal frequencies or speeds decrease motion coherency, with perception shifting to sliding and transparency under these conditions. Even with a choice of gratings that favors a coherent percept, plaids are bi-stable stimuli with alternating episodes of coherency and sliding (Hupé and Rubin 2003). Contrast is not as powerful in modifying coherency, as widely different grating contrast may nevertheless cohere into a global motion. It does, however, influence perceived direction and speed (Stone et al. 1990), presumably because contrast alters component perceived speed and hence the inputs to the motion integration stage (Thompson 1982; Stone and Thompson 1992).

In addition to the exploration of the perceptual organization of plaids, electrophysiological recordings showed that a simple V1 cell selectively respond to the orientation and spatio-temporal component frequency to which it is preferably tuned, but not to the global pattern motion. In contrast, about one third of MT neurons were found to respond to the plaid direction rather than to its component gratings (Movshon et al. 1986; Rodman and Albright 1989). Note that neuronal responses to global motion have also been reported in cat thalamus

(Merabet et al. 1998) and in the pulvinar (Dumbrava et al. 2001) and that inputs to MT bypassing V1 have been described (Sincich et al. 2004).

Psychophysical studies suggested that only neurons tuned to similar spatio-temporal frequencies were combined into a single moving plaid (Adelson and Movshon 1982). These findings were taken as evidence in favor of a linear filtering model in which the motion energy of each grating would be extracted at a first stage by spatio-temporal filters and then selectively combined at a second stage. The possible involvement of non-linearities in motion integration first stemmed from studies seeking for an influence of the intersections that appear with overlapping gratings – also called "blobs" – that are reminiscent of the line-endings or terminators discussed previously. Although a model using linear spatio-temporal filtering should be "blind" to these "blobs," several studies provided evidence that they play a significant role in the motion combination process, as observers seem to rely on the motion of these features in a variety of perceptual tasks (Van Der Berg and Noest 1993). For instance, manipulating the luminance of the gratings' intersections, such that they violate or not the rules of transparency (Stoner and Albright 1990; Vallortigara and Bressan 1991) shifts the percept toward seeing transparent surfaces or global motion, respectively. Others have used unikinematic plaids in which one of the components is stationary, in order to evaluate the contribution of "blobs" to the perceived direction (Gorea and Lorenceau 1990; Masson and Castet 2002; Delicato and Derrington 2005). A two stages model based on spatio-temporal filtering would predict that only the moving component contributes to perceived motion. However, these studies suggested that experimental data could be explained by taking the motion of "blobs" – or non-Fourier motion – into account, thus calling for some non-linearities in analyzing plaid's motion (Wilson and Kim 1994; Van Der Berg and Noest 1993; Bowns 1996).

Studies using RDKs are diverse. In some studies, not described herein, RDKs have been used to look at questions related to cue invariance, showing for instance that form can be recovered from relative motion. The main findings concerned with motion integration of RDKs have been alluded to before. One striking result relevant to this review is the finding that perception shifts from motion repulsion (Marshak and Sekuler 1979) to motion integration (Watamaniuk and Sekuler 1992) when the local reliability and salience of each dot trajectory, a 2D signal, is degraded by imposing a limited life time or a random walk to each dot. This suggests that 2D signals can impose a strong constraint on motion integration and segmentation processes. As for the recovery of contour motion analyzed above, reliable processing of 2D signals seems to drive the segmentation of a moving scene into separate components, while the system defaults to larger integration scale when uncertainty on each dot motion is added – e.g. in the form of motion noise (Lorenceau 1996, Movie 2) or at low dot contrast (Pack and Born 2005).

Similar conclusions stem from studies using multiple contours or gratings distributed across space and partially visible behind occluders, the so-called "aperture stimuli." In many of these studies, the problem of motion integration through combination of 1D component is addressed together with the analysis of 2D junctions that may occur when static occluders partially mask the moving contours, thus creating "spurious" moving terminators at the mask-object junction. This situation of partial occlusion is commonly encountered in a natural environment. One issue is then to understand when and how these signals are classified as spurious and whether their motion influences motion perception. To distinguish line-endings resulting from occlusion and "real" line-ends of objects' contours, Shimojo et al. (1989) introduced the terms "extrinsic" and "intrinsic" that I shall use in the following. A number of examples demonstrate the strong influences of the status of these singularities on motion perception and the need for a classification of these singularities.

In the "chopstick" illusion (Anstis 1990), the crossing of two orthogonal contours translating out of phase along a clockwise trajectory appears to translate in the same clockwise direction, although it is physically moving anticlockwise. Occluding the line terminators' – thus changing their status from intrinsic to extrinsic – changes the percept with the crossing now being perceived as moving anticlockwise (see Movie 3). Shimojo et al. (1989) used a vertical triplet of barber poles each consisting in an oblique grating drifting behind a horizontal rectangle (Fig. 1.5 left). In each barber pole, the perceived motion is along the rectangle longer axis, as in Wallach's demonstrations. Changing the relative disparity between the rectangular apertures and the gratings causes perception to switch to a vertical motion for positive or negative disparities, corresponding to a perception of a unitary surface seen behind, or in front, the three rectangular apertures. Shimojo et al. (1989) accounted for this effect by assuming that extrinsic line-endings motion at the aperture border is discarded from further analysis.

Along a similar line, Duncan et al. (2000) designed a stimulus in which a vertical grating is presented within a diamond aperture (Fig. 1.5 right). In their display, the disparity between the aperture borders and the gratings could be selectively manipulated, such that line-endings distributed along diagonals appeared either near or far relative to the diamond aperture and thus classified either as extrinsic or intrinsic.

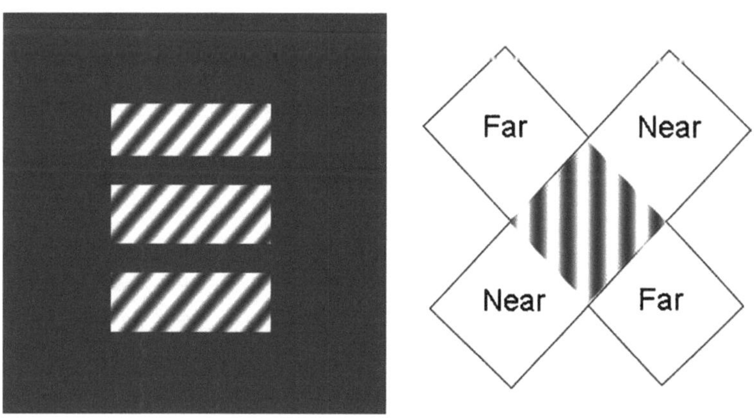

Fig. 1.5 Illustrations of the displays used by Shimojo et al. (1989) in humans (*left*) and by Duncan et al. (2000) in monkeys (*right*), to probe the influence of disparity on motion perception. The gratings are presented at different disparity relative to the background such that the line-ends can appear as intrinsic – belonging to the grating – or as extrinsic – produced by occlusion. With Duncan et al.'s display, the response of MT cells depends on which sides of the square grating are far or near the fixation plane

Under these conditions, perceived drifting direction is "captured" by the intrinsic terminators. The new finding is that recordings from MT neurons show selective responses corresponding to the perceived direction in these different conditions, suggesting that signals from terminators are somehow weighted as a function of their status, extrinsic or intrinsic. Whether this weighting occurs at, or before, the MT stage remains however unclear, although MT neurons are known to be selective to disparity (DeAngelis et al. 1998).

Along the same vein, Rebollo and Lorenceau (unpublished data) measured the effect of disparity on motion integration with Aperture stimuli using outlines of moving shape – diamond, cross and chevron – partially visible at varying depths relative to a background plane made of static dots (Fig. 1.6).

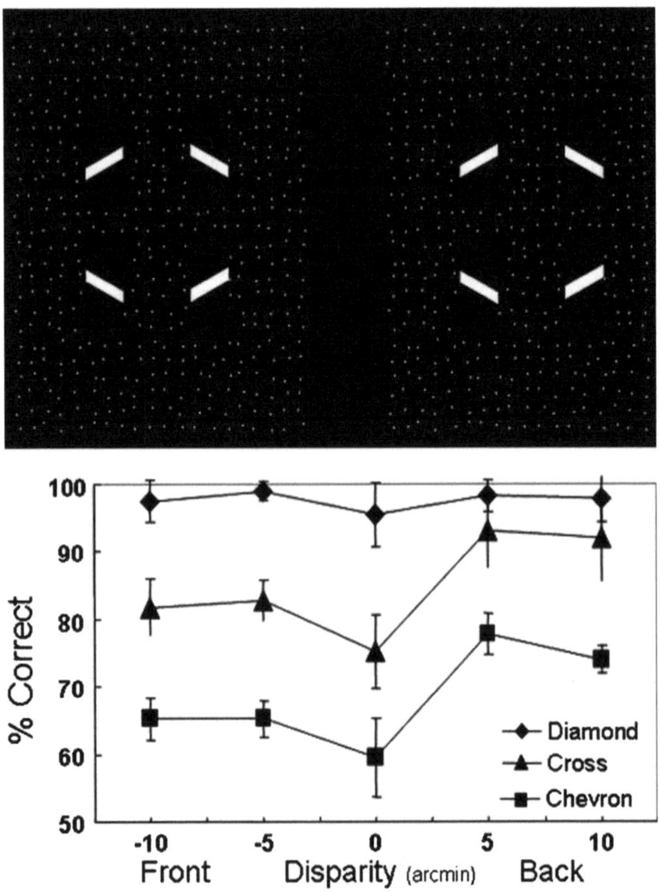

Fig. 1.6 *Top*: Stereo display where moving diamonds are presented at different disparities relative to the fixation plane. *Bottom*: Performance in a clockwise anticlockwise direction discrimination task as a function of disparity for three different shapes. Performance depends on shape but is always worse when the figures and background have the same disparity (Rebollo and Lorenceau unpublished data). See text for details

Using a discrimination task of the global shape motion, they found that, whatever its sign, disparity enhanced motion integration relative to zero disparity, although motion integration was facilitated with negative disparity. An interesting finding is that this effect occurs despite the lack of well defined visible junctions between the plane of fixation and the contour endings (Fig. 1.6 top), suggesting that perceived depth per se rather than local disparity at junctions influences motion integration.

Although these different results suggest that occlusion and disparity "weight" the terminator signals in motion integration, according less weight to extrinsic terminators created by occluders or with negative disparity, a similar effect can be obtained by "blurring" intrinsic line-endings, for instance by introducing motion noise in order to decrease the reliability of line-endings motion (Lorenceau and Shiffrar 1992; Kooi 1993). It thus seems that occlusion and disparity are just two amongst several ways of lowering the weight of terminators in motion integration.

Studying the dynamics of motion integration brings additional insights into the underlying computation. A naïve intuition would consider that the retinal image being initially fragmented in component signals by the mosaic of V1 receptive fields, integration progressively builds into a coherent whole, such that segmentation precedes integration. However, psychophysical data suggest otherwise. Integration appears a fast "automatic" undifferentiated process followed by a slower object-based segmentation. This can be seen in Fig. 1.7 where direction discrimination of global motion – an indirect measure of perceived coherency – is plotted as a function of motion duration for different contrasts of line segments arranged in a diamond or cross shape. Note that under these experimental conditions, line-endings are intrinsic and should therefore be given a strong weight in motion processing as they provide reliable 2D segmentation cues.

As it can be seen, performance increases with increasing motion duration, up to ~800 ms for a diamond shape and around 300 ms for a cross shape. With longer durations, performances for high and low contrast diverge. Notably, with high contrast segments, performance decreases at a long duration while it continues increasing for low contrast segments (also see Shiffrar and Lorenceau 1996). This finding suggests that a contrast dependent competition between integration and segmentation develops over a period of time whose outcome is reflected in psychophysical performance.[2] Why this occurs can be understood within the framework described above for single moving contours: the slow computation of intrinsic line-endings motion, which accounted for the biases in perceived direction of an isolated contour, may also be used to segment the global motion into component signals. Indeed, intrinsic end-points reliably signal discontinuities used to limit the region of space over

[2] One surprising fact is that observers seem to rely on the state of the cerebral networks at the end of the stimulation to give their – highly unreliable – response, although a 'correct' answer – at least relative to the task at hand – is available soon after motion onset.

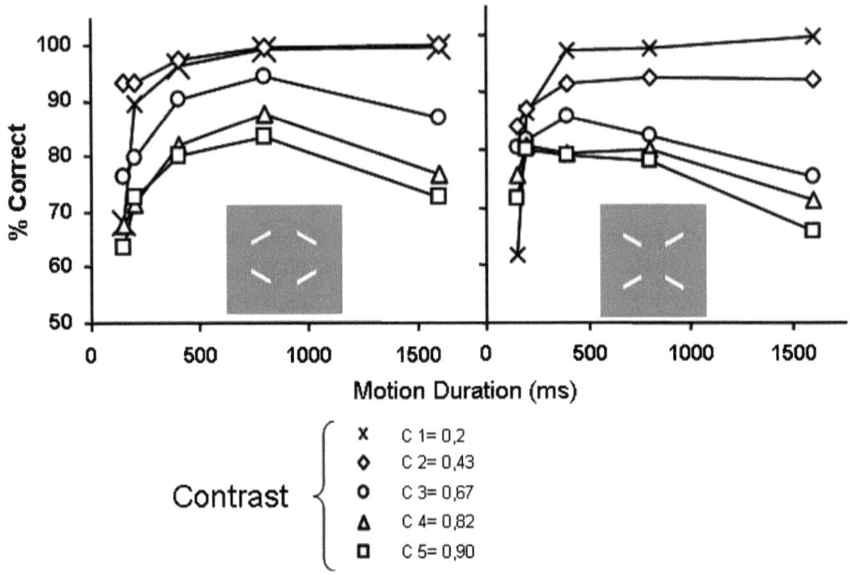

Fig. 1.7 Performance in a clockwise anticlockwise direction, discrimination task as a function of motion duration for a diamond and a cross for five levels of segment luminance. In this display the masks that hide the figures' vertices are of same hue and luminance as the background. Performance reflecting motion integration: performance first increases with motion duration. For longer duration, performance remains high at low segment luminance but decreases for segments at a high luminance (Lorenceau et al. 2003). See text for details

which integration should be functional. One can speculate that fast integration of the responses of V1 – or inputs bypassing V1, see above – direction selective cells to component motion at the MT level is then controlled by a slow segmentation based on signals from line-endings (e.g. from V1 or V2 end-stopped neurons, the latter being involved in junction classification and assignment of border ownership; Qiu et al. 2007) that could involve modulatory inputs to MT. This later idea is supported by the finding that motion segmentation is enhanced after observers were given Lorazepam, a benzodiazepine agonist of $GABA_a$ that potentiates inhibitory neurons (Giersch and Lorenceau 1999). This idea of competitive influences is also supported by the observation that long inspection of these stimuli is accompanied by a bistable perception, with intermittent switches between coherent and incoherent states. In addition, smooth and slow variations of parameters known to modulate motion coherence of "aperture stimuli" – i.e. line-ends or mask luminance – entail perceptual hysteresis such that transitions from coherent to incoherent states and the reverse are not observed for the same parameter values (Movie 4). Such perceptual hysteresis is considered a reliable signature of cooperative/competitive networks (also see Williams and Phillips 1987).

Thus, the reliability, salience and intrinsic/extrinsic status of line endings as well as their perceived depth relative to the fixation plane have a strong, although slow, impact on the integration of motion components distributed across space. Several issues related to the processing of spatial discontinuities of different kind such as end-points, vertices or junctions remain: what is the neuronal mechanism that analyses – and classifies – them? Although these discontinuities are often considered very "local" – at the extreme such singularities are infinitely small – are they also analyzed at low spatial scales? Are they processed at early processing stages like V1 and V2 as suggested by electrophysiological data (Grosof et al. 1993; Peterhans and van der Heydt 1989; Sceniak et al. 1999; Qiu et al. 2007) or do they result from an inference accompanied by (a)modal completion (McDermott and Adelson 2002) involving higher processing stages? In this regard, recent electrophysiological recordings (Yazdanbakhsh and Livingstone 2006) showing that end-stopping is sensitive to contrast polarity brings new insights into the functional properties of end-stopping. One intriguing possibility would be that center and surround interactions in end-stopping are also sensitive to disparity.

Line-ends, junctions, terminators have often been considered "local static" features in the literature. Their role in motion integration has consequently been interpreted as an influence of form. One of the reasons for this assumption is that processing the velocity of all combinations of all possible directions and speeds of singularities would be computationally very demanding (see Löffler and Orbach 1999). However, recent electrophysiolological recordings in monkey suggest that some V1 direction selective cells do process the motion of these singularities (Pack et al. 2003a, b).

1.7 Dorsal Motion and Ventral Form

As pointed out in the introduction, combining motion signals in motion areas is relevant only if the measured signals are bound to the same moving object. Whether this is true requires an analysis of the spatial relationships between local component motions. An assumption common to many models of motion integration is that the moving objects to be analyzed are rigid which somehow relates to their invariant spatial structure. However, the combination of motion components that yields global motion has generally been considered in a velocity space lacking spatial organization where each motion vector representing the direction (polar angle) and speed (vector norm) is considered independently of the underlying moving structure (Adelson and Movshon 1982; Rust et al. 2006). This assumption originates in part from the fact that MT neurons have mostly been studied with extended plaid patterns and RDKs exhibiting a very specific or no spatial structure and the need to design tractable models (but see Grossberg et al. 2001 for modeling including form and motion computation). It also stems from the organization of area MT, exhibiting a columnar organization where close directions – and speeds – are represented in neighboring columns that only present a crude retinotopic organization (Van Essen 1988). Although the antagonistic center-surround organization of many MT receptive fields has been extensively described (Allman et al. 1985; Born and Bradley 2005) and is often proposed to underlie motion segmentation (but see Huang et al. 2007), less is known on the relationships between MT neurons (although long-range connections exist in this area; Levitt and Lund 2002) and it remains unclear whether they may encode the spatial structure of moving objects.[3]

In contrast, neurons in areas distributed along the ventral pathway are not very selective for direction and speed but respond well to polar or concentric spatial organization (Gallant et al. 1993), specific spatial features such as vertices or corners (Pasupathy and Connors 1999) or are selective to more complex arrangements of these features in the infero-temporal cortex of macaque (Tanaka et al. 1991). In man, imaging studies uncovered a cortical region, the lateral occipital complex (LOC), selectively activated for well-structured stimuli and familiar objects (Malachet al. 1995; Kourtzi and Kanwisher 2001).

Whether the spatial organization of the distribution of component motions influences motion integration is worth considering as it could provide additional relevant constraints to segment the distribution of motion signals across space and select those, belonging to the same spatial structure, that should be combined to recover the different motions of objects in a visual scene (Weiss and Adelson 1995; Grossberg et al. 2001). Again, plaids, RDK and "aperture" stimuli have been useful in exploring this issue. Studied in the general framework of a two stage model, where the second MT stage would integrate component motion within a "velocity space" lacking spatial organization, the main novelty is the "intrusion" of form constraints, operating at different spatial scales, that gate motion integration. This influence of form information remains a challenge for most computational models. Overall, the main findings described thereafter are rooted in the Gestalt principles (Koffka 1935) of common fate, similarity, completion and closure. Parallel advances in the analysis of the functional specialization of visual areas provided a new framework for understanding the neural computation underlying motion integration and segmentation.

As plaids and RDKs offer little ways of manipulating spatial structure (with the exception of Glass patterns), this issue has not been thoroughly studied with this type of stimuli. Note, however, that with plaids, the perception of sliding of transparency at small relative gratings' angles, although interpreted as a limit of the motion combination process, could also be seen as a spatial constraint. Similarly RDKs with several fixed spatial distributions of dots, each endowed with a particular velocity, appear as transparent motion of structured surfaces, suggesting that the rigid and invariant spatial relationships between dots are used for motion segmentation. As a matter of fact, an influence of the spatial distribution of dots on motion integration was found by contrasting the capability to integrate motion stimuli made of two clouds of moving dots that were either randomly distributed across space or arranged into a diamond like shape (Lorenceau 1996). Recovering the global motion was better for dots defining a diamond-like shape as compared to a random distribution. Additional studies helped uncovering which spatial characteristics

[3] Note that a different pattern has been proposed for area MST where the selectivity for complex motion – expansion, contraction, rotation – related to the processing of the optic flow field, is supposed to emerge from highly specific, spatially organized, projections from MT cells (Duffy and Wurtz 1995; Koenderink 1986). Indeed, four orthogonal vectors that would share the same representation in a velocity space may define a rotation or an expansion, depending only on their spatial relationships.

Fig. 1.8 Stimuli used in the study of Lorenceau and Zago (1999). Grating patches are presented behind circular apertures. Gratings are different orientation drift sinusoidally out of phase such that integrating their motion yields a perception of a tiled floor translating smoothly along a circular path. At high contrast (*top*) motion integration is difficult but better for L-configurations as compared to T-configurations. At low contrast both configurations appear more rigid and elicit a coherent motion percept. Eccentric viewing conditions facilitate motion integration for both configurations at both contrasts. See Movies 5–8

influence motion integration. Lorenceau and Zago (1999) used a tiled surface of gratings patches forming either L- or T-junctions (Fig. 1.8 and Movies 5–8). Each patch was visible behind a circular aperture that masked the junctions that were thus only virtually present. Although the representation of motion components is the same in a velocity space for both configurations, motion integration was facilitated for L-configurations as compared to T-configurations at high grating contrasts. At a low contrast, motion integration was much easier than at high contrast and the difference between the L and T configurations vanished, suggesting a strong contrast dependency of these "form" constraints.

As for the "Ternus display" described above, one interpretation of these data relies on the idea that "links" between neighboring gratings forming virtual L-junctions have been established, while such links would be weaker or suppressive for virtual T-junctions. This view is also supported by the findings of Lorenceau and Alais (2001, see Movie 9) with aperture stimuli. In this study, recovering the global direction of a collection of rigid geometrical shapes made of identical segments partially visible behind vertical masks was very easy for some shapes – e.g. a diamond – but very difficult for others – e.g. a cross or a chevron

– despite the fact that all component motions had the same representation in a velocity space and very similar frequency spectra.

The mechanisms underlying this influence of form information on motion integration are still unclear. Three possibilities are worth considering.

One relies on the idea that long range connections in area V1, found to underlie contour processing (Field et al. 1993; Kovacs and Julesz 1993; see Hess et al. 2003, for a review), are involved in building a "proto shape" when constraints of good continuity and closure are met, as is it the case for a diamond or for the L-configurations described above. The resulting "neuronal assembly" would then feed the MT stage. This early process would not occur for configurations that do not meet the "good gestalt" criterion and consequently would not be integrated as a whole at the MT stage. However, unless some physiological "signature" or "tagging" of a neuronal ensemble – as for instance the synchronization of neuronal activity (Singer 1995) – is available and can be read-out at further processing stages – or elaborated through recurrent connections – it is unclear what mechanism could "control" motion integration at the MT stage.

A second possibility involves interactions between ventral and dorsal areas. In this scheme, only when component segments are integrated as a rigid shape in ventral areas, e.g. the LOC, would motion integration proceed. Evidence for this account stems from recent fMRI studies where the bi-stability of the "masked diamond stimulus" has been used to identify the regions activated during coherent and incoherent states, as continuously monitored by human observers during long-lasting stimulation (Lorenceau et al. 2006, 2007; Caclin et al. in preparation). With the same distal stimulus, different cortical regions showed state dependent BOLD changes. When the component motions were integrated into a global moving shape, occipital areas (V1, V2) and the LOC were more active than during incoherent states, while the reverse was true in dorsal areas (MT/V5). This pattern of BOLD activity supports the notion of interactions between dorsal and ventral areas during the observation of a bi-stable stimulus, although the precise underlying mechanisms remain unclear. One conceptual account is that of "predictive coding" (Murray et al. 2002), whereby activity in MT/V5 would be reduced whenever the direction and speed of the stimulus can be predicted and anticipated, which is possible during episodes of global motion perception but not during incoherent perceptual states.[4]

It is also possible that feedback from higher stages in the dorsal stream – e.g. from MST or LIP – come into play to modulate the integrative and antagonistic surround of MT neurons, as has been proposed by Huang et al. (2007) to account for their observations of the adaptability of MT receptive field surround to stimulus characteristics. Finally, there is evidence that some STP cells integrate form and motion, at least when stimulated with biological motion stimuli. As each of these proposals corresponds to a specific processing stage – early, medium or high – the whole process may involve all stages in a loop implying feed-forward and feedback computations.

---

[4] In their study, Murray et al. (2002) did not found the same pattern of results as that reported herein, but observed instead a balance of BOLD activity between V1 and the LOC. They do not mention a modulation of MT/V5 activity. The origin of this discrepancy remains unclear and could be related to differences in design and stimulation, or to their limited number of subjects.

Whatever the neuronal mechanisms, it is worth noting that pursuit eye movements, known to be controlled at the MT/MST stage, are strongly constrained by perceptual coherence, indicating that the dorsal pathway has access to a unified representation of component motions also yielding a unified shape, suggesting a least the existence of a shared processing between the ventral and dorsal streams (Stone et al. 2000). An example of the dependency of pursuit on motion coherence is shown in Fig. 1.9 (Lorenceau et al. 2004).

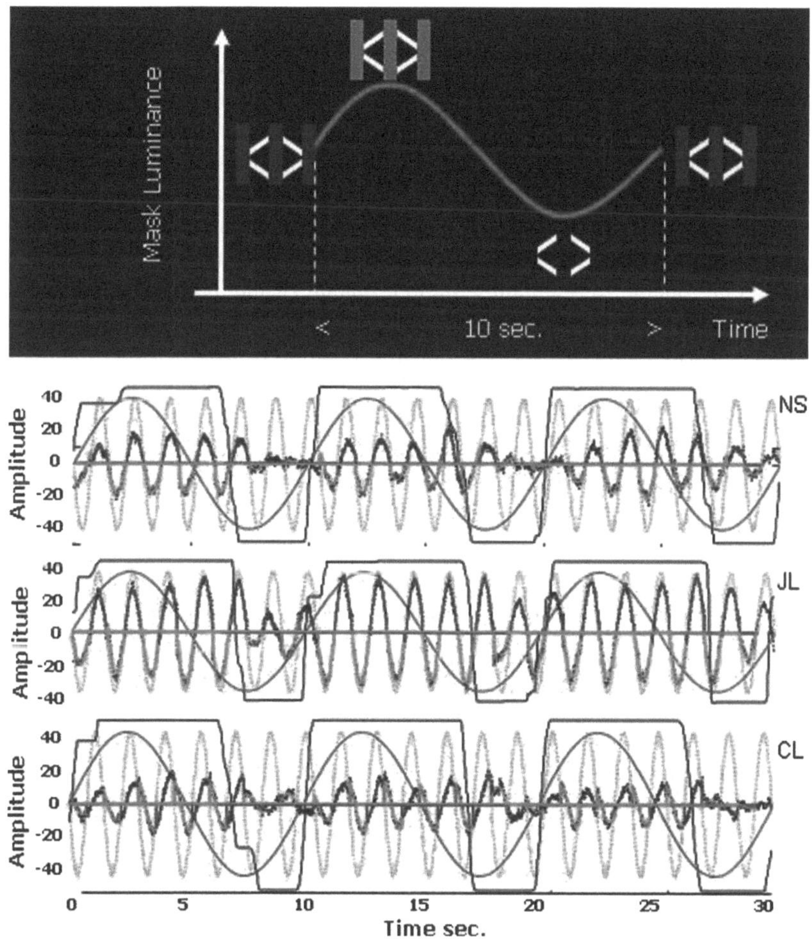

Fig. 1.9 *Top*. Illustration of the display used to study pursuit eye movements recorded during episodes of coherent and incoherent motion. Perceptual transitions were induced by smooth variation of mask luminance while the masked diamond rotated at 1 Hz. Observers were required to actively pursue the diamond's center while reporting their perceptual state with the pen of a tablet. *Bottom*: Results of three observers averaged across three periods of 30 s. *Green/blue* traces show the amplitude of horizontal pursuit eye movements as a function of time. Perceptual states are represented by the *black line*: *upward* for coherent states and *downward* for incoherent states. The *red line* represents mask luminance variations; the *dashed cyan line* shows horizontal stimulus motion. See text for details (*see Color Plates*)

In this experiment, coherence of a "masked diamond" stimulus was modulated by smoothly varying masks luminance (red traces) while a diamond, partially visible behind vertical masks, was rotating at 1 Hz (dashed cyan traces). Observers were asked to actively pursue the diamond's center and to indicate, by moving a pen on a tablet (black traces), the dynamics of their perceptual transitions between coherent and incoherent states. Under these conditions, segments moved up and down with no horizontal component. Thus, horizontal pursuit eye movements should reflect the perceived rather than physical stimulus motion. The results for three observers are shown in Fig. 1.9 (bottom) were horizontal pursuit averaged over three episodes of 30 s and fitted with a sliding sine function (blue/green traces) is plotted as a function of time. The amplitude of horizontal pursuit is large and in phase with stimulus rotation during episodes of coherent movement but is largely reduced or disappears during incoherent states, with a fast decrease of the horizontal pursuit gain after a perceptual switch. Note that the transition points for the two transition types (towards integration or towards segmentation, corresponding to the intersection points between red and black traces) are not identical, reflecting perceptual hysteresis. This hysteresis also exists in eye movement data, showing that observers are unable to maintain a smooth pursuit of the diamond center when a perceived horizontal component is lacking, despite a similar physical motion. Overall, experimental data suggest that the dichotomy between the ventral and dorsal pathways is not as strict as has been previously thought and/or that the assignment of functional properties – processing of form and motion – to these pathways is too schematic. (The observation of widespread responses to motion throughout the visual cortex favors the latter view).

1.8 Eccentric Versus Foveal Motion Integration

One remarkable feature of motion integration is its strong dependency upon the location of the incoming stimulus in the visual field: Central vs. Eccentric viewing conditions. Surprisingly, this dependency has not been the matter of much modeling or electrophysiological investigations despite the fact that for most motion displays used in the studies described above, the competition between motion integration and segmentation seen in central viewing condition is lacking or largely reduced in eccentric viewing conditions. Even for modest eccentricities (~7°) motion components that yield an incoherent percept in central vision blend into a global perceived motion (Lorenceau and Shiffrar 1992; De Bruyn 1997). Such dependency is unlikely to be accounted for by the increasing receptive field size with eccentricity, as the appearance of stimuli presented in central vision is mostly independent of viewing distance – e.g. the retinal size of the stimulus. Moreover, the form constraints described above are released in peripheral vision, such that all spatial configurations that are distinctively processed in central vision appear as having a similar global motion when presented in the periphery (Lorenceau and Alais 2001). The reasons for this dramatic change in the perception of motion are still unclear, but raise questions about the generality of the models aiming at simulating human vision. Several, non-exclusive, possibilities are worth considering. One builds upon the finding that association fields, and presumably the underlying long-range horizontal connections in V1, are absent – or not as dense – for eccentricities above 10° (Field et al. 1993). This fits well with the idea that associations field are involved in shaping the inputs to the motion integration stage. Alternately, the pattern of feedback connectivity which is known to play a role in motion perception (Bullier et al. 2001) may be heterogeneous across the visual field. A third possibility is that the processing of line-ends, which may exert a strong control on whether motion integration should proceed or not, is weakened in the periphery. One may speculate that the property of end-stopping or surround suppression is not homogenously distributed in the visual field, and may instead be restricted to central vision, a suggestion that has some support from electrophysiological studies (Orban 1984). Finally, one cannot exclude the possibility that the effect of eccentricity is related to the ratio of magnocellular to parvocellular cells. One line of research that may shed light on the effect of eccentricity on motion integration is related to "the crowding effect" mostly studied with static stimuli (but see Bex et al. 2003), in which the individuation and accessibility of some basic features are impaired by the presence of contextual stimuli in the target's vicinity.

1.9 Conclusion and Perspectives

In this chapter, I attempted to provide a survey of some experimental works concerned with the integration of form and motion, necessary to elaborate a reliable segmentation of a visual scene into perceptual entities on which recognition and action can rely. Several aspects have been ignored for the sake of clarity. The question of the contribution of mechanisms processing second order motion has not been addressed, mainly because reviews and literature on this topic are already available (Derrington et al. 2004). The question of the analysis of 3D form and motion and the ways the different components of the motion flow: rotation, expansion, etc are analyzed has not been included in this chapter. Let us notice that Rubin and Hochstein (1993) designed "aperture stimuli" with 3D moving shapes. With their displays they report the same dependence of motion integration on the status and reliability of 3D vertices as that described above for 2D translation. However, the processing of motion parallax in structure from motion displays allowing the recovery of 3D form may involve different mechanisms that were not addressed herein. In particular processing motion parallax involves fine estimates of speed, relative speed and speed gradients. Another aspect of motion processing concerns the tight coupling between the perception of motion and the oculomotor behavior and reciprocally the oculomotor behavior, and more generally observer's movement, and their influences on motion perception, whether it concerns perceived direction and speed (Turano and Heidenreich 1999) or the disambiguation of some aspects of the stimulus (see e.g. Wexler et al. 2001; Hafed and Krauzlis 2006). Some of these issues are addressed in other chapters of this book.

Although brief and incomplete, we hope that this overview of recent researches on motion integration provides insights into the mechanisms at work, pointing to a cascade of processes whereby the parsing of moving objects involves numerous intermingled steps recruiting different cortical structures of the visual systems, both in the ventral and dorsal streams. These advances and the progressive identification of the pieces of the puzzle although far from allowing drawing the whole picture, suggests new issues that additional experimental work may uncover in the future. Figure 1.10 provides a schematic representation of circuits underlying form/motion integration together with their functional roles. This schema should definitively not be taken as corresponding to the real computations performed by the brain to recover object's motion but as an attempt to summarize the findings described in this chapter based on our current knowledge of the functional specialization of some visual areas. A large number of studies in macaque monkey and more recently with brain imaging techniques in humans uncovered additional motion areas indicating that the picture is far from the simple one offered here.

Figure 1.10 reads as follows: At an entry stage, neurons in area V1 performs motion detection through limited receptive fields which presumably involves computing motion energy (Adelson and Bergen 1984; Emerson et al. 1992). At this stage each direction selective cell faces the "aperture" problem and only provides crude estimates of local direction. Surround suppression common to many V1 and

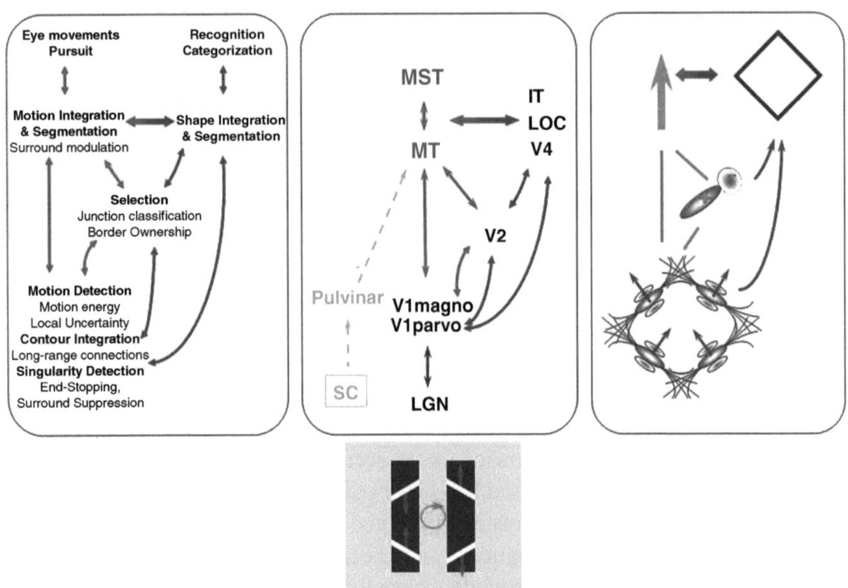

Fig. 1.10 Schematic representation summarizing the results presented in the text. *Left*: Depiction of the processes involved in form/motion integration. *Middle*: Putative areas implementing the perceptual processes. *Right*: Graphical illustration of some mechanisms and perceptual outputs. See text for details

V2 neurons would allow the computation of moving singularities, such as line-endings. At this early stage, processes related to contour integration using long-range horizontal connections and contour segmentation using end-stopped responses would perform the computation of a “proto-shape” implementing some of the gestalt principles, good continuation and closure in particular. This process presumably benefits from feedback from later processing stages (e.g. area V2), but the nature of the neural signature of the resulting neuronal assembly remains to be determined. Area V2 is complex and diverse (see Sincich and Horton 2005). Electrophysiological evidence nevertheless suggests that some sub-structures within area V2 are involved in the assessment of border ownership and in the classification of singularities such as T-junctions, vertices, etc. (Qiu et al. 2007). The central position of area V2, at the cross road between the Ventral and dorsal pathways and the fact that V2 sends projections to the MT/MST complex makes it well suited to gate motion, as well as form, integration. Pooling the responses of V1 direction selective neurons is thought to occur in the MT/MST complex. At motion onset, experimental evidence suggests that pooling is fast and undifferentiated, while motion parsing would process more slowly. There remain, however, uncertainties and debates about the specific computations realized at this stage. They concern the combination rule used to pool motion signals across space but also the functional role of surround suppression that appears more flexible than previously thought and can switch to surround facilitation depending upon the presence and nature of contextual information (Huang et al. 2007). The origin of the modulating influence of the surround influence is still unknown. One intriguing possibility is that they originate from areas processing form information (area V2 and/or areas in the ventral pathway). Oculomotor behavior involves a large network of cortical and sub-cortical areas, not detailed herein (see Krauzlis 2005 for a review). At the cortical level the MT/MST complex is involved in the control of pursuit (Newsome et al. 1986). The observation that pursuit is itself dependent on the perceptual coherence of moving patterns and not solely on the retinal slip (Stone et al. 2000; Stone and Krauzlis 2003) suggests that neural signals related to object motion are present and used at this MT/MST stage.

The parallel computation of shape properties also faces ambiguities and uncertainties related to border ownership, junction classification, stereo “aperture” problem, etc. whose resolutions help motion integration and also benefit from motion processing, e.g. processing of kinetic boundaries, dynamic occlusion (see Shipley and Kellman 1994). It is out of the scope of the present review to detail the processing steps involved in shape processing. Let us just note that V2 and areas distributed within the ventral stream appear to handle shape integration.

One important point to emphasize is that shape and motion integration interact, may be through reciprocal connections between the MT/MST complex and the LOC. Although a functional role of these interactions is the adequate parsing of moving objects in a visual scene, a number of questions still remain. What information is transferred through these interactions? Are they related to intrinsic stimulus characteristics, to expectations and predictions, to attention and decision, to prior knowledge and memory? What kinds of modulating – facilitating, suppressive, recurrent – signals are sent and more importantly how does the system select the neuronal targets of these interactions within the dorsal and ventral areas? Answers to these questions await future experimental and modeling work.

1.10 Supplementary Materials (DVD)

Movie 1 *Dynamics of motion recovery* (file "*1_M1_TiltedLine.avi*"). This movie demonstrates the illusory direction perceived with a single oblique line moving back and forth horizontally. At each direction reversal a brief episode of motion in a direction perpendicular to line orientation can be seen. This effect is attributed to the slow processing of line-endings that carry information relative to the "real" direction of motion (Lorenceau et al. 1993).

Movie 2 *Local salience and motion integration* (file "*1_M2_DotMovDiam.avi*"): This demonstration presents a diamond stimulus made of aligned dots moving with a velocity compatible with a global motion. However, this global rotating motion is seen only when the dot motion salience is decreased by a "motion-noise." Smooth transitions from one extreme (no motion noise) to the other (full motion-noise) yield changes in perceived global motion. Eccentric viewing conditions entail a global motion percept (Lorenceau 1996).

Movie 3 *The "Chopstick" illusion* (file "*1_M3_Chopstick.avi*") illustrates the influence of terminator motion on motion perception. The crossing of two moving lines, strongly depend upon the visibility of their line-ends (Anstis 1990).

Movie 4 *Diamond Integration and Hysteresis* (file "*1_M4_DiamHysteresis.avi*") illustrates the perception of motion integration and segmentation that occur when smoothly varying masks' luminance in the "Masked Diamond" paradigm. In addition, the demo illustrates the phenomenon of hysteresis, a signature of cooperative/competitive mechanisms, whereby the visual system tends to maintain its current state. In this demo the physical parameters corresponding to a perceptual transition from coherent to an incoherent state are different from those corresponding to a perceptual transition from incoherent to a coherent state (Lorenceau et al. 2003).

Movies 5–8 *Tiled moving surfaces* (files "*1_M5_T_Diam_LowC.avi*," "*1_M6_L_Diam_LowC.avi*," "*1_M7_T_Diam_HighC.avi*,", "*1_M8_L_Diam_HighC.avi*"). These four demonstrations illustrate the influence of spatial configuration and contrast on motion integration. At high contrast it is more difficult to perceive a global movement (a translation along a circular trajectory) with the T-like tiled surface as compared to the L-like tiled surface. The global movement is more easily recovered at a low contrast, whatever the spatial configuration. The difference between T and L configurations may reflect the linking of the individual gratings into multiple diamonds for the L configuration, a process that could involve long-range horizontal connections in primary visual cortex (Lorenceau and Zago 1999). Note that eccentric viewing conditions increases coherency for both configurations.

Movie 9 *Shape and motion integration* (File "*1_M9_DiaFormMorph.avi*"): this movie presents a diamond changing into a chevron while rotating along a circular trajectory. Perceiving the global movement is easier when the shape is closed (diamond-like shapes) as compared to a situation when it is not (chevron-like shapes). The effect of shape on motion integration suggests a strong influence of form information on motion perception. Note that the attenuation of the difference between shapes when the stimulus is observed in eccentric vision (Lorenceau and Alais 2001).

References

Adelson EH, & Bergen JE (1985). Spatiotemporal energy models for the perception of motion. Journal of the Optical Society of America 2:284–299

Adelson EH, Movshon JA (1982) Phenomenal coherence of moving visual patterns. Nature 300:523–525

Alais D, Lorenceau J (2002) Perceptual grouping in the Ternus display: evidence for an 'association field' in apparent motion. Vision Res 42:1005–1016

Allman JM, Miezin FM, McGuinness E (1985) Direction and velocity-specific responses from beyond the classical receptive field in the middle temporal visual area (MT). Perception 14:105–126

Anstis SM, (1990) Imperceptible intersections: The chopstick illusion. In A. Blake & T. Troscianko (Eds.), AI and the eye (pp. 105–117). London: Wiley

Barthélemy FV, Perrinet LU, Castet E, Masson GS (2008) Dynamics of distributed 1D and 2D motion representations for short-latency ocular following. Vision Res 48(4):501–522

Bex PJ, Dakin SC, Simmers AJ (2003) The shape and size of crowding for moving targets. Vision Res 43:2895–2904

Bishop PO, Coombs JS, Henry GH (1971) Responses to visual contours: spatiotemporal aspects of excitation in the receptive fields of simple striate neurons. J Physiol (Lond) 219:625

Born RT, Bradley DC (2005) Structure and function of visual area MT. Annu Rev Neurosci 28:157–189

Bowns L (1996) Evidence for a feature tracking explanation of why type II plaids move in the vector sum direction at short durations. Vision Res 36:3685–3694

Bowns L, Alais D (2006) Large shifts in perceived motion direction reveal multiple global motion solutions. Vision Res 46:1170–1177

Bringuier V, Chavane F, Glaeser L, Frégnac Y (1999) Horizontal propagation of visual activity revealed in the synaptic integration field of area 17 neurons. Science 283:695–699

Bulakowski PF, Bressler DW, Whitney D (2007) Shared attentional resources for global and local motion processing. J Vis 7:1–10

Bullier J, Hupé JM, James AC, Girard P (2001) The role of feedback connections in shaping the responses of visual cortical neurons. Prog Brain Res 134:193–204

Cass J, Alais D (2006) The mechanisms of collinear integration. J Vis 6(9):915–922

De Bruyn B (1997) Blending transparent motion patterns in peripheral vision. Vision Res 7:645–648

DeAngelis GC, Cumming BG, Newsome WT (1998) Cortical area MT and the perception of stereoscopic depth. Nature 394:677–680

Delicato LS, Derrington AM (2005) Coherent motion perception fails at low contrast. Vision Res 45:2310–2320

Derrington AM, Allen HA, Delicato LS (2004) Visual mechanisms of motion analysis and motion perception. Annu Rev Psychol 55:181–205

Dobkins A, Zucker SW, Cynader MS (1987) Endstopped neurons in the visual cortex as a substrate for calculating curvature. Nature 329:438–441

Duffy CJ, Wurtz RH (1995) Response of monkey MST neurons to optic flow stimuli with shifted centers of motion. J Neurosci 15:5192–5208

Dumbrava D, Faubert J, Casanova C (2001) Global motion integration in the cat's lateral posterior–pulvinar complex. Eur J NeuroSci 13:2218–2226

Duncan RO, Albright TD, Stoner GR (2000) Occlusion and the interpretation of visual motion: perceptual and neuronal effects of context. J Neurosci 20:5885–5897

Duncker K (1929) Uber induzierts Bewegung. Psychol Forsch 2:180–259 (Translated and condensed as: Induced motion. In: Ellis WD (ed) A source book on gestalt psychology. Humanities Press, New York, 1967)

Emerson RC, Bergen JR, Adelson EH (1992) Directionally selective complex cells and the computation of motion energy in cat visual cortex. Vision Res 32:203–218

Fennema CL, Thompson WB (1979) Velocity determination in scenes containing several moving objects. Comput Graph Image Process 9:301–315

Field DJ, Hayes A, Hess RF (1993) Contour integration by the human visual system: Evidence for a local "association field". Vision Res, 33:173–193

Gallant JL, Braun J, Van Essen DC (1993) Selectivity for polar, hyperbolic, and Cartesian gratings in macaque visual cortex. Science 259:100–103

Georges S, Seriès P, Frégnac Y, Lorenceau J (2002) Orientation dependent modulation of apparent speed: psychophysical evidence. Vision Res 42:2757–2772

Giersch A, Lorenceau J (1999) Effects of a benzodiazepine, Lorazepam, on motion integration and segmentation: an effect on the processing of line-ends? Vision Res 39:2017–2025

Gilbert CD, Wiesel T (1989) Columnar specificity of intrinsic horizontal and corticocortical connections in cat visual cortex. J Neurosci 9(7):2432–2442

Gorea A, Lorenceau J (1991) Directional performance with moving plaids, component-related and plaid-related processing modes coexist. Spatial Vision 5(4):231–252

Grosof DH, Shapley RM, Hawken MJ (1993) Macaque V1 neurons can signal illusory contours. Nature 365:550–552

Grossberg S, Mingolla E, Viswanathan L (2001) Neural dynamics of motion integration and segmentation within and across apertures. Vision Res 41:2521–2553

Hafed ZM, Krauzlis RJ (2006) Ongoing eye movements constrain visual perception. Nat Neurosci 9:1449–1457

Henry GH, Bishop PO (1971) Simple cells of the striate cortex. In: Neff WD (ed) Contributions to sensory physiology. Academic, New York, pp 1–46

Hess RH, Hayes A, Field D (2003) Contour integration and cortical processing. J Physiol – Paris 97:105–119

Huang X, Albright TD, Stoner G (2007) Adaptive surround modulation in cortical area MT. Neuron 53:761–770

Hupé JM, Rubin N (2003) The dynamics of bi-stable alternation in ambiguous motion displays: a fresh look at plaids. Vision Res 43:531–548

Jancke D, Chavane F, Na'aman S, Grinvald A (2004) Imaging cortical correlates of illusion in early visual cortex. Nature 428:423–426

Jazayeri M, Movshon JA (2007) A new perceptual illusion reveals mechanisms of sensory decoding. Nature 446:912–915

Kapadia MK, Ito M, Gilbert C, Westheimer G (1995) Improvement in visual sensitivity by changes in local context: parallel studies in human observers and in V1 of alert monkeys. Neuron 15:843–856

Kapadia MK, Westheimer G, Gilbert CD (2000) Spatial distribution of contextual interactions in primary visual cortex and in visual perception. J Neurophysiol 84:2048–2062

Koenderink JJ (1986) Optic flow. Vision Res 1:161–180

Koechlin E, Anton JL, Burnod Y (1999) Bayesian inference in populations of cortical neurons: a model of motion integration and segmentation in area MT. Biol Cybern 80(1):25–44

Kooi FL (1993) Local direction of edge motion causes and abolishes the barberpole illusion. Vision Res 33:2347–2351
Kourtzi Z, Kanwisher N (2001) Representation of perceived object shape by the human lateral occipital complex. Science 293:1506–1509
Kovacs I, Julesz B (1993) A closed curve is much more than an incomplete one: Effect of closure in figure-ground segmentation. Proc Natl Acad Sci USA 90:7495–7497
Krauzlis RJ (2005) The control of voluntary eye movements: new perspectives. Neuroscientist 11:124–137
Lalanne C, Lorenceau J (2006) Directional shifts in the Barber Pole illusion: effects of spatial frequency, contrast adaptation and lateral masking. Vis Neurosci 23:729–739
Levitt JB, Lund JS (2002) Intrinsic connections in mammalian cerebral cortex. In: Schuez A, Miller R (eds) Cortical areas: unity and diversity. Taylor and Francis, London, UK
Liden L, Pack C (1999) The role of terminators and occlusion cues in motion integration and segmentation: a neural network model. Vision Res 39:3301–3320
Löffler G, Orbach HS (1999) Computing feature motion without feature detectors: a model for terminator motion without end-stopped cells. Vision Res 39:859–871
Lorenceau J (1996) Motion Integration with dot patterns: effects of motion noise and structural information. Vision Res 36:3415–3428
Lorenceau J (1998) Veridical perception of global motion from disparate component motions. Vision Res 38:1605–1610
Lorenceau J, Alais D (2001) Form constraints in motion binding. Nat Neurosci 4:745–751
Lorenceau J, Boucart M (1995) Effects of a static texture on motion integration. Vision Res 35:2303–2314
Lorenceau J, Shiffrar M (1992) The influence of terminators on motion integration across space. Vision Res 2:263–275
Lorenceau J, Shiffrar M, (1999) The linking of visual motion. Visual Cognition, 3–4, vol 6, 431–460
Lorenceau J, Zago L (1999) Cooperative and competitive spatial interactions in motion integration. Vis Neurosci 16:755–770
Lorenceau J, Shiffrar M, Walls N, Castet E (1993) Different motion sensitive units are involved in recovering the direction of moving lines. Vision Res 33:1207–1218
Lorenceau J, Baudot P, Series P, Georges S, Pananceau M, Frégnac Y (2002) Modulation of apparent motion speed by horizontal intracortical dynamics [Abstract]. J Vis 1(3):400a
Lorenceau J, Gimenez-Sastre B, Lalanne C (2003) Hysteresis in perceptual binding. Perception 32, ECVP Abstract Supplement
Lorenceau J, Giersch A, Series P (2005) Dynamics of competition between contour integration and contour segmentation probed with moving stimuli. Vision Res 45:103–116
Majaj N, Smith MA, Kohn A, Bair W, Movshon JA (2002) A role for terminators in motion processing by macaque MT neurons? [Abstract]. J Vis 2(7):415a
Majaj NJ, Carandini M, Movshon JA (2007) Motion integration by neurons in macaque MT is local, not global. J Neurosci 27:366–370
Malach R, Reppas JB, Benson RR, Kwong KK, Jlang H, Kennedy WA, Ledden PJ, Brady TJ, Rosen BR, Tootell RBH (1995) Object-related activity revealed by functional magnetic resonance imaging in human occipital cortex. Proc Natl Acad Sci U S A 92:8135–8139
Marshak W, Sekuler R (1979) Mutual repulsion between moving visual targets. Science 205:1399–1401
Masson GS, Castet E (2002) Parallel motion processing for the initiation of short-latency ocular following in humans. J Neurosci 22:5149–5163
Masson GS, Mestre DR, Stone LS (1999) Speed tuning of motion segmentation and discrimination. Vision Res 39:4297–4308
Masson GS, Rybarczyk Y, Castet E, Mestre DR (2000) Temporal dynamics of motion integration for the initiation of tracking eye movements at ultra-short latencies. Vis Neurosci 17:753–767

Maunsell JHR, Gibson JR (1992) Visual responses latencies in striate cortex of the macaque monkey. J Neurophysiol 68(4):1332–1343
McDermott J, Adelson EH (2004) The geometry of the occluding contour and its effect on motion interpretation. Journal of Vision, 4(10):9, 944–954, http://journalofvision.org/4/10/9/, doi:10.1167/4.10.9
McDermott J, Weiss Y, Adelson EH (2001) Beyond junctions: Nonlocal form contraints on motion interpretation. Perception 30:905–923
Merabet L, Desautels A, Minville K, Casanova C (1998) Motion integration in a thalamic visual nucleus. Nature 396:265–268
Mingolla E, Todd JT, Norman JF (1992) The perception of globally coherent motion. Vision Res 32:1015–1031
Movshon AJ, Adelson EH, Gizzi MS, Newsome WT (1986) The analysis of moving visual patterns. Exp Brain Res 11:117–152
Murray SO, Kersten D, Olshausen BA, Schrater P, Woods DL (2002) Shape perception reduces activity in human primary visual cortex. Proc Natl Acad Sci U S A 99(23):15164–15169
Newsome WT, Dürsteler MR, Wurtz RH (1986) The middle temporal visual area and the control of smooth pursuit eye movements. In: Keller EL, Zee DS (eds) Adaptive processes in visual and oculomotor systems. Pergamon, New York
Nowlan SJ, Sejnowski TJ (1995) A selection model for motion processing in area MT of primates. J Neurosci 15:1195–1214
Orban GA (1984) Neuronal operations in the visual cortex. Springer, New York
Pack CC, Born RT (2001) Temporal dynamics of a neural solution to the aperture problem in visual area MT of macaque brain. Nature 409:1040–1042
Pack CC, Born RT (2005) Contrast dependence of suppressive influences in cortical area MT of alert macaque. J Neurophysiol 93:1809–1815
Pack CC, Born RT, Livingstone MS (2003a) Two-dimensional substructure of stereo and motion interactions in macaque visual cortex. Neuron 37:525–535
Pack CC, Livingstone MS, Duffy KR, Born RT (2003b) End-stopping and the aperture problem: two-dimensional motion signals in macaque V1. Neuron 39(4):671–680
Pack CC, Gartland AJ, Born RT (2004) Integration of contour and terminator signals in visual area MT of alert macaque. J Neurosci 24(13):3268–3280
Pasupathy A, Connor CE (1999) Responses to contour features in Macaque area V4. J Neurophysiol 82:2490–2502
Qian N, Andersen RA, Adelson EH (1994) Transparent motion perception as detection of unbalanced motion signals I. Psychophysics. J Neurosci 14:7357–7366
Qiu FT, Sugihara T, von der Heydt R (2007) Figure-ground mechanisms provide structure for selective attention. Nat Neurosci 10:1492–1499
Rodman HR, Albright TD (1989) Single-unit analysis of pattern-motion selective properties in the middle temporal visual area (MT). Exp Brain Res 75:53–64
Rubin N, Hochstein S (1993) Isolating the effect of one-dimensional motion signals on the perceived direction of moving two-dimensional objects. Vision Res 10:1385–1396
Rust N, Mante V, Simoncelli EP, Movshon JA (2006) How MT cells analyze the motion of visual patterns. Nat Neurosci 9:1421–1431
Sceniak MP, Ringach DL, Hawken MJ, Shapley R (1999) Contrast's effect on spatial summation by macaque V1 neurons. Nat Neurosci 2:733–739
Seriès P, Georges S, Lorenceau J, Frégnac Y (2002) Orientation dependent modulation of apparent speed: a model based on center/surround interactions. Vision Res 42:2781–2798
Seriès PS, Lorenceau J, Frégnac Y (2003) The silent surround of V1 receptive fields: theory and experiments. J Physiol (Paris) 97:453–474
Shiffrar M, Lorenceau J (1996) Increased motion linking across edges with decreased luminance contrast, edge width and duration. Vision Res 36:2061–2068
Shiffrar M, Li X, Lorenceau J (1995) Motion integration across differing image features. Vision Res 35:2137–2146

Shimojo S, Silverman G, Nakayama K (1989) Occlusion and the solution to the aperture problem for motion. Vision Res 29:619–626
Shipley TF, Kellman PJ (1994) Spatiotemporal boundary formation: boundary, form, and motion perception from transformations of surface elements. J Exp Psychol Gen 123:3–20
Simoncelli, Heeger D (1998) A model of neuronal responses in visual area MT. Vision Res 38: 743–761
Sincich LC, Blasdel GG (2001) Oriented axon projections in primary visual cortex of the monkey. J Neurosci 21(12):4416–4426
Sincich LC, Horton JC (2005) The circuitry of V1 and V2: integration of color, form and motion. Annu Rev Neurosci 28:303–326
Sincich LC, Park KF, Wohlgemuth MJ, Horton JC (2004) Bypassing V1: a direct geniculate input to area MT. Nat Neurosci 7(10):1123–1128
Singer W (1995) The organization of sensory motor representations in the neocortex: a hypothesis based on temporal coding. In: C. Ulmita and M. Moscovitch (Eds.) Attention and Performance XV: Conscious and Nonconscious Information processing, MIT Press: Cambridge (Mass.)
Stone GR, Albright TD (1992) Neural correlates of perceptual motion coherence. Nature 358:412–414
Stone LS, Krauzlis RJ (2003) Shared motion signals for human perceptual decisions and oculomotor actions. J Vis 3:725–736
Stone LS, Thompson P (1992) Human speed perception is contrast dependent. Vision Res 32:1535–1549
Stone LS, Watson AB, Mulligan JB (1990) Effects of contrast on the perceived direction of moving plaids. Vision Res 30:619–626
Stone LS, Beutter B, Lorenceau J (2000) Shared visual motion integration for perception and pursuit. Perception 29:771–787
Tanaka K, Saito H, Fukada Y, Moriya M (1991) Coding visual images of objects in the inferotemporal cortex of the macaque monkey. J Neurophysiol 66:170–189
Thompson P (1982) Perceived rate of movement depends on contrast. Vision Res 22:377–380
Turano K, Heidenreich SM (1999) Eye movements affect the perceived direction of visual motion. Vision Res 39:1177–1187
Vaina LM (1989) Selective impairment of visual motion interpretation following lesions of the right occipito-parietal area in humans. Biol Cybern 61:347–359
Vaina LM, Cowey A, Jakab M, Kikinis R (2005) Deficits of motion integration and segregation in patients with unilateral extrastriate lesions. Brain 128:2134–2145
Vallortigara G, Bressan P (1991) Occlusion and the perception of coherent motion. Vision Res 31:1967–1978
Van der Berg AV, Noest AJ (1993) Motion transparency and coherence in plaids: the role of end-stopped cells. Exp Brain Res 96:519–533
Van Essen DC, Maunsell JH, Bixby JL (1981) The middle temporal visual area in the macaque: myeloarchitecture, connections, functional properties and topographic organization. J Comp Neurol 199:293–326
Watamaniuk SNJ, Duchon A (1992) The human visual system averages speed information. Vision Res 32:931–941
Watamaniuk SNJ, Sekuler R (1992) Temporal and spatial integration in dynamic random-dot stimuli. Vision Res 32:2341–2348
Watamaniuk SNJ, Sekuler R, Williams DW (1989) Direction perception in complex dynamic displays: the integration of direction information. Vision Res 29:47–59
Watamaniuk SNJ, Grzywacz NM, McKee SP (1995) Detecting a trajectory embedded in random-direction visual noise. Vision Res 35:65–77
Weiss Y, Adelson EH (1995) Perceptually organized EM: a framework for motion segmentation that combines information about form and motion. MIT Media Laboratory Perceptual Computing Section Technical Report No. 315: ICCV'95
Weiss Y, Adelson EH (2000) Adventures with gelatinous ellipses–constraints on models of human motion analysis. Perception 29(5):543–566

Wexler M, Panerai F, Lamouret I, Droulez J (2001) Self-motion and the perception of stationary objects. Nature 409:85–88
Williams D, Phillips G (1987) Cooperative phenomena in the perception of motion direction. J Opt Soc Am 4:878–885
Williams DW, Sekuler R (1984) Coherent global motion percepts from stochastic local motions. Vision Res 24:55–62
Wilson HR, Kim J (1994) A model for motion coherence and transparency. Vis Neurosci 11:1205–1220
Wilson HR, Ferrera VP, Yo C (1992) A psychophysically motivated model for the two-dimensional motion perception. Vis Neurosci 9:79–97
Yazdanbakhsh A, Livingstone MS (2006) End stopping in V1 is sensitive to contrast. Nat Neurosci 9:697–702
Yo C, Wilson HR (1992) Perceived direction of moving two-dimensional patterns depends on duration, contrast and eccentricity. Vision Res 32:135–147

Chapter 2
Temporal Dynamics of Motion Integration

Richard T. Born, James M. G. Tsui, and Christopher C. Pack

Abstract In order to correctly determine the velocity of moving objects, the brain must integrate information derived from a large number of local detectors. The geometry of objects, the presence of occluding surfaces and the restricted receptive fields of early motion detectors conspire to render many of these measurements unreliable. One possible solution to this problem, often referred to as the "aperture problem," involves differential weighting of local cues according to their fidelity: measurements made near two-dimensional object features called "terminators" are selectively integrated, whereas one-dimensional motion signals emanating from object contours are given less weight. A large number of experiments have assessed the integration of these different kinds of motion cues using perceptual reports, eye movements and neuronal activity. All of the results show striking qualitative similarities in the temporal sequence of integration: the earliest responses reveal a non-selective integration which becomes progressively selective over a period of time. In this chapter we propose a simple mechanistic model based on end-stopped, direction-selective neurons in V1 of the macaque, and use it to account for the dynamics observed in perception, eye movements, and neural responses in MT.

2.1 Temporal Dynamics of Perception and the "Aperture Problem"

Perception is neural computation, and, because neurons are relatively slow computational devices, perception takes time. On the one hand, this sluggish processing is a potential detriment to an animal's survival, and we might expect at least certain

R.T. Born (✉)
Department of Neurobiology, Harvard Medical School, Boston, MA, USA
e-mail: rborn@hms.harvard.edu

U.J. Ilg and G.S. Masson (eds.), *Dynamics of Visual Motion Processing: Neuronal, Behavioral, and Computational Approaches*,
DOI 10.1007/978-1-4419-0781-3_2,

perceptual computations to be highly optimized for speed. On the other hand, the relative slowness of some neural systems may be of benefit to the investigator attempting to understand the circuitry responsible for the computation. Indeed, the temporal evolution of perceptual capacities has been exploited by psychophysicists for many years. By measuring reaction times, limiting viewing times, or using clever tricks such as masking to interrupt perceptual processes at different times, they have gained valuable insights into the nature of successive stages of perceptual computations.

One general theme that has arisen from this body of work is the idea that, when presented with a novel stimulus, perceptual systems first rapidly compute a relatively rough estimate of the stimulus content and then gradually refine this estimate over a period of time. This is demonstrated, for example, by the fact that human observers require less viewing time to recognize the general category to which an object belongs than to identify the specific object (Rosch et al. 1976; Thorpe and Fabre-Thorpe 2001). Similarly, the recovery of stereoscopic depth by comparing images between the two eyes appears to follow a coarse-to-fine progression, with large spatial scales being processed before fine details (Marr and Poggio 1976; Wilson et al. 1991; Rohaly and Wilson 1993, 1994). Furthermore, we will describe in some detail below that the visual motion system uses a similar strategy to compute the direction of motion of objects. Such a strategy may reflect the genuine computational needs of sensory systems – such as the use of coarse stereo matches to constrain subsequent fine ones in order to solve the correspondence problem (Marr et al. 1979) – as well as selective pressures for animals to be able to rapidly initiate behavioral responses, even in the absence of perfect, or detailed, information.

In this chapter, we will consider these issues from the perspective of visual motion perception. A solid object can only be moving in one direction at any given time, yet sampling the motion of small regions of the object can result in disparate estimates of this direction. This constraint on the measurement of motion direction is highly relevant to the visual systems of humans and other animals, in which early visual structures have neurons with small receptive fields. A more concrete way of thinking about the limited receptive field size of these visual neurons is as "apertures," depicted as circles in the inset of Fig. 2.1a. These apertures, in conjunction with the geometry of moving objects, create local motion signals that are frequently ambiguous. For example, if a square-shaped object moves upwards and to the right, a neuron with a small receptive field positioned along one of the object's vertical edges can measure only the rightward component of motion. This measurement is ambiguous, because it is consistent with many possible directions of actual object motion. In general a motion measurement made from a one-dimensional (1D) feature will always be ambiguous, because no change can be measured in the direction parallel to the contour. Only neurons whose receptive fields are positioned over a two-dimensional (2D) feature, such as a *corner* of the square object (often referred to in the literature as a "terminator"), can measure the direction of object motion accurately.

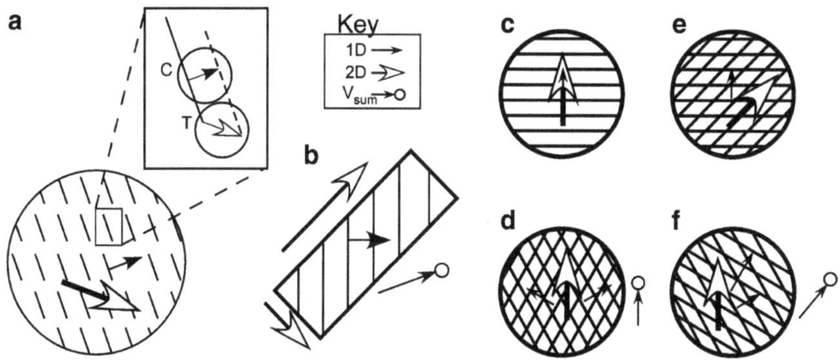

Fig. 2.1 Visual stimuli used to study the dynamics of 1D-to-2D motion. (**a**) Tilted bar-field used by Lorençeau et al. (1993). In this particular example, the 2D direction of motion has a downward component, whereas the 1D direction measured along the contour has an upward component. The inset depicts the situation in greater detail as seen through the apertures of neuronal receptive fields. (**b**) Barber pole in which the direction of grating motion differs by 45° from that of the perceived direction, which is up and to the right (**c**) Single grating. (**d**) Symmetric Type I plaid consisting of two superimposed 1D gratings. (**e**) Unikinetic plaid. Only the horizontal grating moves (*upwards*), but the static oblique grating causes the pattern to appear to move up and to the right. (**f**) Type II plaid in which the perceived direction of the pattern is very different from that of either of the two components or the vector sum. (see also the corresponding movies for each stimulus type)

2.2 Psychophysics of Motion Integration

A large body of experimental and theoretical work has addressed the question of how various local motion measurements are integrated to produce veridical calculations of object motion. Our purpose here is not to review the entire literature (for this, see Pack and Born 2008), but rather to focus on one particular aspect of the computation, namely its temporal dynamics, that may be of particular use in elucidating the neural circuitry that carries it out.

The starting point for this project is the observation that observers make systematic perceptual errors when certain stimuli are viewed for a short amount of time (Lorençeau et al. 1993). That is, the visual system's initial calculations are *not* always veridical. This can be appreciated directly from Movie 1 in which a long, low contrast bar moves obliquely with respect to its long axis. While fixating the red square, most observers see the bar following a curved trajectory, beginning with an upward component that then bends around to the right. In reality the motion is purely horizontal, so this initial upwards component would seem to be a direct manifestation of the aperture problem: of the many direction-selective neurons whose receptive fields would be confined to the bar's contour, those that should respond maximally are those whose preferred direction is up and to the right; hence the mistaken percept.

This phenomenon was explored by Lorençeau et al. (1993), who asked human observers to report the direction of motion of arrays of moving lines similar to those in Movie 1. The lines were tilted either +20° or −20° from vertical, and they moved along an axis tilted either +20° or −20° from the horizontal. Observers were asked to report whether the vertical component of the motion was upwards or downwards using a 2-alternative forced choice procedure. The key aspects of the experimental design were (1) that neither orientation alone nor a combination of orientation and horizontal direction of motion could be used to solve the task and (2) for a given line orientation, the four possible directions of movement produced two conditions in which motion was perpendicular to the orientation of the lines and two in which it was oblique. Importantly, for the two latter conditions, the tilt of the lines would produce "aperture motion" (that is, local motion measured perpendicular to the contours) whose vertical component was opposite to that of the true direction of line motion. For example, for an array of lines tilted 20° to the left of the vertical (counterclockwise), line motion to the right and 20° *downwards* from horizontal would produce aperture motion to the right and 20° *upwards* from the horizontal. Thus, for the two test conditions, insofar as the observers' percepts were influenced by the component of motion perpendicular to line orientation, they should tend to report the wrong direction.

For the control conditions, the observers' reports were accurate under all stimulus conditions. For the test conditions, however, observers often reported the wrong direction of motion, as if their visual systems had been fooled by the aperture problem. For many conditions, the performance was significantly poorer than chance, indicating that the direction of motion was indeed systematically misperceived and not simply difficult to judge. (If the latter had occurred, performance would have been 50% correct.) The Lorençeau group systematically varied three stimulus parameters – line *length*, line *contrast* and the *duration* of stimulus presentation – in order to probe the conditions under which the visual system was most likely to err. The general result was that for arrays of relatively long lines (~3°) at low contrast (<30%) and presented for short durations (~150 ms), observers never reported the true direction of motion. Conversely, as the lines were made shorter, of higher contrast or were viewed for longer durations, performance improved. Although not all possible combinations of these three variables were tested, it was clear that they interacted in relatively predictable ways. Thus, for example, even high-contrast (70%) lines of modest lengths (2.5°) were misperceived by many observers when viewing time was limited to 130 ms. Lowering the contrast to 39% greatly reduced performance for all observers, even for relatively long stimulus presentations (up to 0.5 s).

A similar kind of result was obtained by Yo and Wilson (1992) for the perception of "type II" plaids. In their experiments, the task of the observer was to integrate the motion of two superimposed drifting sinusoidal gratings. By definition, each component grating is a one-dimensional motion stimulus containing directional information only along the axis perpendicular to the orientation of the grating's stripes (Fig. 2.1c). When two such gratings moving in different directions are superimposed, the resulting direction of the plaid motion can be computed in several different ways yielding different possible directions. For certain combinations – referred to as "type II" plaids – the simple vector sum[1] of the two 1D directions produces one predicted direction, whereas an algorithm sensitive to the direction of motion of the 2D features produced by the gratings' intersections, produces a different predicted direction (Fig. 2.1f). The main result of the Yo and Wilson study was that observers reported the vector sum direction for brief stimulus presentations (60 ms) but tended to see the feature direction as viewing time was increased, and, as in the tilted bar experiments, the time-course of the transition was prolonged with gratings of lower contrast.

2.3 Studies of Motion Integration Using Eye Movements

For the perceived direction of tilted bars and type II plaids, the effect of viewing time clearly indicated that motion integration is a dynamic process. Early on, the visual system computes an estimate of motion direction, but it is biased by non-veridical measurements along contours. As time goes by, the correct solution emerges. However, psychophysical judgments are by nature discrete: the stimulus is observed for a fixed amount of time and a single response is given. That single response is presumably the outcome of a dynamic computation, during which various possible solutions are represented in the visual cortex. There is therefore no way to determine whether the observer's response reflects the integrated percept, the most recent mental snapshot or some other way of combining percepts over a period of time. In this respect, psychophysics is not ideal for addressing the issue of temporal dynamics. With respect to visual motion, however, one can monitor other outputs that arguably make use of the same motion processing circuitry: eye movements. In the case of certain eye movements, such as ocular following and smooth pursuit, we are afforded a continuous read-out of the process with no requirement for any kind of conscious judgment on the subject's part. This makes smooth eye movements ideal for studying the dynamics of motion integration, not only in humans but in any animal that can move its eyes. An additional benefit for the cortical physiologist is that both these types of eye movement have been tightly linked to neural signals in the middle temporal (MT or V5) and the medial superior temporal (MST) visual areas of macaque monkeys, (Newsome et al. 1985; Groh et al. 1997; Kawano 1999; Born et al. 2000) thus permitting direct comparison with single unit studies.

A number of labs have availed themselves of eye movement recordings elicited by a variety of visual stimuli that contain 1D and 2D features moving in different directions to study the dynamics of motion integration. As many of these experiments will be described in greater detail elsewhere in this chapter (Chap. 8), we will focus our discussion here to the essential common features and to the

[1] Both the vector sum and the vector average produce resultant vectors with the same direction but different magnitudes. Because we are largely concerned with measurements of direction, we will use the vector sum to refer to both possibilities.

results pertaining to manipulations of contrast and contour length that parallel the psychophysical results described above.

Two different, yet closely related smooth eye movements have been used to probe the 1D-to-2D progression. First, *ocular following* (OF) is a very short-latency, automatic eye movement in response to motion of all, or a large part of, the visual field (Miles and Kawano 1986; Kawano and Miles 1986; Miles et al. 1986). The second type of eye movement is *smooth pursuit*, in which the subject actively tracks the motion of a single target, usually a small spot. Because of the different nature of the visual stimuli used to evoke these two eye movements – large textures for OF vs. single objects for smooth pursuit – they have been suited to probing slightly different features of the computation of motion direction. In particular, OF responses have been evoked by barber poles (Masson et al. 2000) and plaids (Masson and Castet 2002), both essentially textured stimuli that readily lend themselves to the manipulation of stimulus contrast. Smooth pursuit has proven useful for studying the effects of contour length (Born et al. 2006), using single tilted bars (Pack and Born 2001; Born et al. 2006) or objects composed of oblique lines, such as rhombi (Masson and Stone 2002). In the middle ground, where textured patches are reduced in extent and objects are made larger, the distinction between the two types of eye movement becomes blurry and may either cease to exist or reflect a transition over time from OF to pursuit (Masson and Stone 2002).

In any case, the main results from all of the experiments mentioned above have an essential feature in common: the initial eye movement is in the direction of the 1D component of stimulus motion and only some tens of ms later does it reflect the 2D direction. This temporal evolution appears to reflect, at least in part, a difference in the time required to process the two different motion cues. For both types of eye movement, the latency of the 1D response is shorter than that of the 2D response – compare the OF data in Fig. 2.2b with the pursuit data in Fig. 2.3a. Though differing in absolute latency, both show a common relative latency difference between the 1D and 2D responses of approximately 20 ms.

One potential caveat in interpreting the results as a true difference in processing speeds is that, in both cases, the purely 2D component of the eye movement is smaller – the direction difference between the 1D and 2D components of stimulus motion is only 45°, so even the 2D component, when broken down into eye movement coordinates that are either parallel or perpendicular to the 1D direction, has half of its amplitude represented on the 1D axis – and thus may only appear delayed because it takes longer to rise above the baseline plus noise. This explanation is made unlikely, however, by a number of stimulus manipulations in which the relative strengths of the 1D vs. 2D components have been varied. For example, Masson et al. (2000) either masked off the center of the barber pole (a relative decrease of the 1D motion strength) or blurred the edges of the barber pole aperture (a relative decrease of the 2D strength) and found that the respective amplitudes of the two components of the eye movement responses changed in the expected ways, but the latencies of the two remained the same and showed the characteristic 20 ms gap between the 1D and 2D responses (δ in Fig. 2.2b).

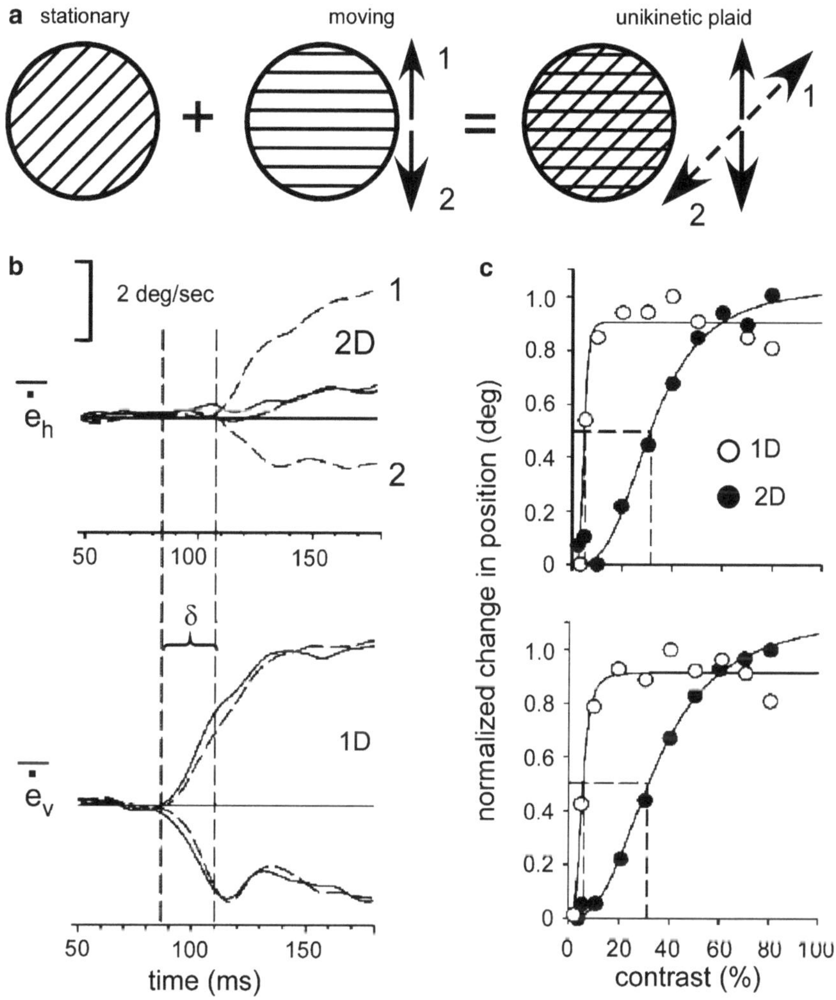

Fig. 2.2 Ocular following elicited by unikinetic plaids from a study by Masson and Castet (2002). (**a**) The unikinetic plaid is composed of an oblique stationary grating superimposed upon a horizontal grating which can move either up (1) or down (2). The resulting motion appears to move along an oblique axis. (**b**) Eye movement responses elicited by either a single horizontal grating (*solid lines*) or the unikinetic plaid (*dashed lines*). In both cases, the 1D motion perpendicular to the horizontal grating is purely vertical and so is manifest as the vertical component of the eye velocity ($\dot{e}_v$). For eye movements evoked by the unikinetic plaid (*dashed lines*), the 1D component seen in the vertical eye velocity traces are identical to those evoked by the grating alone. The 2D component, seen in the horizontal eye velocity traces ($\dot{e}_h$), has a longer latency ($\delta = 20$ ms). (**c**) Contrast sensitivity of the early (1D) and late (2D) components of ocular following elicited by unikinetic plaids in two different human observers. The 1D response shows a very steep dependency on contrast and early saturation, similar to that for magnocellular neurons, whereas the 2D response is shallower and saturates at higher contrasts, more characteristic of parvocellular neurons

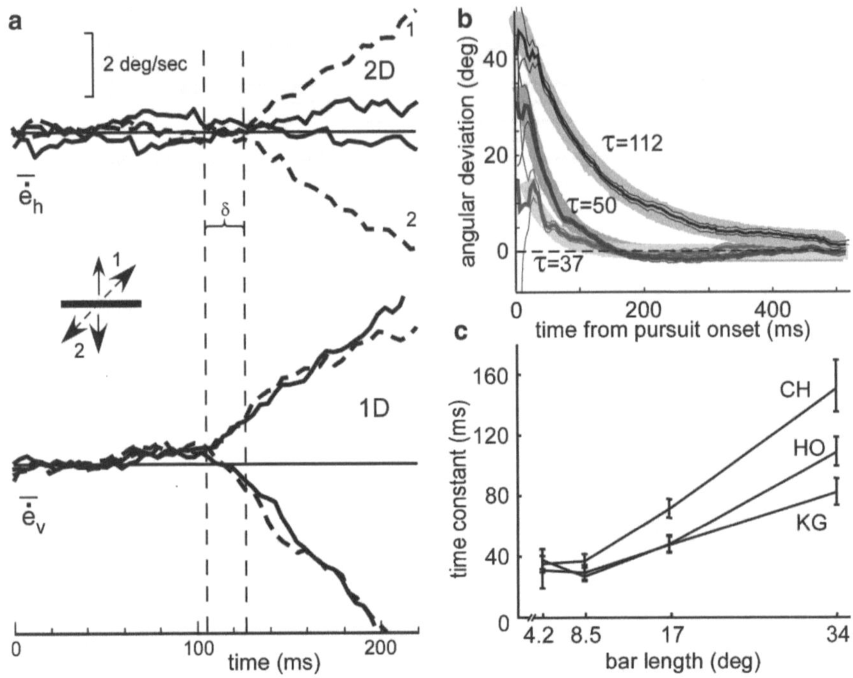

Fig. 2.3 Bar pursuit data from our own lab (Born et al. 2006) displayed so as to be directly comparable to the ocular following experiments of Masson and Castet (2002). (**a**) Monkeys actively pursued a horizontal bar that moved either vertically (solid lines), up (1) or down (2), or obliquely (*dashed lines*), up and to the right (1) or down and to the left (2). In both cases, the eye velocity response to the 1D component is reflected in the vertical component of the eye velocity ($\dot{e}_v$). For the obliquely moving bar, the responses to the 1D component seen in the vertical eye velocity traces are identical to those evoked by the vertically moving bar (*equated for speed*). The 2D responses, seen in the horizontal eye velocity traces ($\dot{e}_h$), show a longer latency (δ=20 ms). This data can be directly compared to that shown in Fig. 2.2b. (**b**) Time course of the angular deviation for tilted bars of different lengths (34, 17 and 4°) along with the best-fitting single exponential decay function for monkey HO. Values for τ are in meter seconds. (**c**) Time constant, τ, as a function of bar length for each of three different monkeys. Panels (**b**) and (**c**) were modified from Fig. 2.4 of Born et al. (2006) (*see Color Plates*)

Similar latency differences were also found for unikinetic plaids (Fig. 2.1e), thus allowing Masson and Castet (2002) to independently measure the effects of varying contrast on the 1D (early) and 2D (late) eye movement responses. The fact that the two responses showed markedly different contrast response functions (Fig. 2.2c) can be taken as further evidence that the neural circuitry underlying the computation of 1D vs. 2D motion is at least partially distinct – either because completely different pathways are involved (Masson and Castet 2002) or, as we will argue below, because the feedback mechanisms responsible for surround suppression or "end-stopping" have a lower contrast sensitivity than do the feed forward center mechanisms. This evidence also argues against the notion that a single mechanism tuned to 1D motion can account for the temporal dynamics of 1D-to-2D motion responses based on longer latencies for frequency components having lower amplitudes (Majaj et al. 2002) .

The other stimulus variable of interest with respect to the psychophysical results is that of contour length. This was explored in some detail by Born et al. (2006), whose results for smooth pursuit agreed, at least qualitatively, with those of Lorençeau and colleagues for perception. Specifically, when monkeys were required to track tilted bars of varying length, the 1D response – measured as the deviation in the direction of the eye movement from the true, 2D direction of bar motion – was both larger in amplitude and more prolonged in duration for longer bars (Fig. 2.3b, c). The angular deviation of pursuit over time was, in general, well described by a single exponential function. In addition to providing a single parameter, τ, that may be useful for comparisons with past and future experiments in perception and physiology, the single-exponential nature of the process may be a clue to the underlying mechanism. In other words, the temporal dynamics observed in these experiments may reflect the gradual accumulation of evidence for a feature that is measured by a single pathway, rather than the integration of two independent pathways.

To summarize up to this point, nearly a dozen different studies using both perceptual reports and smooth eye movements in response to a broad range of visual stimuli reveal a dynamic motion integration process and, while stimulus differences preclude direct quantitative comparisons, the qualitative similarities – particularly with respect to the effects of stimulus contrast and contour length – are striking. On the whole they strongly indicate that all are tapping into common underlying neural circuits. Insofar as this is so, the preceding sections have provided a number of valuable clues to the nature of this circuitry and provided important constraints for the computational models necessary to account for the relationship between neurons and behavior.

2.4 Neural Correlates of Motion Integration

Having observed perceptual and behavioral signatures of a dynamic integration process in vision, one might wonder how these processes are represented in the brain. Although there are many brain areas devoted to visual motion processing, most studies to date have focused on the middle temporal area (MT) of the macaque monkey cortex. Neurons in MT are particularly sensitive to the velocity of visual stimuli and relatively insensitive to other stimulus attributes, such as color and texture. Furthermore, individual MT neurons integrate motion signals over larger regions of visual space than neurons in the primary visual cortex (V1), which provides the bulk of the input to MT. The question of whether the spatial integration observed at the level of single MT neurons displays the sort of temporal dynamics observed perceptually has been the subject of several neurophysiological studies in the last few years.

Pack and Born (2001) recorded the responses of single MT neurons to stimuli that were similar to those used by Lorençeau et al. (1993) to study human perception.

The main differences were that the bars were positioned to fill the receptive field of the neuron under study (Fig. 2.4a) and that they moved somewhat faster than those used in the psychophysical studies. Of course, the output of single neurons can be evaluated on a much finer time-scale than that of human perceptual reports, and the MT results revealed a dynamic integration process that evolved over approximately 60 ms. In agreement with both behavioral and perceptual studies, the earliest responses reflected a non-specific integration that was heavily biased toward the component of motion perpendicular to the orientation of the bars (Fig. 2.4b). In other words, the neurons were initially fooled by the aperture problem, and their responses did not begin to approximate the correct direction of motion until some

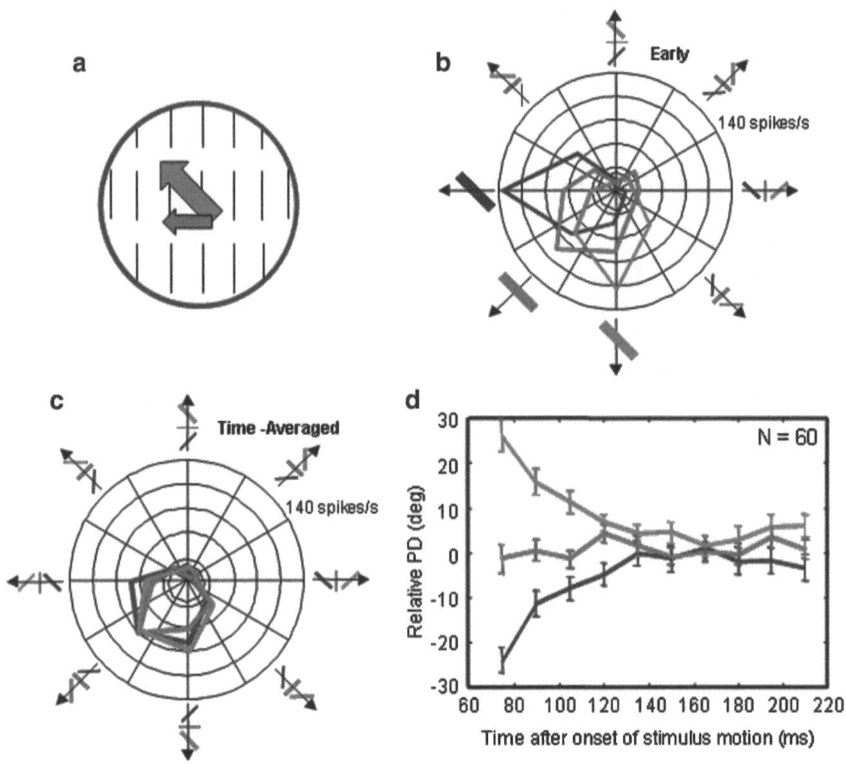

Fig. 2.4 Response of MT neurons to the bar field stimulus. (**a**) The bar field stimulus consisted of rows of bars with the same orientation, filling the receptive field (*blue circle*). When the bars moved obliquely with respect to their orientation (*green arrow*), the component of motion perpendicular to the bar motion (*red arrow*) differed from the true direction by 45°. (**b**) For a single MT neuron, the early part of the response depends on both the orientation and direction of the bar field. This neuron responds best whenever the bar has a left-oblique orientation and a leftward or downward motion component, indicating that it sees only the component of motion perpendicular to the bars. (**c**) The later part of the response depends only on the motion direction. (**d**) The transition from orientation-dependent responses to purely motion-dependent responses is evident in the population of 60 MT neurons. See also movies: *cu085c90.avi, cu085a45.avi, cu085b135.avi*, for a single-cell example of the temporal dynamics (*see Color Plates*)

time later (Fig. 2.4c, d; see also movies: *cu085c90.avi*, *cu085a45.avi*, *cu085b135.avi*). Interestingly, after the dynamics had stabilized, the MT population maintained an average error of roughly 5°, indicating that the motion integration process is not quite perfect, even for stimulus durations of 2,000 ms. A similar residual bias was observed psychophysically (for plaid stimuli) by Yo and Wilson (1992).

The temporal dynamics observed in MT bear at least a superficial similarity to integration processes observed in other aspects of vision. Various studies have found that the earliest responses of visual neurons encode a coarse description of the stimulus features, including orientation (Ringach et al. 1997), stereoscopic depth (Menz and Freeman 2003), faces (Sugase et al. 1999; Tsao et al. 2006), and complex shapes (Hegde and Van Essen 2004; Brincat and Connor 2006). In all cases later responses were linked to more specific details of the stimulus, as evidenced by a narrowing of tuning curves or a decorrelation of population activity in areas such as V1, V2, and IT.

The results on motion processing can be viewed in similar terms, since any resolution of the aperture problem requires the visual system to discern a specific velocity from a large number of candidate velocities. The range of possible velocities would be represented by neurons tuned to the various possibilities, and the neuronal signature of the solution would be a reduction of activity in most of these neurons (which would also entail a decorrelation of population activity). Although this refinement of the velocity representation was implicit in the results of Pack and Born (2001), their result did not elucidate either the underlying mechanism or precisely where in the brain the computation was taking place – it remained possible that the solution to the aperture problem was occurring in some other part of the brain, and that MT was merely reporting the result. An obvious possibility was area V1, since it is the primary source of input to MT.

At a first glance, one might be inclined to rule out the possibility of a resolution of the aperture problem in area V1 a priori, based on the small sizes of the receptive fields found there. Indeed these receptive fields are in effect tiny apertures, and the geometry depicted in Fig. 2.1. guarantees that V1 neurons will be incapable of signaling velocity accurately. Although this assumption is generally valid, there are important exceptions which turn out to be important for understanding the way the brain measures motion.

Consider the situation depicted in Fig. 2.1a. The aperture problem applies to all motion measurements made along the length of each line, but near the ends of the line, velocity can be measured accurately even by neurons with small receptive fields. The reason is that the line-endings or terminators are two-dimensional, and hence permit the measurement of both the perpendicular and parallel components of the bar's velocity. Of course the terminators comprise only a small fraction of each bar's area, and so the existence of these signals does not by itself lead to a solution to the aperture problem. Rather what is needed is a selective integration process that ignores or suppresses the velocities measured along the length of each bar.

A hint at the mechanism by which this selective integration is accomplished was present in the pioneering work of (Hubel and Wiesel 1965). In V1 of the anesthetized cat, they discovered neurons that failed to respond to oriented stimuli extending beyond their central activating regions. Hubel and Wiesel called these neurons "hypercomplex", but this term was eventually replaced by a more descriptive term: "end-stopped neurons". Many end-stopped neurons responded well to terminators, suggesting that in combination with direction selectivity these cells could provide the kind of processing necessary to overcome the aperture problem (see movie: *HWendstop-short.avi*). This possibility was subsequently confirmed by Pack et al. (2003), who showed that direction-selective, end-stopped neurons in the alert macaque were capable of encoding motion direction in a manner that was independent of stimulus orientation. Moreover, this invariance of direction selectivity emerged only after a brief delay – the initial response was influenced by the aperture problem in precisely the way that one would expect based on the geometrical analysis described above (see movie *endstop.avi*). Thus there is good evidence that the dynamic process of motion integration that is observed in MT begins and is partially completed at the level of V1. This possibility is further supported by the observation that the V1 layer that provides the bulk of the projection to MT is also the layer that has the highest prevalence of end-stopped neurons Sceniak et al. (2001).

2.5 A Computational Model of Motion Integration

End-stopping is a specific instance of a more general phenomenon known as surround suppression, which occurs whenever an increase in stimulus size leads to a reduction in neuronal response. Such an inhibitory influence may reflect neuronal mechanisms of normalization, which serve to calibrate the sensitivity of individual neurons to the overall amount of visual stimulation reaching a given visual area. Normalization models have been proposed in various contexts to account for certain nonlinear behaviors that have been observed in V1 neurons (e.g., Heeger 1992). While these models have been successful in explaining many of the V1 results, there has not to our knowledge been a computational account of the interaction of end-stopping and direction selectivity in V1.

A candidate for such a model is depicted in Fig. 2.5a. In this conception the direction selectivity of each cell is based on an implementation of the motion energy model (Adelson and Bergen 1985), with the parameters fully constrained by measurements from macaque V1 neurons (Pack et al. 2006). The end-stopping in each neuron is due to inhibitory input from other V1 neurons with nearby receptive fields, based on the circuit model proposed by Carandini and colleagues (Carandini et al. 1997). This model implements normalization by dividing the activity of each cell by the summed activity of its neighbors. When the receptive fields of the neighboring cells occupy different spatial positions, normalization translates into surround suppression, since large stimuli activate more of the surrounding cells than do small stimuli. Our model extends the proposal of Carandini and colleagues by incorporating a limited spatial range for the normalization pool and a realistic time constant for the surround interactions. Both of these extensions turn out to be important for the ability of the model to account for the dynamic aspects of motion processing.

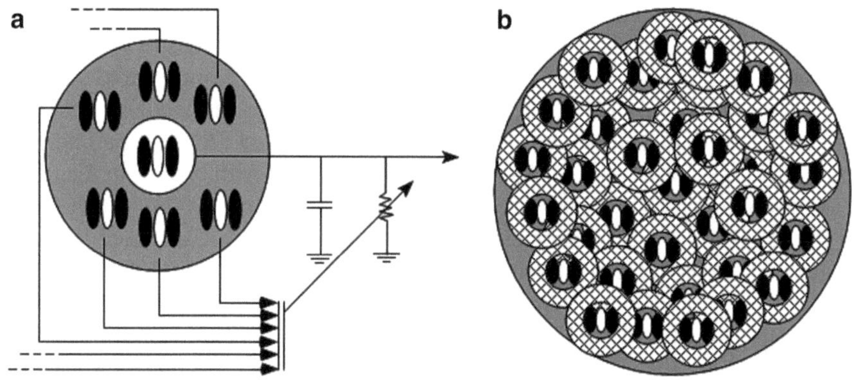

Fig. 2.5 A simple model of motion integration in MT. (**a**) Circuit for a model end-stopped cell, after Carandini et al. (1997). The model is composed of series of identical subunits, each of which is a motion energy detector. The output of the central subunit is modulated by the outputs of several surrounding subunits, and their interaction is characterized by an RC circuit with variable resistance. (**b**) The model MT cell simply sums the outputs of end-stopped cells like those in (**a**), but with receptive fields at different spatial positions

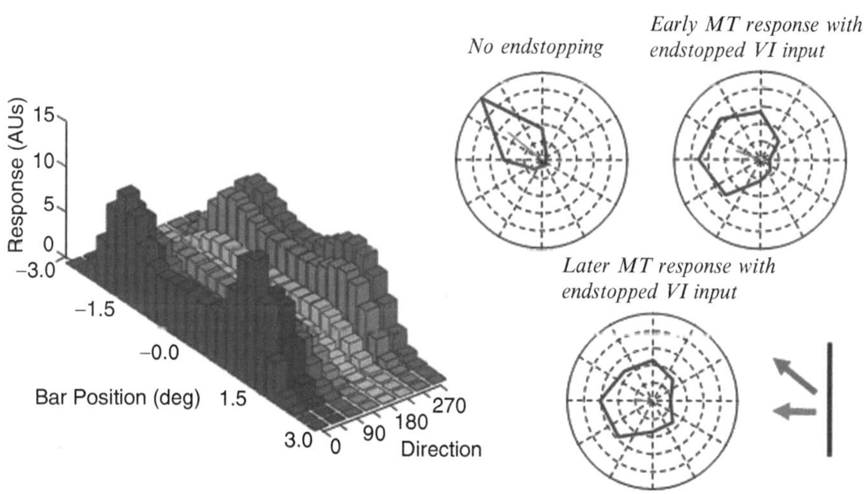

Fig. 2.6 Model output. (**a**). Response of a model V1 cell to a bar moving through different positions and at different orientations. The model responds best to the endpoints of the bar, irrespective of bar orientation, which modulates the overall response level. (**b**). Output of the model MT cell in response to bars oriented 45° with respect to their direction of motion, when V1 endstopping is disabled. The cell normally prefers leftward motion, but in response to the tilted bar stimulus its tuning curve rotates by nearly 45°. (**c**) Early response of the MT neuron with endstopped V1 input. The tuning curve is rotated by more than 20° from its actual preferred direction. (**d**) Later response of the same model MT cell. The tuning curve is centered on leftward motion, with a small residual error of roughly 5°

Figure 2.6a. shows the response of the model to bars at different positions, and it is clear that the neuron responds primarily to the endpoints. This property is invariant with bar orientation, as was observed for real end-stopped neurons

(Orban et al. 1979; Pack et al. 2003). We simulated a population of these end-stopped neurons with receptive fields positioned at different points in visual space. The model MT neuron simply integrated the activity of the population of identical end-stopped V1 neurons (Fig. 2.5b). Figure 2.6c, d shows the output of the MT neuron for the bar-field stimuli used in the Pack and Born (2001) experiment. In this simulation, the bars were oriented at an angle of 45° with respect to the direction of motion, and the model cell preferred a leftward motion. The salient points of the MT data – the accurate measurement of motion and the associated temporal dynamics – are both captured in the model response (Fig. 2.6c, d). The residual bias due to the perpendicular component of the bar velocity is roughly 5°, which is consistent with that observed in MT of alert monkeys (Pack and Born 2001). For comparison, Fig. 2.6b. shows the large errors in the model output when end-stopping in V1 is disabled.

A recently published model (Rust et al. 2006) uses a similar normalization mechanism to account for the responses of MT neurons to plaid stimuli. In this model normalization plays a role similar to that played by end-stopping in the model described above, namely to eliminate spurious motion signals that result from the aperture problem. However, this model lacks temporal dynamics, which would be necessary to provide a full account of the dynamic integration of plaid stimuli seen in MT (Pack et al. 2001; Smith et al., 2005). Furthermore, the normalization mechanism in the Rust et al. (2006) model lacks spatial structure. Indeed the model operates entirely in the velocity domain, so it cannot generate selective responses to endpoints or any other spatial feature of the stimulus.

The Rust et al. (2006) model is primarily an account of the continuum from "component" to "pattern" cells in MT, the latter being generally assumed to be the cells that solve the aperture problem. However, the model shown in Fig. 2.5 is capable of measuring the direction of the bar field stimulus, despite the fact that it would be classified as a "component" cell when tested with plaids; that is, it would respond to the motions of the gratings that make up the plaid, rather than the motion of the entire pattern (simulation results confirmed but not shown). To obtain "pattern" selectivity for plaids, Rust et al. (2006) required a second mechanism, namely a specific pattern of excitatory and inhibitory weights in the feed forward input from V1 to MT. The weighting for "pattern" cells in their model was broader in direction space than that for "component" cells, and it also had inhibitory lobes for directions far from the preferred direction.

Although we have not yet tested this idea, we suspect that the model of Rust et al. (2006) could account for the bulk of the existing data if it used an end-stopping mechanism similar to that shown in Fig. 2.5. This mechanism would provide the temporal dynamics observed in MT with bar fields (Pack and Born 2001), plaids (Pack et al. 2001; Smith et al. 2005), and barber poles (Pack et al. 2004). In combination with the feed forward weighting proposed by Rust et al. (2006), this mechanism would render most cells capable of measuring direction accurately for stimuli that contained terminators, but would limit the proportion of "pattern" cells to those that had broad direction tuning. This might also explain why the proportion of "pattern-like" cells in MT tends to decrease under general anesthesia (Pack et al. 2001), since the tuning bandwidth is also seen to be narrower in anesthetized animals (Pack, Berezovskii, and Born, unpublished observations).

2.6 Conclusions

In summary the existing physiological data demonstrate that MT neurons integrate the motion of complex stimuli gradually over a period of roughly 60 ms after the onset of stimulus motion. When the stimulus contains motion that can be measured accurately by end-stopped V1 neurons, the integration is nearly perfect for all MT neurons. For certain kinds of plaids, the distinction between "component" and "pattern" neurons reflects primarily the variability in tuning bandwidths instantiated by the projection from V1 to MT, with a possible contribution for inhibition from neurons with non-preferred directions (Rust et al. 2006). However, the distinction between "component" and "pattern" cells does not generalize well, as most "component" cells are perfectly capable of integrating pattern motion for other kinds of stimuli. Models that incorporate realistic estimates of the nonlinearities present in V1 are likely to provide a satisfactory account of these data, though a full model with all these characteristics has yet to be implemented. In addition, a weighting of V1 inputs according to spatial and temporal frequency preferences may be helpful in measuring speed (Simoncelli and Heeger 1998). Such a mechanism could easily be incorporated into the framework outlined here.

2.7 Supplementary Materials (CD-ROM)

Movie 1 Illusory motion of a tilted single bar (file "*2_M1_TiltedLines.avi*"). The green bar moves purely from left to right. At the beginning of its sweep, the bar transiently appears to move slightly *upwards* and to the right. The illusory upwards component is due to the aperture problem and the spatially restricted receptive fields of direction-selective neurons at early stages of the visual pathways.

Movies 2–7 Visual stimuli used to study the dynamics of 1D–2D motion (see Fig. 2.1) (files: "*2_M2_Figure1a.avi*", "*2_M3_Figure1b.avi*", "*2_M4_Figure1c.avi*", "*2_M5_Figure1d.avi*, "*2_M6_Figure1e.avi*", "*2_M7.Figure1f.avi*").

1_M2_Figure1a.avi: Tilted bar-field used by Lorençeau et al. (1993). In this particular example, the 2D direction of motion has a downward component, whereas the 1D direction measured along the contour has an upward component. The inset of Fig. 2.1a. depicts the situation in greater detail as seen through the apertures of neuronal receptive fields.

1_M3_Figure1b.avi: Barber pole in which the direction of grating (1D) motion differs by 45° from that of the perceived direction, which is up and to the right.

1_M4_Figure1c.avi: Single horizontal grating moving upwards.

1_M5_Figure1d.avi: Symmetric Type I plaid consisting of two superimposed 1D gratings. The rigid pattern appears to move upwards.

1_M6_Figure1e.avi: Unikinetic plaid. Only the horizontal grating moves (upwards), but the static oblique grating causes the pattern to appear to move up and to the right.

1_M7_Figure1f.avi: Type II plaid in which the perceived direction of the pattern is very different from that of either of the two components or the vector sum.

Movies 8–10. Dynamics of neuronal direction selectivity (files: 2_M8_ *cu085c90.avi, «2_M9_cu085a45.avi», «2_M10_cu085b135.avi»*). A Single MT cell example of the temporal dynamics of the solution of the aperture problem. (Note that this is not the same cell whose data is shown in Fig. 2.4b, c. though the experimental conditions for the different tuning curves were identical.) Each movie shows the dynamics of the neuron's direction tuning for one of the three relative bar orientations:

2_M8_cu085c90.avi corresponds to the control condition (red bars in Fig. 2.4b, c) in which the motion was perpendicular to the bars' orientation;

2_M9_cu085a45.avi represents the tuning curve when the bars have been tilted +45° with respect to their direction of motion (blue bars in Fig. 2.4b, c), and

2_M10_cu085b135.avi is for the bars tilted 45° in the opposite direction (green bars in Fig. 2.4b, c).

Within each movie, the blue line with the open arrowhead indicates the *mean vector* of the neuron's direction tuning curve to the control stimulus, with firing rates averaged over several hundred meter seconds. The mean vector points in the neuron's preferred direction, and its length indicates the width of the tuning curve (the longer the mean vector, the sharper the tuning). The dancing asterisk represents the mean vector during successive 25-ms bins (centered on the time indicated in the upper right corner), and each leaves a filled circle that shows the history over time. The color code indicates the status of the visual stimulus: red, stimulus OFF (i.e. spontaneous activity); green, stimulus ON but stationary; blue, stimulus MOVING. The height of the black line along the y-axis indicates the maximum normalized response. This line shows, for example, that the neuron fires vigorously at stimulus onset (green), but these responses are not direction selective, as shown by the clustering of the green dots around the origin (i.e. length of mean vector is close to zero). Only when the stimulus begins moving do the blue spots move away from the origin, indicative of significant direction tuning. For the +45° (*2_M9_cu085a45.avi*) and −45° (*2_M10_cu085b135.avi*) conditions, the initial preferred direction is systematically deviated according to the predictions of the aperture problem, but then evolves to represent the true (2D) direction over the ensuing 50–75 ms.

Movie 11 Mapping end-stopping in V1 neurons. (file "*2_M11_HWendstop-short.avi*"). This is a truncated version of the movie made originally by David Hubel and Torsten Wiesel depicting the essential feature of an end-stopped neuron, namely that it responds well to a short bar but not at all to a long bar.

Movie 12 Temporal dynamics of end-stopping in V1. (file "*2_M12_endstop.avi*"). Temporal dynamics of end-stopping in V1 of an alert macaque monkey (from Pack et al. 2003). The movie shows the temporal evolution of an end-stopped neuron's receptive field as determined by reverse correlating the neuron's spikes with the position of the center of a long bar. At short correlation delays, the receptive field looks like a solid rectangle, indicating that the neuron responds to the long bar regardless of its location along the receptive field axis. This profile is characteristic of neurons that are *not* end-stopped. Over the following 35 ms, the receptive field profile takes on a dumb-bell shape, indicative of end-stopping. Thus the property of end-stopping requires some time to emerge, and this may account for some or all of the dynamics of MT's solution to the aperture problem (Fig. 2.4 and accompanying movies). For methodological details on the reverse correlation method, see Pack et al. 2003.

Acknowledgments The work discussed in this chapter was supported by NIH Grant EY11379 (RTB). We thank Andrew Zaharia for creating the flash demos of visual stimuli.

References

Adelson EH, Bergen JR (1985) Spatiotemporal energy models for the perception of motion. J Opt Soc Am A 2:284–299

Born RT, Groh JM, Zhao R, Lukasewycz SJ (2000) Segregation of object and background motion in visual area MT: effects of microstimulation on eye movements. Neuron 26:725–734

Born RT, Pack CC, Ponce CR, Yi S (2006) Temporal evolution of 2-dimensional direction signals used to guide eye movements. J Neurophysiol 95:284–300

Brincat SL, Connor CE (2006) Dynamic shape synthesis in posterior inferotemporal cortex. Neuron 49:17–24

Carandini M, Heeger DJ, Movshon JA (1997) Linearity and normalization in simple cells of the macaque primary visual cortex. J Neurosci Nov 1; 17(21):8621–8644

Groh JM, Born RT, Newsome WT (1997) How is a sensory map read Out? Effects of microstimulation in visual area MT on saccades and smooth pursuit eye movements. J Neurosci 17:4312–4330

Heeger DJ (1992) Normalization of cell responses in cat striate cortex. Vis Neurosci 9:181–197

Hegde J, Van Essen DC (2004) Temporal dynamics of shape analysis in macaque visual area V2. J Neurophysiol 92:3030–3042

Hubel DH, Wiesel TN (1965) Receptive fields and functional architecture in two non-striate visual areas (18 and 19) of the cat. J Neurophysiol 28:229–289

Kawano K (1999) Ocular tracking: behavior and neurophysiology. Curr Opin Neurobiol 9:467–473

Kawano K, Miles FA (1986) Short-latency ocular following responses of monkey. II. Dependence on a prior saccadic eye movement. J Neurophysiol 56:1355–1380

Lorençeau J, Shiffrar M, Wells N, Castet E (1993) Different motion sensitive units are involved in recovering the direction of moving lines. Vision Res 33:1207–1217

Majaj N, Smith MA, Kohn A, Bair W, Movshon JA (2002) A role for terminators in motion processing by macaque MT neurons? [Abstract]. J Vis 2(7):415, 415a

Marr D, Poggio T (1976) Cooperative computation of stereo disparity. Science 194:283–287

Marr D, Ullman S, Poggio T (1979) Bandpass channels, zero-crossings, and early visual information processing. J Opt Soc Am 69:914–916

Masson GS, Castet E (2002) Parallel motion processing for the initiation of short-latency ocular following in humans. J Neurosci 22:5149–5163

Masson GS, Rybarczyk Y, Castet E, Mestre DR (2000) Temporal dynamics of motion integration for the initiation of tracking eye movements at ultra-short latencies. Vis Neurosci 17:753–767

Masson GS, Stone LS (2002) From following edges to pursuing objects. J Neurophysiol 88:2869–2873

Menz MD, Freeman RD (2003) Stereoscopic depth processing in the visual cortex: a coarse-to-fine mechanism. Nat Neurosci 6:59–65
Miles FA, Kawano K (1986) Short-latency ocular following responses of monkey. III. Plasticity. J Neurophysiol 56:1381–1396
Miles FA, Kawano K, Optican LM (1986) Short-latency ocular following responses of monkey. I. Dependence on temporospatial properties of visual input. J Neurophysiol 56:1321–1354
Newsome WT, Wurtz RH, Dursteler MR, Mikami A (1985) Deficits in visual motion processing following ibotenic acid lesions of the middle temporal visual area of the macaque monkey. J Neurosci 5:825–840
Orban GA, Kato H, Bishop PO (1979) Dimensions and properties of end-zone inhibitory areas in receptive fields of hypercomplex cells in cat striate cortex. J Neurophysiol 42:833–849
Pack CC, Berezovskii VK, Born RT (2001) Dynamic properties of neurons in cortical area MT in alert and anaesthetized macaque monkeys. Nature 414:905–908
Pack CC, Born RT (2001) Temporal dynamics of a neural solution to the aperture problem in visual area MT of macaque brain. Nature 409:1040–1042
Pack CC, Born RT (2008) Cortical mechanisms for the integration of visual motion. In: Basbaum AI, Kaneko A, Shepherd GM, Westheimer G (eds). The senses: a comprehensive reference, vol 2, vision II Albright TD, Masland R (eds). San Diego Academic Press, pp 189–218
Pack CC, Conway BR, Born RT, Livingstone MS (2006) Spatiotemporal structure of nonlinear subunits in macaque visual cortex. J Neurosci 26:893–907
Pack CC, Gartland AJ, Born RT (2004) Integration of contour and terminator signals in visual area MT of alert macaque. J Neurosci 24:3268–3280
Pack CC, Livingstone MS, Duffy KR, Born RT (2003) End-stopping and the aperture problem: two-dimensional motion signals in macaque V1. Neuron 39:671–680
Ringach DL, Hawken MJ, Shapley R (1997) Dynamics of orientation tuning in macaque primary visual cortex. Nature 387:281–284
Rohaly AM, Wilson HR (1993) Nature of coarse-to-fine constraints on binocular fusion. J Opt Soc Am A 10:2433–2441
Rohaly AM, Wilson HR (1994) Disparity averaging across spatial scales. Vision Res 34:1315–1325
Rosch E, Mervis CB, Gray WD, Johnson DM, Boyes-Braem P (1976) Basic objects in natural categories. Cogn Psychol 8:382–439
Rust NC, Mante V, Simoncelli EP, Movshon JA (2006) How MT cells analyze the motion of visual patterns. Nat Neurosci 9:1421–1431
Sceniak MP, Hawken MJ, Shapley R (2001) Visual spatial characterization of macaque V1 neurons. J Neurophysiol 85:1873–1887
Simoncelli EP, Heeger DJ (1998) A model of neuronal responses in visual area MT. Vision Res 38:743–761
Smith MA, Majaj N, Movshon JA (2005) Dynamics of motion signaling by neurons in macaque area MT. Nat Neurosci 8:220–228
Sugase Y, Yamane S, Ueno S, Kawano K (1999) Global and fine information coded by single neurons in the temporal visual cortex. Nature 400:869–873
Thorpe SJ, Fabre-Thorpe M (2001) Seeking Categories in the Brain. Science 291:260–263
Tsao DY, Freiwal WA, Tootell RB, Livingstone MS (2006) A cortical region consisting entirely of face-selective cells. Science 311:670–674
Wilson HR, Blake R, Halpern DL (1991) Coarse spatial scales constrain the range of binocular fusion on fine scales. J Opt Soc Am A 8:229–236
Yo C, Wilson HR (1992) Perceived direction of moving two-dimensional patterns depends on duration, contrast and eccentricity. Vision Res 32:135–147

Chapter 3
Dynamics of Pattern Motion Computation

Matthew A. Smith, Najib Majaj, and J. Anthony Movshon

Abstract Early in visual processing neurons with small receptive fields can only signal the component of motion perpendicular to the orientation of the contour that passes through them (the "aperture problem"). A moving visual pattern with differently oriented contours can thus elicit neuronal responses that convey conflicting motion cues. To recover the true direction of motion of such a pattern, later visual areas must integrate the different motion cues over space and time. There is extensive evidence which suggests that this integration is not instantaneous – instead it occurs over time and causes profound changes in the perception of direction of motion of some complex moving patterns. To account for such temporal dynamics, previous studies have focused on a two-pathway model of motion perception: a fast pathway to account for the early percept, and a slow one to account for the late percept.

Neurons in macaque area MT are selective for the direction of motion of an object, and their responses appear to be connected directly to the perception of complex motion stimuli in the natural environment. In this chapter, we will discuss neurophysiological data from MT neurons which illustrate how the process of motion perception occurs dynamically. The responses of individual neurons in MT appear to reflect the process by which the primate visual system produces an initial estimate of motion direction and then refines it over time. We will argue that MT neuronal responses are consistent with a single pathway model of motion perception in which temporal dynamics emerge due to two factors: the contrast of elements in the pattern and the time required for the pattern computation.

M.A. Smith (✉)
Center for Neural Basis of Cognition, University of Pittsburgh,
4400 Fifth Avenue, Mellon Institute Room 115,
Pittsburgh, PA, 15213, USA
e-mail: masmith@cnbc.cmu.edu

U.J. Ilg and G.S. Masson (eds.), *Dynamics of Visual Motion Processing: Neuronal, Behavioral, and Computational Approaches*,
DOI 10.1007/978-1-4419-0781-3_3,

3.1 Introduction

In the natural world, our visual experience is not static. Our eyes scan rapidly over the scene before us, fixing on a position for a few hundred milliseconds at a time before moving to the next location. Even within that short fixation, the image on our retina is rarely still – our own bodies and objects in the world around us are in nearly constant motion. Our perception of motion is similarly dynamic, changing over time on a scale of tens to hundreds of milliseconds. Nonetheless, most studies of visual cortex have measured the mean activity of neurons over a period of seconds. In recent years, however, the temporal dynamics of neural response in visual cortex have become subject to increased scrutiny. Perhaps not surprisingly, a number of recent studies have shown extensive changes in the response of visual cortical neurons over time. These dynamics appear to reflect the time course of multiple excitatory and suppressive influences which combine to produce a neuron's response. Comparison of the speed with which our psychophysical performance and physiological response unfold has proved to be an effective tool in understanding motion perception.

3.1.1 Temporal Dynamics in Primary Visual Cortex

Investigation of even the earliest stage of cortical visual processing has revealed significant dynamics in neuronal response. This includes modulation by stimuli which are confined to the receptive field and by those which extend well outside. A consideration of these findings in primary visual cortex (V1) is helpful in understanding the mechanisms of motion perception and the nature of perceptual effects. V1 neurons are tuned to the orientation of a stimulus within their receptive field (RF), and exhibit substantial time-dependent changes in that tuning (Ringach et al. 1997; Ringach et al. 2003; Smith et al. 2006). For small stimuli confined to the receptive field these effects tend to be extremely fast, often occurring with latency equal to or less than the excitatory response onset (Smith et al. 2006). The speed of such phenomena makes it likely that they are generated either by modification of the feedforward input to V1 neurons or very fast computations within the local circuitry in a cortical column.

Stimuli which extend beyond the receptive field, into the non-classical surround, also exert considerable influence on the responses of V1 neurons. In V1, neurons are modulated by a number of oriented stimuli which extend outside the receptive field. Various studies have shown that this effect arrives with some delay after the onset response for a number of stimuli, including fields of bars (Knierim and Van Essen 1992), oriented texture (Lamme 1995; Zipser et al. 1996; Lee et al. 1998), and sinusoidal gratings (Bair et al. 2003; Smith et al. 2006). The timing of surround suppression is one factor that has led most authors to conclude that it originates via feedback to V1 from extrastriate cortex. Angelucci et al. (2002), drawing on data from physiology and neuroanatomy, argued that the spatial scale of surround suppression is well matched to that of feedback circuits from extrastriate cortex to V1.

In searching for neural correlates of perceptual effects in V1, several studies have revealed the presence of dynamics in the response. These effects occur for a number of contextual stimuli presented to an awake animal performing a behavioral task, including curve-tracing (Roelfsema et al. 1998), figure-ground stimuli (Lamme 1995; Lee et al. 1998), illusory contours (Lee and Nguyen 2001), and shape-from-shading (Lee et al. 2002; Smith et al. 2007). The modulation of neuronal response in these paradigms occurs with greater delay than that found for extended iso-orientation stimuli. However, the pattern of neuronal response is similar: it remains normal for some time after the onset, and then after a delay the modulation due to the stimulus context becomes evident.

Finally, there is an additional factor which affects the dynamics of visual processing: the contrast of the stimulus. Specifically, low contrast targets are processed slower than high contrast ones (Albrecht 1995; Carandini et al. 1997; Gawne et al. 1996). This effect is distinct from observations of contextual modulation, in that the latency and magnitude of a neuron's response changes gradually with contrast, and this effect occurs for stimuli confined to the receptive field. Furthermore, the change in latency with contrast can be quite large, spanning up to 100 ms between the lowest and highest contrast stimuli.

3.1.2 Temporal Dynamics in Motion Perception

The relatively small receptive fields and proportion of direction-selective neurons in area V1 make it poorly suited for encoding complex moving stimuli. However, area MT of macaque visual cortex contains a high proportion of neurons which are selective for the direction of stimulus motion (Albright 1984; Movshon et al. 1985; Van Essen et al. 1981; Zeki 1974). Neurons in area MT have also been shown to play an important role in visual motion perception (Britten et al. 1992; Newsome and Paré 1988; Salzman and Newsome 1994). A significant portion of MT neurons encode the true velocity of a stimulus (Perrone and Thiele 2001; Priebe et al. 2003), whereas V1 neurons have independent spatial and temporary frequency responses (Tolhurst and Movshon 1975; Holub and Morton-Gibson 1981; Friend and Baker 1993). Similarly, many MT neurons are capable of decoding the true direction of motion of complex visual patterns such as a plaid stimulus (Fig. 3.2a), composed of two sinusoidal gratings with different orientations (Movshon et al. 1985; Rodman and Albright 1989), a behavior that is not present in V1 neurons that project to MT (Movshon and Newsome 1996). The responses of MT neurons to plaid patterns also vary in a manner consistent with the perceptual phenomenon of motion coherence (Stoner and Albright 1992). However, even though MT neurons have rather precise temporal response properties (Bair et al. 1994), the true direction of pattern motion is not represented in their initial responses. Instead, the encoding of pattern motion direction lags behind the initial estimate of direction by 50–75 ms (Pack and Born 2001; Smith et al. 2005). Taken together, these findings indicate that the detection of complex motion in the visual world is a property which emerges through computation in circuits within area MT.

The dynamics of neural response in area MT are paralleled by psychophysical results in humans. A number of studies have reported that manipulating the components of a complex moving stimulus affects the perception of human observers. When the speed of the component gratings of a plaid stimulus is unequal, observers perceive mostly component motion in brief presentations and pattern motion only after a delay (Yo and Wilson 1992). Similarly, short-latency ocular following responses in humans initially track the motion of a component grating, but later reflect the motion of the pattern (Masson and Castet 2002). A related effect occurs when the aspect ratio of an aperture around a grating stimulus is elongated – observers first track the grating motion and over time are biased toward the aperture's long axis (Masson et al. 2000). Patterns composed of line segments have a similar effect on motion perception. With an array of line segments moving at various angles relative to their orientation, a pair of studies (Castet et al. 1993; Lorenceau et al. 1993) reported that observers were biased by the orientation of the line segments in short observation windows (100–200 ms), but over time they recognize the true motion direction. The initiation of smooth-pursuit eye movements in humans shows a similar bias for stimuli composed of line segments (Masson and Stone 2002), diamonds (Wallace et al. 2005), and a combination of first and second-order motion cues (Lindner and Ilg 2000). The common finding in all of these studies is that observers tend to show an initial bias toward the orientation of components of a pattern, but over the span of 100 ms or more, tend to perceive the true motion direction. The dynamics evident in these results have led to a number of models of how the visual system computes pattern motion. We will consider each of these models in turn.

3.2 Models of Pattern Motion Detection

Consideration of the data from physiological and psychophysical studies has revealed that motion information is processed in at least two stages. The first, likely located in primary visual cortex (V1), extracts basic information (such as orientation) about simple moving patterns from a local region of space. The second stage computes information about the true direction and speed of complex moving patterns by combining inputs from the first stage. Models which strive to explain motion perception typically reflect this two-stage processing in their instantiation.

One such model linearly combines the signals from nonlinear V1 subunits (Heeger et al. 1996; Simoncelli and Heeger 1998), a so-called linear–nonlinear ("L–N") model. This simple model is able to capture many of the properties of direction-selective neurons in macaque area MT, but cannot adequately account for pattern direction selectivity (Simoncelli and Heeger 1998; Mante 2000). A modification of this basic structure, a cascaded L–N model in which the second stage acts on signals from a population of direction-selective units, can accurately decode the motion of complex patterns while maintaining fidelity to the known cortical architecture (Rust et al. 2006). In the cascade model (Fig. 3.1), a stimulus passes through

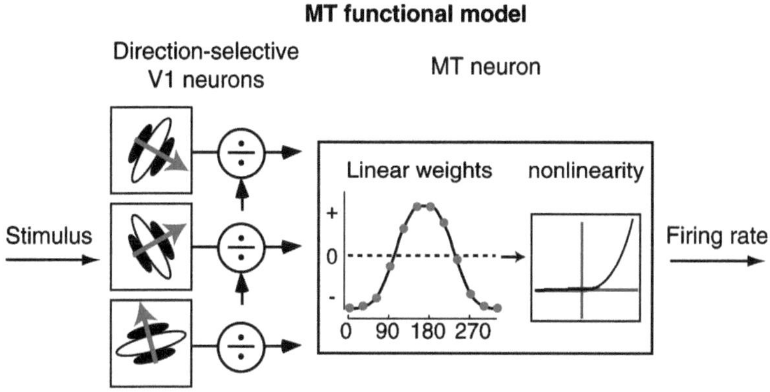

Fig. 3.1 The cascade model

a population of direction-selective V1 neurons and is divisively normalized. The outputs of these model cells feed into a MT neuron which computes their linear weighted sum. The result is converted into firing rate by a nonlinear function, simulating the effect of spike threshold and any additional nonlinear effects which occur post-summation. In the framework of this model, pattern selectivity arises from the recurrent circuit which combines V1 inputs to produce MT neuronal responses. If such a network takes time to stabilize, the selectivity of individual neurons would change over time – first reflecting the simple direction selectivity of input neurons, and later evolving pattern selective responses. One aspect of the cascade model, an orientation-tuned normalization mechanism, may reflect suppressive input from outside the classical receptive field of V1 neurons. This surround suppression is known to occur with some delay after response onset (Bair et al. 2003; Smith et al. 2006), and may lead to the delay in pattern motion computation observed in MT neurons (Smith et al. 2005).

Another class of model, which separately analyzes the contour and terminator information present in a scene (Shimojo et al. 1989; Grossberg and Mingolla 1993; Lorenceau et al. 1993), has also been proposed as a means of decoding motion in complex patterns. A version of this approach uses two parallel pathways (Fourier and non-Fourier), the outputs of which are combined to compute pattern motion (Wilson et al. 1992; Wilson and Kim 1994; Löffler and Orbach 1999). However, the cortical pathways which underlie this model are unknown, and studies of the proposed candidate areas (V2 and V3) do not suggest an important contribution to pattern selectivity (Gegenfurtner et al. 1997; Levitt et al. 1994).

A third approach proposes that neural networks separately process the ambiguous and unambiguous portions of the scene, with the unambiguous locations "filling in" over time (Hildreth 1984; Beutter and Stone 1998). Neural models using recurrent (Lidén and Pack 1999) or feedback (Chey et al. 1997) circuits to decode pattern motion have implemented this proposal. If such a model is tuned so that early responses reflect feedforward signals, and later responses are shaped by recurrent or feedback connections, then the dynamics of pattern motion perception can be replicated.

These three classes of model may provide us with some insight into the dynamics of pattern motion computation. By adjusting the latency or dynamics of the two stages or pathways in each model, it is possible to generate a pattern motion detection system with dynamics which are similar to those shown in experimental studies. It is clear that further exploration of neurophysiological responses is necessary to distinguish the neural mechanisms clearly. We will now describe one series of experiments which aims to explore these mechanisms, and explain the neural basis for our changing perception of complex moving patterns over time.

3.3 Responses of MT Neurons to Plaids

Area MT in the extrastriate cortex of the macaque contains a high proportion of directionally selective neurons (Albright 1984; Movshon et al. 1985; Van Essen et al. 1981; Zeki 1974) and plays an important role in the perception of moving patterns (Britten et al. 1992; Newsome and Paré 1988). When presented with a drifting sinusoidal grating stimulus, the vast majority of MT neurons respond in a direction selective manner (Fig. 3.2b, left). Plaid stimuli, obtained by adding two sinusoidal gratings with different orientations (Fig. 3.2a, right), have been used to demonstrate an important property of some MT neurons which is not present at earlier stages of motion processing. When presented with a plaid stimulus, a directionally selective neuron might respond only to the direction of motion of the component gratings (solid line in Fig. 3.2b, right plot), or it might respond to the true direction of motion of the plaid stimulus (dashed line in Fig. 3.2b, right plot). The former behavior is termed component direction selectivity (CDS) and the latter pattern direction selectivity (PDS).

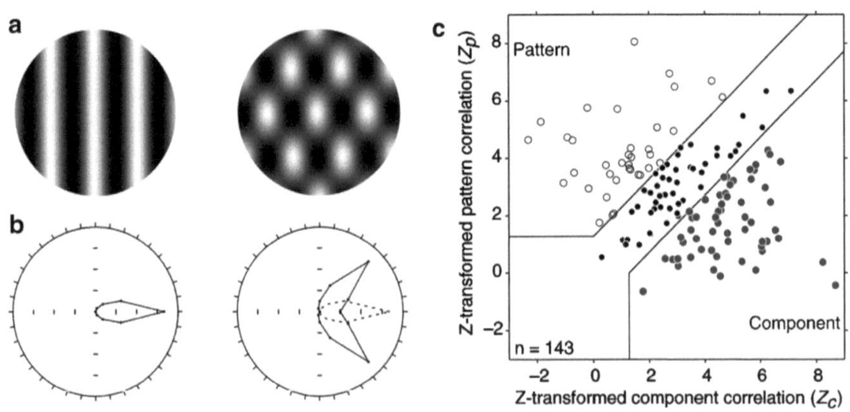

Fig. 3.2 Pattern and component selectivity (modified from Smith et al. 2005)

The classification of a neuron as CDS or PDS is made by comparing its actual tuning curve to a plaid, with two predictions (Fig. 3.2b, right plot) made based on its direction tuning to a single grating. In a population of MT neurons (Fig. 3.2c), 25% were classified as PDS (*white circles*) and 41% were classified as CDS (*gray circles*), with the remainder unclassed (Smith et al. 2005). The solid lines indicate significance boundaries for the classification of neurons as PDS or CDS. V1 neurons signal only the direction of motion of the component gratings (CDS) and not the true pattern direction (Movshon et al. 1985; Movshon and Newsome 1996). This is also true of V1 neurons which project directly to MT (Movshon and Newsome 1996), which is consistent with the idea that pattern motion is computed by circuits within MT.

3.4 Dynamics of MT Neuronal Response

There is evidence from both physiology and psychophysics that the neural representation of complex patterns evolves over tens to hundreds of milliseconds (Pack and Born 2001; Kooi et al. 1992; Lorenceau et al. 1993; Yo and Wilson 1992; Masson and Castet 2002). Since MT neurons are known to play a role in the perception of complex moving patterns, and human observers of plaid stimuli appear to refine their estimate of the direction over time, MT is a natural location in which to look for dynamics of response to such patterns.

Figure 3.3a–d contain scatter plots of pattern and component correlation, computed in the same way as the one shown in Fig. 3.2c. Each panel shows data taken from a small window of time cut out from the full stimulus period. Each point represents one neuron, and the points are colored to indicate the selectivity of that neuron over the entire stimulus period (CDS neurons are *gray circles*, PDS neurons are *white circles*, and unclassed neurons are *black circles*). In the top panel (Fig. 3.3a), only the period from 30 to 50 ms after stimulus onset is included. In this response window, before the onset latency for many MT neurons, there is little or no significant tuning for this measure. In Fig. 3.3, the response window in each row includes an additional 20 ms of data. CDS selective behavior is evident (as indicated by the gray circles that have already crossed the significance line) only 70 ms after stimulus onset (Fig. 3.3b). PDS neurons, however, take longer to show their characteristic behavior. Some neurons reach significance by 90 ms after stimulus onset (Fig. 3.3c), while many others take longer - up to 110 ms (Fig. 3.3d). At this time, 110 ms after stimulus onset, most of the CDS neurons but less than half of the PDS neurons have responded in a way consistent with their final tuning. In the right column (Fig. 3.3e–h), the same analysis is shown using a sliding window. Comparison of the rows reveals that this phenomenon is not due to the reduced noise that comes from averaging over a longer response window. Instead, these scatter plots demonstrate that PDS neurons lag behind CDS neurons in the time it takes them to show their characteristic response.

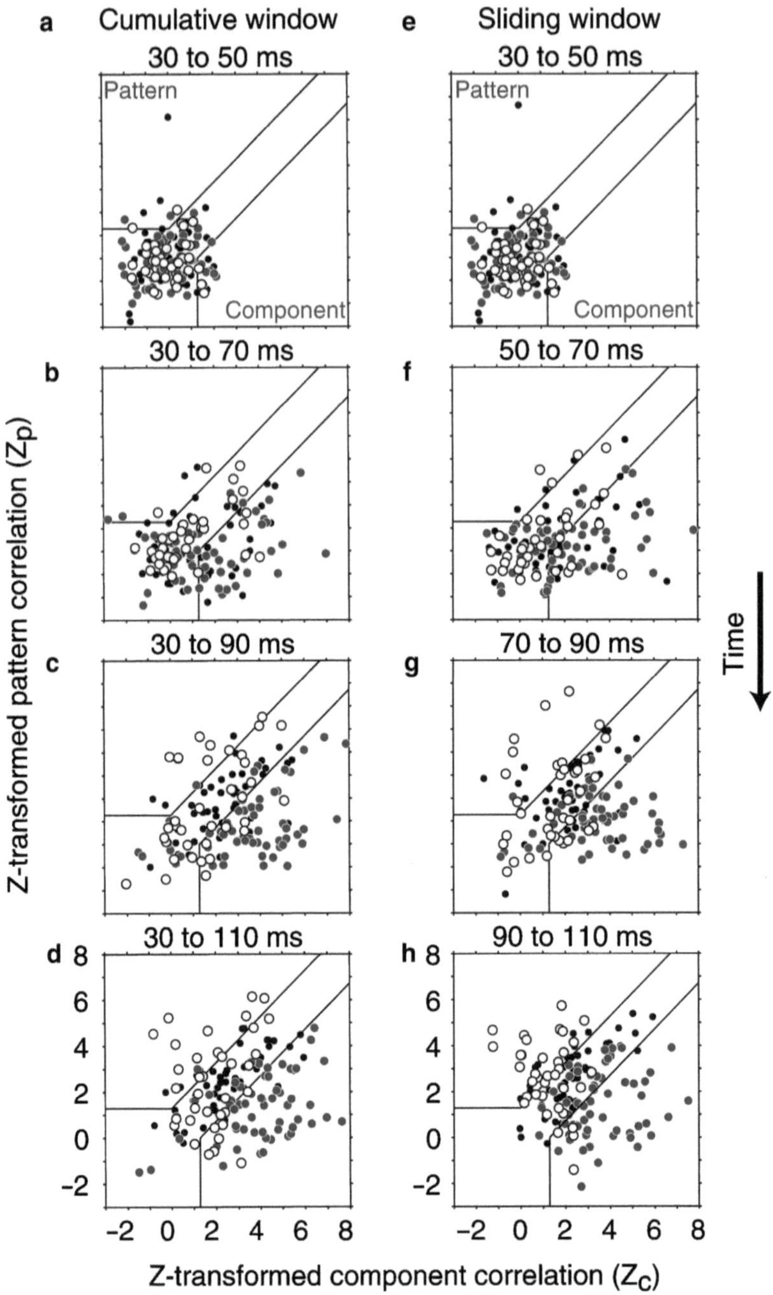

Fig. 3.3 Evolution of pattern and component selectivity for individual neurons

Figure 3.4a shows the evolution of PDS and CDS behavior for the population averages of the three classes of neurons from Fig. 3.2c (Smith et al. 2005). For cells which are eventually labeled as PDS, CDS, or unclassed (based on their response over the full stimulation period), the three lines show the evolution of their pattern and component correlation values over time, starting with the stimulus onset (time after stimulus onset is indicated with the numbers and connected lines). The CDS neurons (*dashed gray* line) cross the significance threshold much earlier in the

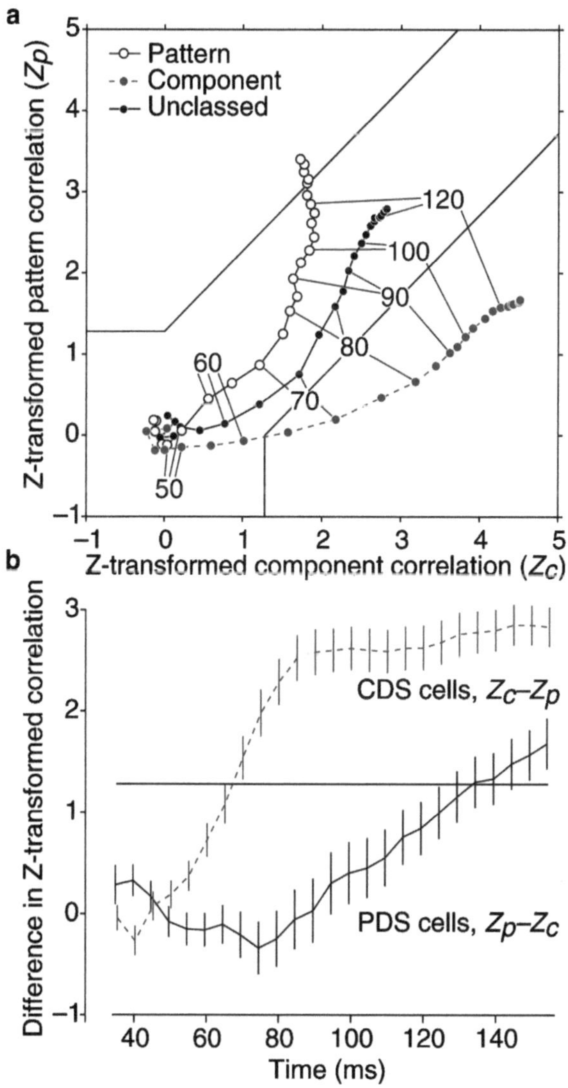

Fig. 3.4 Evolution of pattern and component selectivity in the population average response (modified from Smith et al. 2005)

response than the PDS neurons (*solid gray* line). The difference between the pattern and component indices is shown in Fig. 3.4b, to produce an index of "paternness" for PDS cells (*solid gray* line) and "componentness" for CDS cells (*dashed gray* line). When examining the times at which these two populations cross the significance threshold (horizontal black line), there is a difference of approximately 60–70 ms between the average CDS cell and the average PDS cell.

CDS neurons therefore develop their characteristic response tuning much earlier (60–65 ms) than PDS neurons (125–130 ms), a trend which is evident in the population and also in the responses of individual neurons (Smith et al. 2005). This additional time for pattern direction selectivity to become manifest is considerable, and suggests that circuits more complex than a simple feed-forward network are involved in the computation of pattern motion.

3.5 Relationship Between Bar and Plaid Stimuli

Bar textures moving obliquely to their orientation change apparent direction at low speeds and contrasts. Initially they seem to be moving perpendicular to the orientation of the bar, and over the course of 200–300 ms a human observer's perception tends to shift to the true direction of motion (Castet et al. 1993; Lorenceau et al. 1993). Neurons in macaque area MT exhibit behavior which is analogous – their initial response is based on motion perpendicular to the bar orientation, and their preference later shifts toward the direction in which the terminators move (Pack and Born 2001). These results are consistent with models in which the contours and terminators in a scene are analyzed separately, on different timescales – a rapid signal related to contours or edges and a slower signal related to terminators (Shimojo et al. 1989; Grossberg and Mingolla 1993; Lorenceau et al. 1993; Wilson et al. 1992; Wilson and Kim 1994; Löffler and Orbach 1999).

These models would be able to produce results similar to physiological data through a transition between the two motion signals. If the terminator-related motion signal is processed separately and more slowly than signals related to contours, it is possible to explain physiological and psychophysical observations of dynamic changes in motion perception. There is an alternative interpretation, however, which does not rely on a separate pathway for processing terminator motion. Instead, it may be that contours and terminators only appear to be processed separately due to the well known effect of contrast on visual processing: lower contrast targets are processed slower than high contrast ones (Albrecht 1995; Carandini et al. 1997; Gawne et al. 1996). In order to assess this proposal, we first have to consider a frequency analysis of the bar texture stimuli.

Figure 3.5a shows standard line, dot and bar stimuli in the top row. The motion of the first stimulus (*left*), a series of lines behind an aperture (a rectangular grating), is ambiguous. Motion of this stimulus in any direction is interpreted by an observer (and by MT neurons) as being perpendicular to the orientation of the lines. When there is motion of the second stimulus (*middle*) - a field of dots - it is unambiguous. The third stimulus, a bar texture (*right*), contains elements common

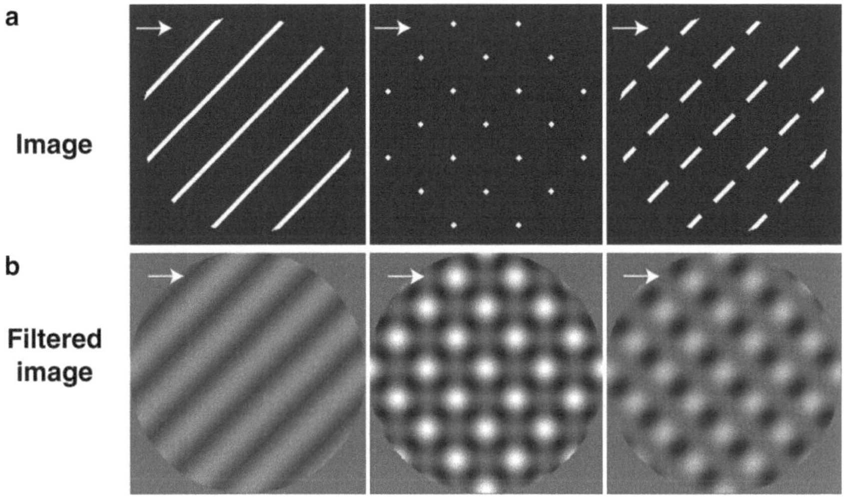

Fig. 3.5 Grating and bar stimuli

to both lines and dots. It is this stimulus which elicits a dynamically changing perception in observers (Castet et al. 1993; Lorenceau et al. 1993) and response in MT neurons (Pack and Born 2001).

The Fourier Transform decomposes any image into a sum of sinusoids of various amplitudes, spatial frequencies, and orientations. In the case of lines, dots, and bars, the Fourier Transform can provide a concrete basis for understanding the commonalities and differences between the stimuli. The first stimulus (*left*), a rectangular grating, has a Fourier amplitude spectrum composed of a sum of sinusoidal gratings of the same orientation but a range of spatial frequencies. In contrast, the dots (*middle*) have a broad amplitude spectrum, consisting of a sum of sinusoidal gratings across a wide range of orientations and spatial frequencies. For both the rectangular grating and dots, however, all the components have similar amplitude (contrast). Although the bar texture (*right*) may at first appear to consist of contours of only one orientation, examination of its Fourier amplitude spectrum reveals that this is not the case. Although it shares features of the lines (a dominant axis of power at one orientation) and dots (broad power), the bar texture's spectrum differs in an important way – the amplitude (contrast) of the components depends on their orientation. Components parallel to the orientation of the bar texture have the highest amplitude, while those at oblique orientations have lower contrast. Changing the contrast of these oblique components has an interesting effect – a decrease in amplitude lengthens the bars until they connect and form continuous lines, while increasing shortens the bar length until they approximate dots.

The bottom row in Fig. 3.5b contains filtered analogs of the corresponding stimuli in the top row. These filtered textures contain only the fundamental spatial frequency components of the original images in Fig. 3.5a. The line stimulus can be approximated by a single grating (*left*) and the dot stimulus by a type of plaid composed of four gratings of equal contrast (*middle*). The filtered bar texture (*right*) is produced by combining the same four constituent gratings but with unequal contrast: a high contrast grating parallel to the bar texture orientation, and three low contrast gratings at +45°, +90° and +135° relative to the high contrast grating. The filtered bar image has no obvious terminators, but nonetheless retains the essential structure of the original bar image. Thus, it is a good stimulus with which to test our hypothesis about the role of contrast in processing bar textures.

3.6 Response to Filtered Bar Textures

If the filtered bar texture shown on the bottom right of Fig. 3.5 is indeed a good approximation to the corresponding bar texture directly above it, then MT neurons should have dynamically evolving responses to this stimulus. Figure 3.6a shows the temporal evolution of direction selectivity in a population of MT neurons to

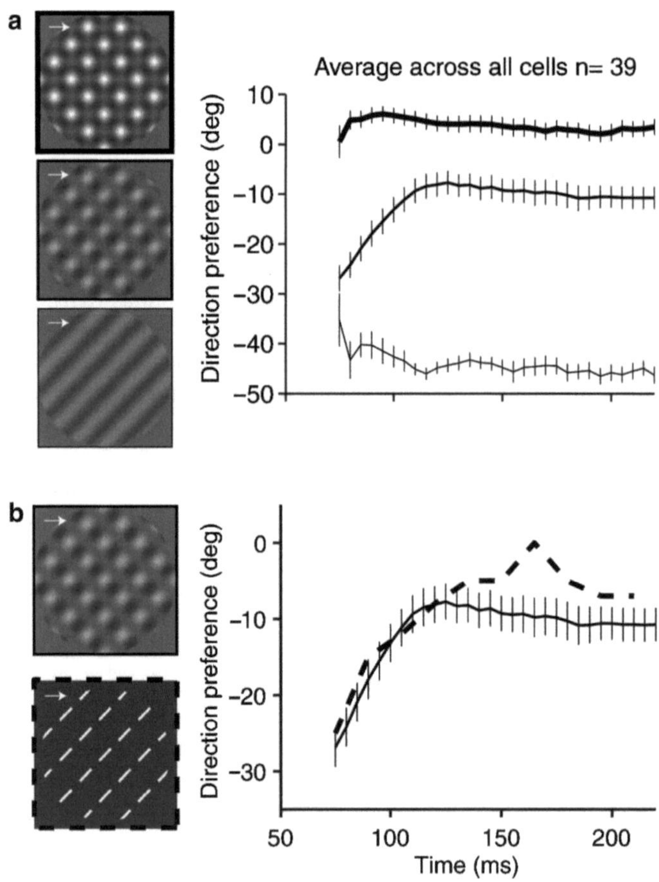

Fig. 3.6 Dynamics of direction selectivity to filtered textures

the three filtered stimuli described above (Majaj 2006). In order to compute the population average, the responses are adjusted so that both the direction preference and onset latency are aligned. When all four components of the filtered stimuli are of equal contrast, the resulting filtered dot texture appears to move to the right, and the neuronal population exhibits consistent direction tuning (*thick black line*). With a single component (a grating), the population of neurons responds prefers the direction orthogonal to the orientation (*thin black line*). These two results are expected based on previous physiological studies of MT and psychophysical studies of human motion perception. For the filtered bar texture, possessing components of different contrast, the direction preference does not remain constant over time. Instead, there is a distinct change in the direction preference over a period extending to 120 ms after stimulus onset (*medium black line*). The early response resembles the direction preference to a single grating; it changes over the next 50 ms, and by 120 ms it stabilizes to resemble the direction preference to the filtered dots.

The change in direction selectivity for filtered bar textures resembles the psychophysical findings of Lorenceau et al. (1993) and physiological results of Pack and Born (2001). Figure 3.6b shows the evolution of direction selectivity for filtered bar textures (middle line from Fig. 3.6a) compared with that for unfiltered bar textures (data replotted from Pack and Born 2001). The dynamics observed in response to the filtered bar texture are remarkably similar to those obtained in response to the texture itself, even though the filtered texture is composed of four sinusoidal gratings and contains no obvious terminators. This suggests that a common mechanism might underlie the dynamics in response to these two stimuli.

3.7 Effects of Contrast on Response Dynamics

The similarity between these two results might be explained by a well known phenomenon in vision. The contrast of a visual stimulus has a powerful effect on the speed of visual processing: neurons respond to low contrast stimuli with a long latency, but high contrast stimuli are processed with a short latency (Albrecht 1995; Carandini et al. 1997; Gawne et al. 1996). The filtered bar texture stimulus is composed of gratings with different contrast – a high contrast grating parallel to the bar texture orientation, and three low contrast oblique gratings. If the different components of the filtered bar texture are processed with different latencies, then the response to the combined stimulus might be expected to exhibit dynamics.

Stimulus contrast affects response magnitude and latency in MT neurons. Figure 3.7 shows this effect for the responses of a single MT neuron. Each row of the figure shows a raster plot of responses to repeated presentations of a drifting grating stimulus in the preferred direction, ranging from 100% contrast (top row) to 10% contrast (bottom row). A histogram, binned at 1 ms precision, of the firing

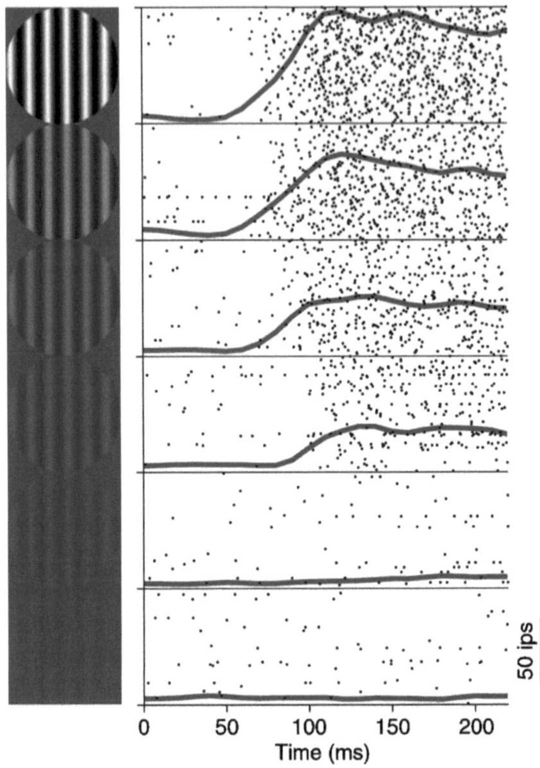

Fig. 3.7 Effect of contrast on response latency

rate over time is shown with a *gray line* in each row. The response latency of this neuron, measured from the response onset or the time to reach peak firing rate, grows as contrast decreases (from the top to bottom rows). In this example neuron, the onset latency changes from approximately 65 to 110 ms as the contrast decreases. The trend evident in the responses shown for the single neuron in Fig. 3.7 is also observed in a larger population of MT neurons (Thiele and Hoffman 1996; Thiele et al. 1999; Majaj 2006). A delay of 100 ms is typical between onset latency at low and high contrast.

In the filtered bar texture, the component gratings may be processed by the visual system with different latencies due to their different contrasts. The highest contrast component, parallel to the orientation of the comparable bar texture, would be processed first. The other components - lower in contrast and at oblique angles relative to the primary component - would be processed with some delay. As an MT neuron integrates the information in all of these components, its direction preference will change over time. The early responses will be dominated by the high contrast grating, but as the lower contrast oblique gratings are processed the later responses will reflect this information.

3.8 Conclusions

In natural vision, our experience of the world is rich with temporal dynamics. Our visual system has evolved to be able to evaluate and interpret this information with the speed necessary to make fast judgments based on our perceptual experience. Nonetheless, the integration necessary to determine the motion direction of a complex visual pattern is not instantaneous. In this chapter, we have described physiological evidence that neurons in macaque area MT have response dynamics which evolve over the first 100–200 ms after a visual stimulus appears. The temporal profile and directional selectivity of these responses parallels results from a well established psychophysical literature based on experiments in human observers.

A number of models have been proposed to explain the dynamics observed in physiological and psychophysical studies. The most common approach has been to propose separate pathways for analyzing different visual features (terminators and contours) – essentially, parallel one- and two-dimensional motion analysis. Here, we have reasoned from data and models that one-dimensional orientation-selective mechanisms can account for these experimental observations. This is not to imply that two-dimensional features are unimportant for visual processing, but rather that a separate pathway for analysis of such features is not necessary to explain dynamics in motion perception. A linear–nonlinear cascade, which incorporates a number of physiologically realistic processes into a functional model, is an alternative approach which provides a good fit to the neuronal and perceptual data.

3.9 Supplementary materials (CD-ROM)

Movie 1 Field of moving upright bars (file « *3_M1_Bars0.avi* ») A field of bars moving perpendicular to their orientation (all of the motion cues in the stimulus are consistent). Videos 1 & 2 are modeled after the stimuli used by Pack and Born (2001).

Movie 2 Field of moving tilted bars (file « *3_M2_Bars45.avi* ») Moving bars with an orientation of +45° relative to the direction of motion. This is the moving version of the stimulus shown in the right panel of Fig. 3.5a. In this case, the local motion signal along the edges of the bars is in conflict with the true direction of motion.

Movie 3 Grating motion (see Fig. 3.5) (file « *3_M3_Filteredlines.avi* ») A single drifting grating stimulus at +45° (Fig. 3.5b, left panel). Because there is only a single orientation present in the stimulus, and no access to terminators (the "aperture problem"), the direction of motion is perceived to be perpendicular to the orientation of the grating.

Movie 4 Filtered dot texture (see Fig. 3.5) (file « *3_M4_Filtereddots.avi* ») The filtered analog of a dot texture, this is a type of plaid stimulus (Fig. 3.5b, middle panel) composed of four gratings of equal contrast. The grating orientations are +0°,

+45°, +90° and +135° relative to the bar texture shown in Movie 2. The result is a filtered dot texture similar to the dot image shown in Fig. 3.5a (middle panel).

Movie 5 Filtered bar texture (see Fig. 3.5) (file « *3_M5_Filteredbars.avi* ») This filtered bar texture (Fig. 3.5b, right panel) is composed of the same four gratings as in Movie 4, but with unequal contrast: a high contrast grating parallel to the bar texture orientation, and three low contrast gratings at +45°, +90° and +135° relative to the high contrast grating.

Movie 6 Filtered bar texture and grating components (file « *3_M6_Texture-components.avi* »). This video shows the filtered bar texture (from Video 5) decomposed into its four constituent gratings and then re-assembled.

References

Albrecht DG (1995) Visual cortex neurons in monkey and cat: effect of contrast on the spatial and temporal phase transfer functions. Vis Neurosci 12:1191–1210

Albright TD (1984) Direction and orientation selectivity of neurons in visual area MT of the macaque. J Neurophysiol 52:1106–1130

Angelucci A, Levitt JB, Walton EJ, Hupé JM, Bullier J, Lund JS (2002) Circuits for local and global signal integration in primary visual cortex. J Neurosci 22:8633–8646

Bair W, Cavanaugh JR, Movshon JA (2003) Time course and time-distance relationships for surround suppression in macaque V1 neurons. J Neurosci 23:7690–7701

Bair W, Koch C, Newsome W, Britten K (1994) Power spectrum analysis of bursting cells in area MT in the behaving monkey. J Neurosci 14:2870–2892

Beutter BR, Stone LS (1998) Human motion perception and smooth eye movements show similar directional biases for elongated apertures. Vision Res 38:1273–1286

Britten KH, Shadlen MN, Newsome WT, Movshon JA (1992) The analysis of visual motion: a comparison of neuronal and psychophysical performance. J Neurosci 12:4745–4765

Carandini M, Heeger DJ, Movshon JA (1997) Linearity and normalization in simple cells of the macaque primary visual cortex. J Neurosci 17:8621–8644

Castet E, Lorenceau J, Shiffrar M, Bonnet C (1993) Perceived speed of moving lines depends on orientation, length, speed and luminance. Vision Res 33:1921–1936

Chey J, Grossberg S, Mingolla E (1997) Neural dynamics of motion grouping: From aperture ambiguity to object speed and direction. J Opt Soc Am A 14:2570–2594

Friend SM, Baker CL Jr (1993) Spatio-temporal frequency separability in area 18 neurons of the cat. Vision Res 33:1765–1771

Gawne TJ, Kjaer TW, Richmond BJ (1996) Latency: another potential code for feature binding in striate cortex. J Neurophysiol 76:1356–1360

Gegenfurtner KR, Kiper DC, Levitt JB (1997) Functional properties of neurons in macaque area V3. J Neurophysiol 77:1906–1923

Grossberg S, Mingolla E (1993) Neural dynamics of motion perception: direction fields, apertures, and resonant grouping. Percept Psychophys 53:243–278

Heeger DJ, Simoncelli EP, Movshon JA (1996) Computational models of cortical visual processing. Proc Natl Acad Sci USA 93:623–627

Hildreth EC (1984) The measurement of visual motion. MIT Press, Cambridge, MA

Holub RA, Morton-Gibson M (1981) Response of visual cortical neurons of the cat to moving sinusoidal gratings: response-contrast functions and spatiotemporal interactions. J Neurophysiol 46:1244–1259

Knierim JJ, Van Essen DC (1992) Neuronal responses to static texture patterns in area V1 of the alert macaque monkey. J Neurophysiol 67:961–980

Kooi FL, DeValois KK, Switkes E, Grosof DH (1992) Higher-order factors influencing the perception of sliding and coherence of a plaid. Perception 21:583–598
Lamme VAF (1995) The neurophysiology of figure-ground segregation in primary visual cortex. J Neurosci 15:1605–1615
Lee TS, Mumford D, Romero R, Lamme VAF (1998) The role of primary visual cortex in higher level vision. Vision Res 38:2429–2454
Lee TS, Nguyen M (2001) Dynamics of subjective contour formation in early visual cortex. Proc Natl Acad Sci USA 98:1907–1911
Lee TS, Yang CF, Romero RD, Mumford D (2002) Neural activity in early visual cortex reflects behavioral experience and higher-order perceptual saliency. Nat Neurosci 5:589–597
Levitt JB, Kiper DC, Movshon JA (1994) Receptive fields and functional architecture of macaque V2. J Neurophysiol 71:2517–2542
Lidén LH, Pack CC (1999) The role of terminators and occlusion cues in motion integration and segmentation: a neural network model. Vision Res 39:3301–3320
Lindner A, Ilg UJ (2000) Initiation of smooth-pursuit eye movements to first-order and second-order motion stimuli. Exp Brain Res 133:450–456
Löffler G, Orbach HS (1999) Computing feature motion without feature detectors: a model for terminator motion without end-stopped cells. Vision Res 39:859–871
Lorenceau J, Shiffrar M, Wells N, Castet E (1993) Difference motion sensitive units are involved in recovering the direction of moving lines. Vision Res 33:1207–1217
Majaj N (2006) Spatial and temporal integration of motion signals in area MT. Thesis, New York University
Mante V (2000) Testing models of cortical area MT. Thesis, Institute of Neuroinformatics, ETH, University of Zurich
Masson GS, Castet E (2002) Parallel motion processing for the initiation of short-latency ocular following in humans. J Neurosci 22:5149–5163
Masson GS, Rybarczyk Y, Castet E, Mestre DR (2000) Temporal dynamics of motion integration for the initiation of tracking eye movements at ultra-short latencies. Vis Neurosci 17:753–767
Masson GS, Stone LS (2002) From following edges to pursuing objects. J Neurophysiol 88:2869–2873
Movshon JA, Adelson EH, Gizzi MS, Newsome WT (1985) The analysis of visual moving patterns. In: Chagas C, Gattass R, Gross C (eds) Pattern recognition mechanisms. Springer, New York, pp 117–151
Movshon JA, Newsome WT (1996) Visual response properties of striate cortical neurons projecting to area MT in macaque monkeys. J Neurosci 16:7733–7741
Newsome WT, Paré EB (1988) A selective impairment of motion perception following lesions of the middle temporal area MT. J Neurosci 8:2201–2211
Pack CC, Born RT (2001) Two-dimensional substructure of MT receptive fields. Nature 409:1040–1042
Perrone JA, Thiele A (2001) Speed skills: measuring the visual speed analyzing properties of primate MT neurons. Nat Neurosci 4:526–532
Priebe NJ, Cassanello CR, Lisberger SG (2003) The neural representation of speed in macaque area MT/V5. J Neurosci 23:5650–5661
Ringach DL, Hawken MJ, Shapley R (1997) The dynamics of orientation tuning in the macaque monkey striate cortex. Nature 387:281–284
Ringach DL, Hawken MJ, Shapley R (2003) Dynamics of orientation tuning in macaque v1: the role of global and tuned suppression. J Neurophysiol 90:342–352
Rodman HR, Albright TD (1989) Single-unit analysis of pattern-motion selective properties in the middle temporal visual area (MT). Exp Brain Res 75:53–64
Roelfsema PR, Lamme VA, Spekreijse H (1998) Object-based attention in the primary visual cortex of the macaque monkey. Nature 395:376–381
Rust N, Mante V, Simoncelli EP, Movshon JA (2006) How MT cells analyze the motion of visual patterns. Nat Neurosci 9:1421–1431

Salzman CD, Newsome WT (1994) Neural mechanisms for forming a perceptual decision. Science 264:231–237
Shimojo S, Silverman GH, Nakayama K (1989) Occlusion and the solution to the aperture problem for motion. Vision Res 29:619–626
Simoncelli EP, Heeger DJ (1998) A model of neuronal responses in visual area MT. Vision Res 38:743–761
Smith MA, Bair W, Movshon JA (2006) Dynamics of suppression in macaque primary visual cortex. J Neurosci 26:4826–4834
Smith MA, Kelly RC, Lee TS (2007) Dynamics of response to perceptual pop-out stimuli in macaque V1. J Neurophysiol 98:3436–3449
Smith MA, Majaj NJ, Movshon JA (2005) Dynamics of motion signaling by neurons in macaque area MT. Nat Neurosci 8:220–228
Stoner GR, Albright TD (1992) Neural correlates of perceptual motion coherence. Nature 358:412–414
Thiele A, Distler C, Hoffman KP (1999) Decision-related activity in the macaque dorsal visual pathway. Eur J Neurosci 11:2044–2058
Thiele A, Hoffman KP (1996) Reaction time and time course of neuronal responses in MT and MST at different stimulus contrasts. Perception 25:ECVP Abstract Supplement
Tolhurst D, Movshon J (1975) Spatial and temporal contrast sensitivity of striate cortical neurones. Nature 257:674–675
Van Essen DC, Maunsell JHR, Bixby JL (1981) The middle temporal visual area in the macaque: myeloarchitecture, connections, functional properties and topographic organization. J Comp Neurol 199:293–326
Wallace JM, Stone LS, Masson GS (2005) Object motion computation for the initiation of smooth pursuit eye movements in humans. J Neurophysiol 93:2279–2293
Wilson HR, Ferrera VP, Yo C (1992) A psychophysically motivated model for two-dimensional motion perception. Vis Neurosci 9:79–97
Wilson HR, Kim J (1994) Perceived motion in the vector sum direction. Vis Neurosci 34:1835–1842
Yo C, Wilson HR (1992) Perceived direction of moving two-dimensional patterns depends on duration, contrast and eccentricity. Vision Res 32:135–147
Zeki SM (1974) Functional organization of a visual area in the posterior bank of the superior temporal sulcus of the rhesus monkey. J Physiol (Lond) 236:549–573
Zipser K, Lamme VA, Schiller PH (1996) Contextual modulation in primary visual cortex. J Neurosci 16:7376–7389

Chapter 4
Multiscale Functional Imaging in V1 and Cortical Correlates of Apparent Motion

Yves Fregnac, Pierre Baudot, Fréderic Chavane, Jean Lorenceau, Olivier Marre, Cyril Monier, Marc Pananceau, Pedro V. Carelli, and Gerard Sadoc

Abstract In vivo intracellular electrophysiology offers the unique possibility of listening to the "synaptic rumor" of the cortical network captured by the recording electrode in a single V1 cell. The analysis of synaptic echoes evoked during sensory processing is used to reconstruct the distribution of input sources in visual space and time. It allows us to infer, in the cortical space, the dynamics of the effective input network afferent to the recorded cell. We have applied this method to demonstrate the propagation of visually evoked activity through lateral (and possibly feedback) connectivity in the primary cortex of higher mammals. This approach, based on functional synaptic imaging, is compared here with a real-time functional network imaging technique, based on the use of voltage-sensitive fluorescent dyes. The former method gives access to microscopic convergence processes during synaptic integration in a single neuron, while the latter describes the macroscopic divergence process at the neuronal map level. The joint application of the two techniques, which address two different scales of integration, is used to elucidate the cortical origin of low-level (non-attentive) binding processes participating in the emergence of illusory motion percepts predicted by the psychological Gestalt theory.

4.1 A Topological Paradox in Comparing Structure and Function

The biological foundations of low-level visual perception in the mammalian brain show an apparent paradox:

On the one hand, the functional specificity of the visual system underlying non-attentive perception seems to be best explained by a parallel cascade of serial filters

Y. Fregnac (✉)
Unité de Neurosciences Intégratives et Computationnelles (UNIC), CNRS UPR 2191,
1 Avenue de la Terrasse Bat 32/33, 91198, Gif-sur-Yvette, France
e-mail: fregnac@unic.cnrs-gif.fr

U.J. Ilg and G.S. Masson (eds.), *Dynamics of Visual Motion Processing: Neuronal, Behavioral, and Computational Approaches*,
DOI 10.1007/978-1-4419-0781-3_4, © Springer Science+Business Media, LLC 2010

from retina to cortex. The synaptic impact of this topographic feed-forward projection is strong and results in multiple point-to-point ordered representations of the retinal periphery onto central target neural structures. The retinofugal projection to the cortex (through the dorsal lateral geniculate nucleus (LGN)), which is responsible for the extraction of contour and motion, forms a continuous homeomorphic mapping of visual space on the layers of the primary visual cortex.

On the other hand, the anatomical architecture along the visual thalamo-cortical pathway includes a profusion of feedback routes, from cortex to thalamus and from higher-order processing stations to primary visual cortex (V1), as well as intrinsic lateral and recurrent connections confined within each processing relay stage.

Thus, the linkage brought about by lateral and feedback connectivity introduces a mismatch with the retinotopic order imposed by the feed-forward projections. How this topological conflict is solved by the whole system is still poorly understood. Nevertheless, the dominance of the feed-forward drive is obvious at the functional level as most thalamic and cortical neurons express in their discharge a "tubular" view of the visual world: spiking responses are evoked only when visual stimuli are present within a small sensory window, defined as the "minimal discharge field". This integration core is limited to 1 or 2° of visual angle in the cat and a few tens of minutes of arc in humans, around the representation of the gaze axis. This reduced average receptive field size is found in spite of the fact that the stellate layer IV cortical neurons, which form the input gate of the cortex, receive most of their synaptic input (>94%) not from the thalamus but from other cortical neurons. These afferent neurons are located in V1 or in other areas (with larger receptive fields), and a large number of them process information in the "silent" surround of the relevant layer IV receptive field (Binzegger et al. 2004). The functional impact of this non-topographic modulation is difficult to measure as its sign (facilitation/depression) and strength depend largely on the contextual conditions used to stimulate the visual system (review in Séries et al. 2003).

In human brain imaging, functional magnetic resonance imaging (fMRI based on the BOLD signal) allows sensory cortical representations to be charted with reduced spatial and temporal precisions, of the order of 1 mm and 1 s, respectively. Two types of approaches, including ad-hoc retinotopic constraints, are classically used: one is based on the study of correlation between stimulus and response while the other takes advantage of the inverse relationship (response→stimulus) by predicting the most likely input on the basis of the target activation pattern. The first approach often utilizes algorithms of multivariate optimization (dependent on stimulus features, such as location, orientation, color, motion, spatial frequency) and associates a multidimensional pseudo-receptive field with each cortical voxel studied. The second approach (often called "brain reading") aims at the classification of cortical activation patterns (distributed across a predefined set of voxels of interest) evoked by classes of stimuli supposedly perceptually "different" (Thirion et al. 2006). Powerful decoding approaches allow the identification and classification of the most likely pattern across new sets of stimuli (not shown during the training of the classifier) (Kay et al. 2008; Haynes and Rees 2006). Notwithstanding the fact that these techniques cannot guarantee the uniqueness of the multivariate decomposition or of the inverse transform, they have been used with remarkable success to show the influence of peripheral information in the processing of foveal information (Lee et al. 2007) and the existence of intra-V1 propagating waves during interocular rivalry (Williams et al. 2008).

These advances notwithstanding, it is obvious that the lack of spatial resolution and the slow kinetics of the hemodynamic signal are poorly adapted to account for perceptual illusions or ambiguities. This is particularly true when the retinal flow results in a percept which is no longer isomorphic to the physical pattern present at the retina, or which is unstable in time. In the spatial domain, even if multivariate analysis indicates distributed changes in the cortical activation patterns with extreme sensitivity, it does not give explicit access, on a voxel by voxel basis, to the read-out of a topographical representation of the percept in the retinotopic cortical space. In the time domain, in spite of sophisticated manipulations of the stimulus phase, or the inclusion of prior knowledge of the perceptual outcome during ambivalent perception, the temporal resolution of the hemodynamic signal is still too slow to track perceptual changes faster than 1 s. However, our everyday experience of the visual world shows that powerful dynamic association mechanisms override the sensory feed-forward imprint of the visual world at the central level and continuously reshape our mental images. These binding mechanisms are expressed in both the primary and secondary cortical areas, and are most likely activated during anesthesia and dream states. Other imaging methods, with higher spatial and temporal resolutions, must be developed to monitor the time-course of dynamic changes in perception (Frégnac 2001).

4.2 Multiscale Imaging

To visualize the sub-threshold functional influence of lateral connections, several experimental approaches can be considered, which require more invasive methods than fMRI (involving craniectomy and pia resection):

The most direct approach is to monitor the spread of evoked activation relayed across the superficial cortical layers, the representation plane of the visual space. This can be done best in vivo by using the voltage-sensitive dye (VSD) imaging technique: CCD cameras have reached a spatial (<50 μm) and temporal precision (<1 ms), compatible with the structural columnar scale and the time-course of synaptic responses. The most adequate stimulation protocol is to use a focal visual stimulus whose features are optimized to trigger the activation of an individual cortical column (or a minimal number of them). For this purpose, one can use a luminance grating patch, of given orientation and spatial and temporal frequency, the size of which is limited to that of visual cortical receptive fields. Such techniques of macroscopic imaging have been applied in areas V1 and V2 of cats and monkeys to track the "horizontal" divergence pattern of the cortical activation process in response to point-like stimuli (Grinvald and Hildesheim 2004).

Rather than looking at the global evoked dynamics of the network, a complementary approach is to address the microscopic organization scale and focus on the synaptic bombardment of a single neuron. Intracellular electrophysiology with sharp electrodes can be used to continuously monitor the membrane potential of a single neuron for several hours, even in vivo (Frégnac and Bringuier 1996; Bringuier et al. 1999). The recorded irregular asynchronous spiking activity is the result of the transient but repeated convergence arising from multiple synaptic sources in the network. For the past 15 years, we have been developing a reverse engineering approach, which allows in principle, retrieval of the effective connection graph in which the cell is embedded at any point in time (Fig. 4.1). The analysis is based on the synaptic rumor recorded in a single cell. Its principle is similar to that of echography in the etymological sense (transcription of echoes), and is referred to as the "functional synaptic imaging" (Frégnac and Bringuier 1996). During sensory activation,

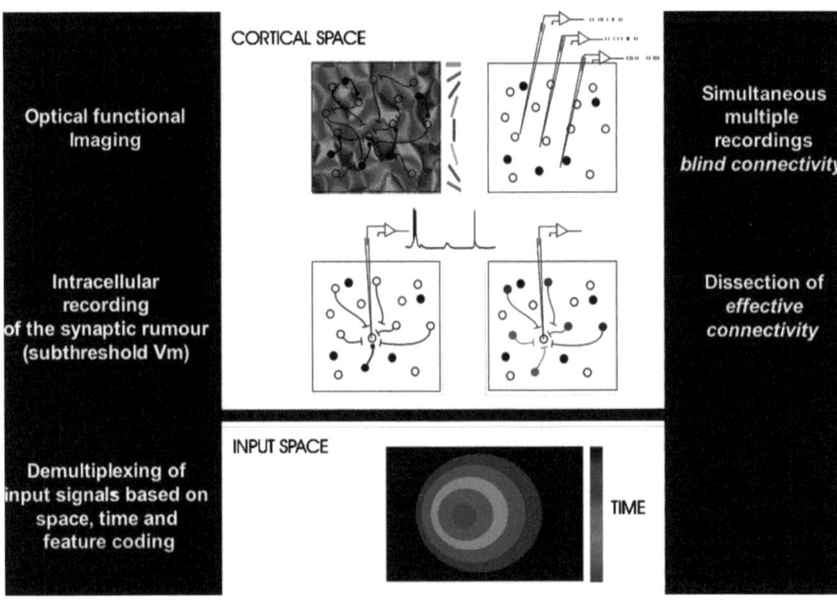

Fig. 4.1 Network dynamics imaging vs. synaptic functional imaging. Three methods for visualizing network dynamics are compared: (1) *top left*, optical functional imaging allows charting of cortical domains of iso-functional preference (color coded) on the basis of metabolic or hemodynamic signals; (2) *top right*, multiple simultaneous extracellular recordings are used to evaluate correlated activity patterns through the blind selection of potentially interconnected neurons; (3) *middle and bottom rows*, the reverse analysis of sub-threshold activity during long-duration intracellular recordings of the same single cell can be used to retrieve the effective network afferent to the recorded cell (see text) (*see Color Plates*)

the cortex is considered a chamber of echoes produced by the thalamo-cortical input. The reverse read-out of the sources is based on the extraction of correlations of synaptic events in space and time with specific stimulus features (orientation, direction, ocular dominance, etc.). It predicts the macroscopic activation of the network in space and time. The success of this de-multiplexing computation relies on the underlying assumption (unproven in the general case) that the input sources are separable in space and their synaptic influence travels in time with similar speed (which is found to be the case for sparse stimulation regimes or during ongoing activity).

4.3 The Electrophysiological Basis of Functional Synaptic Imaging

Since the pioneering work of Hartline (1938), it has been established that the location of receptive field centers is fixed but that the receptive field's extent "however depends upon the intensity and the size of the spot of light used to explore it, and upon the conditions of adaptation". Later work based on electrophysiological recording and dual-stimulation protocols showed, indeed, that the classical receptive field of the cortical neuron is surrounded by a "silent" periphery (or non-classical receptive field (nCRF) review in Séries et al. 2003). Stimulation in the far periphery causes no spiking response, but up- or down-modulates the response to a stimulus presented concurrently in the classical receptive field.

Intracellular recordings give direct access to the sub-threshold responses evoked by peripheral stimuli, which are detected as significant stimulus-locked changes in membrane potential fluctuations (current clamp Bringuier et al. 1999) or conductances (voltage clamp Borg-Graham et al. 1998; Monier et al. 2003). Using clamp intracellular techniques in vivo (70–100 MΩ sharp electrodes with K-acetate or methyl-sulfate), we published almost 10 years ago a seminal study in Science (Bringuier et al. 1999), showing that the spatial extent of the synaptic integration field of cortical V1 neurons extends far beyond the classical discharge field (2- to 20-fold increase in largest elongation, up to 10–15° of solid angle) (Fig. 4.2). The strength of the postsynaptic response evoked by a distal point-like stimulus was found to decrease almost linearly as a function of radial eccentricity (from the RF center), with a space constant of a few degrees of visual angle, and could be described by a Gaussian hill of spatial sensitivity around the RF center. Similarly, the onset of the evoked postsynaptic response showed a linear increase with stimulation eccentricity (relative to the discharge field). Thus, each sub-threshold cortical RF seems to be endowed with dual space-time attributes, i.e. a spatial "hill" of visual sensitivity sitting in a "basin" of temporal delays.

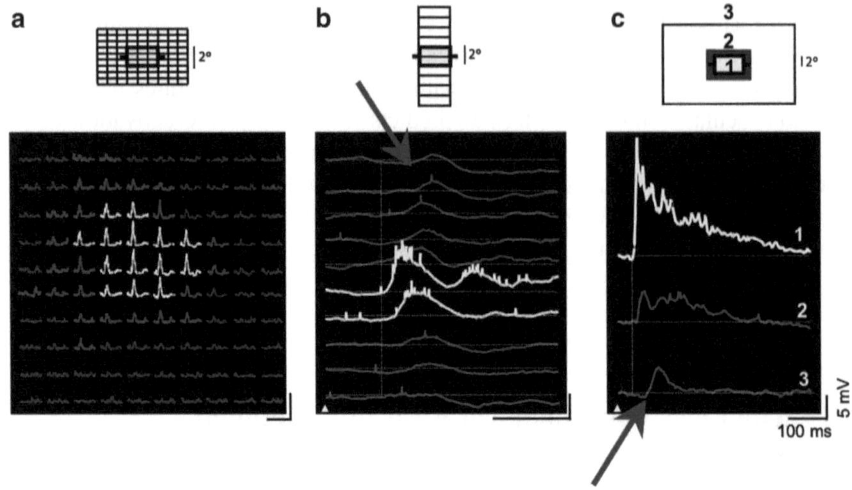

Fig. 4.2 Synaptic integration field and discharge field of visual cortical neurons. **a** to **c**: spatio-temporal maps of sub-threshold (*red*) and spiking responses (*white*) of V1 neurons. (**a**) sparse noise mapping; (**b**) 1-D bar mapping across the width axis of the receptive field; (**c**) co-centric gratings covering either the discharge field (1), the immediate (2) or the far (3) surround the periphery. Note the increase in the spatial extent of the sub-threshold RF with the level of temporal and spatial summation of the stimulus (which increases from **a** to **c**). Note also that the latency of the synaptic response (red arrows) increases as a function of eccentricity from the discharge field center. Adapted from (Bringuier et al. 1999), with permission (*see Color Plates*)

4.4 Reconstruction of Propagation Waves from Synaptic Echoes

Functional synaptic imaging requires reconstruction, first in space and then in cortex, of the source location distribution corresponding to the recorded synaptic echoes produced by the sensory drive. The hypothesis of a traveling wave is made on the assumption of symmetry in exuberant intrinsic connectivity: since V1 is a highly recurrent network, we make the simplifying assumption that each cell is connected reciprocally to any other cell, with identical propagation delays from and to the others (Fig. 4.3, left). In simpler terms, it should take exactly the same time for any given cell to receive a signal from a distant cortical source as to send it back to the same cortical locus. This theoretical shortcut allows us to infer propagation patterns (the cell being seen as a "wave emitter") solely on the basis of the spatio-temporal maps of stimulus-locked synaptic responses recorded in a single cell (the cell being seen as an "echo receiver"). As illustrated in the right panel of Fig. 4.3, the slopes seen in the recorded latency basin (pink dotted lines) suggest that the information received at one point in the cortex through the feed-forward afferents is then propagated radially by the horizontal connectivity to neighboring regions of the visual cortex over a distance that may correspond to up to 10° of visual angle.

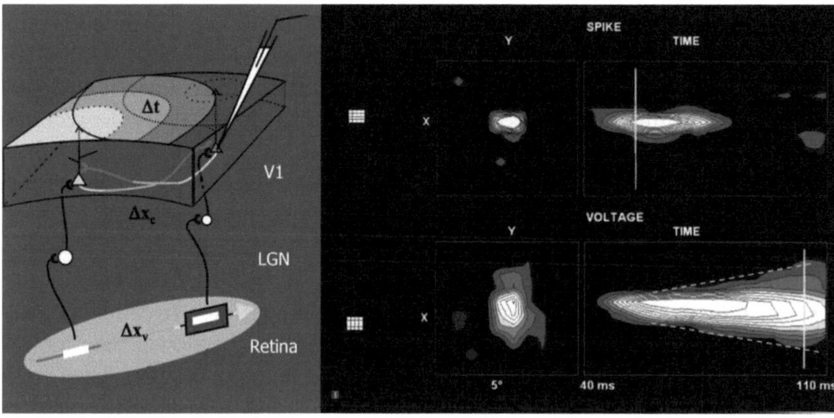

Fig. 4.3 Functional Synaptic Imaging. *Left*: Schematic representation of the hypothesis of reciprocal horizontal connections between two cortical cells (*left*, "sender"; *right* "receiver"). This schema allows to reconstruct a propagating wave (*circles*) from the intracellular measure of evoked latencies of synaptic responses in the "receiver"-cell (*right*, electrode). Δx_v is the eccentricity of the distal stimulus (*white rectangle, no outline*) from the central stimulus shown in the discharge field center (*grey rectangle, black outline*). Δt is the latency between the synaptic response onsets evoked through the two pathways. Δx_c is the intracortical distance between the cortical feed-forward impacts produced by the two stimuli, inferred from the know retinocortical magnification factor (see text for details). *Right*: Spatio-temporal (*left*: X-Y; *right*: X-t) maps of supra-threshold (spike, *upper panels*) and sub-threshold (voltage, *lower panels*) activations in the same V1 cell. The X-Y maps are presented for two specific delays corresponding respectively to the maximal extents of the discharge field (*upper*) and sub-threshold integration field (*lower*). The pink dotted lines show the average speed of the propagation of the reconstructed wave (0.2–0.4 m/s) [see Movie 1] (*see Color Plates*)

These data led to the reconstruction of a propagating wave of visual activity relayed by the horizontal connectivity.

The principle of calculation is straightforward (Fig. 4.3, left): we compare the synaptic effects of two elementary stimuli (white bar), one in the core of the minimal discharge field, and the other in the "silent" surround. The distance between the primary points of the feed-forward impact produced in the cortex by the two stimuli can be predicted on the basis of their relative retinal eccentricity Δx_v and the value of the retino-cortical magnification factor (RCMF). This factor can be measured electrophysiologically (Albus 1975 in cat), by 2-deoxyglucose metabolic labeling (Tootell et al. 1982 in monkey), by intrinsic imaging (Kalatsky and Stryker 2003 in mouse and Basole et al. 2003 in ferret), and by fMRI (Warnking et al. 2002 in humans). Thus, beyond a certain scale of spatial integration (larger than the columnar grain), any distance in the visual space Δx_v can be converted to a distance in the visual cortex Δx_c. The spatial range of the sub-threshold field extent agrees with the anatomical description of 4–7 mm horizontal axons running across superficial layers (Mitchison and Crick 1982) (note that, while the RCMF factor is dependent on the eccentricity from the fovea in primates and humans, this is not so much the case in cats and ferrets where 1° of visual angle corresponds roughly to 1 mm in the cortex within the 10° of the area centralis). Furthermore, the electrophysiological recordings give access to the delay Δt_c between the two synaptic echoes obtained through the feed-forward and the horizontally-mediated pathways. By dividing the inferred cortical distance Δx_c in cortex by the recorded delay Δt_c, an apparent horizontal propagation speed can be computed within the cortical map, in the plane of the layers of V1. The inferred propagation speeds range between 0.02 and 2 m/s, with a peak between 0.1 and 0.3 m/s.

These velocity values have since been confirmed for other sensory cortical structures, such as somatosensory cortex (Moore and Nelson 1998). They are 10 times slower than the X-type thalamic input and feedback propagation from higher cortical areas (2 m/s in Nowak and Bullier 1997) and 100 times slower than the fast Y-pathway (8–40 m/s in Hoffman and Ston Hoffman and Stone 1971). They are in fact within the order of magnitude of conduction speeds measured in vitro and in vivo along non-myelinated horizontal cortical axon fibers (Nowak and Bullier 1997; Hoffman and Stone 1971; Hirsch and Gilbert 1991). In view of the size difference that exists between the sub-threshold receptive field and the discharge field, the propagation most likely involves monosynaptic horizontal connections, although the contribution of rolling waves of postsynaptic activity cannot be entirely excluded. Recent reports based on cortical LFPs triggered on LGN spike activity rule out the possibility that divergence of LGN axons may also contribute to the build-up of the observed latency shifts (Nauhaus et al. 2009, see also Hoffman and Stone 1971).

Thus, we have been able to reconstruct, using electrophysiological recordings for the first time, the propagation of an intracortical wave of visual activation traveling along long-distance horizontal connections. This intracellular study of synaptic echoes showed unexpectedly that the V1 network should not be considered an ordered mosaic of independent "tubular" analyzers, but rather a constellation of wide field integrators, integrating simultaneously input sources arising from much larger regions of visual space than previously thought. Primary visual cortical neurons would thus have the capacity to combine information issuing from different points of the visual field, in a spatio-temporal reference frame centered on the discharge field itself. This ability imposes precise constraints in time and space on the efficacy of the summation process of elementary synaptic responses. Specific functional predictions will be reviewed in the last section of this article.

4.5 Confirmation of Traveling Waves with Voltage Sensitive Dye Imaging

The macroscopic reconstruction of intra-V1 waves on the basis of microscopic echoes, described in the previous section, remains however an extrapolation made between two scales of spatial organization differing by two orders of magnitude (neuron vs. map). Brain imaging methods, particularly adapted to layered structures, whether grown in vitro (Tanifuji et al. 1994) or in organotypic cultures (Carlson and

Coulter 2008) or accessible by direct craniectomy such as cortex in vivo (Grinvald et al. 1994), have been developed in the past 20 years. Intrinsic optical imaging, based on the relative light absorption of oxy- and deoxyhemogloblin, gives a static view of the intrinsic activation pattern averaged over time and trials. VSD techniques, which require an external source of illumination, take advantage of the fast changes in the fluorescence caused by the dye (with a wavelength higher than that of the source) as a function of the state of depolarization of the membrane in which it is incorporated: the higher the change in voltage across the neuronal membrane, the higher the amount of emitted fluorescence. This extrinsic imaging gives an unprecedented view of the state of membrane depolarization in the superficial layers of the cortex, with a time sensitivity close to that of intracellular recordings (Bringuier et al. 1999). Some studies even detect hyperpolarizing « notches » that may be linked to the fast onset of inhibition during sensory activation (Borg-Graham et al. 1998; Monier et al. 2003; Shoham et al. 1999). Since their pioneering study of cortical spread function (Grinvald et al. 1994), the team of Amiram Grinvald has provided detailed quantification of intracortical dynamics evoked by various inducer stimuli (square, annular grating). VSD observations repeated on a large number of cats with a focal light square have shown the slow propagation of an activation wave front, in area 17 or 18, consistent in speed (0.09 m/s) with the conduction velocity of horizontal connections (Jancke et al. 2004).

Many groups have since confirmed the propagation of spontaneous and evoked waves across the cortical laminar planes in visual primary and secondary cortical areas of rodents and higher mammals. A new insight was provided by the team of Per Roland (Roland 2002; Roland et al. 2006; Ahmed et al. 2008), who switched from PET and fMRI in humans to the more invasive VSD techniques in ferret. Per Roland proposed in a seminal review that the activity visualized by VSD techniques represents a waxing and waning of a local depolarized cortical state, best detected in the terminal tuft of the dendrites of layer 2–3 pyramidal cells (Roland 2002). The choice of the ferret was motivated by the ambition to visualize propagating waves, not only the primary visual cortex, but also the secondary visual areas which are concurrently activated by a visual stimulus. The visual paradigm chosen by Per Roland was a simplified figure/ground configuration, and the VSD imaging confirmed the induction of a traveling wave in V1, propagating away from the feed-forward impact evoked by the inducer stimulus. In addition, secondary waves originating from higher cortical areas were shown to bounce back into the primary cortex and invade in a complementary way the representation of the surround of the cue object (Roland et al. 2006).

Three other recent studies have "rediscovered" traveling waves within the visual cortical areas of rat (Xu et al. 2007) and cat (Nauhaus et al. 2009; Benucci et al. 2007). The first study, by Xu and colleagues in the anesthetized rat, reports horizontal waves propagating at slow speed (from 0.05 to 0 .07 m/s), which slow down when crossing the V1-V2 border (corresponding to the vertical meridian representation). It also shows the induction of retro-propagating waves, in agreement with the primary observation of Per Roland's team. The visually evoked waves exhibit stereotyped features and show invariance with the parameters of the drifting grating, such as orientation and temporal or spatial frequency. In contrast, the "ongoing" (spontaneous) waves vary in kinetics. They do not respect area boundaries, and propagate without interruption throughout the entire imaged area (up to 4 mm) more slowly than the evoked activity. It is likely that the two forms of propagating activity are generated by different mechanisms. One possibility for the slowest waves is a polysynaptic column- to-column propagation of "up" states (Angelucci et al. 2002), similar to the "rolling waves" previously reported with calcium imaging in vitro (Tanifuji et al. 1994).

A second study, by the group of Matteo Carandini (Benucci et al. 2007) in cat V1, is based on a clever strategy which exploits the non-linear response properties of certain classes of cortical cells. By using the Fourier transform of the VSD signal evoked by the phase alternation of a static grating or a one-cycle bar, the authors were able to extract simultaneous measurements of the amplitude and phase of the response with respect to the parameters of the visual stimulus. Their method, which yields a high signal/noise ratio and high temporal resolution, takes advantage of the fact that the VSD response oscillates at twice the frequency of a stationary square grating whose contrast is modulated sinusoidally in time. It thus appears in the power spectrum of the VSD signal as a distinctive peak at twice the frequency of the stimulus, i.e., the second harmonic, away from the respiration and heart-beat contamination frequency artifacts. This frequency doubling in the VSD signal is thought by the authors to be generated by the synchronized membrane oscillations in complex cells from the superficial layers of the cortex, but intracellular recordings in our laboratory (Borg-Graham et al. 1998; Monier et al. 2003) show that such behavior is also detectable in the sub-threshold activity of simple cells. By measuring the second harmonic amplitude of the VSD signal, the authors obtained two types of maps, one for orientation preference (using a full-field grating) and the other for retinotopy (using a 1-D bar reduced in width to one high frequency cycle). In the latter case, Benucci and colleagues also measured the phase of the oscillatory response, which corresponds to the temporal delay of the oscillation in each pixel with respect to the inducer stimulus. This phase lag was shown to increase linearly as a function of the lateral distance from the feed-forward impact zone of the bar (Fig. 4.4 shows the amplitude and delay plot in the left bottom corner). These observations confirm that the focal stimulus induced a traveling wave across the cortex, with an apparent speed of propagation estimated at around 0.30 m/s. They fully corroborate the predictions we extracted from our intracellular recordings (Frégnac and Bringuier 1996; Bringuier et al. 1999) in which synaptic responses elicited by stimuli placed far from the center of the receptive field showed increasing delays with the relative eccentricity.

This well publicized account has been recently complemented by an impressive local field potential study, this time applied to the primary cortex of both cat and monkey (Nauhaus et al. 2009). Spike activity and LFPs were sampled simultaneously through 10×10 arrays of electrodes (400 μm separation). It is estimated that the LFP represents a measure of the local postsynaptic activity integrated within a 250 μm radius around the recording sites. During the ongoing activity or under a sparse regime of stimulation at low contrast, the group of Matteo Carandini, in

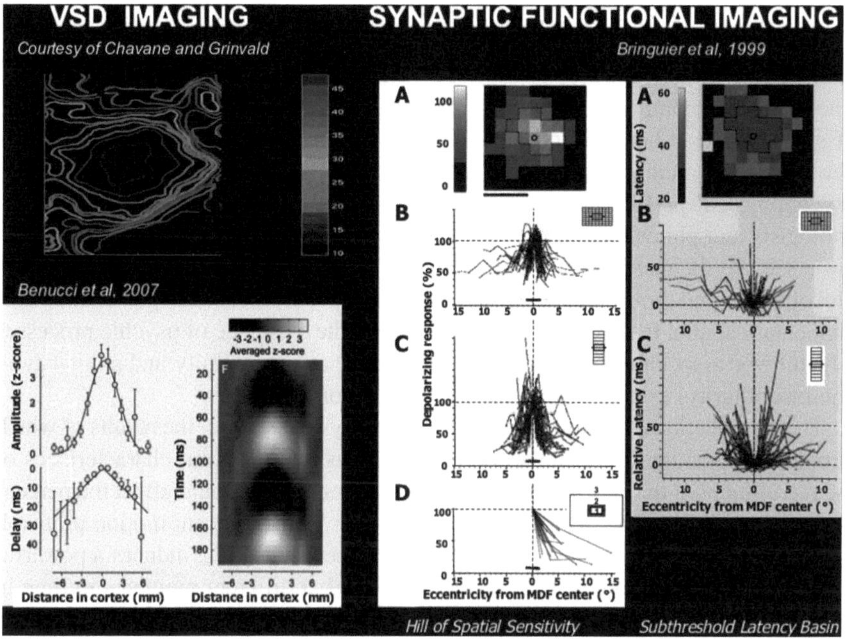

Fig. 4.4 Comparison of spatial and temporal properties of horizontal propagation, using network voltage sensitive dye imaging (*left*) and synaptic functional imaging (*right*). See text for details. *Left panel*, horizontal propagation wave monitored in the superficial layers of cat V1 with voltage sensitive dye imaging (*top left*, with courtesy of F. Chavane and A. Grinvald; *bottom left* (Grinvald et al. 1994). *Right panel*, horizontal propagation inferred with intracellular synaptic functional imaging [10,11,47]. The same propagation speed (0.1–0.3 m/s) is measured by the two imaging techniques (*see Color Plates*)

collaboration with Dario Ringach, confirmed the detection of horizontal waves traveling at low speed (0.20 m/s in monkey and 0.30 m/s in cat). The spatial spread attenuation constant of the propagated waves in each species corresponds quantitatively with the mean extent of horizontal axons (2.1 mm in monkey and 5.8 mm in cat; see also Nowak and Bullier 1997; Angelucci et al. 2002). These authors conclude that "the synaptic inputs to neurons during spontaneous activity can be thought of as the superposition of a myriad of traveling waves originating from individual spikes distributed over an extended region of cortex". They also show conclusively that during the evoked state, the amplitude of spike-triggered LFPs becomes more difficult to ascertain and the waves vanish with contrast increase in the driving stimulus. This suggests that the visibility of intracortical horizontal propagation depends on the operating mode (sparse vs. dense regime) evoked in the cortex and the relative balance between feed-forward and lateral inputs, thus confirming some prior claims (Frégnac and Bringuier 1996; Levitt and Lund 1997; Cannon and Fullenkamp 1993; Polat and Sagi 1993; reviewed in Séries et al. 2003).

4.6 Visualizing Gelstalt Illusions in V1 Cortex

The multiscale comparison of these different imaging techniques opens a new field of study, where it becomes possible to compare the real-time imaging in cortical networks with membrane dynamic recording in single cells on the one hand and psychophysical performance measures on the other. Almost a century ago, psychologists and philosophers proposed a theory of perceptual grouping (the Gestalt laws, see Wertheimer 1912), which predicts the emergence of coherent percepts of global shape and motion from the temporal superposition of static presentations of elementary spatial features. This theory assumes the existence of psychic processes which favor associations in space (according to spatial proximity and similarity in contrast polarity) as well as in time (continuity, common fate).

These predictions inspired a series of psychophysical studies, the results of which strongly support the following working hypothesis: the temporal characteristics of the recruitment of the "horizontal" intracortical connectivity could affect the perception of motion. Among various demonstrators, the "Phi" apparent motion protocol, originally called the "beta phenomenon" by Wertheimer (1912), induces a powerful illusion when the same target is repeatedly flashed at different moments in time in different positions in the visual field ordered along an imaginary trajectory (Fig. 4.5, left). Although at each moment in time the observer only sees a static image, he reports the perception of continuous motion of the same object along the trajectory defined by the "association" path linking the various positions explored in succession. The strength of the percept depends on the complexity of the test stimulus (shape and texture), the duration of the static presentations, the interstimulus interval, and the spatial offset between positions (Anstis et al. 1998). The "line motion" illusion is also based on the same induction process of asynchronous static presentations. In the latter case, the cue feature is a uniform luminance square, followed by a bar of the same luminance, one polar end of which encroaches on the previously flashed square. For adequate interstimulus intervals and presentation durations, the human subject reports a continuous movement of one border, perceived as a smooth morphing of the square into the elongated bar (Hikosaka et al. 1993).

A series of arguments makes it likely that these psychophysical effects are the result of activity waves propagating across V1. At the electrophysiological level, the RF orientation axis of cells that interact through long-range connections are often reported to be co-aligned. Horizontal connections are thought to facilitate the response of cells with similar orientation preference, and to reduce their response otherwise (Roland 2002; Polat et al. 1998; Knierim et al. 1992 reviewed in Séries et al. 2003). A recent local field potential study by the group of Dario Ringach and Matteo Carandini shows the detection of horizontal traveling waves, which most often link cortical loci with the same orientation preference (Nauhaus et al. 2009). Similarly, at the psychophysical level, numerous studies have shown that low-contrast visual contour elements are easier to detect when presented in the context of collinear flankers (Polat and Sagi 1993). If the spatial contextual effect can be easily interpreted in the framework of the "association field" (Field et al. 1993), the temporal determinants of

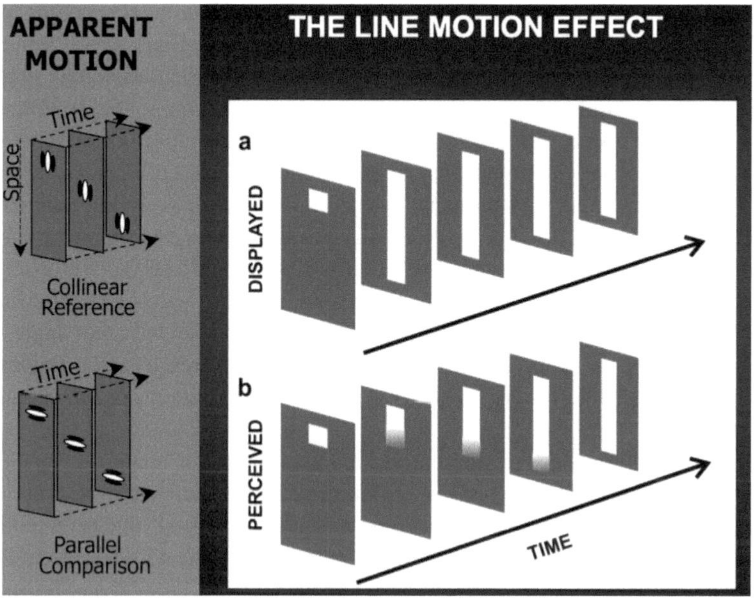

Fig. 4.5 Apparent motion and the hypothesis of the "dynamic association field". *Left*: two-forced choice apparent motion protocol. *Right*: the "dynamic association field" hypothesis; local oriented inputs (Gabor patches) induce a facilitation wave of activity traveling along horizontal connections intra V1. This wave tends to bind proximal receptive fields with co-linear preferred orientations, thus creating a contiguous path of temporal integration. The associative strength of the perceptual effect is maximal when the asynchronous feed-forward sequence produced by joint strokes of apparent motion (*arrow*) travels in phase with the visually evoked horizontal intracortical propagation (*see Color Plates*)

this collinear facilitation have been less explored. By varying the orientation of flanking "contour" elements around their own axis and measuring contrast detection thresholds to a brief foveal target presented at various phases of flanker rotation, a recent psychophysical study showed that the phase-dependency of the pyschophysical effect was compatible with the induction of a facilitatory wave, induced by the flankers and slowly propagating in cortical space at 0.10 m/s (Cass and Alais 2006).

In a collaborative work in our laboratory between psychophysicists (Jean Lorenceau and Sebastien Georges), modelers (Peggy Séries), and electrophysiologists (Pierre Baudot, Frédéric Chavane, Marc Pananceau and Yves Frégnac), we hypothesized from our own intracellular findings (Frégnac and Bringuier 1996; Bringuier et al. 1999; Chavane et al. 2000) that the perception of speed could be differentially modulated during apparent motion sequences of oriented stimuli, either collinear and aligned with respect to the motion axis or at an angle to it. We devised a series of psychophysical experiments in humans that aimed at testing the influence of orientation relative to the motion axis on perceived speed (Georges et al. 2002). Observers were asked to discriminate during a forced choice task between the relative speeds of two apparent-motion (AM) sequences. The elementary test feature was a Gabor patch (an oriented sinusoidal luminance grating whose modulation is weighted by a bi-dimensional Gaussian function) whose luminance profile, spatial frequency, and anisotropy along its main orientation axis precisely reproduce the spatial sensitivity profiles of cortical discharge fields (Daugman 1985). The fixed speed "reference" AM sequence was composed of three identical but positionally offset Gabor patches, whose orientation was collinear to the motion axis. The "comparison" AM sequence was composed of three Gabor patches, this time cross-oriented to the same motion axis, and the global AM speed varied from trial to trial. In contrast with previous results (Castet et al. 1993), we observed a "speedup" illusion: for the same physical speed, a Gabor patch moving along its orientation axis appears much faster than a Gabor patch oriented at an angle to the motion axis. Since the spatio-temporal structures of the "reference" and "comparison" AM sequences used here were identical, it is unlikely that attention can explain the present results.

This psychophysical effect, summarized in Fig. 4.6, may be quantified by the ratio between the speed of the "comparison" collinear sequence and that of the "reference" parallel sequence for which the subject reports equality in speed. The perceptual bias is as strong as threefold in humans, and its strength explains why observers find in more than 80–95% of cases that "parallel" sequences are faster than "horizontal" sequences even if both composite stimuli have the same physical speed. The physical reference speed for which the effect is maximal corresponds to 64°/s in the visual field, which is equivalent for parafoveal tests to a propagation speed of 0.20 m/s in human V1 cortex. Remarkably, and as predicted by considering the difference in the retinocortical magnification factors found respectively in man and cat, the value of speed extrapolated in human cortex is well within the range of those derived from electrophysiological intracellular recordings in cat area 17 (Bringuier et al. 1999; Chavane et al. 2000). Furthermore, the hypothesis that the speed-up is induced when the horizontal wave travels ahead or in phase with the feed-forward inputs is supported both by a computational model (Séries et al. 2002) and by unpublished intracellular observations made in cat primary visual cortex for the same stimulus configurations (Baudot et al. 2000). These different data strongly support our working hypothesis of a dynamic neural "association field" (summarized in the right panel of Fig. 4.5): oriented contours should propagate facilitation across space (co-linearity) as well as across time. In the latter case, synergy is observed when the feed-forward flow travels in phase with the lateral wave evoked by the inducer apparent motion sequence.

A closely related protocol, the "line-motion" illusion (see above), has been applied with voltage sensitive dye imaging in area 18 of the anaesthetized cat, by Amiram Grinvald's team with the participation of one of us (Jancke et al. 2004) (see also Jancke et al. this volume). The aim was to obtain a direct visualization of the spatial spread of the facilitation induced by the cue stimulus. For this purpose, they compared responses to a flashed small square and a long bar alone, and the configuration of the line-motion illusion: a square briefly preceding the bar (Séries et al. 2003). In the associative condition, the VSD pattern demonstrated

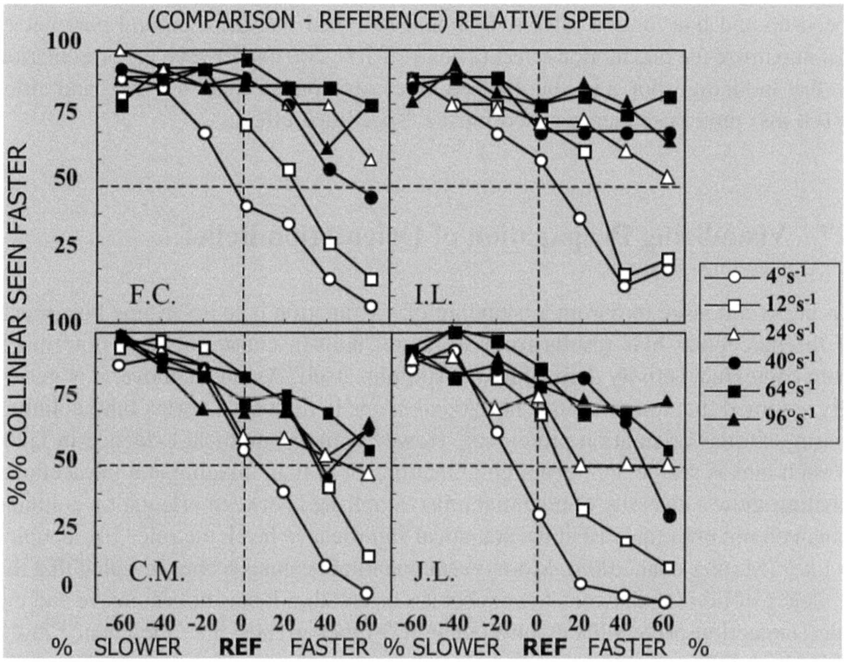

Fig. 4.6 Apparent motion of collinear Gabor elements is seen to be "faster" than that of parallel contours. Comparison of the two-forced choice psychophysical performances when comparing speed between collinear and parallel apparent motion sequences (see text). Each panel represents the individual data for four observers. Each point corresponds to 40 trials. The average proportion of the trials in which "the collinear sequence is perceived as being faster than the parallel sequence" is plotted in ordinates, as a function of the relative speed between the reference (collinear) and the comparison (parallel) apparent motion sequences (see Fig. 4.5, *left panel*). Six reference speeds were used for the collinear sequence (from 4°/s to 96°/s, *right panel*) and each of these were compared to parallel sequences whose speeds varied between +60% and −60% of that of the chosen "reference" collinear sequence. Reprinted with permission from Georges et al. (2002)

the spread of a low-amplitude wave extending far beyond the retinotopic representation of the initial "cue". Again, the observed propagation speed, around 0.10 m/s, was found to be consistent with the conduction velocity reported in our own electrophysiological experiments (Frégnac and Bringuier 1996; Bringuier et al. 1999; Chavane et al. 2000) or visualized with VSD techniques by Amiram Grinvald's team (Grinvald et al. 1994). Furthermore, this work demonstrated that the cue square, even though physically non-moving, induced a propagating wave of cortical depolarization, indistinguishable from the spatio-temporal pattern produced by the continuous motion of the same square, this time moving at 32°/s. Although it is likely that similar low level mechanisms may underlie both the speed-up and line motion effects, it should be noted that the temporal parameters that maximize the line motion effect (a delay of 100–200 ms between the presentation of the inducing spot and the flashed line) are longer than the fast and brief (<100 ms) motion sequences used for the "speed-up" effect.

4.7 Visualizing Propagation of Orientation Belief

An important issue in our understanding of V1 function is to determine how much of the perceptual bias results from structural built-in constraints and how much from contextual activity defined by the stimulus itself. As stated above, it is generally assumed that long-distance horizontal axons in the visual cortex bind columns sharing similar orientation preference. However, the anatomical evidence in favor of such bias is scarce and recent combinations of optical imaging and intracellular labeling show a diversity of potential links established between orientation columns which do not obey, at least at the statistical significance level, the rule "like couples to like" (Monier et al. 2003). More recent quantitative analysis has revealed that the tendency of horizontal axons to connect iso-orientation loci is not exclusive and the interconnection probability is only about 1.5 times greater than the chance level. This bias has been mostly observed for supra-laminar pyramidal neurons. However, inhibitory interneurons and neurons in layer IV or close to pinwheel centers have also been reported to connect lateral orientation columns in a cross-oriented or unselective way (Karube and Kisvarday 2006). As a consequence, at the macroscopic level, the net functional effect cannot be predicted.

In collaboration with the laboratory of Amiram Grinvald, we have recently attempted a multiscale analysis of the visually driven horizontal network activation, using population and single-cell measures of postsynaptic integration. Voltage sensitive dye imaging showed that a local oriented stimulus evoked an orientation selective activity component which remained confined to the feed-forward cortical imprint of the stimulus. Orientation selectivity decreased exponentially along the horizontal spread (space constant ~1 mm). To analyze the local connectivity rules, we also made intracellular recordings to identify the orientation selectivity and preference of converging horizontal inputs onto the same target cell. The combination of imaging and electrophysiological results suggests, somewhat surprisingly, that the horizontal connectivity does not obey iso-orientation rules beyond the hypercolumn scale. In contrast, when increasing the spatial and temporal summation, both optical imaging and intracellular measurements showed the emergence of an iso-orientation selective spread. We conclude that the stimulus cooperativity is a necessary constraint for the emergence of the Gestalt-like binding (Chavane et al. in revision).

This last study, combining network VSD imaging and synaptic functional imaging, shows two contrasting dynamic behaviors of the same network for two distinct stimulus configurations: a single local stimulus does not propagate orientation preference through the long-range horizontal cortical connections whereas stimulation imposing spatial summation and temporal coherence facilitates the build-up of orientation preference propagation. These observations do not forcibly contradict each other. On the one hand, for the local oriented stimulus, the divergent connectivity pattern may facilitate detection of high-order topological properties (e.g. orientation discontinuities, corners, geons). On the other hand, for stimulation protocols involving a larger extent of stimulation, summation of multiple oriented sources in the far "silent" surround can optimize the emergence of iso-orientation preference links. Configurations such as oriented annular stimuli may, for instance, recruit iso-oriented sources collinearly organized with the orientation preference axis of the target column/cell; similar synergy may be obtained when sources, independent of their exact location, share the same motion direction sensitivity as the target grating. Both configurations, which are confounded in annular aperture protocols, correspond to the neural implementation of the Gestalt's continuity and common fate principles, but other more dynamic principles could also emerge from such network configuration.

One research project currently developed within our laboratory (Pedro V. Carelli, Marc Pananceau and Yves Frégnac) aims at the piecewise reconstruction of the "dynamic association field". It takes advantage of our previous observation that apparent motion produced by iso-oriented Gabor stimuli (co-aligned with the motion axis and presented from periphery to center) is more efficient than the simultaneous static presentation of the same stimuli at evoking sub-threshold synaptic responses from the "silent" periphery (Polat and Sagi 1993). Interstimulus intervals of composite stimulation can be chosen to compensate for the latency shift expected between the two retinotopic locations, in such a way as to produce an optimal summation between the two activation waves originating from each local element. A typical protocol is to use a regular paving of visual space by an array of Gabor stimuli, whose central node is optimized to fit the orientation and spatial frequency of the recorded classical receptive field, and use sparse second-order apparent-motion noise to stimulate the visual field. The long-term goal is to extract geometrical and temporal patterns of association which promote or suppress the activation of horizontal (and feedback) connectivity, without altering the feed-forward drive, making these applicable to macroscopic visualization of the horizontal network contribution in human brain imaging.

4.8 Conclusion

This chapter compares two techniques for imaging cortical dynamics during apparent motion, operating at two different scales of integration: the first, macroscopic in nature, is centered on the divergence of activation relayed horizontally across the cortical layers; the second, microscopic, addresses convergence processes at the neuronal level. Their parallel application to the study of the same percept demonstrates the cortical origin of binding processes operating during simple tasks such as pop-out contour integration and motion detection. These processes are low-level and not linked to attention since they are observed in humans during forced choice tasks as well as in the anesthetized mammal. Complementary studies, based on the intracellular electrophysiology, network imaging, and psychophysics, all point to the emergence of cooperative Gestalt-like interactions, when the stimulus carries a sufficient level of spatial and temporal coherence. Above a given activation threshold (yet to be quantitatively defined), a cooperative depolarizing or facilitatory wave becomes detectable in both the primary and secondary visual areas. This wave travels at low speed in the plane of the superficial layers (0.10–0.30 m/s) and becomes anisotropic for oriented inducer stimuli. The physiological features of the spatio-temporal propagation pattern recorded in V1 are highly correlated with the percept reported by the conscious observer and agree with predictions derived from the Gestalt theory. In the two cases of motion illusion reviewed here (apparent motion and line motion), a wave of perceptual binding modulates the integration of feed-forward inputs yet to come: this wave can be seen as the propagation of the network belief of the possible presence of a global percept (the "whole": here, continuous motion of a space-invariant shape) before the illusory percept becomes validated by the sequential presentation of the "parts" (signaled by direct focal feed-forward waves). This neuronal dynamics obeys closely the Gestalt prediction that the emergence of the "whole" should precede in time the detection of the "parts".

These cortical processes result, at the perceptual level, in the propagation of functional biases (binding of contour and motion) which goes beyond the scale of the columnar orientation and ocular dominance network. It remains to be determined whether the correlations we report between perception and horizontal propagation are the result solely of neural processes intrinsic to V1, or whether they reflect the reverberation in V1 of a collective feedback originating from multiple secondary cortical areas, each encoding for a distinct functional representation of the visual field. It may be indeed envisioned that the primary visual cortex plays the role of a generalized echo chamber fed by other cortical areas (visual or not) which participate in the coding of shape and motion in space: accordingly, the waves traveling across V1 would signal the emergence of perceptual coherence when a synergy is reached between the different cortical analyzers. A new challenge is launched at the interface between electrophysiological neurosciences and brain imaging where the genesis and propagation of cognitive beliefs remains to be further understood and explored.

4.9 Supplementary materials (CD-ROM)

Movie 1 Spatio-temporal maps of supra- and sub-threshold activations of a V1 neuron (see **Fig.** 4.3) *(file "4_M1_Spatiotemporalmaps.avi")*. Spatio-temporal (left: X-Y; right: X-t) maps of supra-threshold (spike, upper panels) and sub-threshold (voltage, lower panels) activation in a V1 neuron, recorded intracellularly. The X-Y correlation maps between stimulus location and the postsynaptic event (spike in the upper panel, voltage in the lower panel) are presented as a function of the temporal delay from stimulus onset. The visual stimulation is a sparse (light-dark) noise sequence repeated 15 times for every pixel location. The sub-threshold receptive field extent expands progressively in time and space over several degrees (several millimeters in cortical space) whereas the discharge field remains limited to 1–2° of visual solid angle.

Acknowledgments This work was supported by the CNRS, and grants from ANR (NATSTATS) and the European integrated project FACETS (*FET*- Bio-I3: 015879). This long-lasting line of research has benefited in its realization of the experimental participation of Dr Sebastien Georges in psychophysics, of Dr. Peggy Séries in modeling, and of Julien Fournier, Nazyed Huguet and Drs Alice René, Lyle Graham and Manuel Levy in electrophysiology at UNIC. It has also benefited in the recent years of the scientific collaborations with the laboratory of Pr. Amiram Grinvald (Weizmann Institute, Rehovot, Israel) and the CNRS DyVA team (INCM, Marseille). We thank Drs Andrew Davison and Guillaume Masson for helpful comments.

References

Ahmed B, Hanazawa A, Undeman C, Eriksson D, Valentiniene S, Roland PE (2008) Cortical dynamics subserving visual apparent motion. Cereb Cortex 18(12):2796–2810

Albus K (1975) A quantitative study of the projection area of the central and the paracentral visual field in area 17 of the cat. I. The precision of the topography. Exp Brain Res 24:159–179

Angelucci A, Levitte JB, Walton EJS, Hupé JM, Bullier J, Lund JS (2002) Circuits for local and global signal integration in primary visual cortex. J Neurosci 22:8633–8646

Anstis SM, Verstraten FAJ, Mather G (1998) The motion aftereffect: a review. Trends Cogn Sci 2:111–117

Basole A, White LE, Fitzpatrcik D (2003) Mapping multiple features in the population response of visual cortex. Nature 423:986–990

Baudot P, Chavane F, Pananceau M, Edet V, Gutkin B, Lorenceau J, Grant K, Frégnac Y (2000) Cellular correlates of apparent motion in the association field of cat area 17 neurons. Abstr Soc Neurosci 26:446

Benucci A, Frazor RA, Carandini M (2007) Standing waves and traveling waves distinguish two circuits in visual cortex. Neuron 55(1):103–117

Binzegger T, Douglas RJ, Martin KA (2004) A quantitative map of the circuit of cat primary visual cortex. J Neurosci 24:8441–8453

Borg-Graham LJ, Monier C, Frégnac Y (1998) Visual input evokes transient and strong shunting inhibition in visual cortical neurons. Nature 393:369–373

Bringuier V, Chavane F, Glaeser L, Frégnac Y (1999) Horizontal propagation of visual activity in the synaptic integration field of area 17 neurons. Science 283:695–699

Cannon MW, Fullenkamp SC (1993) Spatial interactions in apparent contrast: individual differences in enhancement and suppression effects. Vision Res 33:1685–1695

Carlson GC, Coulter DA (2008) In vitro functional imaging in brain slices using fast voltage-sensitive dye imaging combined with whole-cell patch recording. Nat Protoc 3(2):249–255

Cass J, Alais D (2006) The mechanisms of collinear integration. J Vis 6(9):915–922

Castet E, Lorenceau J, Shiffrar M, Bonnet C (1993) Perceived speed of moving lines depends on orientation, length, speed and luminance. Vision Res 33:1921–1936

Chavane F, Monier C, Bringuier V, Baudot P, Borg-Graham L, Lorenceau J, Frégnac Y (2000) The visual cortical association field: a Gestalt concept or a physiological entity? J Physiol Paris 94:333–342

Chavane F, Sharon D, Jancke D, Marre O, Frégnac Y, Grinvald A (in revision). Horizontal spread of orientation selectivity in V1 requires intracortical cooperativity. J. Neuroscience
Daugman J (1985) Uncertainty relation for resolution in space, spatial frequency, and orientation optimized two-dimensional visual cortical filters. J Opt Soc Am A2:1160–1168
Field DJ, Hayes A, Hess RF (1993) Contour integration by the human visual system: evidence for a local "association field". Vision Res 33:173–193
Frégnac Y (2001) Le combat des hémisphères. Pour Sci 283:94–95
Frégnac Y, Bringuier V (1996) Spatio-temporal dynamics of synaptic integration in cat visual cortical receptive fields. In: Aertsen A, Braitenberg V (eds) Brain theory: biological basis and computational theory of vision. Springer, Amsterdam, pp 143–199
Georges S, Sèries P, Frégnac Y, Lorenceau J (2002) Orientation dependent modulation of apparent speed: psychophysical evidence. Vision Res 42:2757–2772
Grinvald A, Hildesheim R (2004) VSDI: a new era in functional imaging of cortical dynamics. Nat Rev Neurosci 5(11):874–885
Grinvald A, Lieke EE, Frostig RD, Hildesheim R (1994) Cortical point-spread function and long-range lateral interactions revealed by real-time optical imaging of macaque monkey primary visual cortex. J Neurosci 14:2545–2568
Hartline HK (1938) The response of single optic nerve fibers of the vertebrate eye to illumination of the retina. Am J Physiol 121:400–415
Haynes JD, Rees G (2006) Decoding mental states from brain activity in humans. Nat Rev Neurosci 7:523–534
Hikosaka O, Miyauchi S, Shimojo S (1993) Focal visual attention produces illusory temporal order and motion sensation. Vision Res 33:1219–1240
Hirsch JA, Gilbert CD (1991) Synaptic physiology of horizontal connections in the cat's visual cortex. J Neurosci 11:1800–1809
Hoffman KP, Stone J (1971) Conduction velocity of afferents to cat visual cortex: a correlation with cortical receptive field properties. Brain Res 32:460–466
Jancke D, Chavane F, Naaman S, Grinvald A (2004) Imaging cortical correlates of illusion in early visual cortex. Nature 428:423–426
Kalatsky VA, Stryker MP (2003) New paradigm for optical imaging: temporally encoded maps of intrinsic signal. Neuron 38(4):529–545
Karube F, Kisvarday ZF (2006). Bouton distribution of deep-layer spiny neurons on the functional maps in cat visual cortex. FENS Forum Abstr 3:179.14.
Kay KN, Naseralis T, Prenger RJ, Gallant JL (2008) Identifying human natural images from brain activity. Nature 452:352–355
Knierim JJ, Van Essen DC (1992) Neuronal responses to static texture patterns in area V1 of the alert macaque monkey. J Neurophysiol 67:961–980
Lee S, Blake R, Heeger DJ (2007) Hierarchy of cortical responses underlying binocular rivalry. Nature Neurosci 10(8):1048–1054
Levitt JB, Lund JS (1997) Contrast dependence of contextual effects in primate visual cortex. Nature 387:73–76
Mitchison G, Crick F (1982) Long axons within the striate cortex: their distribution, orientation, and patterns of connection. Proc Natl Acad Sci U S A 79:3661–3665
Monier C, Chavane F, Baudot P, Graham L, Frégnac Y (2003) Orientation and direction selectivity of excitatory and inhibitory inputs in visual cortical neurons: a diversity of combinations produces spike tuning. Neuron 37:663–680
Moore CI, Nelson SB (1998) Spatio-temporal subthreshold receptive fields in the vibrissa representation of rat primary somatosensory cortex. J Neurophysiol 80:2882–2892
Nauhaus I, Busse L, Carandini M, Ringach DL (2009) Stimulus contrast modulates functional connectivity in visual cortex. Nature Neurosci 12:70–76
Nowak LG, Bullier J (1997) The timing of information transfer in the visual system. In: Rockland KS, Kaas JH, Peters A (eds) Extrastriate visual cortex in primates. New York, Plenum, pp 205–241

Polat U, Sagi D (1993) Lateral interactions between spatial channels: suppression and facilitation revealed by lateral masking experiments. Vision Res 33:993–999
Polat U, Mizobe K, Pettet MW, Kasamatsu T, Norcia AM (1998) Collinear stimuli regulate visual responses depending on cell's contrast threshold. Nature 391:580–584
Roland PE (2002) Dynamic depolarisation fields in the cerebral cortex. Trends Neurosci 25:183–190
Roland PE, Hanazawa A, Undeman C, Eriksson D, Tompa T, Nakamura H, Valentiniene S, Ahmed B (2006) Cortical feedback depolarization waves: a mechanism of top-down influence on early visual areas. Proc Natl Acad Sci U S A 103(33):12586–12591
Séries P, Lorenceau J, Frégnac Y (2003) The silent surround of V1 receptive fields : theory and experiments. J Physiol Paris 97(4–6):453–474
Séries P, Georges S, Lorenceau J, Frégnac Y (2002) Orientation dependent modulation of apparent speed: a model based on the dynamics of feed-forward and horizontal connectivity in V1 cortex. Vision Res 42:2781–2797
Shoham D, Glaser DE, Arieli AI, Kenet T, Wijnbergen C, Toledo Y, Hildesheim R, Grinvald A (1999) Imaging cortical dynamics at high spatial and temporal resolution with novel blue voltage-sensitive dyes. Neuron 24(4):791–802
Tanifuji M, Sugiyama T, Murase K (1994) Horizontal propagation of excitation in rat visual cortical slices revealed by optical imaging. Science 266:1057–1059
Thirion B, Diuchesnay E, Hubbard EM, Dubois J, Poline J-B, LeBihan D, Deheane S (2006) Inverse retinotopy: inferring the visual content of images from brain activation patterns. Neuroimage 33:1104–1116
Tootell RB, Silverman MS, Switkes E, De Valois RL (1982) Deoxyglucose analysis of retinotopic organization in primate striate cortex. Science 218:902–904
Warnking J, Dojat M, Guerin-Dughe A, Delon-Martin C, Olympieff S, Richard N, Chehikian A, Segebarth C (2002) FMRI retinotopic mapping-step by step. Neuroimage 17:1665–1683
Wertheimer M (1912) Experimentelle Studien über das Sehen von Beuegung. Z Psychol Physiol Sinnesorg 61:161–265
Williams MA, Baker C, Op De Beeck HP, Shim WM, Dang S, Triantafyllou C, Kanwisher N (2008) Feedback of visual object information to foveal retinotopic cortex. Nature Neurosci 11:1439–1445
Xu W, Huang X, Takagaki K, Wu JY (2007) Compression and reflection of visually evoked cortical waves. Neuron 55(1):119–129

Chapter 5
Stimulus Localization by Neuronal Populations in Early Visual Cortex: Linking Functional Architecture to Perception

Dirk Jancke, Fréderic Chavane, and Amiram Grinvald

Abstract In primary visual areas any local input is initially transmitted via horizontal connections giving rise to a transient peak of activity with spreading surround. How does this scenario change when the stimulus starts to move? Psychophysical experiments indicate that localization is different for stationary flashed and moving objects depending on the stimulus history. We here demonstrate how successively presented stimuli alter cortical activation dynamics. By a combination of electrophysiological and optical recordings using voltage-sensitive dye we arrive at the conclusion that sub-threshold propagating activity pre-activates cortical regions far ahead of thalamic input. Such an anticipatory mechanism may contribute in shifts of the perceived position as observed for the flash-lag effect and line-motion illusion in human psychophysics.

5.1 Introduction

How are local stimuli represented across primary visual cortex? Even small visual objects activate large populations of neurons that form extended and densely inter connected excitatory and inhibitory circuits acting on millisecond time scales. The dynamic balance between these antagonist circuits thereby determines each cells' output and hence the amount of information being transmitted further downstream to form perceptual events.

In this chapter we review how the primary visual cortex represents position of small stimuli that are briefly flashed or moved. Neuronal representations of appearing visual objects, the sudden onset of motion, or tracking of moving objects are necessarily confronted with the need of processing accurately space and time. Localization of visual objects must therefore deal with neuronal delays.

We further address the question in, how far activity in early cortical areas correlates to visual perception without the need of top-down contributions due to

D. Jancke (✉)
Department of Neurobiology, Ruhr-University Bochum, ND7/72, Bochum, 44780, Germany
e-mail: jancke@neurobiologie.rub.de

U.J. Ilg and G.S. Masson (eds.), *Dynamics of Visual Motion Processing: Neuronal, Behavioral, and Computational Approaches*,
DOI 10.1007/978-1-4419-0781-3_5, © Springer Science+Business Media, LLC 2010

voluntary action. In particular we focus on well-known visual illusions (flash-lag, line-motion) that are assumed to involve early cortical mechanisms in motion processing. The reported experiments are based on measurements in visual areas 17 and 18 in anaesthetized and paralyzed preparations. Thus the obtained data reflect "automatic" cortical computation without involvement of attentional processes.

Within our framework we link neuronal signals derived from extracellular and voltage-sensitive dye recordings to psychophysical phenomena in order to unravel the relationship between architecture and function of early visual cortex.

5.2 Neuronal Population Dynamics in Visual Space

A small local stimulus evokes widespread cortical activity recruiting thousands of densely interconnected neurons through horizontal connectivity. Figure 5.1 sketches a snapshot of synaptic activity in response to a flashed square measured by voltage-sensitive dye imaging. The fluorescent dye signal reports net neuronal depolarization across the cortical surface, with emphasis on superficial layers. By definition, single neurons fire action potentials more strongly when activated within their receptive field centers (Fig. 5.1, red circles). In contrast, neurons that are stimulated outside their receptive field (RF) may receive input by long-range horizontal connections but do not reach firing thresholds (lower trace Fig. 5.1b).

5.2.1 The Concept of a Population Receptive Field

Taking into account that many neurons with overlapping RFs, are activated when a small stimulus is presented (Fig. 5.1c), we applied a population approach derived from single unit recordings. In this approach, each neuron participates

Fig. 5.1 (continued) (see symbolized spike traces) but are activated subthreshold due to long-range horizontal synaptic input. (**c**) Many neurons with densely overlapping receptive fields (RFs) are stimulated when a small square is presented (colored ellipses sketch their projection in visual field coordinates). (**d**) Time resolved population representation of a square that was flashed in the middle axis of the sampled space. Activity builds-up coherently ~30 ms after stimulus onset, reaches its maximum after ~50 ms, and decays homogenously centered on stimulus location. (**e**) Seven squares within a fixed reference frame (*white outline*) were presented independently of the neurons' individual RF location. Distributions of activity were obtained by pooling the spiking responses of 178 neurons (recordings in cat area 17). Each cell's normalized firing rate in response to the stimuli was mapped to individual RF-centers, resulting in distributions of activity. The responses were interpolated with a Gaussian, with Sigma matching the average RF widths. In summary, each neuron contributes to the entire population activity by its firing rate dependent on the location of its RF-center relative to the stimulus. Time averages of 30–80 ms after stimulus onset are shown (flash duration 25 ms). Note that the peaks of activity are closely around each of the stimulus positions. Thus the applied population approach enables sampling of visual space smaller than average RF sizes (size of squares was 0.4°). (**f**) Population representations derived by means of OLE for all seven elementary stimuli used (time average, 30–80 ms). The optimization procedure allowed the individual peaks to be accurately aligned with the position of each stimulus (*see Color Plates*)

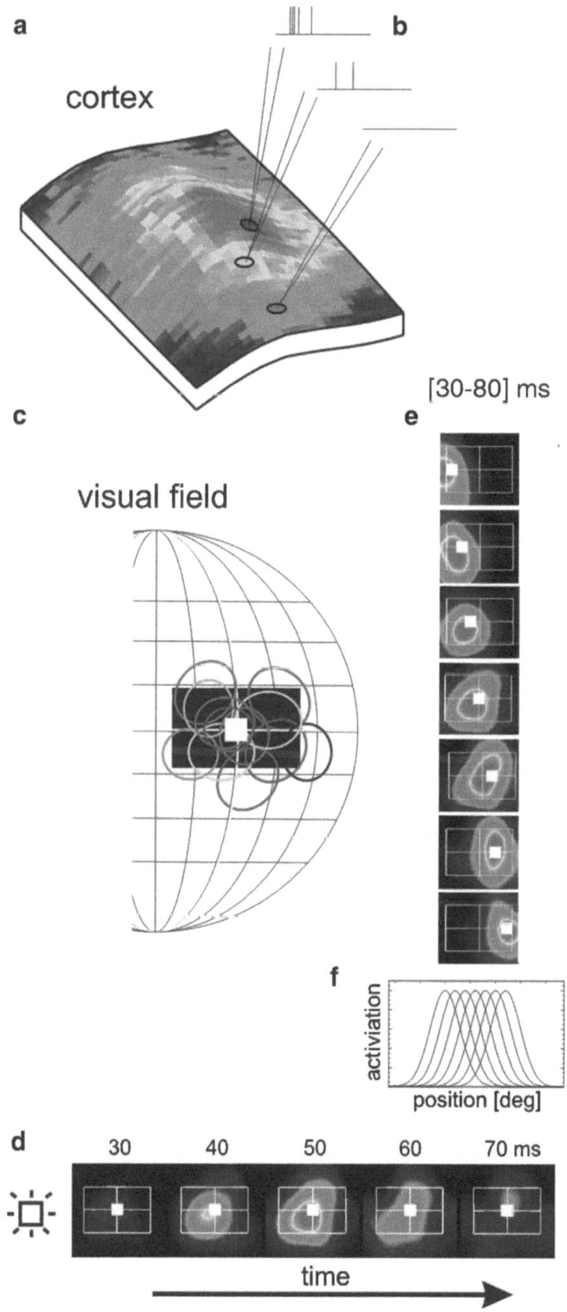

Fig. 5.1 Widespread cortical activity in response to small local stimuli. (**a**) A portion of visual cortex is sketched overlaid with an activity pattern that was evoked by a flashed square. Warm colors indicate high-amplitude activity revealed by voltage-sensitive dye imaging. (**b**) Extracellular measurements at three different recordings sites lead to different results: Whereas neurons that are located around the center of activation reach spiking thresholds, neurons further away do not

in a reconstructed population activity built within a fixed reference frame, here a visual position. Thus, all neurons contribute to the population representation in visual space coordinates dependent on their firing rates and relative RF positions (Jancke et al. 1999, 2004b; Jancke 2000).

Our concept is a straightforward consequence of the observation that a large number of differently tuned neurons are activated, after even the simplest form of sensory stimulation or motor output. For any stimulation, some of these neurons are optimally activated, but the majority sub-optimally, having receptive fields positioned along the edges of a local visual stimulus; for example Fig. 5.1a, c. In addition, under natural viewing conditions stimuli are arbitrarily distributed across many RFs with highly diverse spatio-temporal properties (Szulborski and Palmer 1990; Gegenfurtner and Hawken 1996; Fitzpatrick 2000; Dinse and Jancke 2001a, b; for a similar approach in the somatosensory system see Nicolelis et al. 1998).

5.2.2 RF-Derived Population Representations of Stimulus Position

We first applied a straightforward interpolation procedure in which each cell "votes" with its firing rate for the center of its RF into the population code. Each cells' RF center was quantitatively assessed, the resulting RF-profiles were smoothed, and the RF-centers were defined at the location of maximal amplitude.

Subsequently, seven squares were flashed within a fixed reference frame (Fig. 1c, gray area; Fig. 5.1d, e, white rectangle) regardless of the respective RF position, and the cell responses to these stimuli were normalized to the maximum. Thus each cell's normalized response to a given square fed the population representation at their RF-center position. Population responses were then smoothed with a Gaussian (width=0.6°). The built-up and decay of the population response of 178 neurons is presented in Fig. 5.1d in response to a central stimulation, showing a gradual and well-localized peak of activity centered on the position of the given square (each frame depicts a 10 ms time step).

Time averages of population representations of the seven flashed stimuli are depicted in Fig. 5.1e (white squares). The individual distributions of activity in response to squares at different locations can be regarded as the summed activity profile within a population's receptive field (PRF[1]). The extent of the PRF is larger than the retinotopic stimulus representation, and reflects the average RF size and scatter in cat area 17 (Albus 1975). However, our population procedure allows reconstruction of stimulus position in resolutions much smaller than individual RF sizes.

[1] We used this term for spiking distributions of population activity (Jancke et al. 1999, 2004b). Originally, the term was introduced for LFP recordings (Victor et al. 1994), and most recently applied for voxel analysis of fMRI data (Dumoulin and Wandell 2008).

5.2.3 OLE-Derived Population Representations of Stimulus Position

As an alternative to the RF-derived interpolation procedure, and in order to construct the PRF in a well defined mathematical way, we employed an Optimal Linear Estimator (OLE). This technique, originally developed to estimate a single value of an encoded physical quantity (Salinas and Abbott 1994), is based on a Bayesian theoretical framework (Dayan and Abbott 2001). We used an extension of this method that enabled us to estimate entire distributions of population activity across visual space (Jancke et al. 1999, 2004b; Jancke 2000; Jazayeri and Movshon 2006).

The method is based on two ideas. First, the population distribution is generated as a linear superposition of a set of basis functions, one such function for each neuron. Each neuron's basis function is multiplied with the current firing rate of the neuron. Second, a template function for the distribution of population activity was defined as a Gaussian centered at each of the seven reference stimulus positions. Its width (0.6°) in visual space approximately matched the average RF profile of all neurons measured (Jancke et al. 1999). The basis function of each neuron was determined such that for the seven reference stimuli the reconstructed population distribution approximated the template function optimally. For this optimization, we used the mean firing rates within the time interval around maximum activity, 40–65 ms after the onset of the stimulus. However, the exact size of the integration window was not critical for the estimation procedure.

Figure 5.1f depicts distributions of population activity in the specified time window with high precision in visual space. Note that all seven flashes were fairly represented by 0.4° shifts of activity profiles across the PRF. This reconstruction was used as a tool for the next step, the calculation of motion trajectories across the PRF.

5.2.4 Representation of Moving Stimuli across the PRF

To extend the estimation to moving stimuli, the basis function that each neuron contributes to the seven reference stimuli was held fixed, but was now multiplied with the firing rate of that neuron in response to a stimulus moving with a particular velocity.

For all speeds and both directions tested, Fig. 5.2 depict space-time diagrams of OLE-derived population distributions within the time intervals that revealed significant propagation of spiking activity. The diagrams show coherent peaks of activity tracking each moving stimulus across the PRF. Dependent on speed we observed a spatial lag between the current stimulus position and the activity peaks (compare Fig. 5.4), which were most evident at high stimulus speeds (see red and blue lines). The profiles of the PRF showed some variability in response amplitudes and scatter

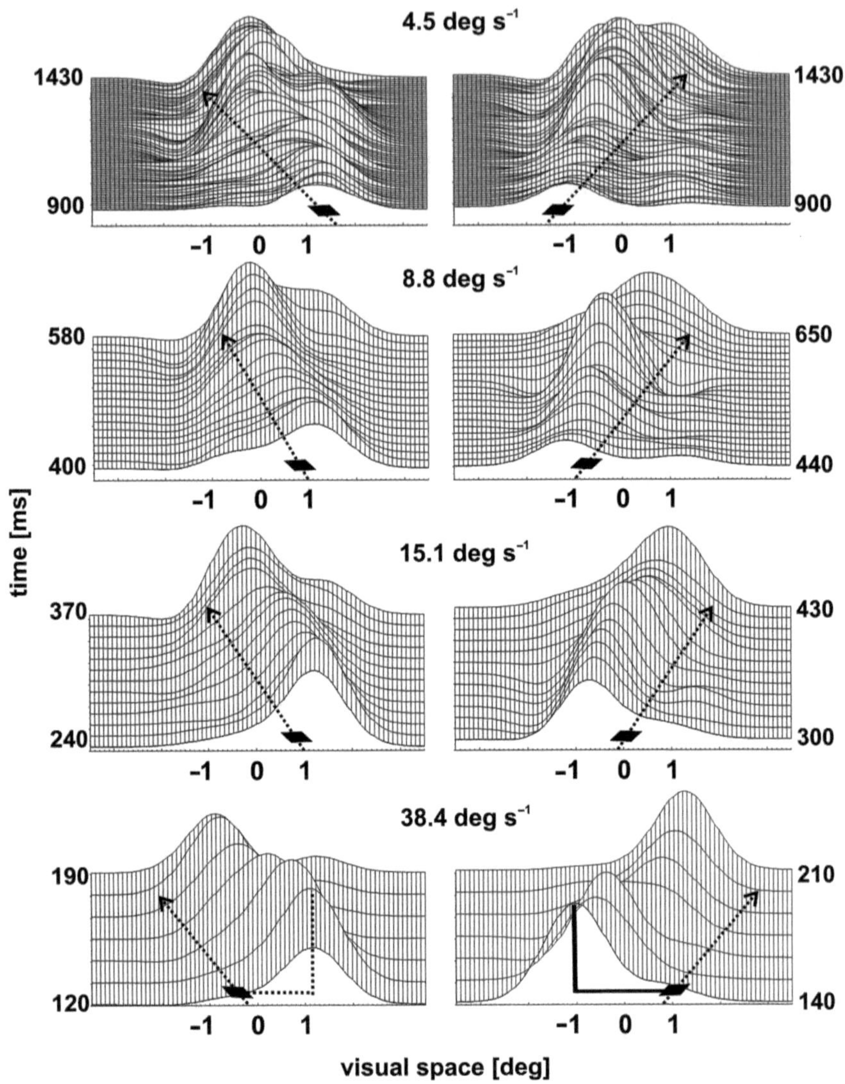

Fig. 5.2 Motion trajectories across the population receptive field (PRF). Space-time diagrams of population activity representing squares that moved with different speeds and directions (stimulus shown in *black*, *arrows* sketch trajectories). Profiles were derived by OLE. The *x*-axis depicts visual space in degree, zero indicating mid-point of trajectories. Starting position of movement was at ±3.2°. *Left column* shows propagation reflecting peripheral-central motion; *right column* shows opposite direction. The *y*-axis resolves 10 ms time steps within the time interval in which propagation of activity occurred. Activity was normalized for each speed separately. The different tilt angles of the space-time plots arise from activity peaks matching stimulus speed and direction. *Red and blue lines* in the *bottom row* sketch the spatial lag between current stimulus position and peak location within the PRF (cf. Fig. 5.4). (Modified from Jancke et al. ©2004b, by permission from Wiley-Blackwell)

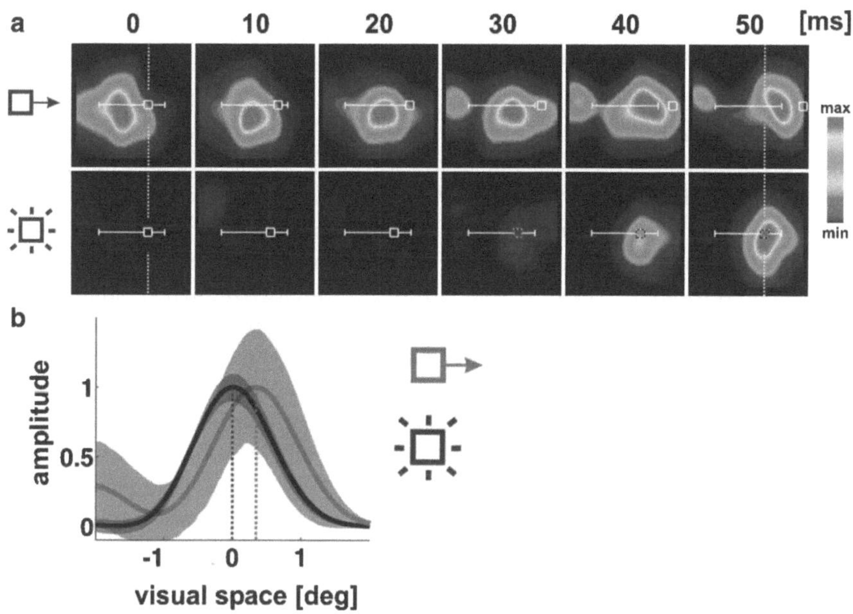

Fig. 5.3 The flash lags: impact of shorter latency for motion compared to a flash on the representation of position. (**a**) *Upper row*: moving square (38.4°s⁻¹). *Lower row*: flashed square (outlined in white; 25 ms on; shown stippled after stimulus off). Population representations are shown in time frames of 10 ms from *left* to *right*. The sampled space covered 2.8° of the central visual field (*horizontal white line*, cf. Fig. 5.1). *Color bar* indicates level of population activity; data were normalized separately for each stimulus condition. The moving stimulus started 3.2° outside the sampled visual space tracked by a peak of population activity that has been evoked some time before beyond the sampling region. At time zero stimulus position was identical for the moving and the flashed square (compare *vertical pointed lines*). During the next time steps activity in response to the flash emerged while the moving peak continued propagating. When activity for the flashed stimulus reached its maximum (50 ms), the moving peak already passed the mutual flash position due to its faster processing. (**b**) The spatial offset shown in (**a**) was confirmed applying the OLE approach in combination with a bootstrap analysis ($n = 1{,}000$). The time window shown depicts activity 50–60 ms after presentation of the flashed stimulus. Activity was separately normalized to the mean of each stimulus condition. *Gray shaded* areas show 99% confidence intervals for the flashed stimulus (*bright gray*; *black line* shows mean), respectively for the moving square (*bright red*; *red line* shows mean). A clear spatial offset ($p < 0.00001$) can be seen between the peaks of both conditions due to reduced time-to-peak latencies in response to motion (*vertical dotted line*).(Modified from Jancke et al. ©2004b, by permission from Wiley-Blackwell) (*see Color Plates*)

in peak positions. These fluctuations were due to irregularities in cell sampling and of no significance ($p > 0.05$, bootstrap, $n = 1{,}000$). Thus, our approach depicts propagating population activity on a fine spatial scale, resolving shifts of activity peaks in visual space much smaller than the RF sizes (see Fig. 5.3a upper row, for a motion trajectory derived from the RF interpolation procedure).

5.2.5 The Flash-Lag Effect: Differences between Flashed and Moving Stimuli

Judging correctly the location of moving objects is of crucial importance for evading obstacles or predators or for catching prey. This task would be almost impossible, particularly for high speeds, if the relevant information is delayed due to neural conduction and processing times. To overcome this problem, compensatory mechanisms have evolved that allow for anticipation of the path of motion.

Psychophysically, a moving and a flashed stimulus presented aligned, are perceived as being displaced: Surprisingly, the moving stimulus appears ahead of the flash (Hazelhoff and Wiersma 1924). One explanation for this "flash-lag" effect is that the visual system is predictive and extrapolates the position of a moving stimulus into the future (Nijhawan 1994). Alternatively, the "latency difference" model assumes that the visual system processes moving objects more rapidly compared to single flashes (Purushothaman et al. 1999; Kirschfeld and Kammer 1999; Whitney et al. 2000; for review and other explanations see Eagleman and Sejnowski 2000; Krekelberg and Lappe 2001). In none of those studies, however, were neurophysiological correlates examined. We therefore used our population approach for comparison between population representations of a single flash and a moving square.

The upper row in Fig. 5.3a illustrates the central portion of a motion trajectory of a square moving at a speed of $38.4°s^{-1}$. At the onset of the flash (lower row, time zero), the moving square is located at the same position as the flashed square. Clearly, due to neural delay times, cortical population activity for the flash at that time is at baseline. In contrast, for the moving square, a propagating peak of activity is already observed that had been evoked previously since the stimulus trajectory started outside the PRF (see supplementary movie 1). Five steps later (50 ms), the moving stimulus is at a new position, tracked by the peak of population activity with a spatial lag. At that time, activity for the flash has reached its maximum without changing position, thus representing faithfully the initial location of the stimulus. Applying the OLE procedure, Fig. 5.3b depicts the same situation as shown in Fig. 5.3a, 50 ms, after presentation of the flash. We found the peak representing the moving stimulus approximately 0.33° ahead of the peak representing the flash.

5.2.6 Latency Differences – Spatial Offset – Directional Asymmetry

Assuming equal processing times for both flashed and moving stimuli, the population representations should be localized at identical positions. Instead, activity peaks elicited by motion and the flash showed a significant spatial offset, i.e. the moving square was represented ahead of the flash, indicating shorter latencies for motion due to the past history created by the moving stimulus.

To compare latencies evoked by single flashes with those for moving stimuli, we used the spatial lag method introduced by Bishop et al. (1971) for analysis of single cell latencies. This method is based on the fact that the time-to-peak of activity depends on stimulus speed. As our approach, produces continuous propagation of activity across the PRF (Figs. 5.2 and 3a) rather than single discharge peaks we modified the original method: Along the entire trajectory across the PRF, we calculated the average spatial lag between current stimulus position and its representing activity peak. This mean spatial lag was plotted as a function of stimulus speed (Fig. 5.4) indicating that the spatial lag increased linearly with stimulus speeds.

The slope of the regressions revealed latencies of 38 ms for the peripheral-central ($r=0.99$) and 42 ms ($r=0.99$) for central-peripheral direction. The spatial lag was significantly shorter than estimated from time-to-peak latencies in response to flashed stimulation (~54 ms, mean peak latency for single flashes across the PRF; bootstrap).

The spatial lag was also dependent on the direction of motion. Peripheral-central movement's latencies were smaller compared to the opposite direction, particularly for higher speeds (see Fig. 5.2). With decreasing speed, this asymmetry became less significant due to the increasing positional scatter ($p<0.01$ for 8.8°s^{-1}; $p>0.05$ for 4.5°s^{-1}). For a slow speed of 4.5°s^{-1}, reduced latency for peripheral-central motion

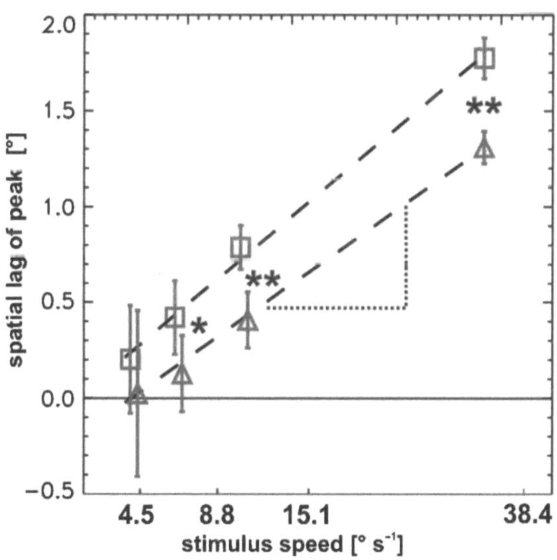

Fig. 5.4 Calculation of peak latencies for moving stimuli: speed-dependence of the spatial lag. The average spatial lag between the propagating peak of population activity and the actual position of the moving stimulus increased with stimulus speeds. Data points show mean spatial lags for each speed and direction (bootstrap, $n=1{,}000$). *Blue squares* indicate centro-peripheral movement, red triangles indicate opposite motion direction. Latencies were calculated by linear regression. Latency for centro-peripheral motion was 42 ms, and 38 ms for peripheral-central direction (significance of direction difference: **=$p<0.00001$; *=$p<0.01$). (Modified from Jancke et al. ©2004b, by permission from Wiley-Blackwell) (*see Color Plates*)

led to a match between the peak of population activity and actual stimulus position as indicated by a spatial lag of nearly zero.

Spatial asymmetries for representations of moving objects (Mateeff and Hohnsbein 1988; Müsseler and Aschersleben 1998) might be the result of active mechanisms, compensating neural delays for one direction on the cost of delays in the opposite direction (Jensen and Martin 1980). Van Beers et al. (2001) proposed a number of putative mechanisms underlying differences in localization for foveopetal and foveofugal motion. These mechanisms include temporal asymmetries in neural delays, and a partial asymmetric spatial expansion of the retinal representation, both comparable to our findings. On the other hand, these authors provided evidence that when shifting gaze, the central nervous system is able to compensate for localization errors by sensor motor integration to maintain position constancy, maybe by taking advantage of these internally generated erroneous position signals.

Taken together we showed that in cat area 17 small moving squares of light could be represented as propagating peaks of activity across a "population receptive field" in which latencies for motion were significantly shorter than expected from the response to flash stimuli.

In terms of spike rates of single cell RFs, such behavior can be interpreted as dynamic and asymmetric enlargement of RF-sizes. As a consequence, RFs are shifted opposite to motion direction (Fu et al. 2004) causing neurons to respond with shorter delay times. Likewise, one might interpret such a shift as spatial phenomena: RF boundaries that were not responsive when mapped with flashed stimuli, become responsive when a stimulus moves, and RFs are therefore "pulled" towards an approaching moving stimulus (Pulgarin et al. 2003).

5.2.7 Motion Anticipation in Early Visual Cortex: A Novel Achievement in the Representation of Stimulus Position?

On the single cell level, neural latencies vary within a wide range of delay times. Generally, neurons that were sensitive to high stimulus speeds were found to have also short latencies for stationary light bars (Duysens et al. 1982). Comparing responses to flashed and moving slits of light, only few cells showed reduced latencies (Bishop et al. 1971). A stimulus dark edge evoked responses in advance of the discharge coming from the light edge (Bishop et al. 1971, 1973). Also, LGN neurons were found to respond with shorter delays to moving compared to flashed light bars (Orban et al. 1985). Part of these controversial findings may result from the relative wide bin sizes used for analyzing, which makes it difficult to detect small changes in latencies at the single cell level. Furthermore, the spatial-lag method is critically sensitive to response variability of single cells. The population approach used here however, transforms the various spatio-temporal dynamics of single cell activity into homogeneous activity patterns at the population level leading to a qualitative processing step from microscopic to mesoscopic levels (Freeman 2000; Dinse and Jancke 2001a). As a consequence, the population approach allows for densely sampled and fine-scaled analysis of activity trajectories across a representative neural population.

Anticipatory mechanisms have also been reported for populations of retinal ganglion cells (Berry et al. 1999). The authors found that nonlinear contrast-gain control together with spatially extended receptive fields and a biphasic temporal response account for population activity that travels near the leading edge of a moving bar. Translating their data from millimeter retina into visual field coordinates (cf. DeVries and Baylor 1997; Hughes 1971), the retinal compensatory mechanisms were limited to stimulus speeds of approximately $5°s^{-1}$ which is in accordance with our result for peripheral-central motion direction. While we found a reduction of latencies in primary visual cortex up to $40°s^{-1}$, latencies for this range of speeds have not been investigated for the retina.

5.2.8 Latency Differences May Contribute to the Flash-Lag Effect

There is an extensive ongoing discussion about the nature of flash-lag effect, which has been studied under a large variety of experimental designs (Metzger 1932; MacKay 1958; Nijhawan 1994; Purushothaman et al. 1999; Krekelberg and Lappe 1999, 2001; Kirschfeld and Kammer 1999; Eagleman and Sejnowski 2000; Krekelberg et al. 2000; Sheth et al. 2000; Whitney et al. 2000). However, the neural substrates underlying this effect remain unknown.

The flash-lag effect has also been reported with no retinal motion indicating that extra-retinal information can be used to alternatively derive motion information (Schlag et al. 2000). Moreover, the flash-lag phenomenon not only applies to motion, but to other stimulus dimensions as well, such as color (Sheth et al. 2000). In any case, while not designed to mimic a particular psychophysical experiment – our experimental set-up corresponds to the traditional continuous motion protocol (Hazelhoff and Wiersma 1924) – the presented data revealed ~16 ms difference in latency between a flash and a moving stimulus corresponding to a 30% reduction in processing time. Reduced latencies of 10 ms for moving stimuli ($15°s^{-1}$) have recently been reported for neurons in primate V1 (Ceriatti et al. 2007). For the flash-lag effect described by Eagleman and Sejnowski (2000), the stimulus moved at $360°s^{-1}$ rotation angle leading to a displacement of about 5°, which translates into a delay of 14 ms, thus within the same range as found in our study. On the other hand, latency differences obtained in various paradigms commonly range between 40 and 80 ms (Krekelberg and Lappe 2001) most likely involving additional mechanisms of downstream cortical areas (but see next section). Furthermore, compensation of neural processing times is not restricted to the perceptual domain. It has recently been demonstrated that pointing movements towards the final position of a moving target were directed beyond its vanishing point suggesting that for goal-directed tasks, sensorimotor integration is critical for compensation of neural latencies (Kerzel and Gegenfurtner 2003).

Interestingly, a recent study investigated the flash-lag effect using two stimuli that moved towards the position of the flash. The authors hypothesized that two moving objects that would collide should "add their pre-activations" (Maiche et al. 2007). Indeed, it was found that such a stimulus configuration lead to even shorter latencies resulting in an increase of the flash-lag effect.

5.3 Neuronal Population Dynamics in Cortical Space

We hypothesized that pre-activation resulting from preceding stimulus displacements along the trajectory lead to an increased excitability, and thus to higher probabilities to fire action potentials when the stimulus moves across the PRF. Long-range horizontal connections may constitute a possible cortical substrate for such pre-activation as the spreading sub-threshold activity (Grinvald et al. 1994; Bringuier et al. 1999) extends far beyond the classical spiking RF (Allman et al. 1985; cf. Fig. 5.1a). However, extracellular recordings as shown in the previous paragraphs provide no information about the accompanying intracellular events (see chapter Fregnac et al. in this book). To identify those cortical mechanisms involved in motion processing, we employed voltage-sensitive dye (VSD) imaging (Cohen et al. 1978; Grinvald et al. 1984) that allows sensitive monitoring of the changes in membrane potential across neuronal populations with high spatial and temporal resolutions (Shoham et al. 1999).

5.3.1 Optical Imaging: Recording of Widespread Cortical Activity in Real-Time

VSD imaging makes use of a fluorescent dye that incorporates into neuronal membranes. Dependent on the depolarization of the membrane potential in which it is inserted, the dye changes its properties: the higher the change in voltage across the neuronal membrane the higher the amount of emitted fluorescence. These optical signals are then detected by a highly sensitive camera system (Grinvald and Hildesheim 2004).

5.3.2 Sub-threshold Preactivation by Long-Range Horizontal Connections

To visualize how motion trajectories are represented across the cortical surface, small squares were moved with different speeds (16, 32, $64°s^{-1}$). Figure 5.5a depicts cortical VSD responses in cat area 18 evoked by small squares moving downwards in the visual field. In all cases, we observed an initial fast spread (white contour) that

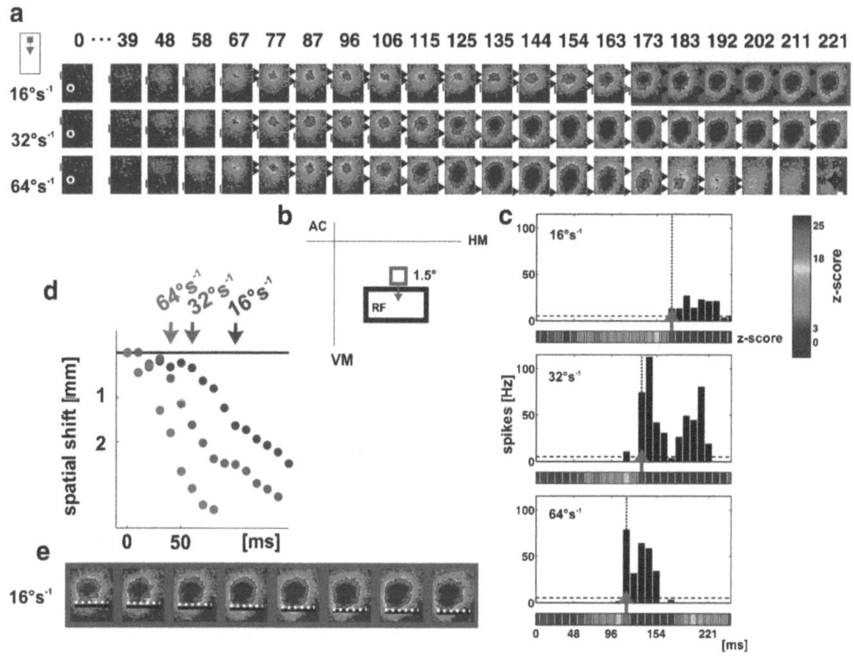

Fig. 5.5 Imaging cortical dynamics in response to moving squares. (**a**) Time courses of evoked cortical activity are shown in single frames (scale bar at *bottom right* 1 mm). Time after stimulus onset is indicated at the *top*. From *top to bottom row*: moving squares presented at different speeds, 16, 32, and 64°s^{-1}. *Green vertical lines* indicate the onset of stimuli and their estimated trajectory along posterior–anterior axis (M, medial; P, posterior). Extracellular recording (multi units) was performed simultaneously at the cortical location marked by *white circles* at frame zero (see (**b**) for receptive field (RF) of the recorded neurons. Spiking activity propagates with different speeds in anterior direction (*lower black arrows*). (**c**) To verify that the approaching wave front of VSD activity corresponds to spiking threshold, Post Stimulus Time Histograms (PSTHs) of spike events were derived. Horizontal stippled line depicts significance (z-score = 2, 30 repetitions). As the electrode was placed outside the retinotopic representation of the square's starting position, no spikes are evoked at the onset of motion, although the dye signal reports subthreshold spread of activity (*bottom row* shows actual z-score levels at the electrode position). Spikes are detected, however, as soon as activity reaches the spiking threshold at the position of the electrode (see *red arrows* in PSTH, and in **a**). With increasing stimulus speed, shifts in latency between the PSTHs can be seen because spiking level reached the electrode earlier for higher speeds. Thus squares moving at different speeds evoke consecutive spikes at a fixed electrode position. (**d**) Spatial shift of the wave front of spiking activity, slopes indicate speed of propagation across the cortex. Increasing hue of green depicts increasing stimulus speed. (**e**) Magnified section of imaging frames shown in (**a**), *upper row*. Note that the wave front of spiking activity (*white stippled lines*) closely matches the projected location of the stimulus (*black lines*) indicating "on-line" tracking of motion (see text). (Modified and reprinted by permission from Macmillan Publishers Ltd: Nature (Jancke et al. ©2004a) (*see Color Plates*)

was followed by emerging activity, which started to propagate a gradually drawn-out anterior in the direction of the induced motion (lower black arrows in Fig. 5.5a; green dots in d).

In order to detect regions that are activated at spiking levels we placed a single electrode at a given cortical position (white circles in 1st frames) and multi-unit recordings were performed simultaneously to VSD imaging in order to map the neurons' RF location in visual space. Subsequently, the start of the stimulus was positioned above the marked RF area (Fig. 5.5b). Moving the square evoked spikes as soon as the stimulus entered the RF. Dependent on stimulus speed however, significant spiking activity was detected at different times resulting in shifts of spiking response onset (Fig. 5c). Importantly, this procedure enabled to relate the amplitude of the dye signal directly to spiking threshold. As can be seen by the traces at the bottom of Fig. 5.5c, high levels of the dye signal (black contours in Fig. 5.5a) consistently reported the onset of spiking activity at different times for individual stimulus speeds (red arrows). Thus, the stimulus trajectories were characterized by asymmetrically propagating waves of spiking activity moving according to stimulus speeds (Fig. 5.5d).

Note that anterior to the spiking wave front, significant sub-threshold activity could be detected (white contours). We propose that such spreading activity may provide a basis for faster spiking as it pre-activates cortical areas ahead of the motion trajectory. Figure 5.5e depicts a larger scaled section of evoked activity in response to a square that was moved at $16°s^{-1}$ (compare Fig. 5.5, top row). The black bars mark the estimated center of the actual stimulus position projected onto the cortical coordinates (Albus 1975; Tusa et al. 1979). The white dotted lines show the cortical position of the propagating spiking wave front. As both lines closely match their position frame by frame, we observe an "on-line" tracking of a moving stimulus translating into ~50 ms reduction of latencies. For further increasing stimulus speeds however, spiking activity lagged behind stimulus projections.

Compared to area 17, cat area 18 seems to have more capacities in latency reduction, probably due to its preference for processing of higher temporal frequencies. On the other hand, if we consider the wave front (see transition from blue to green color in Fig. 5.3a, upper row; supplementary movie 1, and Fig. 5.2) instead of the peak of the population representations, full compensation of latencies could be claimed even for the PRF approach in area 17. At significant spiking wave front levels ($p<0.01$, bootstrap) we calculated a spatial lag of nearly zero (0.006°) for a stimulus speed of $38°s^{-1}$. Thus depending on the read-out cortical mechanisms, different degrees of latency compensation could be achieved.

Note also that reduced latency does not occur instantaneously but needs time to build-up. Only when the stimulus trajectory has been established, continuative stimulation leads to the observed catch-up effect of the spiking wave front (Fig. 5.5a, upper row).

5.3.3 Propagating Waves of Cortical Activity: The Line-Motion Illusion

Almost a century ago, Gestalt psychologists (Wertheimer 1912; Kenkel 1913) made the striking observation that non-moving stimuli can give rise to motion perception (Kanizsa 1951). Assuming that the facilitatory effect of the cue spread gradually in space and time (Posner et al. 1980; Sagi and Julesz 1986; Polat et al. 1998), the sensation of motion evoked by a subsequently flashed bar during the line-motion illusion could result from sequential suprathreshold activation, thus mimicking real motion drawing away from the local cue (Hikosaka et al. 1993; von Grünau et al. 1996a); Fig. 5.6a, b; see supplementary movie 2).

To explore cortical mechanisms underlying the "line-motion" illusion we examined, in cortical area 18 of the anaesthetized cat, responses to a flashed small square and a long bar alone, and the configuration of the line-motion illusion: a square briefly preceding the bar (Jancke et al. 2004a). Figure 5.6c depicts the cortical VSD response evoked by the small square alone. As the VSD signal is sensitive primarily to synaptic potentials rather than to spikes (Sterkin et al. 1999; Grinvald et al. 1999; Petersen et al. 2003), we searched here for an indirect way to delineate the spiking regions in the resulting optical maps (Albus 1975;

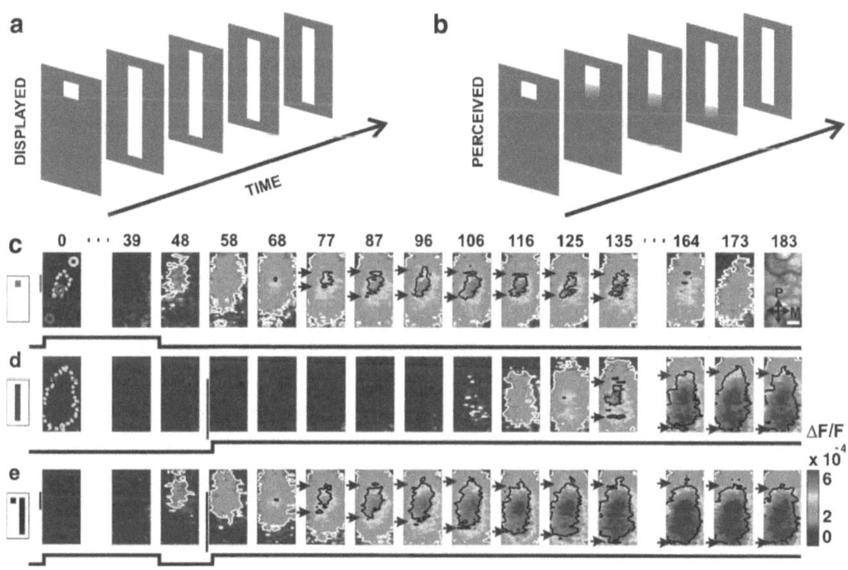

Fig. 5.6 Representations of stationary stimuli. (**a/b**) The line-motion paradigm. (**a**) Square ('cue') presented before a bar stimulus. (**b**) Subjects report illusory line-drawing (see supplementary movie 2). (**c–e**) Patterns of evoked cortical activity as a function of time. *Yellow dotted* contours approximate retinotopic representation of the stimuli; *coloured circles* indicate extracellular recording sites. *White contours* delimit low-amplitude activity (significance level; $p<0.05$). The cortical area imaged is shown at upper right. Scale bar, 1 mm; P, posterior; M, medial. *Vertical lines* indicate estimated position of the stimuli (sizes: 1.5° square, 1.5°×6° bar) along posterior-anterior axis. Time in milliseconds after stimulus onset is shown at the *top*. Stimulation period is depicted at the bottom of each row. Colour scale indicates averaged fractional changes in fluorescence intensity ($\Delta F/F$). Stimuli: (**c**) Flashed small square. (**d**) Flashed bar. (**e**) Line-motion paradigm. Average of 22 trials. (Modified and reprinted by permission from Macmillan Publishers Ltd: Nature (Jancke et al. ©2004a) (*see Color Plates*)

Tusa et al. 1979). As has been shown in the previous section, spiking regions should be located in the area of high VSD amplitudes. We therefore analyzed the dynamic behavior of evoked low- and high-amplitude activity separately. Again, low-amplitude activity spread far beyond the retinotopic representation of the square at ~0.09 ms^{-1}, consistent with conduction velocity along horizontal connections (white contour, Grinvald et al. 1994; Bringuier et al. 1999; Benucci et al. 2007). In contrast, high-amplitude activity (delimited by black contours) showed negligible lateral extension (Fig. 5.6c, arrows), after an initial ~20 ms period of spread. Single-unit recordings (colored circles; Fig. 5.6c, first frame) confirmed that spiking activity evoked by the square was limited to the high-amplitude area. The second stationary stimulus, the bar, flashed alone 60 ms later, yielded a similar finding (Fig. 5.6d). High-level activity delineated a circumscribed region representing the elongated shape of the bar almost at once (compare upper and lower black arrows).

Next, we investigated the spatio-temporal pattern of activity evoked by the line-motion paradigm (Fig. 5.6e). In contrast to the conditions in which either the square or the bar was flashed alone, the spike discharge zone did not remain stable but was gradually drawn out towards the end of the cortical bar representation (lower arrows). Thus, although these two stimuli were stationary, the anterior portion of the high-activity region propagated similarly to moving stimuli (Fig. 5.6a), suggesting, that cortical correlates of illusory line-motion were directly visualized here (see supplementary movie 3).

These findings demonstrate the effect of spatio-temporal patterns of sub-threshold synaptic potentials on cortical processing and the shaping of perception. While psychophysical studies also implicate top-down involvement of higher cortical areas (Hikosaka and Miyauchi 1993; von Grünau and Faubert 1994; Shimojo et al. 1997), others argue for essential bottom-up, i.e. stimulus induced mechanisms, occurring at early processing stages (Hikosaka et al. 1993; Steinman et al. 1995; von Grünau et al. 1996b).

We suggest that the characteristics of sub-threshold spread in early visual areas as shown here, are basic properties of the primary visual cortex involved in motion induction, and thus, cortical correlates of the bottom-up mechanism underlying the line-motion illusion without need of attention (Cavanagh et al. 1989, von Grünau et al. 1996b; Kawahara et al. 1996; Downing and Treisman 1997). It has been reported that when multiple inducers (or different stimulus features like color, contrast, or texture) are used there is evidence for a second top-down component that operates on a slower timescale (Hikosaka and Miyauchi 1993; von Grünau et al. 1996b; Nakayama and Mackeben 1989). Furthermore, the illusory line-motion can be modulated by attention (Downing and Treisman 1997), or can be voluntarily induced by non-retinotopic mechanisms (Hikosaka and Miyauchi 1993) even across sensory modalities (Shimojo et al. 1997, but see Downing and Treisman 1997). These top-down processing and attentional/attention seeking mechanisms remain to be explored in behaving subjects, by bringing together psychophysics and functional imaging.

5.4 Conclusions

Visual illusions such as the flash-lag and the line-motion illusions reveal discrepancies between the physically present world outside and internal *re*-presented sensation. What can we learn about the brain when studying such phenomena? Of course, the brain did not evolve its enormous complexity in order to deceive in every- day life. The task the visual system has to solve is to generate and interpret internally meaningful patterns of neuronal activity that enable successful interaction in natural environments. Visual illusions may therefore appear as artificial extremes of what the neuronal system has evolved to handle. The brain uses its tremendous flexibility - that still cannot be achieved by technical devices - in order to navigate and survive within rapidly changing visual input. In this framework, adaptation, stability that preserves flexibility, and prediction, are the essential capacities that have to be acquired throughout ongoing individual learning and evolutionary processes.

5.4.1 Anticipation and Integration: Important Capacities of the Brain

What are then advantages of producing far reaching sub-threshold waves of activities? Apart from well-documented context dependent lateral perceptual modulation (Polat and Sagi 1993; Polat et al. 1998), we believe that the spatio-temporal pattern of cortical activity waves is important in motion integration (Jancke 2000) and prediction. Any sudden appearing object (e.g. a prey, a predator – or a car nowadays) is potentially about to move (Dragoi et al. 2002). Sub-threshold cortical activity that spreads far ahead of the actual stimulus representation may therefore "prepare" the cortex for an object's putative trajectory. Such a strategy can save neuronal processing times. When actual movement occurs, neurons that represent position ahead of the stimulus are already close to the firing threshold and can react more rapidly to motion onset.

However, there is still a gap of knowledge about how timing information provided by visual cortical neurons map to perception (Seriès et al. 2002). Along the visual pathway various predictive (Nijhawan 1994; Rao and Ballard 1999) and integrative mechanisms (Krekelberg and Lappe 1999; Eagleman and Sejnowski 2000) starting from the retina (Berry et al. 1999) up to sensorimotor transformation (Kerzel and Gegenfurtner 2003) are involved in processing of motion. It remains an open question how the representation of moving stimuli in the primary visual cortex, in particular its reduced response latencies as reported here, contribute to the processing of object position in higher brain areas.

In summary we have showed that neuronal populations at the first cortical processing stage, the primary visual cortex, incorporate signatures of neuronal motion initiation and anticipation. Once a stimulus continues to move, we demonstrated that sub-threshold cortical activity spreads ahead of retino-thalamic input leading to reduction of neuronal delay times as compared to a stationary flash. Depicting neuronal activity in stimulus space, we observed that reduced delay times cause shifts of RFs opposite to motion direction. This effect does not fully explain but may contribute to the well-known "flash-lag" phenomena: a moving and a flashed stimulus presented aligned, are perceived with spatial offset in which the moving stimulus appears ahead.

We suggest that preceding stimulation evoke pre-activation via long-range horizontal axons and increase the postsynaptic potential close to firing thresholds, thus preparing the ground for reduced latencies for subsequent stimuli. Such a cortical mechanism seems to be also involved in the line-motion phenomena, in which a bar, flashed at once after a small dot, is perceived as being gradually drawn-out. Here, illusory motion is seen despite stationary input, thus uncovering the underlying propagation of far spreading cortical activity.

Integration and boosting of propagating sub-threshold activity may therefore be a fundamental cortical mechanism for initiation of motion processing in response to spatially close and temporally successive visual input.

5.5 Supplementary Materials (CD-ROM)

Movie 1 Population representation of a moving square stimulus in cat area 17 (*file 5_M1_population.avi*). Representation of a moving square across the population receptive field (PRF). The PRF was derived from pooled activity of 178 neurons recorded in cat area 17. The RFs of all neurons densely cover a region along 2.8° in the central visual field (horizontal white line). Each neuron contributes to the overall activity distribution dependent on its relative RF location (see text). The white outlined square shows the stimulus (0.4°) as it moves across the visual field (speed = 38.4°s^{-1}). Spiking population activity forms a propagating peak that tracks the stimulus with a spatial lag due to neural delay times.

Movie 2 The line-motion stimulus (file *5_M2_linemotion.avi).* The line-motion illusion. The movie presents four different configurations of the line-motion illusion, consisting of pre-cueing small squares at four different positions followed by a stationary bar. Although the bar is presented at once, it is perceived as being drawn out from the position of the small squares.

Movie 3 V1 population response to the line motion (*file 5_M3_linemotionV1.avi).* Imaging cortical correlates of illusion in the visual cortex. Slow-motion video of the cortical representation of the line-motion paradigm in cat area 18 (posterior = top, medial = right, cf. Fig. 5.6e). Optically detected activity within 5.5 × 2.9 mm cortical surface is shown (sampling rate 9.6 ms, stimulus onset at frame zero). The first frames show a rapid spread of low amplitude activity (light blue, green, and yellow) followed by high-level activity (red) that gradually propagates towards the end of the cortical bar representation thus reporting motion. Blue vertical line indicates estimated position and sizes of the stimuli along posterior-anterior axis (1.5° square, 6° bar; horizontal line approximates position along medial-lateral axis). Color bar indicates averaged fractional changes in fluorescence intensity ($\Delta F/F$, 22 trials). White contour delimit significance level, black contour encircles the spike discharge zone, see text).

Acknowledgments D.J. supported by Minerva Foundation, BMBF, Deutsche Forschungsgemeinschaft (Scho 336/4-2 and Di 334/5-1,3). F.C. supported by Marie Curie EU fellowship. A.G. supported by the Grodetsky Center, Goldsmith, Korber & ISF Foundations, BMBF/MOS, and NIH 1R01-EB00790-01 grants.

References

Albus K (1975) A quantitative study of the projection area of the central and the paracentral visual field in area 17 of the cat I. The precision of the topography. Exp Brain Res 24:159–179

Allman J, Miezin F, McGuiness E (1985) Stimulus specific responses from beyond the classical receptive field: neurophysiological mechanisms for local-global comparisons in visual neurons. Annu Rev Neurosci 8:407–430

Benucci A, Frazor RA, Carandini M (2007) Standing waves and traveling waves distinguish two circuits in visual cortex. Neuron 55:103–117

Berry MJ II, Brivanlou IH, Jordan TA, Meister M (1999) Anticipation of moving stimuli by the retina. Nature 398:334–338

Bishop PO, Coombs JS, Henry GH (1971) Responses to visual contours: Spatio-temporal aspects of excitation in the receptive fields of simple striate neurones. J Physiol (London) 219:625–657

Bishop PO, Coombs JS, Henry GH (1973) Receptive fields of simple cells in the cat striate cortex. J Physiol (London) 231:31–60

Bringuier V, Chavane F, Glaeser L, Frégnac Y (1999) Horizontal propagation of visual activity in the synaptic integration field of area 17 neurons. Science 283:695–699

Cavanagh P, Arguin M, von Grünau M (1989) Interattribute apparent motion. Vision Res 29:1197–1204

Ceriatti C, Botelho EP, Soares JGM, Gattass R, Fiorani M (2007) Shorter latencies to moving stimuli in primate V1 single units. Soc Neurosci Abstracts 37:279.22

Cohen LB, Salzberg BM, Grinvald A (1978) Optical methods for monitoring neuron activity. Annu Rev Neurosci 7:171–182

Dayan P, Abbott LF (2001) Theoretical neuroscience: computational and mathematical modeling of neural systems. Cambridge, MIT Press

DeVries SH, Baylor DA (1997) Mosaic arrangement of ganglion cell receptive fields in rabbit retina. J Neurophysiol 78:2048–2060

Dinse HR, Jancke D (2001a) Time-variant processing in V1: From microscopic (single cell) to mesoscopic (population) levels. Trends Neurosci 24:203–205

Dinse HR, Jancke D (2001b) Comparative population analysis of cortical representations in parametric spaces of visual field and skin: a unifying role for nonlinear interactions as a basis for active information processing across modalities. Prog Brain Res 130:155–173

Dragoi V, Sharma J, Miller EK, Sur M (2002) Dynamics of neuronal sensitivity in visual cortex and local feature discrimination. Nat Neurosci 5:883–891

Dumoulin SO, Wandell BA (2008) Population receptive field estimates in human visual cortex. Neuroimage 39:647–660

Downing PE, Treisman AM (1997) The line-motion illusion: Attention or impletion? J Exp Psychol Hum Percept Perform 23:768–779

Duysens J, Orban GA, Verbeke O (1982) Velocity sensitivity mechanisms in cat visual cortex. Exp Brain Res 45:285–294

Eagleman DM, Sejnowski TJ (2000) Motion integration and postdiction in visual awareness. Science 287:2036–2038

Fitzpatrick D (2000) Seeing beyond the receptive field in primary visual cortex. Curr Opin Neurobiol 10:438–443

Freeman WJ (2000) Mesoscopic neurodynamics: from neuron to brain. J Physiol (Paris) 94:303–322

Fu YX, Shen Y, Gao H, Dan Y (2004) Asymmetry in visual cortical circuits underlying motion-induced perceptual mislocalization. J Neurosci 24:2165–2171

Gegenfurtner KR, Hawken MJ (1996) Interaction of motion and color in the visual pathways. Trends Neurosci 19:394–400

Grinvald A, Anglister L, Freeman JA, Hildesheim R, Manker A (1984) Real-time optical imaging of naturally evoked electrical activity in intact frog brain. Nature 308:848–850

Grinvald A, Lieke E, Frostig R, Hildesheim R (1994) Cortical point-spread function and long-range lateral interactions revealed by real-time optical imaging of macaque monkey primary visual cortex. J Neurosci 14:2545–2568

Grinvald A, Shoham D, Shmuel A, Glaser D, Vanzetta I, Shtoyerman E, Slovin H, Arieli A (1999) In vivo optical imaging of cortical architecture and dynamics. In: Wind-Horst U, Johansson H (eds) Modern techniques in neuroscience research. Springer, New York, pp 893–969

Grinvald A, Hildesheim R (2004) VSDI: a new era in functional imaging of cortical dynamics. Nat Rev Neurosci 5:874–885

Hazelhoff FF, Wiersma H (1924) Die Wahrnehmungszeit [The sensation time]. Zeitschrift für Psychologie 96:171–188

Hikosaka O, Miyauchi S (1993) Voluntary and stimulus induced attention detected as motion sensation. Perception 22:517–526

Hikosaka O, Miyauchi S, Shimojo S (1993) Focal visual attention produces illusory temporal order and motion sensation. Vision Res 33:1219–1240

Hughes A (1971) Topographical relationships between the anatomy and physiology of the rabbit visual system. Doc Ophthalmol 30:33–159

Jancke D, Erlhagen W, Dinse HR, Akhavan AC, Giese M, Steinhage A, Schöner G (1999) Parametric population representation of retinal location: neuronal interaction dynamics in cat primary visual cortex. J Neurosci 19:9016–9028

Jancke D (2000). Orientation formed by a spot's trajectory: a two-dimensional population approach in primary visual cortex. J Neurosci 20 RC86:1–6

Jancke D, Chavane F, Naaman S, Grinvald A (2004a) Imaging correlates of visual illusion in early visual cortex. Nature 428:423–426

Jancke D, Erlhagen W, Schöner G, Dinse HR (2004b) Shorter latencies for motion trajectories than for flashes in population responses of primary visual cortex. J Physiol (London) 556:971–982

Jazayeri M, Movshon JA (2006) A new perceptual illusion reveals mechanisms of sensory decoding. Nat Neurosci 446:912–915

Jensen HJ, Martin J (1980) On localization of moving objects in the visual system of cats. Biol Cybern 36:173–177

Kanizsa G (1951). Sulla polarizzazione del movimento gamma. Archiva Psicologica Neurologica Psichiatrica 3:224–267

Kawahara J, Yokosawa K, Nishida S, Sato T (1996) Illusory line motion in visual search: attentional facilitation or apparent motion? Perception 25:901–920

Kenkel F (1913) Untersuchungen über den Zusammenhang zwischen Erscheinungsgröße und Erscheinungsbewegung bei einigen sogenannten optischen Täuschungen. Zeitschrift für Psychologie 67:358–449

Kerzel D, Gegenfurtner KR (2003) Neuronal processing delays are compensated in the sensorimotor branch of the visual system. Curr Biol 13:1975–1978

Kirschfeld K, Kammer T (1999) The Fröhlich effect: a consequence of the interaction of visual focal attention and metacontrast. Vision Res 39:3702–3709

Krekelberg B, Lappe M (1999) Temporal recruitment along the trajectory of moving objects and the perception of position. Vision Res 39:2669–2679
Krekelberg B, Lappe M, Whitney D, Cavanagh P, Eagleman DM, Sejnowski TJ (2000) The position of moving objects. Science 289:1107a
Krekelberg B, Lappe M (2001) Neuronal latencies and the position of moving objects. Trends Neurosci 24:335–339
MacKay DM (1958) Perceptual stability of a stroboscopically lit visual field containing self-luminous objects. Nature 181:507–508
Maiche A, Budelli R, Gomez-Sena L (2007) Spatial facilitation is involved in flash-lag effect. Vision Res 47:1655–1661
Mateeff S, Hohnsbein J (1988) Percepual latencies are shorter for motion towards the fovea than for motion away. Vision Res 28:711–719
Metzger W (1932) Versuch einer gemeinsamen Theorie der Phänomene Fröhlichs und Hazelhoffs und Kritik ihrer Verfahren zur Messung der Empfindungszeit. Psychologishe Forschung 16:176–200
Müsseler J, Aschersleben G (1998) Localizing the first position of a moving stimulus: The Fröhlich effect and an attention-shifting explanation. Percept Psychophys 60:683–695
Nakayama K, Mackeben M (1989) Sustained and transient components of focal visual attention. Vision Res 29:1631–1647
Nicolelis MA, Ghazanfar AA, Stambaugh CR, Oliveira LM, Laubach M, Chapin JK, Nelson RJ, Kaas JH (1998) Simultaneous encoding of tactile information by three primate cortical areas. Nat Neurosci 1:621–630
Nijhawan R (1994) Motion extrapolation in catching. Nature 370:256–257
Orban GA, Hoffmann KP, Duysens J (1985) Velocity selectivity in the cat visual system. I. Responses of LGN cells to moving bar stimuli: a comparison with cortical areas 17 and 18. J Neurophysiol 54:1026–1049
Petersen C, Grinvald A, Sakmann B (2003) Spatiotemporal dynamics of sensory responses in layer 2/3 of rat barrel cortex measured in vivo by voltage-sensitive dye imaging combined with whole-cell voltage recordings and neuron reconstructions. J Neurosci 23:1298–1309
Polat U, Sagi D (1993) Lateral interactions between spatial channels: suppression and facilitation revealed by lateral masking experiments. Vision Res 33:993–999
Polat U, Mizobe K, Pettet MW, Kasamatsu T, Norcia AM (1998) Collinear stimuli regulate visual responses depending on cell's contrast threshold. Nature 391:580–584
Posner MI, Snyder CRR, Davidson BJ (1980) Attention and the detection of signals. J Exp Psychol 109:160–174
Pulgarın M, Nevado A, Guo K, Robertson RG, Thiele A, Young MP (2003) Spatio-temporal regularities beyond the classical receptive field affect the information conveyed by the responses of V1 neurons. Soc Neurosci Abstracts 33:910.16
Purushothaman G, Patel SS, Bedell HE, Ogmen H (1999) Moving ahead through differential visual latency. Nature 396:424
Rao RPN, Ballard DH (1999) Predictive coding in the visual cortex: a functional interpretation of some extra-classical receptive field effects. Nat Neurosci 2:79–87
Sagi D, Julesz B (1986) Enhanced detection in the aperture of focal attention during simple discrimination tasks. Nature 321:693–695
Salinas E, Abbott LF (1994) Vector reconstruction from firing rates. J Comput Neurosci 1:89–107
Schlag J, Cai RH, Dorfman A, Mohempour A, Schlag-Rey M (2000) Extrapolating movement without retinal motion. Nature 403:38–39
Seriès P, Georges S, Lorenceau J, Frégnac Y (2002) Orientation dependent modulation of apparent speed: a model based on the dynamics of feed-forward and horizontal connectivity in V1 cortex. Vision Res 42:2781–2797
Sheth BR, Nijhawan R, Shimojo S (2000) Changing objects lead briefly flashed ones. Nat Neurosci 3:489–495
Shimojo S, Miyauchi S, Hikosaka O (1997) Visual motion sensation yielded by non-visually driven attention. Vision Res 37:1575–1580

Shoham D, Glaser DE, Arieli A, Kenet T, Wijnbergen C, Toledo Y, Hildesheim R, Grinvald A (1999) Imaging cortical dynamics at high spatial and temporal resolution with novel blue voltage-sensitive dyes. Neuron 24:791–802

Steinman BA, Steinman SB, Lehmkuhle S (1995) Visual attention mechanisms show a center-surround organization. Vision Res 35:1859–1869

Sterkin A, Arieli A, Ferster D, Glaser DE, Grinvald A (1999) Real time optical imaging in cat visual cortex exhibits high similarity to intracellular activity. Abstract of the 5th IBRO World Congress of Neuroscience, p 122

Szulborski RG, Palmer LA (1990) The two-dimensional spatial structure of nonlinear subunits in the RFs of complex cells. Vision Res 30:249–254

Tusa RJ, Rosenquist AC, Palmer LA (1979) Retinotopic organization of areas 18 and 19 in the cat. J Comp Neurol 185:657–678

van Beers RJ, Wolpert DM, Haggard P (2001) Sensorimotor integration compensates for visual localization errors during smooth pursuit eye movements. J Neurophysiol 85:1914–1922

Victor JD, Purpura K, Katz E, Mao B (1994) Population encoding of spatial frequency, orientation, and color in macaque V1. J Neurophysiol 72:2151–2166

von Grünau M, Faubert J (1994) Intra-attribute and interattribute motion induction. Perception 23:913–928

von Grünau M, Racette L, Kwas M (1996a) Measuring the attentional speed-up in the motion induction effect. Vision Res 36:2433–2446

von Grünau M, Dube S, Kwas M (1996b) Two contributions of motion induction: a preattentive effect and facilitation due to attentional capture. Vision Res 36:2447–2457

Wertheimer M (1912) Experimentelle Studien über das Sehen von Bewegung. Zeitschrift für Psychologie 61:161–265

Whitney D, Murakami I, Cavanagh P (2000) Illusory spatial offset of a flash relative to a moving stimulus is caused by differential latencies for moving and flashed stimuli. Vision Res 40:137–149

Chapter 6
Second-Order Motion Stimuli: A New Handle to Visual Motion Processing

Uwe J. Ilg and Jan Churan

Abstract This chapter addresses three different aspects related to visual motion processing by means of second-order motion stimuli. The first question discussed is whether there is a need for separate mechanisms underlying the execution of action and perception elicited by these motion stimuli. Second, light is shed on the neuronal responses to second-order motion stimuli recorded from the middle temporal (MT) and medial superior temporal (MST) areas. While neuronal responses are recorded, the monkeys performed a psychophysical task and reported the direction of stimulus movements. Third and final, the perception of biological motion in man and monkeys and its relationship to second-order motion is addressed.

6.1 Definition of First-Order and Second-Order Motion

As the content of this book demonstrates, the visual processing of motion has been extensively investigated in a strong alliance between psychophysical, electrophysiological, and theoretical approaches. Many models proved to be successful in accounting for the perception of moving patterns in which the luminance profile is shifted across space. These moving patterns are called first-order or Fourier motion. One specific model for the detection of such Fourier motion is the elementary motion detector of the correlation type (EMD), commonly referred to as Reichardt-detector, which uses the cross-correlation of the spatio-temporal luminance distribution (Reichardt 1987). This model is functionally equivalent to some other operators widely discussed in the psychophysical literature which are called energy models (Adelson and Bergen 1985; van Santen and Sperling 1985).

U.J. Ilg (✉)
Department of Cognitive Neurolgy, Hertie-Institute of Clinical Brain Research, Tuebingen, Germany
e-mail: uwe.Ilg@uni-tuebingen.de

U.J. Ilg and G.S. Masson (eds.), *Dynamics of Visual Motion Processing: Neuronal, Behavioral, and Computational Approaches*,
DOI 10.1007/978-1-4419-0781-3_6, © Springer Science+Business Media, LLC 2010

Neither of these models, however, is appropriate to explain motion perception elicited by non-Fourier or second-order motion stimuli (Chubb and Sperling 1988) in which the motion signal is defined by another stimulus quality other than luminance (e.g. flicker, contrast, or texture). In second-order stimuli, the mean luminance distribution is not shifted at all or is shifted in a direction different from the perceived one.

Two specific cases of second-order motion stimuli will be considered in this chapter. They are presented in the form of random dot kinematograms (RDKs), i.e., stimulus sequences of black and white dots distributed randomly in space. In drift-balanced motion (**dbm**), a region of static dots is embedded in a background of dynamic noise, in which all dots are replaced between successive frames of the RDK in a random fashion. When the region in which the dots do not move is shifted, there is no net displacement of the luminance distribution (Chubb and Sperling 1988; Lelkens and Koenderink 1984). Imagine you are watching a field of wheat from above in which a rabbit is running. You cannot see the rabbit itself; you can only see the random movement of the spears caused by the rabbit. This describes a **dbm** stimulus outside the laboratory.

Second, in theta motion (**tm**), the region of the traveling object is defined by dots moving coherently in a different direction as the object itself (Zanker 1993). You can observe **tm** under natural condition if the crowd in a filled soccer stadium starts the famous Laola wave. Each hand is moving upwards, but the wave itself moves horizontally. The **tm** stimulus offers the possibility of differentiating local from global motion signals. The movements of the individual dots represent a local motion signal, whereas the movement of the entire **tm** stimulus gives a global motion signal.

These two second-order motion stimuli are now compared to two conventional first-order motion stimuli. First, in the simplest case, a black bar is moving across the screen, which is filled with dynamic noise in the background. Since this stimulus provides motion energy on a rather coarse spatial scale, which can be readily extracted by simple EMDs, it will be labeled here as coarse Fourier motion or luminance defined motion (**lm**). Trivially, this luminance-defined target can also be perceived if it is stationary. Second, when a group of random dots is moving coherently in front of the dynamic noise background, the motion signal has to be extracted by EMDs on a finer spatial scale, and thus the stimulus is labeled as Fourier motion (**fm**). An **fm** stimulus can be observed in nature if a perfectly camouflaged animal starts to move. The motion signal, either first- or second-order, provides an input to structure-from-motion mechanisms according to gestalt theory (Koffka 1935; Wertheimer 1923). Successful processing of any of these motion signals is an important task since the motion signal is able to uncover even perfectly adapted camouflage which is frequently achieved by adaptation of skin color and luminance to the background.

The space-time histograms of **fm**, **dbm,** and **tm** stimuli are shown in Fig. 6.1. On the x-axis of these histograms, the luminance of a given row in the stimulus is shown. The change in luminance over time is given on the y-axis. However, it is much easier to see these stimuli on the attendant DVD.

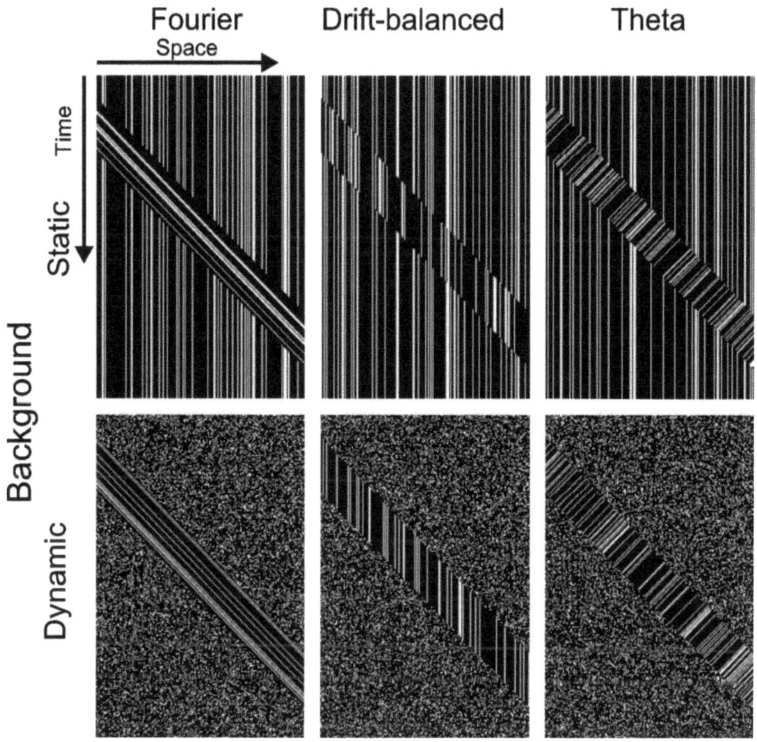

Fig. 6.1 Space-time-diagrams of the motion stimuli: Fourier motion, drift-balanced motion, and theta motion. An exemplary horizontal line of the stimulus is displayed as it develops over time. The *upper row* shows the motion stimuli presented in front of a stationary background, the *lower row* gives the stimuli presented in front of a dynamic background. Each stimulus can also be found on the DVD

6.2 Mechanisms Underlying Motion Perception and Motion-Dependent Action

An interesting question is whether there are separate or common mechanisms of visual motion processing for perception of second-order motion and for the generation of smooth-pursuit eye movements (SPEM). Influenced by the concept that the cortical visual system consists of two subsystems, a ventral stream responsible mainly for object identification (what) and a dorsal stream underlying the localization of objects (where) (Ungerleider and Mishkin 1982), it was proposed that there are also separate visual pathways for perception and action (Goodale and Milner 1992). Initially, this proposal was supported by the observation that a patient suffering from a visual-form agnosia showed very poor perception of objects. However, the patient was able to perform a manual grasping task on different objects correctly

(Goodale et al. 1991). In other words, the patient's ability to localize and grasp the target was normal (i.e. normal action), whereas her ability to perceive the form of the target (i.e. disturbed perception) was absent.

To address specifically the question whether separate systems are used for perception of visual motion and for the generation of SPEM, several experiments were performed in the past. In these experiments, the subject's perception was examined on the one hand using psychophysical techniques. On the other, the identical stimulus was used to elicit eye movements. A very typical stimulus in these studies consisted of an RDK with variable ratio of coherently moving dots superimposed onto randomly moving dots displayed within a stationary aperture (Newsome and Pare 1988). This variable ratio constituted the coherence of the motion signal. If all dots were moving to the right (100% coherence), the subject clearly perceived rightward movement. If all dots were moving randomly (0% coherence), the perception of the subject was at chance level. The reports of the subjects constituted the psychometric function, which described the chance of perceiving the correct direction of motion as a function of the percentage of coherently moving dots. At the same time, the eye movements of a subject exposed to these moving dots were recorded and analyzed. This analysis yielded the oculometric function of the subject. To determine this function, the direction of the elicited eye movements was determined within a time window of 500 ms as a function of the coherence of the moving dots. In order to analyze only the smooth eye movements, saccades were eliminated carefully. The obtained psychometric and oculometric functions were convincingly similar (Krauzlis and Adler 2001). In addition, a cue which signaled the direction of subsequent stimulus movement, affected perception and eye movements similarly (Krauzlis and Adler 2001). A common mechanism of visual motion processing for perception and generation of eye movements was also suggested by the similarity of oculometric and psychometric functions obtained when moving plaids were used (Krauzlis and Stone 1999), in case of the directional biases for elongated apertures (Beutter and Stone 1998), and when the effects of motion coherence (Beutter and Stone 2000) and motion integration (Stone et al. 2000) were addressed.

Another attempt to reveal whether perception and oculomotor control share a common mechanism of visual motion processing consists of the use of second-order motion stimuli as targets for smooth pursuit eye movements. Psychophysical experiments have shown that human subjects (Chubb and Sperling 1988; Zanker 1993) and monkeys (Churan and Ilg 2001) are able to detect and discriminate different types of second-order (**dbm** and **tm** as described earlier) correctly. So are human subjects able to track a moving stimulus defined by second-order motion? We asked our subjects to pursue moving first-order (**fm**) and second-order (**dbm**, **tm**) stimuli. During steady-state pursuit, the eyes followed precisely the object independent of its type (Butzer et al. 1997). Similar results were obtained from rhesus monkeys (Ilg and Churan 2004).

However, the onset of the eye movements revealed a strong dependency on the stimulus type: The latency of the initial saccades was shortest (209 ms) for the luminance defined stimulus while second-order motion stimuli revealed a significantly longer saccadic latency of 260 ms (Butzer et al. 1997). In agreement with these results, psychophysical experiments on motion perception have shown, that second-order motion stimuli have to be presented longer than first-order motion stimuli for successful direction discrimination (Derrington et al. 1993; Yo and Wilson 1992). This similarity in the dependence of latency and stimulus duration on the type of motion stimulus also indicates common visual motion processing for perception and action.

In contrast to the steady-state pursuit, initiation of pursuit depends massively on Fourier motion. As Fig. 6.2 shows, when the Fourier component and the second-order component of the stimulus are moving in opposite directions (as in the **tm** stimulus), the eyes start to move in the direction of the Fourier component even when the later steady-state pursuit follows the second-order motion. In case of **dbm**, the acceleration of the eyes is clearly reduced, and the latency is increased for this subject and stimulus condition. However, for the grand average across all subjects used in this study, the pursuit onset latency did not depend on stimulus type (at least for the small and slowly moving target) (Lindner and Ilg 2000). The importance of the first-order or Fourier component for pursuit initiation was also documented recently by the missing fundamental stimulus that elicited an ocular following

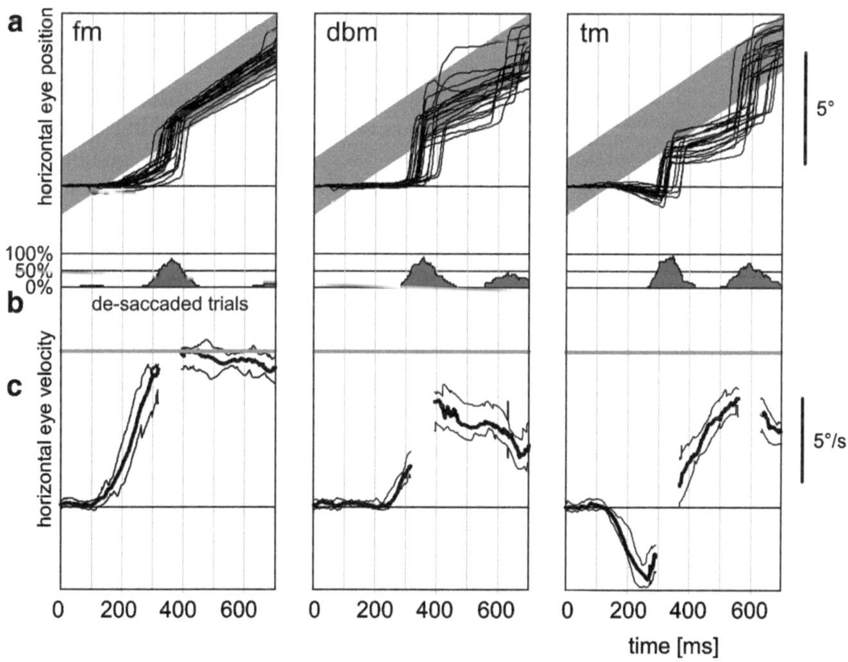

Fig. 6.2 Initiation of smooth-pursuit eye movement elicited by Fourier motion (**fm**), drift-balanced motion (**dbm**), and theta motion (**tm**). (**a**) shows single trial eye position traces of a typical subject together with the stimulus position indicated as grey band. Time zero represents the target motion onset. (**b**) gives the saccade histograms. (**c**) shows the median and 1st and 3rd quartile of the de-saccaded eye velocity profiles (modified from (Lindner and Ilg 2000))

response in the opposite direction of the pattern movement (see chapter 7 by Frederick Miles) (Sheliga et al. 2005). In addition, the dependency of pursuit initiation on the Fourier motion signal is very similar to the directional error during pursuit initiation obtained by tilted contours (see chapters by Masson et al and Born). When instantaneous pursuit velocity was compared with a low-level and a high-level speed estimation task, it emerged that pre-saccadic pursuit reflects the output of a low-level, motion energy based mechanism whereas post-saccadic pursuit is dominated by a high-level, position-tracking mechanism (Wilmer and Nakayama 2007). Obviously, pursuit initiation is dominated by need-for-speed characteristics. It is more important to initialize pursuit as fast as possible as to direct the eyes in the correct direction of target movement.

When motion perception and generation of smooth-pursuit is compared, it is very close to the examination of the effects of the motion after-effect (MAE) or waterfall illusion. It is well known that prolonged viewing of motion elicits an apparent motion in the opposite direction (see example on DVD). However, second-order motion stimuli are much less able to elicit a strong MAE compared to first-order motion [for review see (Mather et al. 1998). So we asked whether prolonged viewing of motion affects not only motion perception but also the initiation of pursuit eye movements. We ran psychophysical experiments in which three subjects had to determine their point of subjective stationarity (PSS) of a random dot pattern presented for 200 ms. If the stimulus was preceded by either 2 or 4 s of motion at 12° per second, PSS was shifted approx. 12 min of arc per second in the adapted direction. In other words, if the adaptation was towards the left, the test stimulus had to move at 12 min/s leftwards to be perceived as being stationary. When we performed pursuit experiments with the same three subjects, pursuit onset latency depended on the direction of adaptation: 171 ms if adaptation was in same direction, 163 ms if adaptation was in opposite direction of the pursuit target. This difference was statistically significant (ANOVA, $p=0.0043$). Interestingly, initial saccade latency, initial acceleration, or steady state pursuit gain were not affected by the adaptation. Very similar findings were obtained for pursuit initiation in monkeys when post-saccadic eye velocity was analyzed (Gardner et al. 2004). So our own data together with data from the literature suggest that adaptation to motion affects motion perception in a similar manner as the initiation of smooth-pursuit.

In conclusion, the results of several experiments with respect to visual motion processing suggest that there is no need to assume separate mechanisms underlying the execution of smooth-pursuit eye movements and motion perception.

6.3 Neuronal Substrate of Second-Order Motion Processing

It is widely accepted that motion processing in primates is a cortical feature. This is different from other vertebrates which possess directionally selective ganglion cells, i.e. motion processing takes place within the retina (for review see (Wassle 2004). Within the hierarchy of the primate visual system, V1 neurons are the first stage which displays directional selectivity (Hubel and Wiesel 1968). Most likely these directionally selective neurons provide the direct input (Movshon and Newsome 1996) to the middle temporal (MT) area, located on the posterior bank of the superior temporal sulcus (STS) (Allman and Kaas 1971; Dubner and Zeki 1971). Directionally selective neurons in STS are not confined to MT but also dominate neighboring medial superior temporal area (MST), located on the floor and the anterior bank of the sulcus (Maunsell and van Essen 1983; Ungerleider and Desimone 1986). Although neurons in area MST are also sensitive to retinal image motion, they show a number of features which distinguish them from the kind of pure elementary motion detector neurons characterizing area MT.

Before comparing neuronal responses to first-order and second-order motion recorded from rhesus monkeys, the experiments addressing these processes in human subjects are shortly described. It was shown that different types of second-order stimuli could be easily recognized by human subjects. Additionally, subjects were able to discriminate the motion direction as well as execute SPEM to second-order motion stimuli in the same way as for first-order stimuli (Butzer et al. 1997). On the other hand simple motion-energy detectors were not able to detect second-order motion. Therefore the question arises, whether first-order and second-order stimuli are processed by the same or by different neuronal mechanisms.

Some psychophysical observations suggested a similarity in the processing of the two types of stimuli. Smith and colleagues (Smith et al. 1994) have shown that the thresholds for detection of orientation and direction of motion are very similar for second-order stimuli, which is also true for first-order stimuli (Green 1983; Watson et al. 1980). Nishida (Nishida 1993) observed that perception of Phi motion (Anstis 1970) can be produced similarly by switching the second-order properties on a display while keeping the luminance constant, instead of inverting the luminance profile of the display.

In contrast, other investigations also emphasized the differences in the perception of the two types of motion. It was shown, that subjects required more time to discriminate the direction of motion for second-order stimuli (Derrington et al. 1993). Other experiments have demonstrated that adaptation to first- or second-order motion, respectively, influenced the perception only for the motion type which was adapted (Nishida et al. 1997). Further, Harris and Smith (Harris and Smith 1992) have shown that second-order motion stimuli were not able to elicit an optokinetic nystagmus. Although the evidence from psychophysical experiments in favor of or against a common processing mechanism was mixed, other evidences were raised from physiological investigations. Patzwahl and colleagues (Patzwahl et al. 1993) reported that visually evoked potentials differ for first-order and second-order stimuli. In a clinical study, one patient with lesions in areas V2 and V3 was described to have a selective impairment of first-order motion perception but a preserved perception of second-order motion stimuli (Vaina et al. 1999; Vaina et al. 1998). More differences in the processing of first-order and second-order stimuli were found using functional magnetic resonance imaging (fMRI). Area V3 was shown to be stronger activated by second-order motion than by first-order (Smith et al. 1998). Another area was found in the extrastriate cortex, which responded selectively to motion defined contours and could therefore also be involved in second-order motion processing (area KO: Van Oostende et al. 1997).

Taken together, the vast majority of findings suggest a separate processing of first-order and second-order motion. However, the different processing pathways may converge at some point to establish a common percept of motion. It was suggested, that this converging point for a *'Form-cue invariant motion processing'* might be the area MT (Albright 1992), since neurons were found in this area, which responded to different types of first-order and second-order motion stimuli.

In order to examine the neuronal substrate of second-order motion processing in single-unit recordings, it is necessary to document that rhesus monkeys perceive these stimuli the same way as human subjects do. One way to prove that monkeys are able to perceive second-order stimuli is the execution of smooth-pursuit eye movements. Only if the monkeys are able to process the motion signal, are they able to execute SPEM to moving second-order objects. As we found, monkeys are able to pursue different types of second-order motion stimuli (Ilg and Churan 2004). However, we could not use SPEM during our recordings from areas MT and MST since it is known that neurons in MST respond to eye movements (Ilg et al. 2004). Therefore, we performed a direction discrimination task as behavioral control during the recordings. During fixation of a stationary target, a moving stimulus was presented. After a first-order or second-order stimulus was presented moving in one of two alternative directions, the monkeys had to perform a saccade towards one of two simultaneously presented saccade targets. The selected target signaled the perceived direction of stimulus movement. Figure 6.3a shows the responses of a typical MST neuron to the presentation of **fm**, **dbm,** and **tm** stimuli in preferred and non-preferred direction.

We recorded data from 106 neurons in areas MT and MST all of which responded significantly to the **fm** stimulus. As for the neuron whose response is shown in Fig. 6.3a, 58 neurons have shown a significant response to the **dbm** stimulus. We did not find any other visual response properties which correlated with the ability of a neuron to code for **dbm;** neither did we find any clustering of neurons responsive to **dbm.** This finding supports results reported by others (Albright 1992; O'Keefe and Movshon 1998) using similar types of motion stimuli.

For **tm** stimulus, the example neuron apparently inverted its preferred direction. This means that the neuron did not respond to the global motion component of the **tm** stimulus, but responded only to the first-order motion component which was in the opposite direction to the object motion. So local motion signals determined the

Fig. 6.3 (continued) direction discrimination task (**a**). The neuronal activity is shown as raster display and as spike density function (kernel width 25 ms). The vertical dotted lines inform about the size of the receptive field. Note that the monkey discriminated correctly the motion direction in each single trial. (**b**) shows the directional index (DI) calculated from the responses of all 106 neurons to **fm**, **dbm,** and **tm**. Since the preferred direction to tm was apparently inverted, the DI elicited by tm is negative. To facilitate comparison, the dashed bar shows the inverted value in this condition. (**c**) shows the initial eye acceleration of the monkey tracking the **fm**, **dbm**, or **tm** stimulus, respectively (modified from (Ilg and Churan 2004))

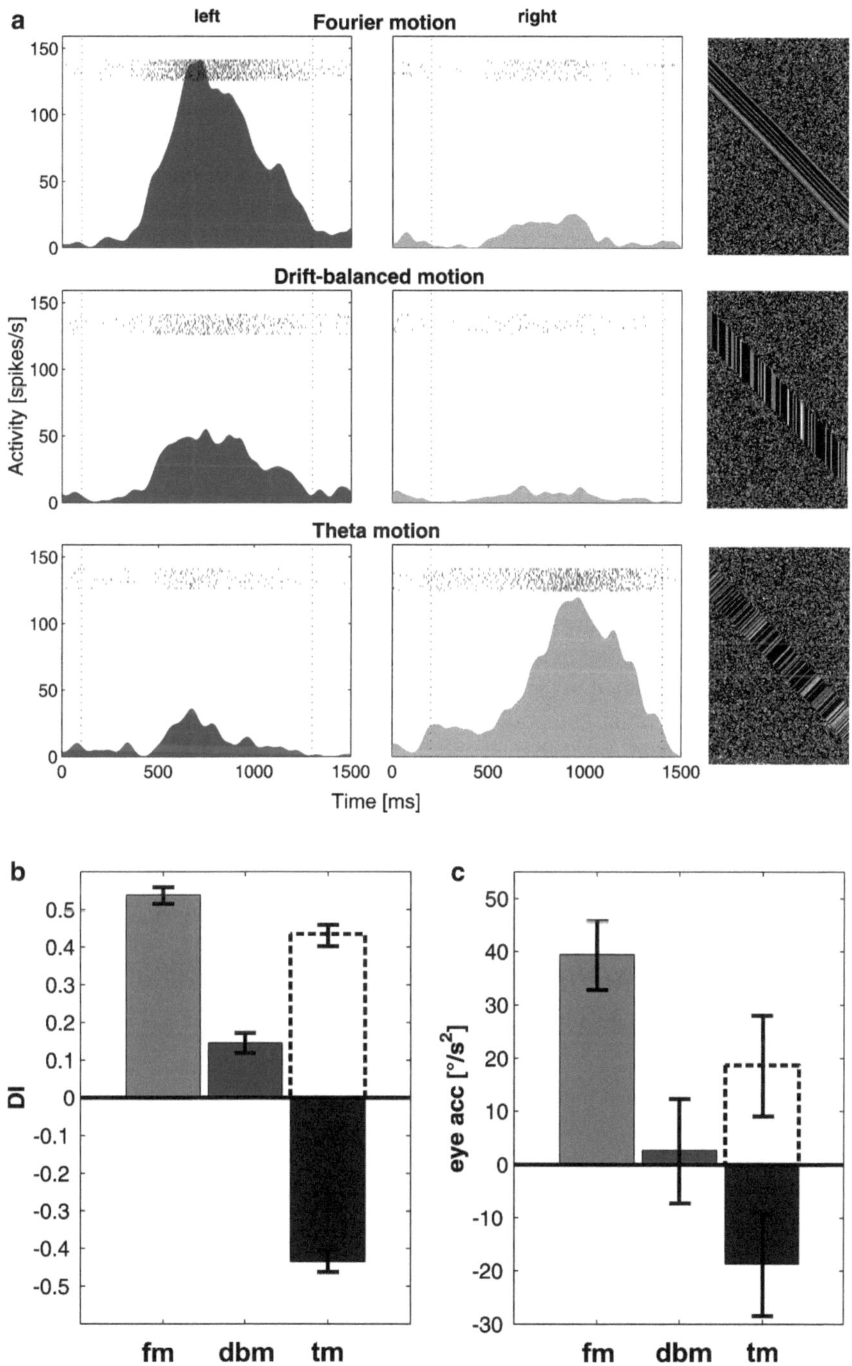

Fig. 6.3 Response of a typical neuron recorded from area MST to leftward (preferred) and rightward (non-preferred) first-order, drift-balanced, and theta motion while the monkey was performing a

neuronal responses recorded from MT and MST and not the global motion signal. Movshon and colleagues came to the same conclusion using a slightly different stimulus which consisted of Garbor patches (Majaj et al. 2007).

When the responses to **fm** and **tm** are compared, one might assume that they are simply mirror inverted. However, the directional selectivity of our entire population of neurons expressed as DI[1] is larger for **fm** as compared to **tm** (see Fig. 6.3b). Parallel to the directional selectivity of the neuronal responses, the initial eye acceleration during smooth-pursuit directed towards a **tm** stimulus is smaller as compared to **fm** (see Fig. 6.3c). So both, neural activity and eye movements were not totally determined by raw retinal image motion signals but reflected to a certain extent interactions between local and global motion signals. These interactions are possibly quite simple: the life time of the individual dots is much shorter in the **tm** stimulus compared to the **fm** stimulus. However, the idea of interactions between the processing of first-order and second-order motion was documented by a recent study, too. Experiments performed in anaesthetized marmorsets showed that responses in V1, V2, and V3 to first-order motion stimuli were modulated by second-order stimuli presented simultaneously (Barraclough et al. 2006a). The fact that second-order motion modulates the response to first-order motion suggests that processing of both types of motion depends on common mechanisms.

Areas MT and MST are retinotopically organized. Even if single neurons in these areas do not respond to the object motion component in the tm stimulus, this motion is still represented in the neuronal population as a moving '*wave of activity*' caused by the first-order component. It is close at hand to imagine that this wave can be implemented by coherent discharges across populations of neurons. According to a two-stage model of second-order processing (Zanker 1993), MT and MST represent the first stage in which the first-order component of the motion is coded. It remains for future studies to determine which area upstream of MT and MST displays responses to code for object motion.

6.4 Is There a Multimodal Representation of Motion (in MT and MST)?

Since it has been shown that human subjects can perform residual SPEM towards a moving sound source or moving tactile stimulus (Berryhill et al. 2006; Hashiba et al. 1996), and the activity of neurons in MT and MST is closely related to the execution of SPEM, it is possible to speculate that the neuronal activity in areas MT and MST might code in general for a moving stimulus, independent of its modality. In a series of experiments, we asked whether MT and MST neurons were activated similarly by a moving visual and a moving auditory stimulus. It is important to note that after some training, monkeys were easily able to report the direction of the

[1] $\mathbf{DI} = 1 - \text{activity}_{\text{non-preferred direction}}/\text{activity}_{\text{preferred direction}}$.

moving acoustic stimulus. To avoid technical problems related to the mechanics of a moving loudspeaker, we applied apparent motion stimuli. In a linear arrangement of 48 loudspeakers (spacing of 1°), subsequent speakers were activated every 50 ms resulting in an apparent velocity of 20° per second.

However, no response to the moving sound source was observed as indicated by the responses of a typical MST neuron depicted in Fig. 6.4. We performed this experiment on 32 neurons recorded from areas MT and MST of three different monkeys. Each neuron gave a very clear and directionally selective response to a moving visual stimulus. Not a single neuron gave a response to the moving sound source. Despite this lack of neuronal response, the monkeys responded to almost every presentation of the stimulus correctly. This absence of response was rather surprising compared

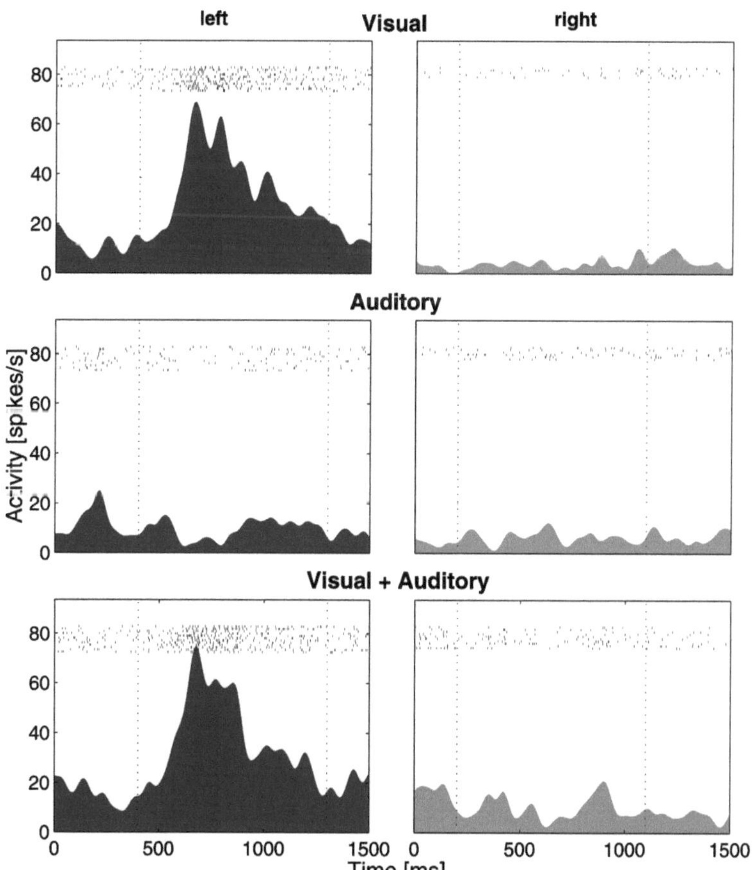

Fig. 6.4 Response of a typical MST neuron to visual, auditory, and audio-visual apparent motion. The neuronal activity is shown as raster display and spike density function (kernel width 25 ms). Vertical dotted lines give the size of the receptive field. The monkey performed a direction discrimination task while data was recorded; all trials showed correct responses (modified from (Ilg and Churan 2004))

to other electrophysiological findings. With respect to multi-sensory processing of visual and auditory information, it has been shown that neurons recorded from the superior temporal sulcus (STS), especially from its anterior parts, responded to stimulation in both modalities (Benevento et al. 1977; Bruce et al. 1981).

Another line of evidence suggesting general coding builds on results obtained from symbolic motion cues. After a monkey had learned to understand a static symbol (e.g. an arrow) as a cue representing a specific motion direction, directionally selective neurons in area MT responded to the presentation of this cue. In contrast, before the training, neurons did not respond to the presentation of the cue (Schlack and Albright 2007). This acquired responsiveness observed in MT resembles the responsiveness of neurons recorded from the lateral intraparietal area (LIP) to auditory stimulation. These neurons responded to an auditory stimulus only if this stimulus was used as a saccade target. This auditory response was only observable after training while in naïve monkeys, there was no response to auditory stimulation (Grunewald et al. 1999).

Taken together, our results clearly confirm previous findings indicating that areas MT and MST are not the final stages in the cortical processing of motion. Areas higher up in the hierarchy of visual motion processing such as the anterior and posterior part of the superior temporal poly-sensory area (STPa and STPp) (for overview see (Felleman and van Essen 1991), area 7A, the ventral intraparietal area (VIP), and the cortical vestibular areas (e.g. as the parieto-insular vestibular cortex PIVC and area 2v) may be involved in a truly multi-modal, general, and form-cue invariant representation of a moving object. VIP may be another hot candidate for multi-modal motion processing since neurons recorded here displayed responses to somatosensory, vestibular, and visual stimulation (Bremmer et al. 2002; Duhamel et al. 1998). In addition, VIP stands as example for multi-sensory area since its existence was also shown in the human brain by means of functional brain imaging techniques (Bremmer et al. 2001).

6.5 Dynamic Flicker in the Background Reduces Perception and Directional Selectivity

So far, neither the neuronal responses to second-order motion stimuli nor the responses to moving sound sources suggest a strong relationship between perception of these moving stimuli and firing rates observed in MT and MST. In contrast to our results, there are numerous reports in the literature emphasizing the important role of areas MT and MST for motion perception. It has been shown that single-unit activity corresponds to motion perception (Bisley and Pasternak 2000; Britten et al. 1996; Celebrini and Newsome 1994), experimental lesions yielded a clear deficit in motion perception (Newsome and Pare 1988; Rudolph and Pasternak 1999), and finally, micro-stimulation affected motion perception (Bisley et al. 2001; Celebrini and Newsome 1995; Ditterich et al. 2003).

A very recent brain imaging study showed that the BOLD response of human MT is closely matched to the perceived direction of an ambiguous stimulus whereas earlier visual areas (V1 to V4v) matched the physical presence of motion (Serences and Boynton 2007). So it is rather surprising that our data presented so far shows a rather weak correlation between motion perception and firing rates in MT and MST. However, we were also able to document the strong relation between neuronal activity in MT and MST and motion perception when we slightly changed the paradigm. Instead of presenting the moving stimulus in front of a static background, we added a dynamic background. This offers the possibility to examine the influence of the temporal structure of the background on motion perception and single-unit responses.

First, we asked whether the temporal structure of the background influences motion perception. With respect to human subjects, we were able to show that the discrimination thresholds (as measured by the duration of stimulus presentation necessary to obtain 75% correct responses in the direction discrimination task) increased with increasing flicker density of the background (see Fig. 6.5a). With respect to our monkeys, we found that the percentage of correct responses dropped if the stimulus was presented in front of a dynamic background compared to a static background (see Fig. 6.5b). Parallel to this reduction, the reaction time increased. The effect of flicker on sensory and sensorimotor performance has been shown psychophysically in several earlier investigations (Baccino et al. 2001; Macknik et al. 1991; Scase et al. 1996). It has been reported that flicker increases the perceived velocity of slowly moving random dot patterns; this effect was named `*temporal capture*' (Treue et al. 1993). However, at higher stimulus velocities, no influence of flicker was reported (Zanker and Braddick 1999).

Second, we asked whether the responses from MT and MST are also modulated by the temporal structure of the background. It was shown earlier that MT neurons respond weaker to a transparent stimulus (RDK which contains motion of dots in preferred and non-preferred direction) compared to an unidirectional stimulus (Snowden et al. 1991). In our experiments, we used a random-dot background with the same average luminance as the moving stimulus, in which the dots were replaced in every frame constituting a dynamic flicker. Since a dynamic background contains motion energy in all directions, the neuronal responses to moving stimuli presented in front of a dynamic background are reduced compared to responses elicited by stimuli moving across a stationary background. We quantified the sharpness of the directional tuning of each neuron by d' calculated upon the responses to stimuli moving in preferred and non-preferred direction, respectively.[2] Figure 6.5c shows that the directional selectivity of neuronal responses was in fact reduced by the presence of a dynamic background.

The aforementioned modulation of neuronal responses in visual cortex due to second-order motion (Barraclough et al. 2006a) can also be explained by additional flicker provided by the second-order motion stimulus.

[2] $\mathbf{d'} = 2\times(\text{mean activity}_{\text{pref}} - \text{mean activity}_{\text{non-pref}})/(\text{std activity}_{\text{pref}} + \text{std activity}_{\text{non-pref}})$.

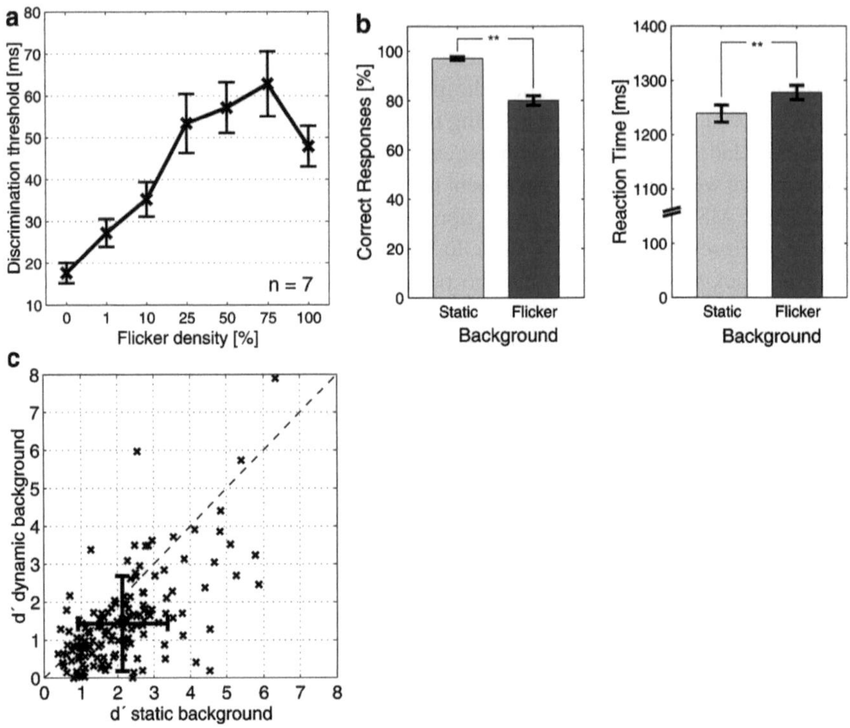

Fig. 6.5 Effects of dynamic background on perception and neuronal responses. (**a**) shows the dependency of the discrimination thresholds from seven subjects on the flicker density. 0% flicker density represents the static background condition. The subjects' perception of a moving stimulus is affected by different strength of flicker in the background. (**b**) gives the percentage of correct responses as well as the reaction time of a monkey performing the direction discrimination task (DDT) with static and dynamic background. (**c**) The directional discriminability of the neuronal tuning expressed as d' is shown for static and dynamic background conditions. The cross gives the mean d' for static and dynamic background together with standard deviations (modified from (Churan and Ilg 2002))

6.6 Complex Perception of Structure from Motion: Biological Motion

In the beginning of this chapter, we addressed the issue that motion information provides input to the perception of object form known as Gestalt (Koffka 1935; Wertheimer 1923). In the simplest case, the form of the object is defined by coherent motion. In addition, the perception of depth is influenced by motion parallax which is the retinal image motion caused by self-motion. This motion parallax strongly depends on the distance between object and observer (Braunstein 1962). More complex is the perception biological motion (Jansson and Johansson 1973; Johansson 1976, 1975). Even if a human walker is reduced to only 10 points (point light walker PLW), it elicits a very vivid perception as long as the walker is in motion. It is possible to determine the sex of the walker, the executed action, or its emotional state (for review see (Troje 2002). Despite this very vivid perception, the Gestalt perception immediately vanishes if the motion display stops. Similarly, if only still pictures of the PLW are shown, a naïve subject cannot perceive the walker immediately as he can for the moving display. For more details see chapter 14 by Martin Giese and colleagues.

6.6.1 Perception of Biological Motion by Man and Monkey

Using the method of preferential looking, it was found that even 4 to 6-month old infants preferred PLW compared to a noise pattern (Fox and McDaniel 1982). The authors speculated that the ability to perceive biological motion might be largely intrinsic rather than acquired slowly through experience. If that is true, the next question arises whether the perception of biological motion is restricted to the human species. Specifically, we asked whether rhesus monkeys are able to perceive biological motion and whether the monkeys' perception follows similar pattern as in humans.

The point light walker (PLW) used in our experiments had the form of a leftward or rightward walking human which consisted of 10 points The data set was kindly provide by Martin Giese. The PLW had a total height of 10° and one gait cycle had the duration of 600 ms. We trained the monkey on direction discrimination of a luminance defined bar initially. After a successful training (performance higher as 80%), we displayed the PLW moving to the left or right, respectively, and trained the monkey again in direction discrimination of the PLW. After success in the second stage of training, we displayed the PLW without horizontal displacement. Figure 6.6a gives the amount of correct responses for stimulus presentations from 100 to 900 ms.

So there is experimental evidence that the monkey was able to perceive the human walker since he was able to report correctly the direction of movement from the PLW walking in place. If the monkey's perception is similar to the perception of human subjects, it should be possible to degrade this perception by small modifications of the stimulus. From humans it is known that the perception of biological motion can be specifically degraded if the stimulus is presented upside down (Pavlova and Sokolov 2000). We asked the monkey to report the direction of PLW presented upright and upside down. As Fig. 6.6b shows, the monkey's performance dropped down substantially in case of the inverted PLW. Another possibility of degrading the perception of the PLW was to modulate the trajectory of each of the 10 dots. We simply added position noise to trajectory each dot. The performance of the monkey dropped with increasing amplitude of noise as shown in Fig. 6.6c. Finally, we added a variable horizontal displacement to PLW walking originally in place. If the displacement was in the opposite direction of the PLW, human observers described the stimulus as Michael Jackson's moon-walk (see DVD). The performance of the monkey depended on this translational component of velocity up to 7° per second and returned to normal values for decreasing translational components (see Fig. 6.6d).

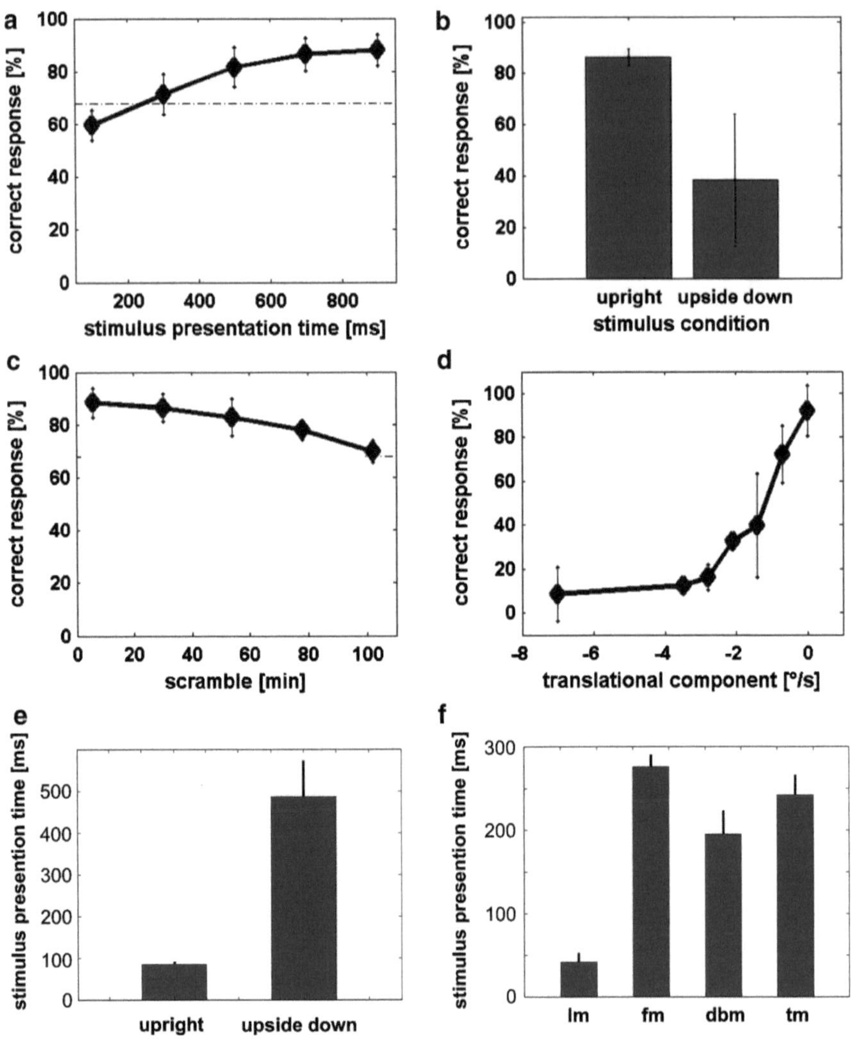

Fig. 6.6 Details of perception of biological motion, (**a**) to (**d**) for a monkey, (**e**) and (**f**) for human subjects. (**a**) gives the percentage of correct responses dependent on duration of stimulus presentation. The dash-dotted line represents the perceptual threshold at 68% correct responses. (**b**) shows the percentage of correct responses for the upright and inverted presentation of the PLW; stimulus presentation was 500 ms. **C** gives details of the relationship between the percentage of correct responses and the amount of stimulus disturbance. PLW presentation was 500 ms. (**d**) shows the dependence of the response on the horizontal displacement of the stimulus (500 ms presentation). (**e**) shows the mean perceptual thresholds of 9 human observers for upright and upside down presented PLW defined by luminance. (**f**) shows the perceptual thresholds of the same human observers for PLW defined by luminance, first-order motion, drift-balanced motion, and theta motion, respectively. In each diagram, mean and standard errors are shown. Note that all stimuli can be found on the DVD.

Taken together, our data strongly support the idea that monkeys are able to perceive biological motion, with similar characteristics as a human observer. This result shows the validity of the monkey model for human perception of biological motion. It in particular increases the relevance of the single-unit responses from the anterior part of STS in monkeys using biological motion stimuli (Oram and Perrett 1994). Neurons in this part of the monkey brain also respond to still images of humans in motion (Barraclough et al. 2006b).

6.6.2 Biological Motion Based on Second-Order Motion

Recently, it was suggested that the perception of biological motion depended primarily on form and just to a lesser extent on motion information (Beintema et al. 2006; Beintema and Lappe 2002). One of the arguments in favor of the form information is the perception of biological motion disappearing if the stimulus is presented upside down as mentioned above (Pavlova and Sokolov 2000). In order to shed some light on the form versus motion discrepancy, we asked whether biological motion can also be perceived if the individual dots of a PLW are not defined by luminance, but by second-order motion. If the dots are not defined by luminance, motion processing is a necessary first step in the processing of the biological motion stimulus.

We used the same stimuli as in the monkey experiments described above. Each of the 10 patches constituting the PLW had the size of 10 by 10 pixels (approx 1°). Each patch could be either defined by luminance, Fourier motion, drift-balanced motion, or theta motion, respectively. The PLW was always displayed without any horizontal displacement, i.e. the PLW was walking in place.

First, we investigated the importance of the appropriate orientation of the stimulus. As shown in Fig. 6.6e, human subjects were able to report the gait direction of a luminance defined PLW when it was presented only for approximately 100 ms. If the stimulus was inverted, the threshold increased to approx. 500 ms. In the next experiment, every single point of the PLW was defined either by luminance, coherent dot motion (first-order), drift-balanced motion, or theta motion (see Fig. 6.6f). Note that all stimuli can also be seen on the DVD. A Kruskal-Wallis test revealed that the influence of the stimulus type on the perceptual threshold was highly significant ($p=0.00006$). When we applied a post-hoc test, only the differences in thresholds obtained from the luminance defined stimulus were significantly different from the thresholds obtained for all other stimuli ($p<0.027$).

The fact that human subjects are able to perceive a PLW defined by motion cues emphasizes the importance of a motion pre-processor for the processing of biological motion stimuli. Without this early motion processing, no input signal for a form analyzing mechanism would be available. So motion processing seems to be a necessary first step for the perception of biological motion.

6.7 Supplementary Materials (CD-ROM)

Movies 1-4 Motion stimuli presented in front of a stationary background (files "*6_M1_movie_stat_lm.avi*", "*6_M2_movie_stat_fm.avi*", "*6_M3_movie_stat_dbm.avi*", "*6_M4_movie_stat_tm.avi*")

6_M1_movie_stat_lm.avi: a luminance defined (i.e. black) rectangle moves from left to right

6_M2_movie_stat_fm.avi: a fourier motion defined rectangle moves from left to right

6_M3_movie_stat_dbm.avi: a drift balanced motion defined rectangle moves from left to right

6_M4_movie_stat_tm.avi: a theta motion defined rectangle moves from left to right

Movies 5-8 Motion stimuli presented in front of a dynamic background (files *"6_M5_movie_dyn_lm.avi", "6_M6_movie_dyn_fm.avi", "6_M7_movie_dyn_dbm.avi", "6_M8_movie_dyn_tm.avi")*

6_M5_movie_dyn_lm.avi: a luminance defined (i.e. black) rectangle moves from left to right

6_M6_movie_dyn_fm.avi: a fourier motion defined rectangle moves from left to right

6_M7_movie_dyn_dbm.avi: a drift balanced motion defined rectangle moves from left to right

6_M8_movie_dyn_tm.avi: a theta motion defined rectangle moves from left to right

Movie 9 Motion after-effect (file "*6_M9_movie_mae.avi*") Demonstration of the motion after-effect.

Movies 10-13 Point-light walker presented in front of a stationary background (files *"6_M10_walker_stat_lm.avi","6_M11_walker_stat_fm.avi","6_M12_walker_stat_dbm.avi", "6_M17_walker_dyn_tm.avi")*

6_M10_walker_stat_lm.avi: each dot of the walker is defined by color (i.e. black)

6_M11_walker_stat_fm.avi: each dot of the walker is defined by Fourier motion

6_M12_walker_stat_dbm.avi: each dot of the walker is defined by drift balanced motion

6_M13_walker_stat_tm.avi: each dot of the walker is defined by theta motion

Movies 14-17 Point-light walker presented in front of a stationary background (files *"6_M14_walker_dyn_lm.avi", "6_M15_walker_dyn_fm.avi", "6_M16_walker_dyn_dbm.avi", "6_M17_walker_dyn_tm.avi")*

6_M14_walker_dyn_lm.avi: each dot of the walker is defined by color (i.e. black)

6_M15_walker_dyn_fm.avi: each dot of the walker is defined by Fourier motion

6_M16_walker_dyn_dbm.avi: each dot of the walker is defined by drift balanced motion

6_M17_walker_dyn_tm.avi: each dot of the walker is defined by theta motion

Movies 18 A walker with opposite linear displacement (file "*6_M18_moonwalker.avi*") Point light walker added to a linear displacement in the opposite direction

References

Adelson EH, Bergen JR (1985) Spatiotemporal energy models for the perception of motion. J Opt Soc Am A 2:284–299

Albright TD (1992) Form-cue invariant motion processing in primate visual cortex. Science 255:1141–1143

Allman JM, Kaas JH (1971) A representation of the visual field in the caudal third of the middle tempral gyrus of the owl monkey (Aotus trivirgatus). Brain Res 31:85–105

Anstis SM (1970) Phi movement as a subtraction process. Vis Res 10:1411–1430

Baccino T, Jaschinski W, Bussolon J (2001) The influence of bright background flicker during different saccade periods on saccadic performance. Vis Res 41:3909–3916

Barraclough N, Tinsley C, Webb B, Vincent C, Derrington A (2006a) Processing of first-order motion in marmoset visual cortex is influenced by second-order motion. Vis Neurosci 23:815–824

Barraclough NE, Xiao D, Oram MW, Perrett DI (2006b) The sensitivity of primate STS neurons to walking sequences and to the degree of articulation in static images. Prog Brain Res 154:135–148

Beintema JA, Georg K, Lappe M (2006) Perception of biological motion from limited-lifetime stimuli. Percept Psychophys 68:613–624

Beintema JA, Lappe M (2002) Perception of biological motion without local image motion. Proc Natl Acad Sci USA 99:5661–5663

Benevento LA, Fallon J, Davis BJ, Rezak M (1977) Auditory–visual interaction in single cells in the cortex of the superior temporal sulcus and the orbital frontal cortex of the macaque monkey. Exp Neurol 57:849–872

Berryhill ME, Chiu T, Hughes HC (2006) Smooth pursuit of nonvisual motion. J Neurophysiol 96:461–465

Beutter BR, Stone LS (1998) Human motion perception and smooth pursuit eye movements show similar directional biases for elongated apertures. Vis Res 38:1273–1286

Beutter BR, Stone LS (2000) Motion coherence affects human perception and pursuit similarly. Vis Neurosci 17:139–153

Bisley JW, Pasternak T (2000) The multiple roles of visual cortical areas MT/MST in remembering the direction of visual motion. Cereb Cortex 10:1053–1065

Bisley JW, Zaksas D, Pasternak T (2001) Microstimulation of cortical area MT affects performance on a visual working memory task. J Neurophysiol 85:187–196

Braunstein ML (1962) The perception of depth through motion. Psychol Bull 59:422–433

Bremmer F, Klam F, Duhamel JR, Ben Hamed S & Graf W (2002) Visual-vestibular interactive responses in the macaque ventral intraparietal area (VIP). *European Journal of Neuroscience* 16: 1569-1586., 2002.

Bremmer F, Schlack A, Shah NJ, Zafiris O, Kubischik M, Hoffmann K, Zilles K, Fink GR (2001) Polymodal motion processing in posterior parietal and premotor cortex: a human fMRI study strongly implies equivalencies between humans and monkeys. Neuron 29:287–296

Britten KH, Newsome WT, Shadlen MN, and Celebrini S (1996) A relationship between behavioral choice and the visual responses of neurons in macaque MT. Vis Neurosci 13:87–100

Bruce C, Desimone R, Gross CG (1981) Visual properties of neurons in a polysensory area in superior temporal sulcus of the macaque. J Neurophysiol 46:369–384

Butzer F, Ilg UJ, Zanker JM (1997) Smooth-pursuit eye movements elicited by first-order and second-order motion. Exp Brain Res 115:61–70

Celebrini S, Newsome WT (1995) Microstimulation of extrastriate area MST influences performance on a direction discrimination task. J Neurophysiol 73:437–448

Celebrini S, Newsome WT (1994) Neuronal and psychophysical sensitivity to motion signals in extrastriate area MST of the macaque monkey. J Neurosci 14:4109–4124

Chubb C, Sperling G (1988) Drift-balanced random stimuli: a general basis for studying non-fourier motion perception. J Opt Soc Am 5:1986–2007

Churan J, Ilg UJ (2002) Flicker in the visual background impairs the ability to process a moving visual stimulus. Eur J Neurosci 16:1151–1162

Churan J, Ilg UJ (2001) Processing of second-order motion stimuli in primate middle temporal area and medial superior temporal area. J Opt Soc Am A 18:2297–2306
Derrington AM, Badcock DR, Henning GB (1993) Discriminating the direction of second-order motion at short stimulus durations. Vis Res 33:1785–1794
Ditterich J, Mazurek ME, Shadlen MN (2003) Microstimulation of visual cortex affects the speed of perceptual decisions. Nat Neurosci 6:891–898
Dubner R, Zeki SM (1971) Response properties and receptive fields of cells in an anatomically defined region of the superior temporal sulcus in the monkey. Brain Res 35:528–532
Duhamel JP, Colby CL, Goldberg ME (1998) Ventral intraparietal area of the macaque: congruent visual and somatic response properties. J Neurophysiol 79:126–136
Felleman DJ, van Essen DC (1991) Distributed hierarchical processing in the primate cerebral cortex. Cereb Cortex 1(1–47):1991
Fox R, McDaniel C (1982) The perception of biological motion by human infants. Science 218:486–487
Gardner JL, Tokiyama SN, Lisberger SG (2004) A population decoding framework for motion aftereffects on smooth pursuit eye movements. J Neurosci 24:9035–9048
Goodale MA, Milner AD (1992) Separate visual pathways for perception and action. Trends Neurosci 15:20–25
Goodale MA, Milner AD, Jakobson LS, Carey DP (1991) A neurological dissociation between perceiving objects and grasping them. Nature 349:154–156
Green M (1983) Contrast detection and direction discrimination of drifting gratings. Vis Res 23:281–289
Grunewald A, Linden JF, Andersen RA (1999) Responses to auditory stimuli in macaque lateral intraparietal area I Effects of training. J Neurophysiol 82:330–342
Harris LR, Smith AT (1992) Motion defined exclusively by second-order characteristics does not evoke optokinetic nystagmus. Vis Neurosci 9:565–570
Hashiba M, Matsuoka T, Baba S, Watanabe S (1996) Non-visually induced smooth pursuit eye movements using sinusoidal target motion. Acta Otolaryngologica Suppl 525:158–162
Hubel DH, Wiesel TN (1968) Receptive fields and functional architecture of monkey striate cortex. J Physiol (London) 195:215–243
Ilg UJ, Churan J (2004) Motion perception without explicit activity in areas MT and MST. J Neurophysiol 92:1512–1523
Ilg UJ, Schumann S, Thier P (2004) Posterior parietal cortex neurons encode target motion in world-centered coordinates. Neuron 43:145–151
Jansson G, Johansson G (1973) Visual perception of bending motion. Perception 2:321–326
Johansson G (1976) Spatio-temporal differentiation and integration in visual motion perception. An experimental and theoretical analysis of calculus-like functions in visual data processing. Psychol Res 38:379–393
Johansson G (1975) Visual motion perception. Sci Am 232:76–88
Koffka K (1935) Principles of gestalt psychology. Lund Humphries, London
Krauzlis RJ, Adler SA (2001) Effects of directional expectations on motion perception and pursuit eye movements. Vis Neurosci 18:365–376
Krauzlis RJ, Stone LS (1999) Tracking with the mind's eye. Trends Neurosci 22:544–550
Lelkens AMM, Koenderink JJ (1984) Illusory motion in visual displays. Vis Res 24:1083–1090
Lindner A, Ilg UJ (2000) Initiation of smooth-pursuit eye movements to first-order and second-order motion stimuli. Exp Brain Res 133:450–456
Macknik SL, Fisher BD, Bridgeman B (1991) Flicker distorts visual space constancy. Vis Res 31:2057–2064
Majaj NJ, Carandini M, Movshon JA (2007) Motion integration by neurons in macaque MT is local, not global. J Neurosci 27:366–370
Mather G, Verstraten F, Anstis S (1998) The motion aftereffect. The MIT Press, Cambridge, MA
Maunsell JHR, van Essen DC (1983) The connections of the middle temporal visual area (MT) and their relationship to a cortical hierarchy in the macaque monkey. J Neurosci 3:2563–2586

Movshon JA, Newsome WT (1996) Visual response properties of striate cortical neurons projecting to area MT in macaque monkeys. J Neurosci 16:7733–7741
Newsome WT, Pare EB (1988) A selective impairment of motion perception following lesions of the Middle Temporal Visual Area (MT). J Neurosci 8:2201–2211
Nishida S (1993) Spatiotemporal properties of motion perception for random-check contrast modulations. Vis Res 33:633–645
Nishida S, Ashida H, Sato T (1997) Contrast dependencies of two types of motion aftereffect. Vis Res 37:553–563
O'Keefe LP, Movshon JA (1998) Processing of first- and second-order motion signals by neurons in area MT of the macaque monkey. Vis Neurosci 15:305–317
Oram MW, Perrett DI (1994) Responses of anterior superior temporal polysensory (STPa) neurons to "biological motion" stimuli. J Cogn Neurosci 6:99–116
Patzwahl DR, Zanker JM, Altenmueller EO (1993) Cortical potentials in the humans reflecting the direction of object motion. NeuroReport 4:379–382
Pavlova M, Sokolov A (2000) Orientation specificity in biological motion perception. Percept Psychophys 62:889–899
Reichardt W (1987) Evaluation of optical motion information by movement detectors. J Comp Physiol A 161:533–547
Rudolph K, Pasternak T (1999) Transient and permanent deficits in motion perception after lesions of cortical areas MT and MST in the macaque monkey. Cereb Cortex 9:90–100
Scase MO, Braddick OJ, Raymond J (1996) What is noise for the motion system? Vis Res 36:2579–2586
Schlack A, Albright TD (2007) Remembering visual motion: neural correlates of associative plasticity and motion recall in cortical area MT. Neuron 53:881–890
Serences JT, Boynton GM (2007) The representation of behavioral choice for motion in human visual cortex. J Neurosci 27:12893–12899
Sheliga BM, Chen KJ, Fitzgibbon EJ, Miles FA (2005) The initial ocular following responses elicited by apparent-motion stimuli: Reversal by inter-stimulus intervals. Vis Res 46:979–992
Smith AT, Greenlee MW, Singh KD, Kraemer FM, Hennig J (1998) The processing of first- and second-order motion in human visual cortex assessed by functional magnetic resonance imaging (fMRI). J Neurosci 18:3816–3830
Smith AT, Hess RF, Baker CL Jr (1994) Direction identification thresholds for second-order motion in central and peripheral vision. J Opt Soc Am A 11:506–514
Snowden RJ, Treue S, Erickson RG, and Andersen RA (1991) The response of area MT and V1 neurons to transparent motion. J Neurosci 11(9):2768–2785
Stone LS, Beutter BR, Lorenceau J (2000) Visual motion integration for perception and pursuit. Perception 29:771–787
Treue S, Snowden RJ, Andersen RA (1993) The effect of transiency on perceived velocity of visual patterns: a case of "temporal capture". Vis Res 33:791–798
Troje NF (2002) Decomposing biological motion: a framework for analysis and synthesis of human gait patterns. J Vis 2:371–387
Ungerleider LG, Desimone R (1986) Cortical connections of visual area MT in the macaque. J Comp Neurol 248:190–222
Ungerleider LG, Mishkin M (1982) Two cortical visual systems. In: Ingle DJ (ed) Analysis of visual behavior. MIT Press, Cambridge, MA, pp 549–586
Vaina LM, Cowey A, Kennedy D (1999) Perception of first- and second-order motion: separable neurological mechanisms? Hum Brain Mapp 7:67–77
Vaina LM, Makris N, Kennedy D, Cowey A (1998) The selective impairment of the perception of first-order motion by unilateral cortical brain damage. Vis Neurosci 15:333–348
Van Oostende S, Sunaert S, Van Hecke P, Marchal G, Orban GA (1997) The kinetic occipital (KO) region in man: an fMRI study. Cereb Cortex 7:690–701
van Santen JP, Sperling G (1985) Elaborated Reichardt detectors. J Opt Soc Am A 2:300–321
Wassle H (2004) Parallel processing in the mammalian retina. Nat Rev Neurosci 5:747–757

Watson AB, Thompson PG, Murphy BJ, Nachmias J (1980) Summation and discrimination of gratings moving in opposite directions. Vis Res 20:341–347
Wertheimer M (1923) Untersuchungen zur Lehre von der Gestalt II. Psychologische Forschung 4:301–350
Wilmer JB, Nakayama K (2007) Two distinct visual motion mechanisms for smooth pursuit: evidence from individual differences. Neuron 54:987–1000
Yo C, Wilson HR (1992) Perceived direction of moving two-dimensional patterns depends on duration, contrast and eccentricity. Vis Res 32:135–147
Zanker JM (1993) Theta motion: a paradoxical stimulus to explore higher order motion extraction. Vis Res 33:553–569
Zanker JM, Braddick OJ (1999) How does noise influence the estimation of speed? Vis Res 39:2411–2420

Part II
Active Vision, Pursuit and Motion Perception

Chapter 7
Motion Detection for Reflexive Tracking

Frederick A. Miles and Boris M. Sheliga

Abstract The moving observer who looks in the direction of heading experiences radial optic flow, which is known to elicit horizontal vergence eye movements at short latency, expansion causing convergence and contraction causing divergence: the Radial Flow Vergence Response (RFVR). The moving observer who looks off to one side experiences linear flow, which is known to elicit horizontal version eye movements at short latency: the Ocular Following Response (OFR). Although the RFVR and OFR are very different kinds of eye movement and are sensitive to very different patterns of global motion, they have very similar local spatiotemporal properties. For example, both responses are critically dependent on the Fourier composition of the motion stimuli, consistent with early spatio-temporal filtering prior to motion detection, as in the well-known energy model of motion analysis. When the motion stimuli are sine-wave gratings, the two responses share a very similar dependence on the spatial frequency and contrast of those gratings, and even the quantitative details are very similar. When the motion consists of a single step ("two-frame movie") then a brief inter-stimulus interval results in the reversal of both responses, consistent with the idea that both are mediated by motion detectors that receive a visual input whose temporal impulse response function is strongly biphasic. Further, when confronted with two sine-wave gratings that differ slightly in spatial frequency and have competing motions, both responses show nonlinear dependence on the relative contrasts of those two gratings: when the two sine waves differ in contrast by more than about an octave then the one with the higher contrast completely dominates the responses and the one with lower contrast loses its influence: winner-take-all. It has been suggested that these nonlinear interactions result from mutual inhibition between the low-level mechanisms sensing the motion of the different competing harmonics. Lastly, single unit recordings and local lesions in monkeys strongly suggest that both types of eye movements are

---

F.A. Miles (✉)
Laboratory of Sensorimotor Research, National Eye Institute/NIH,
Bldg 49 Rm 2A50, Bethesda, MD, 20892, USA
e-mail: fam@lsr.nei.nih.gov

U.J. Ilg and G.S. Masson (eds.), *Dynamics of Visual Motion Processing: Neuronal, Behavioral, and Computational Approaches*,
DOI 10.1007/978-1-4419-0781-3_7, © Springer Science+Business Media, LLC 2010

mediated by neurons in the MT/MST region of the cerebral cortex that are sensitive to global optic flow. We will argue that these various findings are all consistent with the idea that the RFVR and OFR acquire their different global properties at the level of MT/MST, where the neurons respond to large-field radial and linear optic flow, and their shared local properties from a common earlier stage, the striate cortex, where the neurons respond to the local motion energy.

7.1 Two Different Kinds of Reflexive Eye Movement That Use Visual Motion

This chapter is concerned with two kinds of eye movements that are elicited at short latencies by large-field visual motion and function to help stabilize the gaze of the moving observer. The global structure of the visual motion conforms to the patterns of optic flow experienced by the moving observer who undergoes pure translation and determines which kind of eye movement is elicited. The overall pattern of the optic flow experienced during translation consists of radial streams of images emerging from a focus of expansion straight ahead and disappearing into a focus of contraction behind, cf., the lines of longitude on a globe. The *direction* of flow at any given point depends solely on the motion of the observer but the *speed* of the flow also depends on the 3-D structure of the visual surroundings, being inversely proportional to the viewing distance at that location. Thus, nearby objects move across the field of view much more rapidly than more distant ones: motion parallax (Gibson 1950, 1966). However, given the eyes' restricted fields of view, the pattern of motion actually seen depends on where the eyes are pointing relative to the direction of heading.

The moving observer who looks in the direction of heading sees a radially expanding pattern of flow and, as objects that lie ahead get closer, he/she converges his/her two eyes in order to keep the two foveas aligned on those objects utilizing a number of depth-tracking mechanisms. The mechanism of interest here is the so-called *Radial-Flow Vergence Response* (RFVR), which senses the radial flow and generates vergence (disconjugate) eye movements at ultra-short latency, <60 ms in monkeys and <80 ms in humans (Busettini et al. 1997; Inoue et al. 1998; Kodaka et al. 2007; Yang et al. 1999). When radial optic flow is applied to large random dot patterns, expansion causes convergence – consistent with compensation for forward motion of the observer – and contraction causes divergence – consistent with compensation for backward motion of the observer.

The moving observer who looks off to one side sees a laminar pattern of optic flow and, as nearby objects pass by, he/she tracks them with both eyes utilizing a number of conjugate tracking mechanisms. The one of interest here is the so-called *Ocular Following Response* (OFR), which senses the laminar flow and generates version (conjugate) eye movements at ultra-short latency (Barthélemy et al. 2006; Busettini et al. 1991; Masson and Castet 2002; Masson et al. 2000, 2001, 2002a, b; Miles and Kawano 1986; Miles et al. 1986a, b).

Clearly, these two oculomotor reflexes – the RFVR and the OFR – respond to very different kinds of global optic flow – radial and laminar – and generate two very different kinds of eye movement – *vergence* that alters the angle between the two lines of sight and thereby changes the distance to the plane of fixation, and *version* that alters the eccentricity of the two eyes together and thereby shifts gaze within the plane of fixation. Vergence (Vg), which is given by the difference in the positions of the two eyes [L – R], and version (Vs), which is given by the average position of the two eyes [(L + R)/2], are orthogonal representations and provide a complete description of binocular eye movements, so that the positions of each eye can be reconstructed from them. Thus, adopting the convention that rightward movement is positive, increases in convergence are also positive, and L = Vs + Vg/2 while R = Vs – Vg/2.

Although the RFVR and OFR utilize very different global patterns of optic flow (see Miles et al. 2004, for recent review), this chapter will concentrate on the properties of the local-motion detectors mediating these two reflexes, which we will argue are very similar, perhaps even the same. One important feature of all the experiments that will be reviewed is that they describe only the *initial open-loop eye movements*, that is, the eye movements generated within two reaction times. The reason for this is that the eye movements during this time are the direct result of the visual processing that occurred prior to response onset. Our general thesis is that these initial eye movements provide a powerful probe for investigating the early cortical processing of visual motion.

7.2 Responses to First-Order Motion Energy

Recent studies manipulated the Fourier composition of the visual stimuli used to elicit the OFR and the RFVR (Kodaka et al. 2007; Sheliga et al. 2005a, b, 2006b), employing a variety of spatial patterns including a square wave lacking the fundamental, which is the so-called missing fundamental (*mf*) stimulus. As first pointed out by Adelson (1982), the *mf* stimulus has the special property that, when advanced in ¼-wavelength steps, its harmonics all shift ¼ of their respective wavelengths, the $4n+1$ harmonics (like the 5th, 9th, etc.) in the forward direction and the $4n-1$ harmonics (like the 3rd, 7th, etc.) in the backward direction. Importantly, the amplitude of the ith harmonic of the *mf* stimulus is proportional to $1/i$, so that the major Fourier component is the 3rd harmonic. It has been known for some time that when *mf* stimuli in the form of 1-D grating patterns are moved in successive ¼-wavelength steps, the direction of perceived motion is often opposite to the actual motion (Adelson 1982; Adelson and Bergen 1985; Baro and Levinson 1988; Brown and He 2000; Georgeson and Harris 1990; Georgeson and Shackleton 1989). It is generally argued that 1st-order-motion detectors are responsible for the perception here and that these detectors are not sensing the motion of the raw images (or their features) but rather the motion energy in a spatially filtered version of the images, so that the perceived motion depends critically on the harmonic composition of the spatial stimulus and especially the principal Fourier component, the 3rd harmonic. Note that when the *mf* stimulus shifts ¼ of its (fundamental) wavelength, the 3rd harmonic shifts ¾ of its wavelength in the same (forward) direction. However, a ¾-wavelength *forward* shift of a sine wave is exactly equivalent to a ¼-wavelength *backward* shift and, because the brain gives greatest weight to the nearest-neighbor matches (spatial aliasing), the perceived motion is generally in the backward direction: see Fig. 7.1. On the other hand, subjects sometimes perceive motion in the correct direction and this is generally attributed to higher-order detectors sensitive to the motion of specific features in the image. These observations are consistent with many others indicating that there are (at least) two neural

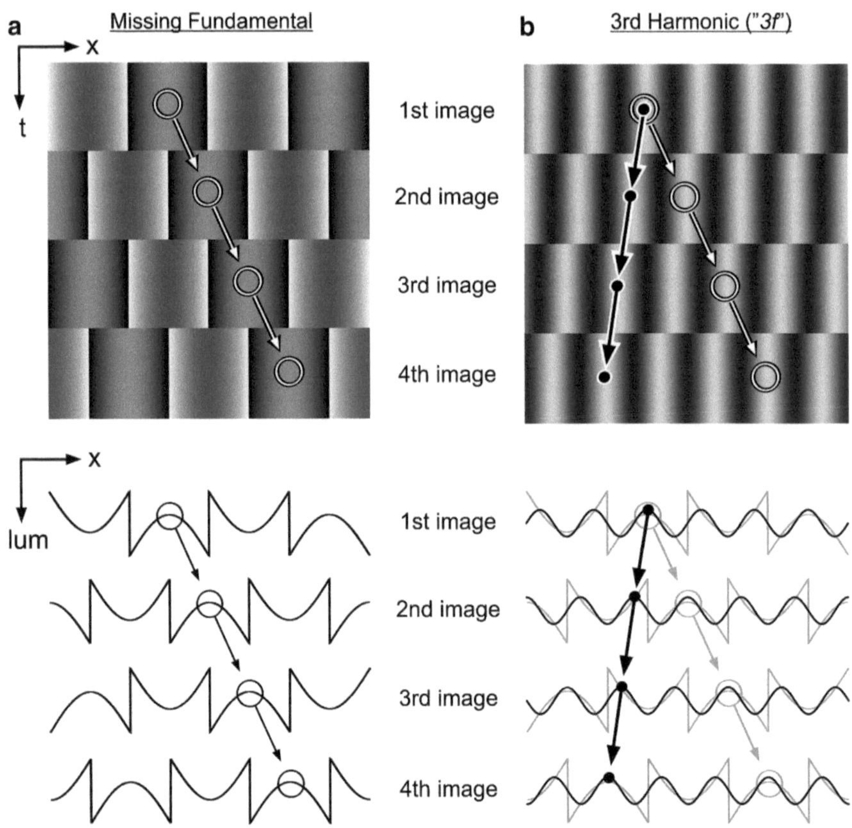

Fig. 7.1 The 1-D vertical *mf* stimulus grating and its 3rd harmonic. When the *mf* stimulus undergoes successive ¼-wavelength steps to the right (**a**), its 3rd harmonic undergoes ¾-wavelength steps to the right (**b**). *Upper panels* show horizontal slices through the stimuli at successive points in time (*x–t* plot) and *lower traces* show luminance as a function of horizontal spatial position (*x–lum* plot) after each step. The ¾-wavelength rightward steps of the 3rd harmonic (*gray circles* linked by *gray arrows* in (**b**) cannot be distinguished from ¼-wavelength leftward steps (*black dots* linked by *black arrows* in (**b**). In fact, when a pure sinusoid with the wavelength of the 3rd harmonic undergoes such steps it is invariably perceived to move leftwards, indicating that the brain gives greatest weight to the nearest matching images. After Chen et al. (2005), with permission (Wiley-Blackwell Publishing Ltd)

mechanisms by which we can sense visual motion.[1] The distinguishing characteristics of these mechanisms are sometimes controversial, and various descriptors have been applied to them: "short-range" vs. "long-range" (Braddick 1974), "1st-order" vs. "2nd-order" (Cavanagh and Mather 1989), "Fourier" vs. "non-Fourier" (Chubb and Sperling 1988), "passive" vs. "active" (Cavanagh 1992), and "energy-based" vs. "feature-based" or "correspondence-based" (Smith 1994).

Quarter-wavelength steps applied to 1-D *mf* grating stimuli elicit initial OFRs in the backward direction, i.e., in the direction of motion of the 3rd harmonic rather than the direction of motion of the overall pattern (Chen et al. 2005; Sheliga et al. 2005a). An example of such a response is shown in Fig. 7.2 (see trace labeled, *mf*).

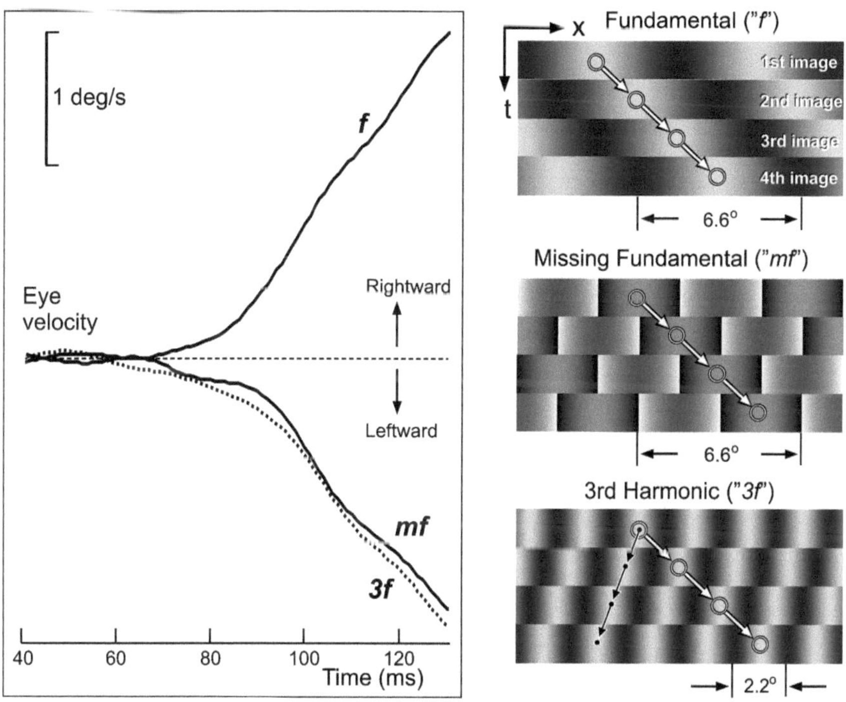

Fig. 7.2 The initial horizontal OFRs resulting from successive rightward steps applied to various 1-D vertical grating patterns (sample data for one subject). Trace *mf*: the initial OFR generated when the *mf* stimulus (wavelength, 6.6°) underwent successive ¼-wavelength rightward steps (1.65°). Trace *f*: the initial OFR when steps of the same magnitude (1.65°) and direction (rightward) were applied to pure sine-wave gratings that had the same spatial frequency as the fundamental (i.e., wavelength, 6.6°). Trace *3f*: the initial OFR when steps of the same magnitude (1.65°) and direction (rightward) were applied to pure sine-wave gratings that had the same spatial frequency and contrast (8%) as the principal Fourier component (3rd harmonic) of the *mf* stimulus. Note that time on the abscissa starts 40 ms after stimulus onset, and the *mf* and *3f* response profiles are almost identical. The cartoons at the right show *x–t* plots of the three stimuli, which all underwent the same motion steps (indicated by the *circles* and *white arrows*). After Chen et al. (2005), with permission (Wiley-Blackwell Publishing Ltd)

[1] Lu and Sperling (1995, 1996, 2001) postulate *three* different mechanisms by which we sense motion.

It is important to note that when ¼-wavelength steps are applied to 1-D gratings with a *pure sinusoidal* luminance profile the OFRs are always in the direction of those shifts (see *f* and *3f* traces in Fig. 7.2), indicating that the motion detectors mediating the OFR give greatest weight to the nearest-neighbor matches. In fact, the OFRs to the *mf* stimuli were very similar to those when the same steps were applied to a pure sinusoid with the spatial frequency and contrast of the 3rd

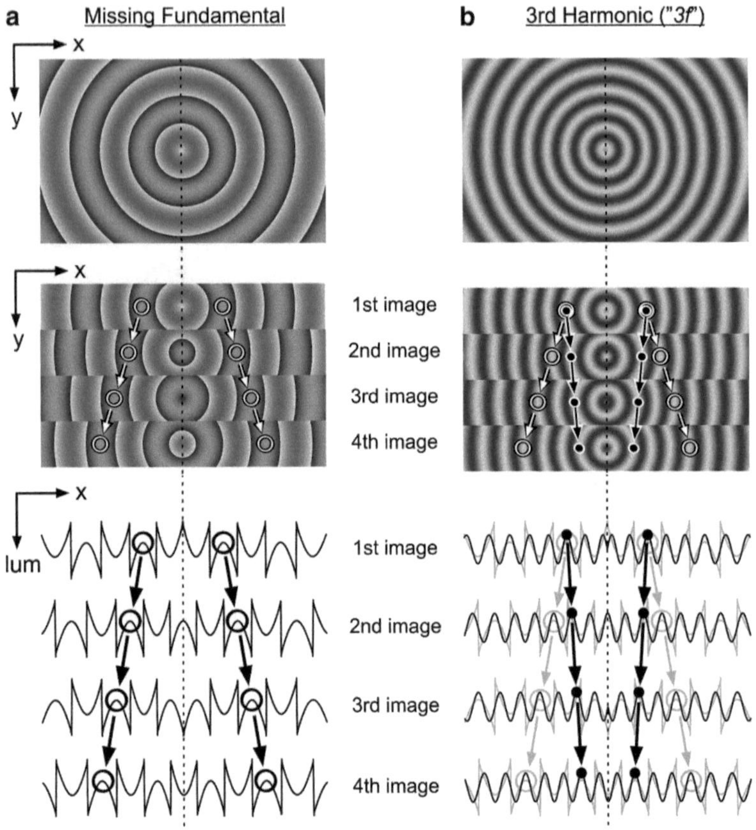

Fig. 7.3 The radial *mf* stimulus and its 3rd harmonic. When the *mf* stimulus undergoes successive ¼-wavelength expansion steps (**a**), its 3rd harmonic undergoes ¾-wavelength expansion steps (**b**). *Upper panels* show *x–y* luminance, indicating the appearance of the patterns on the screen at a given moment. *Middle panels* show horizontal slices through the centers of the *x–y* luminance plots after successive ¼-wavelength expansions of the *mf* pattern. *Lower traces* show horizontal cross sections of the luminance profile through the center of the stimulus. The ¾-wavelength expansion steps of the 3rd harmonic (*gray circles* linked by *gray arrows* in (**b**)) cannot be distinguished from ¼-wavelength contraction steps (*black dots* linked by *black arrows* in (**b**)). In fact, when a concentric pattern with a sinusoidal radial luminance profile as in (**b**) undergoes such steps it is invariably perceived to contract, indicating that the brain gives greatest weight to the nearest matching images. Furthermore, the RFVRs elicited by stimuli like those in (**a**) and (**b**) are divergent (data not shown), consistent with contracting radial flow. After Kodaka et al. (2007)

harmonic: compare the *mf* and *3f* traces in Fig. 7.2. These findings are consistent with the idea that the OFR is mediated by local-motion detectors sensitive to 1st-order motion, such as those in the well-known *energy model* of motion analysis (Adelson and Bergen 1985; van Santen and Sperling 1985; Watson and Ahumada 1985). Further support for this comes from the clear reversal of OFRs with "1st-order reverse-phi motion", one of the hallmarks of an energy-based mechanism (Masson et al. 2002a).

In an analogous study on the RFVR (Kodaka et al. 2007), *mf* stimuli were arranged in concentric circles whose *radial* luminance modulation was that of a square wave with a missing fundamental and these patterns were subject to motion consisting of successive ¼-wavelength *radial* steps: see Fig. 7.3. Once more it is important to note that when successive ¼-wavelength radial shifts are applied to concentric patterns with a *pure sinusoidal* radial luminance profile the RFVRs conform to those seen with random-dot patterns – expansion steps cause convergence and contraction steps cause divergence – indicating that the local-motion detectors mediating the RFVR also give greatest weight to the nearest-neighbor matches. Analogous to the OFR, the RFVRs when ¼-wavelength steps were applied to the radial *mf* stimulus were invariably reversed, so that expansion steps resulted in divergence and contraction steps resulted in convergence, and closely resembled the RFVRs elicited when the same radial steps were applied to concentric patterns with a pure sinusoidal luminance profile whose spatial frequency and contrast were the same as those of the 3rd harmonic (not shown). In sum, these data indicate that the local-motion detectors mediating both the OFR and the RFVR are sensitive to the 1st-order motion energy in the stimulus.

7.3 Non-Linear Interactions with Opponent Motion: Winner-Take-All (WTA)

Subsequent studies of the OFR that also used 1-D *mf* stimulus gratings examined the effect of *selectively* reducing the contrast of the principal Fourier component, the 3rd harmonic, while leaving the contrasts of the other harmonics unchanged (Sheliga et al. 2006c). This revealed the existence of powerful nonlinear interactions between the mechanisms sensing the various competing harmonics: as the contrast of the 3rd harmonic was reduced below that of the next most prominent harmonic, the 5th, then, as expected, the OFR reversed direction (because the 5th is a $4n+1$ harmonic whereas the 3rd is a $4n-1$ harmonic). However, surprisingly, once the contrast of that 3rd harmonic fell to less than ½ the contrast of the 5th harmonic then further reductions in its contrast had no impact, as though the influence of that 3rd harmonic had been suppressed by the 5th harmonic, which was now the principal Fourier component and dominated the OFR. In the example data shown in Fig. 7.4a, the 5th harmonic of the *mf* stimulus had a contrast of 20%, and selectively reducing the contrast of the 3rd harmonic from 10 to 1% had almost no impact (closed circles in Fig. 7.4a), whereas the equivalent drop in the contrast

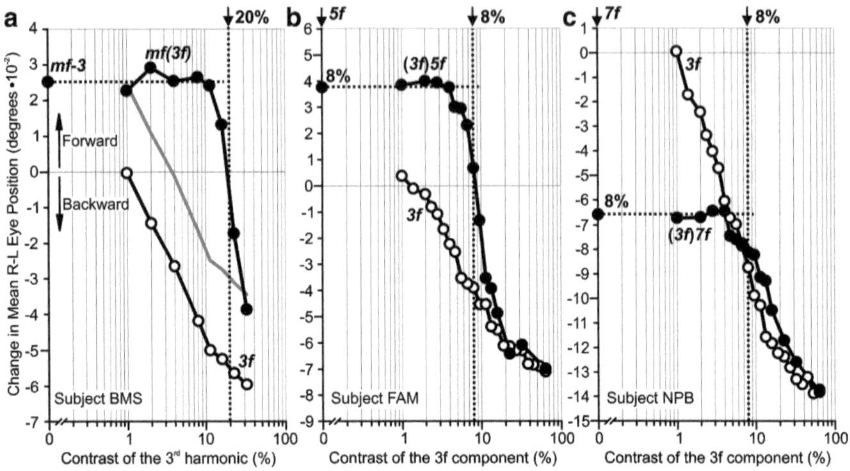

Fig. 7.4 Evidence for nonlinear interactions between the mechanisms sensing competing motions (sample response measures for one subject). (**a**) The initial OFRs to the *mf* stimuli: dependence on the contrast of the 3rd harmonic; plots show the OFR elicited by *mf* stimuli when the contrast of the 3rd harmonic was varied selectively while the contrasts of all other harmonics were held constant at the level they had when the 3rd harmonic was maximal, i.e., 32% (*closed circles*, labeled *mf(3f)*); plots also show the dependence on contrast of the OFRs to pure *3f* stimuli alone (*open circles*); the response to the *mf* stimulus that completely lacks the 3rd harmonic (*mf-3* stimulus) is plotted on the vertical axis (*filled circle* and extrapolated *horizontal dashed line*); also shown are the simulated OFRs based on the vector sum of the responses to the *mf-3* and *3f* stimuli (*gray continuous line*); the contrast of the 5th harmonic (20%) is indicated in vertical dotted line; the *mf(3f)* data are all plotted with respect to the contrast of the 3rd harmonic. (**b**) The initial OFRs to the competing *3f* and *5f* stimuli: dependence on the contrast of the *3f* component when the contrast of the *5f* component was fixed at 8% (*closed circles*, labeled *(3f)5f*); plots also show the OFR elicited by pure *3f* stimuli (*open circles*), and pure *5f* stimuli with 8% contrast (*closed circle* on the vertical axis and extrapolated *horizontal dashed line*); the *(3f)5f* data are all plotted with respect to the contrast of the *3f* component. (**c**) The initial OFRs to the competing *3f* and *7f* stimuli: dependence on the contrast of the *3f* component when the contrast of the *7f* component was fixed at 8% (*closed circles*, labeled *(3f)7f*); plots also show the OFR elicited by pure *3f* stimuli (*open circles*), and pure *7f* stimuli with 8% contrast (*closed circles* on the vertical axis and extrapolated *horizontal dashed line*). The *(3f)7f* data are all plotted with respect to the contrast of the *3f* component. Sample data from Sheliga et al. (2006c)

of a pure sinusoidal stimulus that had the same spatial frequency and underwent the same steps as that 3rd harmonic had a dramatic impact on the OFR (open circles labeled *3f* in Fig. 7.4a). This suggests that the neural channels carrying the information about the competing harmonics are mutually antagonistic so that if one harmonic has a contrast significantly greater than all others then it will tend to prevail over its competitors.

This idea was investigated further by restricting the moving stimuli to just two competing sine waves equivalent to the 3rd and 5th harmonics of the *mf* stimulus, so that their motions were in *opposite* directions. In the example data shown in Fig. 7.4b, the *5f* component always had a contrast of 8%, and increasing the contrast of the *3f* component from 1 to 4% had almost no impact (closed circles) whereas the equivalent increase in the contrast of the pure *3f* stimulus alone had a dramatic impact on the OFR (open circles). Also, when the contrast of the *3f* component was more than twice that of the *5f* component (i.e., >16%) then the *5f* component was now without influence and the responses now approximated those to the pure *3f* stimulus alone. Systematically changing the contrast at which the *5f* component was fixed indicated that the critical factor was the ratio of the contrasts of the competing gratings: when of similar contrast both were effective (vector sum/averaging), but when the contrast of one was less than about ½ that of the other then the one with the higher contrast became dominant and the one with the lower contrast became ineffective: Winner-Take-All (WTA).

Analogous studies on the RFVR (Kodaka et al. 2007) used concentric circular patterns whose radial luminance modulation was that of two superimposed sine waves with spatial frequencies in the ratio 3:5. One grating underwent contracting steps and the other expanding steps, effectively mimicking the competing motions of the 3rd and 5th harmonics of the *mf* stimuli, and when the contrast of one exceeded that of the other, on an average, by a factor of almost two then the one with the higher contrast dominated the RFVRs and the one with lower contrast lost its influence (WTA).

This nonlinear behavior of the OFR and RFVR was attributed to mutual inhibition between the neural channels sensing the competing stimuli (cf., Ferrera 2000; Ferrera and Lisberger 1995, 1997; Recanzone and Wurtz 1999). One important issue is the spatial extent of these postulated inhibitory connections and this was recently investigated by recording the initial horizontal OFRs when horizontal motion in the form of successive ¼-wavelength steps was applied in opposite directions to *3f* and *5f* 1-D vertical sine-wave gratings that were each confined to horizontal strips extending the full width of the display (45°) but only 1–2° high (Sheliga et al. 2007a). The initial OFRs again showed strong dependence on the relative contrasts of the competing gratings and when these gratings were coextensive (i.e., overlapping) this dependence was always highly nonlinear, showing WTA behavior, exactly as with the full-screen overlapping gratings used in the previous OFR study. However, a vertical gap of 1° between the competing gratings was sufficient to completely eliminate the nonlinear interaction and OFRs now approximated the vector sum of the responses to each grating stimulus alone. Thus, the nonlinear interactions responsible for the WTA outcome were strictly local, indicating that the postulated inhibitory connections do not extend much beyond the confines of the visual stimuli.

The postulated mutual inhibition between channels subserving opposite directions of motion is often termed, "motion opponency", and has substantial supporting evidence from psychophysical studies (Levinson and Sekuler 1975; Mather and Moulden 1983; Qian et al. 1994; Stromeyer et al. 1984; van Santen and Sperling 1984; Zemany et al. 1998), functional magnetic resonance imaging (Heeger et al. 1999), and single unit recordings in area MT (Bradley et al. 1995; Mikami et al. 1986; Qian and Andersen 1994; Rodman and Albright 1987; Rust 2004; Snowden et al. 1991) and area V1 (Rust 2004; Rust et al. 2005). Interestingly, Rust (2004) concluded that MT inherited motion opponency from V1.

From the functional viewpoint, it has been suggested that motion opponency will improve noise immunity and increase directional selectivity (Born and Bradley 2005; Qian et al. 1994). Also, in recent neuronal models of motion processing, motion opponency makes an important contribution to the *pattern selectivity* evident in some MT neurons (Rust et al. 2006). The study that demonstrated WTA behavior in the OFR (Sheliga et al. 2006c) argued that the strong preference given to the images with higher contrast would give objects in the plane of fixation an advantage: because of accommodation, the retinal images of objects in the plane of fixation will tend to be better focused – and hence tend to have higher contrasts – than those of objects in other depth planes. It was pointed out that this would be in line with earlier studies, which showed that when random-dot stimuli are used, the OFR is effectively disabled by binocular disparities of more than a few degrees (Masson et al. 2001; Yang et al. 2003; Yang and Miles 2003), suggesting that the motion detectors mediating the OFR are also disparity selective and that, in everyday conditions, these reflexes will have a strong preference for objects in the immediate vicinity of the plane of fixation and will tend to ignore objects in other depth planes. This same reasoning could be applied to the RFVR but, as pointed out by Kodaka et al. (2007), it is not clear how favoring images moving in the plane of fixation would necessarily operate to this system's advantage.

7.4 Non-Linear Interactions with Component Motion: WTA and Normalization

The OFR study of Sheliga et al. (2006c) that reported WTA behavior with opponent motion also included experiments with two competing 1-D sine waves that were equivalent to the 3rd and 7th harmonics of the *mf* stimulus so that their motions were in the *same* direction, here termed *component* motion. In the example data shown in Fig. 7.4c, the *7f* component always had a contrast of 8%, and increasing the contrast of the *3f* component from 1 to 4% had almost no impact (closed circles) whereas the equivalent increase in the contrast of the pure *3f* stimulus alone had a dramatic impact on the OFR (open circles). Also, when the contrast of the *3f* component exceeded twice that of the *7f* component (i.e., >16%) then the *7f* component was now without influence and the responses now approximated those to the *3f* stimulus alone. Systematically changing the contrast at which the *7f* component was fixed again indicated that the critical factor was the ratio of the contrasts of the competing gratings: when of similar contrast both were effective (vector sum/averaging), but when the contrast of one was less than about ½ that of the other then the one with the higher contrast became dominant and the one with the lower contrast became ineffective: Winner-Take-All (WTA). When the two gratings were each confined to horizontal strips only 1–2° high this nonlinear interaction was still very robust when the two gratings were overlapping (Sheliga et al. 2007a). However, unlike the situation with the *3f* and *5f* stimulus strips, separating the *3f* and *7f* grating strips by a vertical gap of up to 8° (the largest separation tried) reduced the nonlinear interaction somewhat but did not eliminate it and OFRs were still far short of the linear sum of the responses to each grating alone. The suggestion here is that the inhibitory interactions generally postulated to account for the WTA behavior are again very local but there are also more global inhibitory interactions resembling the divisive normalization often described in visual-motion-sensitive neurons in the cortex (Britten and Heuer 1999; Carandini and Heeger 1994; Carandini et al. 1997; Heeger 1992; Heuer and Britten 2002; Simoncelli and Heeger 1998).

This postulated global normalization was recently examined further by recording the horizontal OFRs to successive ¼-wavelength steps applied to a single 1-D vertical sine-wave grating that could occupy the full monitor screen (45° wide, 30° high) or a number of horizontal strips, each 1° high and extending the full width of the display (Sheliga et al. 2008). These strips were always equally spaced vertically, and increasing the number of strips could reduce the response latency by up to 20 ms, so the magnitude of the initial OFRs was estimated from the change in eye position over the initial open-loop period measured with respect to response onset. A single (centered) strip (covering 3.3% of the screen) always elicited robust OFRs, and 3 strips (10% coverage) were sufficient to elicit the maximum OFR. Further increasing the number of strips to 15 (50% coverage) had little impact, i.e., responses had asymptoted, and further increasing the coverage to 100% (full screen image) actually *decreased* the OFR so that it was now less than that elicited with only 1 strip. In this experiment, the gratings always had the same contrast, and in a second experiment, the contrast of the gratings could be fixed at one of four levels: the OFR showed essentially the same pattern of dependence on the number of strips (i.e., screen coverage) at any given contrast but, significantly, the lower the contrast, the lower the level at which the response asymptoted. This indicated that the asymptote was *not* due simply to the passive achievement of some intrinsic upper limit in the magnitude of the eye movement or the underlying motion signals ("ceiling effect"). Rather, this asymptote was seen as the result of an active process consistent with the normalization attributed to global divisive inhibition among cortical neurons cited in the previous paragraph.

Sheliga et al. (2008) attributed the *decrease* in the OFR when the image filled the monitor screen to the increased continuity of the gratings arguing that it would favor the local inhibitory surround mechanisms over the central excitatory ones (cf., Barthélemy et al. 2006). Direction-selective neurons with powerful inhibitory surrounds are commonplace in cortical area MT, which is a major source of the motion signals reaching MST, a region known to be critical for the genesis of the OFR (Takemura et al. 2007). Some MT neurons have antagonistic surrounds whose preferred direction of motion is the same as that at the center, rendering these neurons sensitive to local-motion contrast and insensitive to wide-field motion: see Born and Bradley (2005) for recent review. Sheliga et al. (2008) suggested that it is because of such neurons that introducing spatial discontinuities increases the OFR – even while decreasing the area stimulated by the motion – by reducing the activation of the antagonistic surrounds. This study indicates that robust OFRs can be elicited by much smaller motion stimuli than are commonly used and strongly suggests that this is because of divisive normalization and inhibitory surround mechanisms. Ideally, the responses of an ocular tracking mechanism to motion of a given speed and direction should be insensitive to the physical characteristics of the moving images and these new data indicate that, for a given contrast, the initial OFRs are independent of the size of the stimulus over a five-fold range (10–50% coverage). Over this range, there is clear vector averaging, exactly the sort of behavior one expects of a system subject to divisive normalization. Sheliga et al. (2008) suggested that these effects are mediated by the same mechanism that is responsible for contrast gain control whereby the OFR saturates at relatively low contrast, ~30% (Masson and Castet 2002; Sheliga et al. 2005a).

A crucial feature of the study of Sheliga et al. (2008) was that the stimuli were in effect seen through elongated apertures aligned with the axis of motion and hence were inherently broadband. Moving images confined to stationary circular apertures, as in the study of Barthélemy et al. (2006), become increasingly high-pass when the aperture is reduced in diameter, compromising the low spatial frequencies that are preferred by the OFR. Thus, the effects of the aperture here are less to do with its area than with its spatial-frequency bandwidth, which depends on the length of the aperture along the axis of motion. Many other studies have examined the so-called smooth pursuit tracking responses to single small moving spots that are obviously not confined to a stationary window, but these pursuit responses have latencies that are generally at least twice that of the OFR (e.g., Heinen and Watamaniuk 1998).

7.5 Dynamics: The Biphasic Temporal Impulse Response

Recent studies using two-frame movies, i.e., single steps, to elicit OFRs and RFVRs showed that brief ISIs (10–100 ms) reversed the initial direction of these responses (Kodaka et al. 2007; Sheliga et al. 2006a). Sample data showing this effect for the OFR can be seen in Fig. 7.5a. These reversals are reminiscent of the oft-reported reversal of perceived motion by brief ISIs that has generally been attributed to the temporal dynamics of the early visual pathway and, in particular, to the negative phase of the well-known biphasic temporal impulse response function of the human visual system (Pantle and Turano 1992; Shioiri and Cavanagh 1990; Strout et al. 1994; Takeuchi and De Valois 1997; Takeuchi et al. 2001). In this scheme, the polarity of the visual responses reaching the underlying motion detectors is assumed to undergo reversal during the ISI, so that the neural representation of the 2nd image – whose appearance marks the onset of motion – is matched to a representation of the 1st image that has undergone (transient) reversal during the ISI. This 180° phase shift in the neural representation of the 1st image means that the ¼-wavelength difference between the 1st and 2nd stimuli would be seen as a 90° phase shift in one direction when there is no ISI and as a 90° phase shift in the opposite direction when there is a brief ISI.

The consensus from the above-mentioned psychophysical studies was that with ISIs of less than ~100 ms the perceived motion depended on 1st-order energy-based

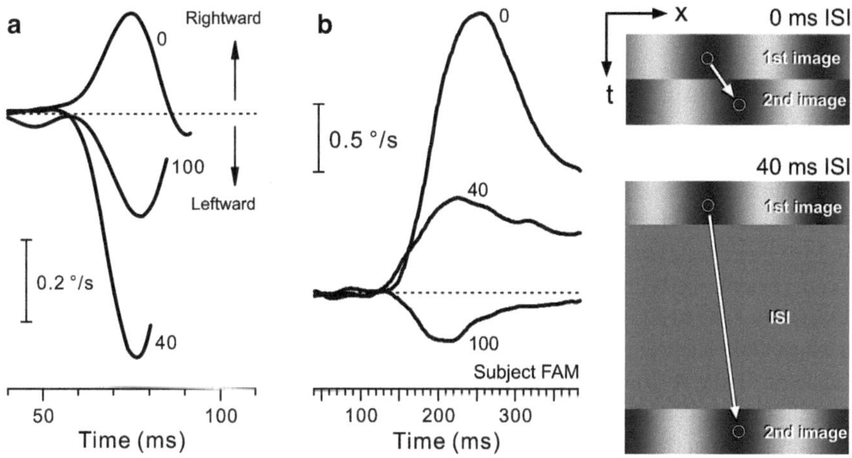

Fig. 7.5 The initial horizontal OFRs elicited by two-frame movies (single ¼-wavelength rightward steps) applied to 1-D vertical gratings: dependence of mean eye velocity response profiles on an intervening luminance-matched period of gray, the ISI (one subject). (**a**) Photopic conditions. (**b**) Scotopic conditions. The ISIs, indicating the time interval between the disappearance of the 1st image and the appearance of the 2nd image, were 0, 40 or 100 ms (numbers on the traces). Note that time on the abscissa starts 40 ms after the appearance of the 2nd image. Upward deflections of the traces denote rightward eye movements. *Dotted lines* indicate zero eye velocity. Contrast was always 32%. Sample traces from Sheliga et al. (2006a). The cartoons at the right show *x–t* plots of the stimuli when the ISI was 0 ms (above) and 40 ms (below)

mechanisms, whereas any perceived motion with longer ISIs depended on higher-order feature-based mechanisms. Thus, the reversal of the OFR and RFVR with short ISIs is a further support for mediation by detectors sensitive to 1st-order motion energy.

The dependence of the initial OFR on the ISI has also been shown to be very sensitive to the mean luminance level. Thus, the strong reversal of the initial OFR with ISIs of 10–60 ms seen in Fig. 7.5a was obtained under photopic conditions and the reversed OFRs actually reached much higher velocities than the non-reversed OFRs with 0-ms ISI. However, under scotopic conditions, reversal occurred only with ISIs $\geq$ 60 ms and these reversed OFRs were always appreciably weaker than the non-reversed OFRs with 0-ms ISI: see Fig. 7.5b, for which the luminance was below the human cone threshold. That the dependence of OFRs on the ISI shifted from biphasic to more monophasic with dark adaptation accords with the changes in the human modulation transfer function from band-pass to low-pass in the frequency domain and from biphasic to monophasic in the time domain (Kelly 1961, 1971a, b; Roufs 1972a, b; Snowden et al. 1995; Swanson et al. 1987). Note that the reversal of perceived motion with intermediate ISIs (30–90 ms) reported by Takeuchi and De Valois (1997) was obtained under photopic viewing conditions and these workers also showed that the reversal was reduced at low luminance. In fact, when the retinal illuminance was reduced below cone threshold, Takeuchi and De Valois (1997) found that ISIs no longer reversed perceived motion.

7.6 Neural Mediation

There is extensive evidence from monkeys that the OFR and RFVR are cortically mediated despite their ultra-short latency. Bilateral lesions of the medial superior temporal area of the cortex (MST) result in major impairments of both reflexes (Takemura et al. 2002, 2007) and there is extensive data from single unit recordings indicating that neurons in this region discharge in relation to the visual stimuli used to drive these reflexes. Thus, MST is specialized for the processing of optic flow (for recent review, see Wurtz 1998) and has long been known to contain neurons that are selectively sensitive to radial optic flow patterns such as those used to evoke RFVRs at ultra-short latencies (Duffy 2000; Duffy and Wurtz 1991a, b, 1995, 1997a, b, c; Lagae et al. 1994; Saito et al. 1986; Tanaka et al. 1986; Tanaka and Saito 1989). Kawano and colleagues have shown that there are neurons in MST that discharge in relation to the earliest OFR responses, their temporal profiles even reproducing the irregularities in the temporal profiles of the OFRs (Kawano et al. 1994; Takemura et al. 2000; Takemura and Kawano 2006). This cortical region is thought to rely heavily on magnocellular pathways, which are so named because they include the magnocellular layers of the LGN (Livingstone and Hubel 1987, 1988; Maunsell et al. 1990; Merigan and Maunsell 1990; Schiller et al. 1990). The contrast-dependence of the OFR in monkeys (Miles et al. 1986a) and humans (Masson and Castet 2002; Sheliga et al. 2005a), and of the RFVR in humans (Kodaka et al. 2007), closely resemble that in the magnocellular pathway, which is characterized by saturation at relatively low contrast levels (Kaplan and Shapley 1982). Recordings from monkeys also indicate that, at scotopic luminance levels, vision is dominated by rod inputs to magnocellular-projecting retinal ganglion cells (Lee et al. 1997; Purpura et al. 1988), consistent with the finding of Sheliga et al. (2006a) that the OFR continues to operate even at very low luminance and contrast levels. Lesions and electrophysiological studies in monkeys strongly suggest that the OFR is mediated by projections from MST to the dorsolateral pons, which then projects to the ventral paraflocculus, a region of the cerebellum well known for its involvement with the generation of tracking eye movements (see Takemura and Kawano 2002, for review).

7.7 A Window onto the Processing of Visual Motion in the Human Striate Cortex?

Earlier studies suggested that the OFR and RFVR are synergistic reflexes[2] that combine to assist in the visual stabilization of the gaze of the moving observer and pointed out a number of shared features in addition to their ultra-short latency, such

[2] We have not mentioned a 3rd reflex, the Disparity Vergence Response, that is also thought to be a member of this family, because it responds to binocular disparity rather than motion. This reflex shares many fundamental properties with the OFR and RFVR, including dependence on 1st-order (disparity) energy (Sheliga et al. 2006b), and WTA behavior when competing (disparity) stimuli are used (Sheliga et al. 2007b).

as post-saccadic enhancement, dependence on the preëxisting vergence angle, and – in monkeys at least – mediation by MST: for review see Miles (1998), Miles et al. (2004) and Takemura et al. (2007). More recently, Kodaka et al. (2007) showed that the fundamental spatiotemporal characteristics of the OFR and RFVR – such as their dependence on contrast, spatial frequency and an ISI, as well as the nonlinear interactions that are evident with competing motions – were very similar, quantitatively as well as qualitatively. Kodaka et al. (2007) suggested that these two very different kinds of eye movements share these basic spatiotemporal properties because they are mediated by the same low-level, local-motion detectors. As pointed out above, work on monkeys strongly implicates the MST area of cortex in the genesis of the RFVR and OFR, and this area is known to receive major inputs from area MT (Maunsell and van Essen 1983; Ungerleider and Desimone 1986), which receives a direct projection from direction-selective neurons in V1 (Movshon and Newsome 1996). Of particular interest is that recent authors have suggested that neurons in MT inherit their local-motion selectivity from neurons in V1 (e.g., Born and Bradley 2005; Churchland et al. 2005; Movshon and Newsome 1996; Priebe et al. 2006; Rust 2004; Rust et al. 2006). This raises the possibility that the local spatiotemporal properties of the MST neurons mediating both the RFVR and the OFR directly reflect the local motion energy computed by V1 direction-selective neurons. Thus, even though the MST neurons mediating these two reflexes must have very different global properties – preferring radial vs. linear optic flow, respectively – they nonetheless probably share the same local spatiotemporal characteristics. One especially attractive feature of these two reflexes is that many of their basic characteristics are well captured by simple mathematical functions with only two free parameters (e.g., dependence on log spatial frequency is Gaussian, dependence on contrast is well described by the Naka–Rushton equation, and dependence on the relative contrast of two competing motions is well described by a Contrast-Weighted-Average model) and these quantitative characterizations generally show little inter-subject variability. Thus, although the OFR and RFVR are *motor* responses, they directly reflect the detailed properties of the low-level *sensory* detectors mediating those responses and effectively provide a quantitative window onto the early cortical processing of visual motion, perhaps as early as striate cortex.

Acknowledgments This research was supported by the Intramural Research Program of the National Eye Institute at the NIH.

References

Adelson EH (1982). Some new motion illusions, and some old ones, analysed in terms of their Fourier components. Invest Ophthalmol Vis Sci, 34 (Suppl.):144 (Abstract)

Adelson EH, Bergen JR (1985) Spatiotemporal energy models for the perception of motion. J Opt Soc Am A 2:284–299

Baro JA, Levinson E (1988) Apparent motion can be perceived between patterns with dissimilar spatial frequencies. Vision Res 28:1311–1313

Barthélemy FV, Vanzetta I, Masson GS (2006) Behavioral receptive field for ocular following in humans: dynamics of spatial summation and center-surround interactions. J Neurophysiol 95:3712–3726
Born RT, Bradley DC (2005) Structure and function of visual area MT. Annu Rev Neurosci 28:157–189
Braddick O (1974) A short-range process in apparent motion. Vision Res 14:519–527
Bradley DC, Qian N, Andersen RA (1995) Integration of motion and stereopsis in middle temporal cortical area of macaques. Nature 373:609–611
Britten KH, Heuer HW (1999) Spatial summation in the receptive fields of MT neurons. J Neurosci 19:5074–5084
Brown RO, He S (2000) Visual motion of missing-fundamental patterns: motion energy versus feature correspondence. Vision Res 40:2135–2147
Busettini C, Masson GS, Miles FA (1997) Radial optic flow induces vergence eye movements with ultra-short latencies. Nature 390:512–515
Busettini C, Miles FA, Schwarz U (1991) Ocular responses to translation and their dependence on viewing distance. II. Motion of the scene. J Neurophysiol 66:865–878
Carandini M, Heeger DJ (1994) Summation and division by neurons in primate visual cortex. Science 264:1333–1336
Carandini M, Heeger DJ, Movshon JA (1997) Linearity and normalization in simple cells of the macaque primary visual cortex. J Neurosci 17:8621–8644
Cavanagh P (1992) Attention-based motion perception. Science 257:1563–1565
Cavanagh P, Mather G (1989) Motion: the long and short of it. Spat Vis 4:103–129
Chen KJ, Sheliga BM, FitzGibbon EJ, Miles FA (2005) Initial ocular following in humans depends critically on the fourier components of the motion stimulus. Ann N Y Acad Sci 1039: 260–271
Chubb C, Sperling G (1988) Drift-balanced random stimuli: a general basis for studying non-Fourier motion perception. J Opt Soc Am A 5:1986–2007
Churchland MM, Priebe NJ, Lisberger SG (2005) Comparison of the spatial limits on direction selectivity in visual areas MT and V1. J Neurophysiol 93:1235–1245
Duffy CJ (2000) Optic flow analysis for self-movement perception. Int Rev Neurobiol 44:199–218
Duffy CJ, Wurtz RH (1991a) Sensitivity of MST neurons to optic flow stimuli. I. A continuum of response selectivity to large-field stimuli. J Neurophysiol 65:1329–1345
Duffy CJ, Wurtz RH (1991b) Sensitivity of MST neurons to optic flow stimuli. II. Mechanisms of response selectivity revealed by small-field stimuli. J Neurophysiol 65:1346–1359
Duffy CJ, Wurtz RH (1995) Response of monkey MST neurons to optic flow stimuli with shifted centers of motion. J Neurosci 15:5192–5208
Duffy CJ, Wurtz RH (1997a) Medial superior temporal area neurons respond to speed patterns in optic flow. J Neurosci 17:2839–2851
Duffy CJ, Wurtz RH (1997b) Multiple temporal components of optic flow responses in MST neurons. Exp Brain Res 114:472–482
Duffy CJ, Wurtz RH (1997c) Planar directional contributions to optic flow responses in MST neurons. J Neurophysiol 77:782–796
Ferrera VP (2000) Task-dependent modulation of the sensorimotor transformation for smooth pursuit eye movements. J Neurophysiol 84:2725–2738
Ferrera VP, Lisberger SG (1995) Attention and target selection for smooth pursuit eye movements. J Neurosci 15:7472–7484
Ferrera VP, Lisberger SG (1997) The effect of a moving distractor on the initiation of smooth-pursuit eye movements. Vis Neurosci 14:323–338
Georgeson MA, Harris MG (1990) The temporal range of motion sensing and motion perception. Vision Res 30:615–619
Georgeson MA, Shackleton TM (1989) Monocular motion sensing, binocular motion perception. Vision Res 29:1511–1523
Gibson JJ (1950) The perception of the visual world. Houghton Mifflin, Boston
Gibson JJ (1966) The senses considered as perceptual systems. Houghton Mifflin, Boston

Heeger DJ (1992) Normalization of cell responses in cat striate cortex. Vis Neurosci 9:181–197
Heeger DJ, Boynton GM, Demb JB, Seidemann E, Newsome WT (1999) Motion opponency in visual cortex. J Neurosci 19:7162–7174
Heinen SJ, Watamaniuk SN (1998) Spatial integration in human smooth pursuit. Vision Res 38:3785–3794
Heuer HW, Britten KH (2002) Contrast dependence of response normalization in area MT of the rhesus macaque. J Neurophysiol 88:3398–3408
Inoue Y, Takemura A, Suehiro K, Kodaka Y, Kawano K (1998) Short-latency vergence eye movements elicited by looming step in monkeys. Neurosci Res 32:185–188
Kaplan E, Shapley RM (1982) X and Y cells in the lateral geniculate nucleus of macaque monkeys. J Physiol (Lond) 330:125–143
Kawano K, Shidara M, Watanabe Y, Yamane S (1994) Neural activity in cortical area MST of alert monkey during ocular following responses. J Neurophysiol 71:2305–2324
Kelly DH (1961) Visual response to time-dependent stimuli. I. Amplitude sensitivity measurements. J Opt Soc Am 51:422–429
Kelly DH (1971a) Theory of flicker and transient responses. I. Uniform fields. J Opt Soc Am 61:537–546
Kelly DH (1971b) Theory of flicker and transient responses. II. Counterphase gratings. J Opt Soc Am 61:632–640
Kodaka Y, Sheliga BM, Fitzgibbon EJ, Miles FA (2007) The vergence eye movements induced by radial optic flow: some fundamental properties of the underlying local-motion detectors. Vision Res 47:2637–2660
Lagae L, Maes H, Raiguel S, Xiao D-K, Orban GA (1994) Responses of macaque STS neurons to optic flow components: a comparison of areas MT and MST. J Neurophysiol 71:1597–1626
Lee BB, Smith VC, Pokorny J, Kremers J (1997) Rod inputs to macaque ganglion cells. Vision Res 37:2813–2828
Levinson E, Sekuler R (1975) Inhibition and disinhibition of direction-specific mechanisms in human vision. Nature 254:692–694
Livingstone M, Hubel DH (1988) Segregation of form, color, movement, and depth: anatomy, physiology, and perception. Science 240:740–749
Livingstone MS, Hubel DH (1987) Psychophysical evidence for separate channels for the perception of form, color, movement, and depth. J Neurosci 7:3416–3468
Lu ZL, Sperling G (1995) The functional architecture of human visual motion perception. Vision Res 35:2697–2722
Lu ZL, Sperling G (1996) Three systems for visual motion perception. Curr Dir Psychol Sci 5:44–53
Lu ZL, Sperling G (2001) Three-systems theory of human visual motion perception: review and update. J Opt Soc Am A 18:2331–2370
Masson GS, Busettini C, Yang D-S, Miles FA (2001) Short-latency ocular following in humans: sensitivity to binocular disparity. Vision Res 41:3371–3387
Masson GS, Castet E (2002) Parallel motion processing for the initiation of short-latency ocular following in humans. J Neurosci 22:5149–5163
Masson GS, Rybarczyk Y, Castet E, Mestre DR (2000) Temporal dynamics of motion integration for the initiation of tracking eye movements at ultra-short latencies. Vis Neurosci 17:753–767
Masson GS, Yang D-S, Miles FA (2002a) Reversed short-latency ocular following. Vision Res 42:2081–2087
Masson GS, Yang D-S, Miles FA (2002b) Version and vergence eye movements in humans: open-loop dynamics determined by monocular rather than binocular image speed. Vision Res 42:2853–2867
Mather G, Moulden B (1983) Thresholds for movement direction: two directions are less detectable than one. Q J Exp Psychol A 35:513–518
Maunsell JH, Nealey TA, DePriest DD (1990) Magnocellular and parvocellular contributions to responses in the middle temporal visual area (MT) of the macaque monkey. J Neurosci 10:3323–3334

Maunsell JH, van Essen DC (1983) The connections of the middle temporal visual area (MT) and their relationship to a cortical hierarchy in the macaque monkey. J Neurosci 3:2563–2586
Merigan WH, Maunsell JH (1990) Macaque vision after magnocellular lateral geniculate lesions. Vis Neurosci 5:347–352
Mikami A, Newsome WT, Wurtz RH (1986) Motion selectivity in macaque visual cortex. I. Mechanisms of direction and speed selectivity in extrastriate area MT. J Neurophysiol 55:1308–1327
Miles FA (1998) The neural processing of 3-D visual information: evidence from eye movements. Eur J Neurosci 10:811–822
Miles FA, Busettini C, Masson GS, Yang D-S (2004) Short-latency eye movements: evidence for rapid, parallel processing of optic flow. In: Vaina LM, Beardsley SA, Rushton S (eds) Optic flow and beyond. Kluwer Academic Press, Dordrecht, pp 79–107
Miles FA, Kawano K (1986) Short-latency ocular following responses of monkey. III. Plasticity. J Neurophysiol 56:1381–1396
Miles FA, Kawano K, Optican LM (1986) Short-latency ocular following responses of monkey. I. Dependence on temporospatial properties of visual input. J Neurophysiol 56:1321–1354
Movshon JA, Newsome WT (1996) Visual response properties of striate cortical neurons projecting to area MT in macaque monkeys. J Neurosci 16:7733–7741
Pantle A, Turano K (1992) Visual resolution of motion ambiguity with periodic luminance- and contrast-domain stimuli. Vision Res 32:2093–2106
Priebe NJ, Lisberger SG, Movshon JA (2006) Tuning for spatiotemporal frequency and speed in directionally selective neurons of macaque striate cortex. J Neurosci 26:2941–2950
Purpura K, Kaplan E, Shapley RM (1988) Background light and the contrast gain of primate P and M retinal ganglion cells. Proc Natl Acad Sci USA 85:4534–4537
Qian N, Andersen RA (1994) Transparent motion perception as detection of unbalanced motion signals. II. Physiology. J Neurosci 14:7367–7380
Qian N, Andersen RA, Adelson EH (1994) Transparent motion perception as detection of unbalanced motion signals. I. Psychophysics. J Neurosci 14:7357–7366
Rodman HR, Albright TD (1987) Coding of visual stimulus velocity in area MT of the macaque. Vision Res 27:2035–2048
Recanzone GH, Wurtz RH (1999) Shift in smooth pursuit initiation and MT and MST neuronal activity under different stimulus conditions. J Neurophysiol 82:1710–1727
Roufs JAJ (1972a) Dynamic properties of vision. I. Experimental relationships between flicker and flash thresholds. Vision Res 12:261–278
Roufs JAJ (1972b) Dynamic properties of vision. II. Theoretical relationships between flicker and flash thresholds. Vision Res 12:279–292
Rust NC (2004) Signal transmission, feature representation and computation in areas V1 and MT of the macaque monkey. Doctoral dissertation, New York University, Dissertation Abstracts International 65/09-B, 4444
Rust NC, Mante V, Simoncelli EP, Movshon JA (2006) How MT cells analyze the motion of visual patterns. Nat Neurosci 9:1421–1431
Rust NC, Schwartz O, Movshon JA, Simoncelli EP (2005) Spatiotemporal elements of macaque v1 receptive fields. Neuron 46:945–956
Saito H, Yukie M, Tanaka K, Hikosaka K, Fukada Y, Iwai E (1986) Integration of direction signals of image motion in the superior temporal sulcus of the macaque monkey. J Neurosci 6:145–157
Schiller PH, Logothetis NK, Charles ER (1990) Role of the color-opponent and broad-band channels in vision. Vis Neurosci 5:321–346
Sheliga BM, Chen KJ, FitzGibbon EJ, Miles FA (2005a) Initial ocular following in humans: a response to first-order motion energy. Vision Res 45:3307–3321
Sheliga BM, Chen KJ, FitzGibbon EJ, Miles FA (2005b) Short-latency disparity vergence in humans: evidence for early spatial filtering. Ann NY Acad Sci 1039:252–259
Sheliga BM, Chen KJ, FitzGibbon EJ, Miles FA (2006a) The initial ocular following responses elicited by apparent-motion stimuli: Reversal by inter-stimulus intervals. Vision Res 46:979–992

Sheliga BM, FitzGibbon EJ, Miles FA (2006b) Short-latency disparity vergence eye movements: a response to disparity energy. Vision Res 46:3723–3740

Sheliga BM, FitzGibbon EJ, Miles FA (2007a) Competing image motions: evidence for local and global nonlinear interactions. 2007 Neuroscience Meeting Planner. Program No. 337.314. Society for Neuroscience, San Diego, CA

Sheliga BM, FitzGibbon EJ, Miles FA (2007b) Human vergence eye movements initiated by competing disparities: evidence for a winner-take-all mechanism. Vision Res 47:479–500

Sheliga BM, FitzGibbon EJ, Miles FA (2008) Human ocular following: evidence that responses to large-field stimuli are limited by local and global inhibitory influences. Prog Brain Res 171:237–243

Sheliga BM, Kodaka Y, FitzGibbon EJ, Miles FA (2006c) Human ocular following initiated by competing image motions: evidence for a winner-take-all mechanism. Vision Res 46: 2041–2060

Shioiri S, Cavanagh P (1990) ISI produces reverse apparent motion. Vision Res 30:757–768

Simoncelli EP, Heeger DJ (1998) A model of neuronal responses in visual area MT. Vision Res 38:743–761

Smith AT (1994) Correspondence-based and energy-based detection of second-order motion in human vision. J Opt Soc Am A 11:1940–1948

Snowden RJ, Hess RF, Waugh SJ (1995) The processing of temporal modulation at different levels of retinal illuminance. Vision Res 35:775–789

Snowden RJ, Treue S, Erickson RG, Andersen RA (1991) The response of area MT and V1 neurons to transparent motion. J Neurosci 11:2768–2785

Stromeyer CF, Kronauer RE, Madsen JC, Klein SA (1984) Opponent-movement mechanisms in human vision. J Opt Soc Am A 1:876–884

Strout JJ, Pantle A, Mills SL (1994) An energy model of interframe interval effects in single-step apparent motion. Vision Res 34:3223–3240

Swanson WH, Ueno T, Smith VC, Pokorny J (1987) Temporal modulation sensitivity and pulse-detection thresholds for chromatic and luminance perturbations. J Opt Soc Am A 4:1992–2005

Takemura A, Inoue Y, Kawano K (2000) The effect of disparity on the very earliest ocular following responses and the initial neuronal activity in monkey cortical area MST. Neurosci Res 38:93–101

Takemura A, Inoue Y, Kawano K (2002) Visually driven eye movements elicited at ultra-short latency are severely impaired by MST lesions. Ann N Y Acad Sci 956:456–459

Takemura A, Kawano K (2002) Sensory-to-motor processing of the ocular-following response. Neurosci Res 43:201–206

Takemura A, Kawano K (2006) Neuronal responses in MST reflect the post-saccadic enhancement of short-latency ocular following responses. Exp Brain Res 173:174–179

Takemura A, Murata Y, Kawano K, Miles FA (2007) Deficits in short-latency tracking eye movements after chemical lesions in monkey cortical areas MT and MST. J Neurosci 27:529–541

Takeuchi T, De Valois KK (1997) Motion-reversal reveals two motion mechanisms functioning in scotopic vision. Vision Res 37:745–755

Takeuchi T, De Valois KK, Motoyoshi I (2001) Light adaptation in motion direction judgments. J Opt Soc Am A 18:755–764

Tanaka K, Hikosaka K, Saito H, Yukie M, Fukada Y, Iwai E (1986) Analysis of local and wide-field movements in the superior temporal visual areas of the macaque monkey. J Neurosci 6:134–144

Tanaka K, Saito H (1989) Analysis of motion of the visual field by direction, expansion/contraction, and rotation cells clustered in the dorsal part of the medial superior temporal area of the macaque monkey. J Neurophysiol 62:626–641

Ungerleider LG, Desimone R (1986) Cortical connections of visual area MT in the macaque. J Comp Neurol 248:190–222

van Santen JP, Sperling G (1984) Temporal covariance model of human motion perception. J Opt Soc Am A 1:451–473

van Santen JP, Sperling G (1985) Elaborated Reichardt detectors. J Opt Soc Am A 2:300–321
Watson AB, Ahumada AJ (1985) Model of human visual-motion sensing. J Opt Soc Am A 2:322–341
Wurtz RH (1998) Optic flow: a brain region devoted to optic flow analysis? Curr Biol 8:R554–R556
Yang D-S, FitzGibbon EJ, Miles FA (1999) Short-latency vergence eye movements induced by radial optic flow in humans: dependence on ambient vergence level. J Neurophysiol 81:945–949
Yang D-S, FitzGibbon EJ, Miles FA (2003) Short-latency disparity-vergence eye movements in humans: sensitivity to simulated orthogonal tropias. Vision Res 43:431–443
Yang D-S, Miles FA (2003) Short-latency ocular following in humans is dependent on absolute (rather than relative) binocular disparity. Vision Res 43:1387–1396
Zemany L, Stromeyer CF, Chaparro A, Kronauer RE (1998) Motion detection on flashed, stationary pedestal gratings: evidence for an opponent-motion mechanism. Vision Res 38:795–812

Chapter 8
When the Brain Meets the Eye: Tracking Object Motion

Guillaume S. Masson, Anna Montagnini, and Uwe J. Ilg

Abstract To accurately track a moving object of interest with appropriate smooth eye movements, the brain needs to reconstruct a single velocity vector describing the global motion of this object. Because of the aperture problem (see Chap. 1), the visual system must integrate piecewise local information from either elongated edges and contours or particular features such as corners and texture elements. Here, we show that investigating smooth eye movements unveil several dynamical properties of this visual motion integration stage. Signals are weighted according to their uncertainties. The integration is highly dynamical - eye movements being always launched first in the simplest, linear (vector sum) prediction. Tracking trajectories are then progressively adjusted to match the object trajectory after 200 ms of pursuit. Such strategy is immune to higher factors such as prediction about incoming 2D target trajectory. On the contrary, mixing retinal and extra-retinal signals become important later during the pursuit to accommodate partial or total object motion occlusion for instance. We propose a framework computing and representing object motion through two recurrent loops (V1-MT and MST-FEF, respectively), with area MST playing the role of gear. Such architecture would accommodate two important constraints of motor behavior: quick reaction to a new visual event and utilization of extra-retinal information to smooth out transient changes in the image such as those occurring when an object moves in a crowded environment.

8.1 Introduction

While most of our knowledge on smooth pursuit eye movements comes from experiments where subjects were asked to track a small red spot moving over a dark background (see Lisberger et al. 1987), human and non-human primates face much

G.S. Masson (✉)
Institut de Neurosciences Cogntiives de la Méditerranée, CNRS and Université de la Méditerranée, 31 Chemin Joseph Aiguier, 13402, Marseille, France
e-mail: guillaume.masson@incm.cnrs-mrs.fr

U.J. Ilg and G.S. Masson (eds.), *Dynamics of Visual Motion Processing: Neuronal, Behavioral, and Computational Approaches*,
DOI 10.1007/978-1-4419-0781-3_8, © Springer Science+Business Media, LLC 2010

more complex situations in natural environments. Tracking eye movements are used to stabilize the image of the object of interest onto the retinas so that its properties can be scrutinized with high-resolution spatial vision mechanisms. To do so, the brain must compute a reliable and dynamical velocity signal that fully describes the trajectory of the selected object despite its complex shape, its own movement kinematics (think of a spinning ball), and the large variations in the retinal input due to sudden changes in illumination levels or the presence of other occluding and reflecting surfaces. Several recent pieces of work reveal that such a nearly perfect sensori-to-motor transformation relies on complex neuronal dynamics where inflow estimation is constantly updated by using not only the different visual cues available from the retinal image but also the internal representations of object properties and trajectories (see Krauzlis and Stone 1999; Ilg 2002; Masson 2004 for reviews).

The time course of smooth pursuit eye movements reflects the neuronal solution for this intricate computation of object motion trajectory. Recent studies have suggested that different motion processing might control different aspects of pursuit. Wilmer and Nakayama (2007) have brought evidence supporting the idea that low-level motion signals (motion-energy-based) influence the earliest phase of pursuit initiation, before the occurrence of the first catch-up saccade. This is consistent with many other studies on both voluntary (e.g. Lindner and Ilg 2000; Hawken and Gegenfurtner 2001; Priebe et al. 2001) and reflexive tracking (Masson et al. 2002; Sheliga et al. 2005; see Chaps. 7 and 11) in primates. Both pursuit onset and first acceleration phase seem to be tightly coupled with low-level motion mechanisms that sense spatio-temporal changes in the retinal image. On the other hand, an independent high-level motion signal (position-tracking) can influence post-saccadic accuracy of steady-state tracking. The same, later part of tracking responses can also be influenced by higher factors such as training (Madelain and Krauzlis 2003) or attention (Madelain et al. 2005). Thus, the overall process of computing the exact motion signals that have to be converted into motor commands is considerably more complex than originally thought (see Ilg 2002; Masson 2004; Krauzlis 2004, for reviews). Here, we will show that such complicated process can be dissected out within a single, experimental, and theoretical framework such as 2D object motion computation, in particular its part devoted to motion signals integration. By using a set of very simple visual stimuli as illustrated in Fig. 8.1b–d, we can address the various levels of motion integration and separate them in time, thanks to the temporal dynamics of the neural solutions of fundamental questions such as the aperture problem or features binding (Fig. 8.1a). Moreover, the same line-drawing objects offer an opportunity to unveil the role of mid-level vision rules such as motion coherency or object segmentation (e.g. Beutter and Stone 2000) as well as higher-order, cognitive processes such as attention, anticipation, or prediction (see Ilg 2002; Masson 2004). Lastly, there is now a large bulk of experimental evidences obtained at the physiological (e.g. Pack and Born 2001), psychophysical (e.g. Lorenceau et al. 1993), behavioral (e.g. Masson and Stone 2002), and computational (e.g. Mingolla 2003; Berzhanskaya et al. 2007; Bayerl and Neumann 2004; Weiss et al. 2002) levels using these

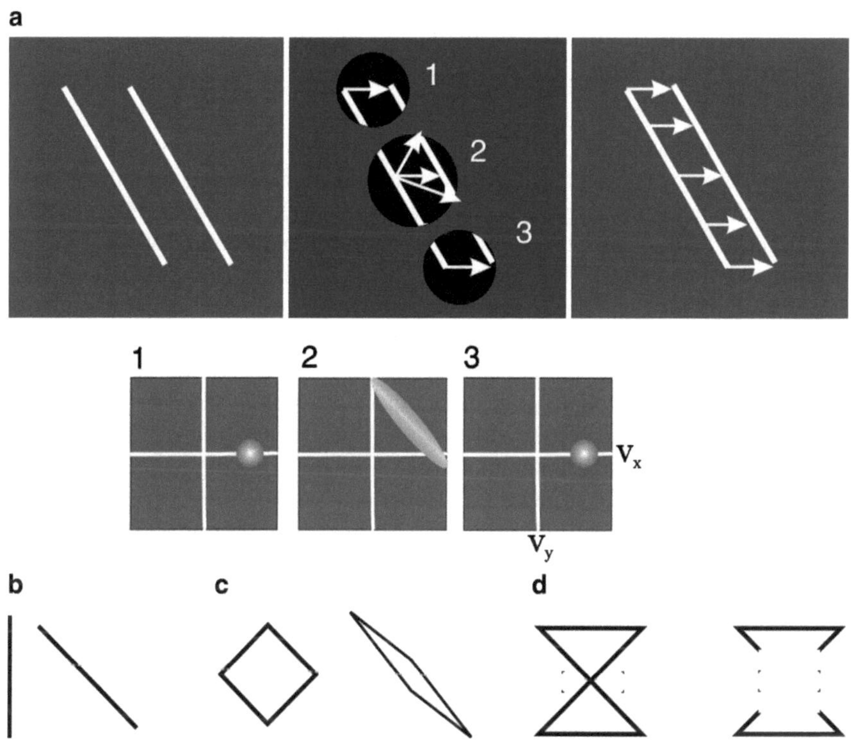

Fig. 8.1 Line-drawing stimuli for investigating the dynamics of biological visual motion. (**a**) Illustration of the aperture problem. The leftward image shows two successive positions of a single, tilted line drifting to the right. The second image illustrates the aperture problem. The same two positions are sensed through small apertures localized along the edge or at line-ends. Lower plots are velocity likelihood distributions sensed in three different image locations. At location 2, motion information is highly ambiguous at the central aperture since an infinite number of velocity vector are compatible with the change in position of the uni-dimensional edge. This is illustrated by an elongated Gaussian distribution drawn in the velocity space and indicating all possible velocities being consistent with the edge translation. On the contrary, translation of 2D features such as line-ending is highly non-ambiguous when seen through a small aperture (locations 1 and 3). In velocity space, motion information is described as a single Gaussian blob centered onto the actual 2D velocity (i.e. horizontal). The different examples of line-drawing objects are illustrated in plots b–d. (**b**) Upright and tilted lines where 1D and 2D motion direction are either aligned or 45° separated, respectively. (**c**) Upright and tilted diamonds. Vector average of the four 1D edge motions is aligned with actual object velocity for upright diamond. However, with tilted diamonds, vector average prediction is 44° away from the actual 2D trajectory. (**d**) Real and imaginary targets, illustrated together with the electronic window checking eye position relative to object center position during smooth pursuit. With illusory targets, center of the four diagonal edges must be inferred. The height of the target was 20°; the height of the blanked area in the center in case of the imaginary figure was 12.°

stimuli in human and non-human primates, so that an unified framework can be drawn. We want herein to elaborate such a framework in the specific context of pursuit eye movements.

8.2 From Following to Pursuing: Tracking Initiation and Visual Motion Dynamics

Pursuit eye movements are primarily driven by retinal image motion. Early studies using small moving spots have made the seminal observation that smooth pursuit is mainly controlled using target velocity (Rashbass 1961; Carl and Gellman 1987). This target velocity signal is reconstructed from a rapid pooling of local motion information over a considerable part (up to 10–20°) of the visual field (Heinen and Watamaniuk 1998; Pola and Wyatt 1985). Such considerable spatial summation area is at the root of a motion integration mechanism used to improve the neural estimation of target direction and speed (e.g. Verghese and Stone 1995) and also overcome an intrinsic limitation of early visual pathways - the aperture problem (Wallach 1935). Each neuron of early visual stages such as primary visual cortex has a limited access to the image input due to both its small receptive field (i.e. <0.5° in foveal vision for primate V1: Hubel and Wiesel 1974; Dow et al. 1981; ~1–2° in foveal vision for macaque MT: Albright and Desimone 1987) and its broadly tuned direction selectivity (for a recent review, see Lennie and Movshon 2005). As the image of the object of interest will usually cover up to several degrees of visual field, local visual signals are most often ambiguous. In particular, elongated edges moving across a small receptive field will generate a 1-dimensional (1D) change in luminance distribution which is inherently ambiguous since it is compatible with a large number of 2-dimensional (2D) translations of the object (see Chap. 1). Given the properties of early motion detectors, motion direction orthogonal to the edge orientation will be most probably locally encoded as shown both by neuronal responses of MT neurons to elongated bars (Albright 1984; Pack and Born 2001) and the bias observed in the perceived direction of moving tilted bars (e.g. Wallach 1935; Lorenceau et al. 1993; Castet et al. 1993; Castet and Wuerger 1997). Motion integration is the key mechanism by which different local motion measurements are selectively combined within larger receptive fields to reconstruct the global, 2D velocity (i.e. speed and direction) of the object (see Braddick 1993 for a review). With 2D surfaces, different edge motions can be integrated to reconstruct this global motion using the Intersection of Constraints Rule (Fennema and Thompson 1979; Adelson and Movshon 1982). However, with more simple objects such as a single bar, other motion information must be extracted to solve the aperture problem. Neurons whose receptive fields are positioned over 2D features such as objects' corners, endpoints ("terminators"), or texture elements can measure their direction of motion accurately (see Fig. 8.1a). These local measurements must be integrated with the other, more ambiguous signals to reconstruct the 2D trajectory of the object. From a computational point of view, this integration stage poses two problems that are often overlooked. First, some computational rules must be used to automatically combine 1D and 2D local measurements. Several linear (vector sum or averaging) or non-linear (winner-take-all) rules have been proposed (Wilson et al. 1992; Nowlan and Sejnowski 1995; Löffler and Orbach 1999). We will show that the earliest part of tracking responses most probably follows the vector average direction while later parts are driven by a nonlinear estimate of global motion. Second, non ambiguous 2D motion must be propagated along the edges so that at some point in time, all local motion measurements will be coherent with the translation of a rigid object (Mingolla 2003; Bayerl and Neumann 2004, 2007). This would indicate that all direction-selective neurons activated sequentially along the object trajectory have solved the aperture problem (Pack and Born 2001).

The spatio-temporal dynamics of motion integration can have important consequences for tracking objects in a crowded environment (Masson and Stone 2002). Reversely, smooth pursuit is a powerful tool to probe the temporal dynamics of the neural solution for the aperture problem. This problem has been recently tackled by several studies and a coherent picture emerges for both human (Masson and Stone 2002; Wallace et al. 2005; Montagnini et al. 2006) and non-human (Pack and Born 2001; Born et al. 2002, 2006) primates. More important, these studies allow us to relate the temporal dynamics of behavioral and neuronal levels. Neuronal aspects of 2D motion integration are addressed elsewhere in this Book (Chap. 2 by Born, Tsui and Pack), and we will herein focus on the behavioral and computational approaches. The links between perceptual and motor aspects of motion integration are discussed by Lorenceau (Chap. 1) and Hafed and Krauzlis (Chap. 13). A key aspect of all these pieces of work is that very similar, but not identical stimuli have been used, offering a unique integrative perspective.

The following key points can sketch the basic properties of motion integration in the context of smooth pursuit in primates. First, when presented with single tilted (±45°) lines, initial pursuit direction is always strongly biased toward motion direction orthogonal to bar orientation (see movies 1–4) (Pack and Born 2001; Masson and Stone 2002, Born et al. 2006; Montagnini et al. 2006). Similar observations have been made for reflexive tracking elicited with single gratings presented behind tilted apertures (i.e. barber-poles), albeit at shorter latency in both humans (Masson et al. 2000) and monkeys (Barthélemy et al. *unpublished observations*). On average, deviations up to 25–30° away from the actual bar motion direction were observed in humans (Montagnini et al. 2006) and monkeys (Born et al. 2006). Using line-drawing objects with four edges (i.e. diamonds, see Fig.8.1c), Masson and colleagues showed that initial pursuit direction was nearly aligned with the average of the 4 velocity vectors normal to the edges (Masson and Stone 2002; Wallace et al. 2005). The largest biases (~40°) were observed with low contrast and fast moving targets - that is with high noise levels when considering motion detectors with small receptive fields (Wallace et al. 2005).

Nevertheless, a key difference between initial tracking error and misperception of object motion direction (see Lorenceau et al. 1993; Castet et al. 1993) is that only the former is always observed, including at maximum target contrast and with fast speed target motion presented in the fovea. Moreover, biases in initial tracking direction are seen in every trial, albeit of variable amplitude as shown by the distributions illustrated in Fig. 8.2a. We superimposed distributions of initial tracking error observed with different targets of same contrast and speed - an upright line target where 1D local and 2D local/global motion directions are aligned; a tilted

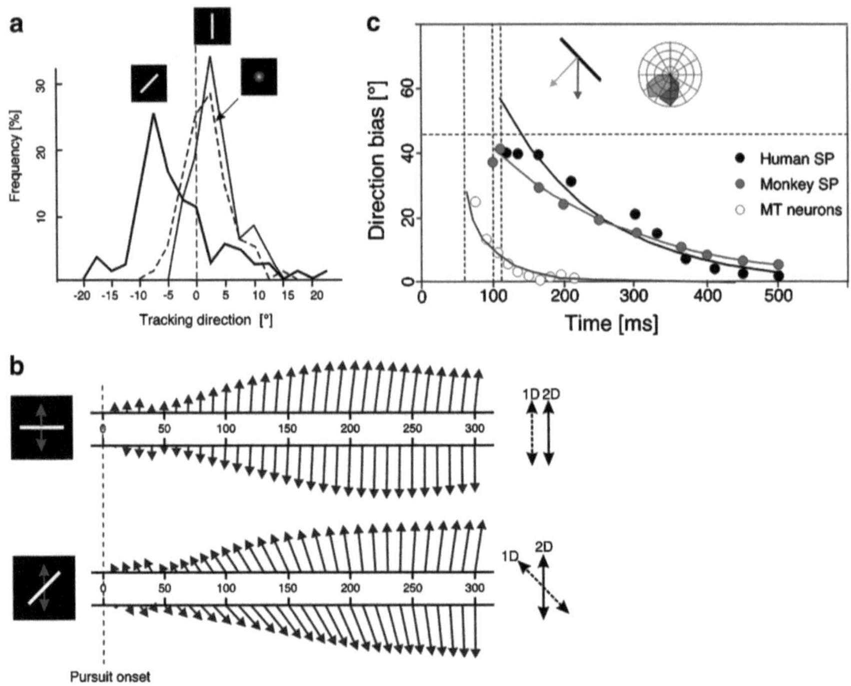

Fig. 8.2 Dynamics of a biological solution for the aperture problem. (**a**) Frequency distribution of initial tracking direction errors, for each type of motion stimuli: a small blob, an upright and a 45° tilted line. Trials for leftward and rightward 2D motion have been mixed. (**b**) Instantaneous eye velocity vector, over 300 ms of pursuit for lines moving either along horizontal or oblique axis. With the former, pursuit is initiated in the direction orthogonal to the bar, the actual 2D velocity. With tilted bars, smooth pursuit is initiated along the oblique direction, normal to bar orientation and is then slowly rotated to be aligned with the vertical axis after more than 200 ms of movement. See movie 2. (**c**) Temporal dynamics of the direction bias measured in response to single moving bars, for MT neurons (replotted from Pack and Born 2001), monkey (from Born et al. 2006), and human (Montagnini et al. 2007) pursuit.

line where 1D and 2D motion are 45° away from each other, and a small Gaussian blob. One can see that large variability in pursuit direction can be observed with moving bars, illustrating the influence of ambiguous local motion measurements on the initial phase of pursuit initiation. This result is reminiscent of the observation made earlier by Castet et al. (1999) that global motion computation must be considered in a probabilistic sense, where the visual machinery infers direction and speed from noisy measurements of the different motion cues present in the image and their relative weighting (Weiss et al. 2002). It is also evident that direction of ocular responses to tilted bars has a uni-modal distribution, shifted towards the 1D motion direction (vertical dotted lines). In short, initial bias in pursuit initiation is a very robust phenomenon, highly consistent on a trial-by-trial basis.

Once tracking has been initiated in a direction away from the actual object motion direction, how long does it take before this initial tracking error is fully corrected? The time course of tracking direction reflects, at least in part the temporal dynamics of motion integration. Such dynamics have been investigated under two different aspects. The first question to address is when the different motion signals are made available to the integration stage. The fact that, with full contrast targets, initial smooth pursuit direction is not perfectly aligned with the perpendicular prediction indicates that some influence of motion signals other than 1D cues are already present at pursuit onset - that is ~100 ms after motion onset. Using large motion stimuli and optimal conditions for fastest ocular movement initiation (see Miles and Sheliga, Chap. 11), Masson and colleagues have been able to dissect out the different latencies of 1D and 2D local motion signals. With both barberpoles (Fig. 8.3) and plaid motions (see Chap. 2), they showed that in humans, reflexive ocular following eye movements are initiated in the direction perpendicular to grating orientation, at latency ~85 ms. It was only 20 ms later that tracking direction started to deviate towards global, pattern motion direction, thereby signaling when 2D-related motion cues were becoming available to the motion integration stage (Masson et al. 2000; Masson and Castet 2002; Masson 2004).

Such 20 ms latency difference remains constant over a large range of total stimulus contrast (Masson and Castet 2002; Barthélemy et al. 2008) and is not affected by the relative position of either 1D or 2D cues within the visual field (Masson et al. 2000). Masson and coworkers have replicated these findings in macaque monkeys where ocular following is initiated with latency ~55 ms (Miles et al. 1986). The earliest influence of 2D cues can be seen only ~10–15 ms after response initiation. Altogether, these results demonstrate that 1D and 2D local motion cues are extracted with different latencies. They are consistent with a delayed 2D motion computation, using for instance end-stopping properties in area V1 (Hubel and Wiesel 1965; Pack et al. 2003; Pack et al., 2004). In the same vein, Smith et al. (2005) found that pattern-motion selectivity takes about 20 ms more to emerge than component-motion selectivity in a large population of MT neurons tested with plaid motions. This computation must involve a local combination of motion signals but it remains unclear how and at which stage (V1 or MT) such computation is done (Majaj et al. 2007). Regardless of the mechanism used by the primate brain to extract 2D local motion cues, our results suggest that 1D and 2D motion signals are not available at the same time for computing global motion. The tiny, but largely significant latency difference might explain the initial tracking direction error seen for smooth pursuit eye movements. At latency ~100–120 ms, both 1D and 2D motion signals are now available, albeit with different weights. Several studies have suggested that earliest 2D motion perception (Yo and Wilson 1992) or tracking initiation (Masson et al. 2000; Masson and Stone 2002) were best predicted in a direction near the vector average of 1D and 2D motion (Fig. 8.3c). Moreover, the influence of contrast or speed upon initial tracking direction might reflect the relative weight given to these different vectors in the averaging computation (Wallace et al. 2005). We will return to this when considering the various models of 2D motion integration.

The second aspect of temporal dynamics is the time course of tracking error correction. Pack and Born (2001) found that MT neurons change their preferred direction over ~60 ms when presented with a set of tilted lines drifting across their

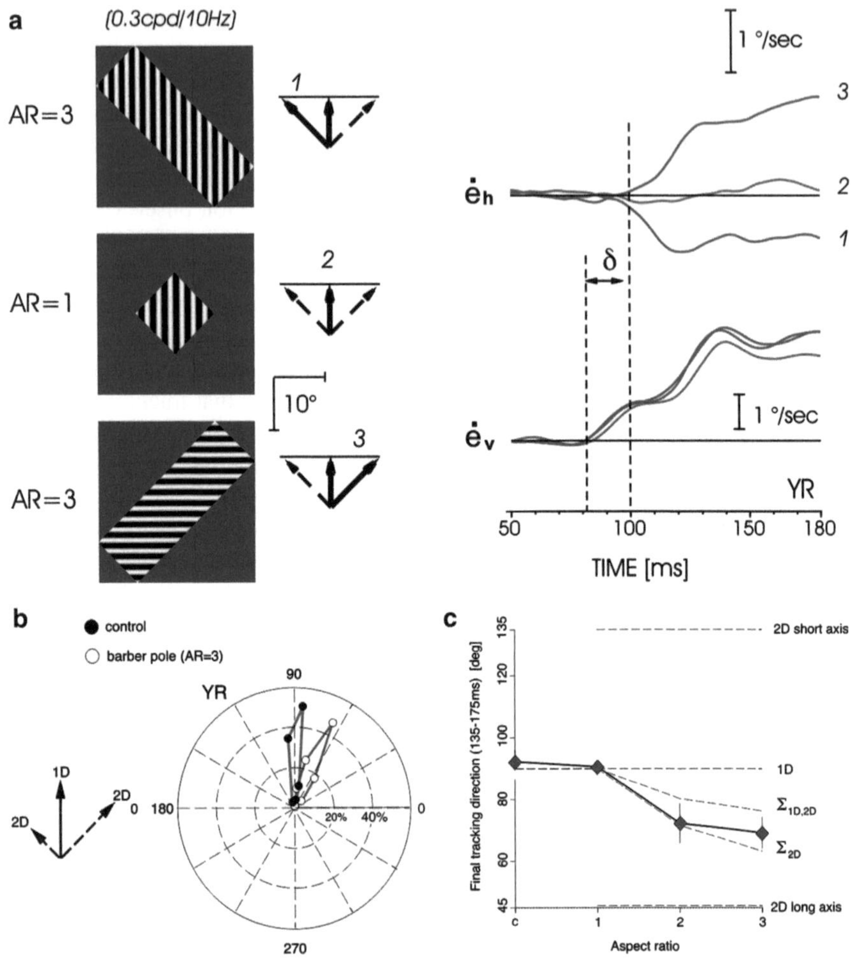

Fig. 8.3 Early dynamics of 2D motion integration. (**a**) A barber pole containing a horizontal grating moving upward is presented behind tilted apertures of aspect ratio (AR) either 1 or 3. With AR = 1, vector average of all 1D (grating) and 2D (line-endings) motion cues point towards the grating drifting direction (i.e. upward). With AR = 3, the vector average solution is biased towards the long axis of the aperture, so that global motion direction is diagonal along the upward-leftward axis (case 1) or the upward-rightward axis (case 2). Right-hand plot shows mean horizontal and vertical eye velocities for each case. Control condition (AR = 1) is shown in *blue*: reflexive tracking is initiated in the vertical direction at the usual latency of ~85 ms. No horizontal deflection is seen. The same vertical eye movements are initiated in cases 1 and 3. However, after a delay of 20 ms, eye velocity deviates towards either rightward (case 3) or leftward direction, indicating the influence of line-endings motion. (**b**) Direction distributions of late part of ocular following (i.e. ~150 ms after motion onset). Ones can see that an average 30° deviation is found with barber-pole of AR = 3, roughly consistent across all trials. (**c**) Mean tracking direction at the end of the open-loop period for a control condition (upright square aperture: 0) and different aspect ratios. *Dotted lines* show the predicted directions from the vector average of either 2D motions alone or 1D and 2D motions together. Modified from Masson et al. (2000). (*see Color Plates*)

receptive field. At response onset, direction selectivity of MT neurons was aligned to the motion direction orthogonal to the edges. Then, direction selectivity gradually changed over time to become insensitive to edge orientation and, therefore, to be aligned with the actual 2D trajectory of the stimulus. In the same study, Pack and Born (2001) measured how pursuit direction evolves over time when presented with a similar stimulus (a single, tilted bars) and found that pursuit slowly converge onto target direction in ~200 ms. Similar dynamics were found in more detailed studies, in both humans (Masson and Stone 2002; Wallace et al. 2005) and monkeys (Born et al. 2006). Wallace and colleagues used a double exponential function to estimate both the peak direction error and its timing, and fit the time course of tracking direction errors. The reduction of direction error over time is well rendered by an exponential decay and therefore we have fitted this function to both the neuronal data and the monkey and human pursuit data, taken from these previously published studies (Fig. 8.2c). All data were aligned with respect to visual motion onset to capture the relative dynamics of motion integration at different levels. The comparison indicates that, in macaques, neural and behavioral time constants are of ~60 and ~100 ms, respectively. Moreover, time constants were nearly identical between monkey and human pursuit data. Overall, this indicates that pursuit eye movements most certainly reflect the dynamics of the neural solution of the aperture problem. Area MT is critical for the initiation of smooth pursuit in monkeys: neuronal responses led eye movements and population activity encodes retinal speed and acceleration (Lisberger and Movshon 1999; Priebe and Lisberger 2004). The first 100 ms of neuronal activity convey most of the direction information that is needed for driving smooth pursuit (Osborne et al. 2004). Given that very small numbers of spikes are emitted by MT neurons during this first hundred of ms, the neural mechanism that decides when and how to initiate smooth pursuit must do so on the basis of a rather limited number of spikes (Osborne et al. 2004). This suggests that the temporal window for integrating motion direction information is rather small and therefore the changes in neural pattern of activity such as those seen by Pack and Born (2001) will be passed almost unchanged to downstream oculomotor pathway for generating motor commands. The low pass properties of the oculomotor plant can then explain the differences between the neural and behavioral dynamics. These experiments once more demonstrate the interest of tracking responses for probing visual motion computation (Lisberger et al. 1987; Miles 1998; Masson 2004). There is however one aspect of the behavioral temporal dynamics which has not been taken into account so far. Pursuit system acts as a negative feedback loop. Wallace et al. (2005) have shown that correction of initial pursuit direction error began to be corrected before the closing of the oculomotor loop (see Fig. 8.4a: time-to-peak of direction error occurs well before the closing of the oculomotor loop, as indicated by the vertical dotted line). This further suggests that dynamics of eye movements largely reflect the neuronal dynamics. However, a full description of the oculomotor responses to ambiguous motion will need to consider the properties of the oculomotor loop and how they influence the dynamical solution of the aperture problem (see Montagnini et al. 2007). Hafed and Krauzlis (2006;

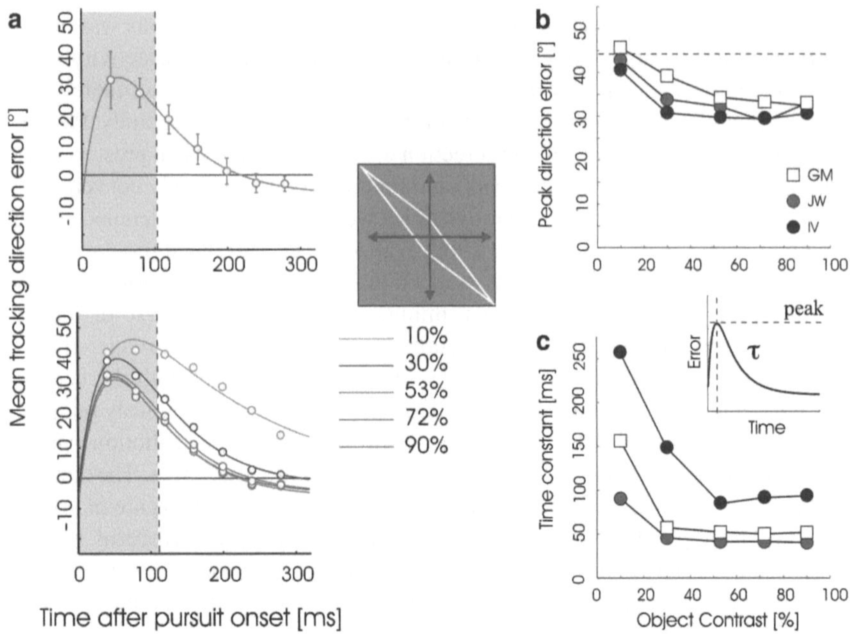

Fig. 8.4 Dependence of tracking direction errors on visual properties of moving object. (**a**) Temporal dynamics of the tracking direction error, obtained for rightward motion of a tilted line with different contrast. *Upper plot* shows the dynamics at full contrast (i.e. white bars against a black background). *Lower plots* illustrate the results with targets at contrast ranging from 10 to 90%. (**b**) Dependency of initial tracking direction error upon target contrast, for three subjects. (**c**) Relationship between time constant of the tracking direction error reduction and target contrast. Modified with permission from Wallace et al. (2005). (*see Color Plates*)

see Chap. 13) have shown that pursuit itself plays a role in integrating edges into a coherent global diamond motion. They however only consider steady-state tracking. Future work shall investigate how long it takes before eye movements behavior interacts with perceptual motion integration.

The temporal dynamics of the computation solving the aperture problem varies with several parameters of the visual stimuli. In particular, for low contrast targets, Wallace et al. (2005) found both a higher initial bias in pursuit direction and a slower decay of the tracking error (Fig. 8.4b-c). As mentioned above, faster target speed resulted in larger tracking errors. These results were interpreted as further evidence that target speed is analyzed by MT neurons of rather small receptive fields. Low contrast and high speed both result in smaller signal-to-noise ratio and thus a smaller contribution of 2D features motion is expected. These results are consistent with the facts that properties such as end-stopping in area V1 (Sceniak et al. 1999) and center-surround interactions in both V1 and MT areas (Sceniak et al. 1999; Pack et al. 2005) vanish at low contrast.

We stated above that initial pursuit direction is most certainly based on a vector average computation of 1D and 2D motion direction. This prediction is however extremely difficult to be verified quantitatively with single line-drawing objects, since the relative weights of 1D edges and 2D features are difficult to compute (Wallace et al. 2005; Montagnini et al. 2007). Several attempts have however been made to qualitatively test this aspect of motion integration. Increasing the number of 2D local cues not only strongly reduces the initial bias but also accelerates the correction process (Wallace et al. 2005). Inversely, lengthening the pursued titled bars induced a longer time course for reducing pursuit direction error down to the 2D trajectory (Born et al. 2006). It however remains unclear if a slower decay of direction errors reflects the time needed to propagate 2D motion information along the edge or a weaker contribution of these 2D features, due to their more eccentric location. In humans, we also found that longer bars resulted in stronger biases in initial pursuit direction. These results were used to constrain a Bayesian model of motion integration, in which a dynamical decision about the target motion direction must be dynamically computed using the only available noisy pieces of evidence - edges and line-endings motion (Montagnini et al. 2007).

8.3 Modeling the Temporal Dynamics in an Inferential, Recurrent Framework

Motion integration is often seen as an inferential process where the brain attempts to estimate the most probable 2D motion trajectory from noisy and ambiguous inputs that are represented within populations of local motion detectors (Simoncelli et al. 1991; Weiss and Fleet, 2002). Simoncelli and colleagues have proposed a Bayesian model rendering the key aspects of motion integration as measured psychophysically (Weiss et al. 2002; Stocker and Simoncelli 2006). Using probabilistic representations of 1D motion cues (i.e. motion likelihoods in the velocity space) and a *Prior* distribution favoring slow speeds (i.e. a Gaussian distribution centered at zero speeds), this model simulates the bias observed in perceived direction towards the normal velocity (or, equivalently a vector average of the different 1D cues) at low contrast and slow speeds, with both line-drawing objects (Lorenceau et al. 1993) and plaids (Yo and Wilson 1992). The same model has been used to explain the lower initial eye acceleration of pursuit to small targets of low contrast or high spatial frequency (Priebe and Lisberger 2004). Very recently, Barthélemy et al. (2008) have shown that the initial contrast dynamics of 1D- and 2D-driven components of ocular following responses can be explained by the distributed representations of 1D and 2D motion cues within the same inference framework.

The Bayesian model proposed by Weiss et al. (2002) cannot explain the temporal dynamics of motion integration. Moreover, the actual contribution of 2D motion cues was rather unclear. In the original version of their model, the time course of perceived motion was modeled as lowering the variance of motion likelihoods, therefore reducing the effect of the *Prior* distribution. Time course of perceived direction thus mimics the changes observed when increasing target contrast. As a consequence, the model fails to account for the large initial bias observed at full contrast with smooth pursuit

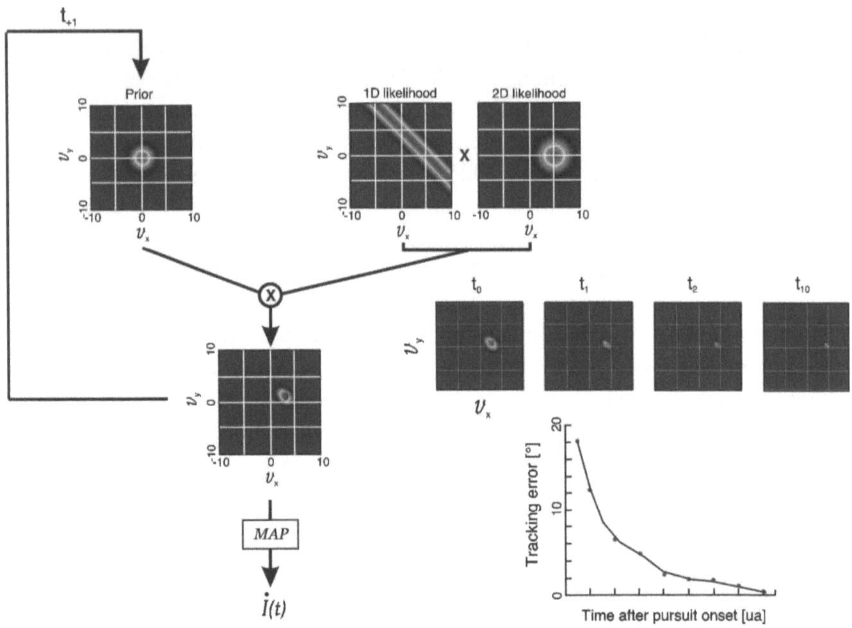

Fig. 8.5 A dynamical Bayesian model of motion integration. Ambiguous (i.e. edge) and non-ambiguous (i.e. features) motion information are represented as independent likelihood distributions in the velocity space. Following the earlier models by Weiss et al. (2002) and Barthélemy et al. (2008), these likelihoods are multiplied with a Prior distribution favoring slow speeds, to give the A Posteriori distribution of target velocity. The maximum of this distribution (MAP) is selected to give the best estimate of target velocity $\dot{I}(t)$ that can be used by the pursuit system. Such distributed A posteriori representation is then used as Prior information for the next iteration, assuming that likelihoods are left unchanged. Such recurrent processing is accumulating evidence about 2D target motion and therefore, the A posteriori distribution gradually evolves over time. Several time steps are illustrated (from t_0 to t_3), as well as the complete time course of tracking direction error. This model mimics two important aspect of the experimental data illustrated in Fig. 8.2: the initial tracking direction error and its time course (modified from Montagnini et al. 2007). (*see Color Plates*)

eye movements, unless 2D cues are ignored. Moreover, one needs to account for the varying time-course observed with parameters such as contrast or bar length (Wallace et al. 2005; Born et al. 2006). Instead, we have explored such a dynamical version of the Bayesian model at a more formal level, where the *Prior* distribution is iteratively updated (Montagnini et al. 2007). Instead of arbitrarily changing the *Prior* distribution, we decided to update it using the *A Posteriori* distribution obtained from the preceding computing step. By doing so, the best estimate of target velocity that is extracted from a *Maximum-A-Posteriori* (MAP) computation gradually converges towards the actual 2D target trajectory (Fig. 8.5). Such recurrent Bayesian model implements one particular type of adaptive filtering known as Kalman filters where noisy inputs are gradually estimated by adapting internal filter parameters from recurrent outputs (Anderson and Moore 1979).

The model parameters - in particular the distributions of velocity likelihoods - were estimated from eye movements data collected in humans in response to moving tilted lines of various orientations, length, and contrast. As already stated above, one critical problem of the model is to represent the weighted contribution of 1D and 2D cues, and thus their relative variances. We attempted to solve this problem by comparing eye acceleration of smooth pursuit responses to either very elongated bars (i.e. "pure" 1D stimuli) or small luminance blobs (i.e. "pure" 2D inputs). A second critical aspect is to estimate the relative contribution of the different noise sources (i.e. sensory or motor). Following an earlier suggestion by Osborne et al. (2005), we assumed that most of the variability found in initial ocular responses reflects the noise level within the biological motion processing. Lastly, in the first version of the model we did not attempt to implement the oculomotor negative feedback loop and in particular two key aspects - its consequences onto the input retinal image (i.e. on the 1D and 2D velocity likelihoods) and the low-pass properties of the oculomotor plant. Future work will incorporate these aspects. Overall, we may assume that the MAP computation used to extract target velocity from probability distributions is performed within a spatial, not a retinal frame of reference.

Montagnini et al. (2007) demonstrated that the model can account for both the initial pursuit direction and the time course of direction tracking errors. Quantitatively, the relative contribution of 1D and 2D likelihoods, relative to prior distribution was estimated from the variance of responses observed with two perifoveal dots (i.e. "pure" 2D motion input) or one very long upright bar (i.e. "pure" 1D motion input). This was done for several target speeds. The estimates of the variance of the prior and the two independent likelihood distributions are estimates of hidden variables, which are considered as characterizing the internal inferential processes underlying motion integration. Importantly, because these variables are fully constrained by experimental data, they provide a first general validation of the model. Indeed, the estimated prior variance turned out to be roughly constant – as expected for a prior – across a threefold increase in target speed or in the stimulus log-contrast. Second, our estimates for the 1D and 2D likelihood variance for different target speed provide evidence of a monotonic increase of the variance with target speed. This functional relationship was not predicted by the original Bayesian model of 2D motion perception (Weiss et al. 2002), but was proposed in a later study fitting the model to psychophysical data for speed discrimination (Stocker and Simoncelli 2006). It also appears to be reasonable in the light of Weber–Fechner's law. Finally, our model provided an estimate of the tracking error across several iterations of the updating routine. With only two free parameters (one to adapt the model arbitrary iteration step to the experimental sampling period in milliseconds and one to account for a global energy-scaling factor), the model predictions were in good agreement with the time-course of the experimentally measured tracking error across different target speed and contrast values.

Dynamical Bayesian models are one particular type of adaptive filters, such as Kalman filters (Kalman 1960). Implementation of a Kalman filter by means of recurrent networks has been proposed for estimating motion flows and computing sensorimotor transformations from time-varying inputs (e.g. Rao 2004; Denève

2004; Denève et al. 2007). A similar recurrent architecture was proposed by Bayerl and Neumann (2004, 2007) to solve the aperture problem and then to reconstruct veridical 2D retinal flows for simple objects. This model suggests that temporal dynamics of the neural solution for solving this aperture problem, such as found by Pack and Born (2001) can be explained by the dynamics of the recurrent processing between areas V1 and MT. Overall, the output of motion integration computation is fed back to the motion detection stage (presumably V1) to resolve local motion ambiguities. However, for the implementation of such recurrent networks, several unknown information about the dynamics of motion integration in macaque cortex is required such as the exact role of MT feedback or the contribution of other local feedbacks such as one recurrent signal about local luminance distribution (i.e. form information) to control the diffusion of motion information within the V1-MT network (Tlapale et al. 2007). Similar interactions between local form and motion cues have been proposed on psychophysical (Lorenceau et al. 1993; Lorenceau and Alais 2001; see Chap. 1) and physiological (e.g. Pack and Born 2001; Hueng et al. 2007) levels and are used in the dynamical models developed by Grossberg and colleagues (Chang et al. 1998; Berzhanskaya et al. 2007; see Chap. 15).

8.4 Mixing Retinal and Non-Retinal Cues for Solving the Aperture Problem

One striking aspect of misperceptions due to the aperture problem is their genuine and automatic aspect. As said above, initial bias in pursuit initiation is seen on every trial in both humans and monkeys, albeit with some variability in its magnitude (Wallace et al. 2005; Born et al. 2006). Montagnini et al. (2006) have investigated whether cognitive factors can reduce or even eliminate it. The short cut answer is no! First, cueing subjects about object's shape or orientation does not reduce the initial bias towards the vector average direction (see also Wallace et al. 2005). This is consistent with previous psychophysical findings that static viewing of complex shapes does not help in resolving their ambiguous motion (Lorenceau and Alais 2001). Second, initial tracking errors remained remarkably stable over more than two hundred trials. We have recently extended these results in macaque monkeys and found that repetition of tilted bar motions over more than 1,000 trials cannot eliminate it. Similar results were reported by Born et al. (2006). Thus, learning a small set of bars orientation and directions of motion cannot overcome the aperture problem.

In the preceding experiments, motion was always randomized across at least two directions. When a single motion direction is presented within one block of trials, anticipatory tracking eye movements are seen in both humans (Kowler et al. 1985; Freyberg and Ilg 2008) and monkeys (Missal and Heinen 2004; de Hemptinne et al. 2006). This anticipatory pursuit is characterized by a slow eye velocity buildup, starting about 200 ms before target motion onset, in the direction of the predicted retinal slip. Such anticipatory eye movements depend upon neuronal signals originating from the pursuit areas within the frontal lobe, in particular the frontal eye field (FEF) (MacAvoy et al. 1991) and the supplementary eye field (SEF) (Missal and Heinen 2004).

That predictive mechanisms are used to maintain accurate pursuit eye movements despite the rather long delay of the oculomotor feedback loop and the fluctuations present in the retinal input, has already been demonstrated, in particular when using periodical target motion (Becker and Fuchs 1985; Barnes 1993; Kettner et al. 1996). Moreover, anticipatory smooth responses have been shown to reflect cognitive expectation about upcoming target motion, based on recent target motion history (Kowler and Steinman 1979a, 1979b; de Hemptinne et al. 2006). We asked how these predictive mechanisms would affect motion integration. First, we ran blocks of trials in which bar orientation and motion direction were kept constant so that 2D trajectory became highly predictable. As expected, anticipatory responses were seen as early as 200 ms before target motion onset, for both upright and tilted bars (Fig. 8.6a). With tilted bars moving horizontally, slow anticipatory build-up was seen only in the horizontal eye velocity profiles, indicating a pure horizontal anticipatory pursuit. Target motion onset resulted in strong eye acceleration at normal pursuit latency. However, these visually-driven responses were along the oblique axis for tilted bars and were not different from the biased responses observed in the unpredictable conditions. In other words, anticipatory responses were driven along the 2D motion direction whereas visually-driven responses were still initiated toward the 1D motion direction. We concluded that predictive signals about 2D target velocity cannot be used to solve the aperture problem at motion onset. In the Bayesian framework introduced above, this would indicate that the initial *a priori* assumption is not influenced by either the recent history of target motion or by predictive signals. At pursuit onset, this information may be used to control the internal gain of visual-motor transformation (see Heinen et al. 2005; Kodaka and Kawano 2003; Tabata et al. 2005, 2006) but does not seem to be used by the visual mechanisms involved in solving the aperture problem.

If predictive signals about actual target trajectory cannot be used at pursuit initiation, the question remained open to see whether these internal representations can be used during steady-state eye movements, that is, once the aperture problem has been solved. Masson and Stone (2002) investigated this question using a moving tilted diamond (Fig. 8.6b). When set into motion, these tilted diamonds elicited strong transient bias towards the vector average direction. Once tracking direction was perfectly aligned with the actual target trajectory, the pursued line disappeared during 100 ms on half of the trials. A drop in pursuit eye velocity was observed, so that when the target reappeared a small velocity error had to be measured and corrected (Becker and Fuchs 1985). A smooth acceleration of the eye was always seen but only along the 2D target trajectory (i.e. purely along the vertical direction in the examples showed in Fig. 8.6b). A transient bias toward the 1D motion direction was never observed indicating that the estimation of retinal image velocity at reappearance was independent of the shape of the object. The absence of tracking direction error cannot be explained by the slow retinal image velocity at target reappearance ($<2°$ per second) since similar target speed presented at initiation did produce tracking biases (Masson and Stone 2002; Wallace et al. 2005). These results suggest that the

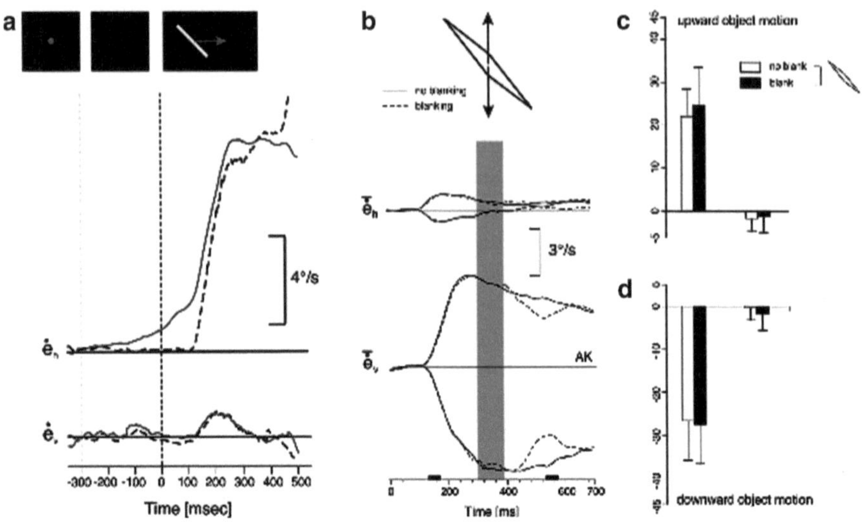

Fig. 8.6 Interaction between target visual motion and trajectory prediction in humans. (**a**) Smooth pursuit responses to a moving tilted bar, moving rightward. Trials started with a red fixation dot, followed by rightward motion of a tilted bar, presented after a 200 ms gap. Rightward target motion was presented either interleaved with leftward motions (*broken curves*) or blocked (300 consecutive trials). When leftward and rightward motions were interleaved, pursuit onset arose with a latency of ~100 ms and presented the vertical bias seen with tilted targets. When target motion was fully predictable (i.e. block of rightward motion), anticipatory pursuit was clearly seen during the gap period. Anticipatory pursuit started at target fixation offset and built-up until ~100 ms after target motion onset, where the visually-driven response was initiated. Clearly, anticipatory pursuit was purely horizontal and left intact the transient bias present in the visually-driven component (adapted from Montagnini et al. 2006). (**b**) Pursuit of a tilted diamond was initiated along the oblique direction, corresponding to the vector average prediction. After 300 ms, pursuit direction was correctly aligned with the vertical target motion (continuous lines). On half of the trials, pursuit was briefly (100 ms) blanked. Target disappearance resulted in a decrease in eye velocity (broken line). This tracking error was corrected ~100 ms after target re-appearance, indicating the contribution of retinal target image motion. However, such visually-driven correction was always purely vertical: the directional bias was never observed. (**c**, **d**) Quantitative estimates of tracking direction errors measured either after target first appearance or after target reappearance (measurement windows are illustrated as black short horizontal bars in (**b**)). No difference was observed between trials with or without blanking, indicating that corrections of pursuit velocity are based upon 2D target motion (from Masson and Stone 2002).

aperture problem is solved only once during a single pursuit event and the motion integration stage is able to stabilize this solution despite large, transient fluctuations in the retinal inputs.

Control of pursuit eye movements intermingles low- and high-level cues to maintain gaze onto the target (Masson and Stone 2002; Churchland et al. 2003; Madelain and Krauzlis 2003). For instance, Churchland et al. (2003) showed that the decrease in steady-state eye velocity in monkeys observed during a transient disappearance of a visual target was lesser when the occluder was visible than when the target was simply extinguished for the same duration. This suggests that visual cues about the spatial layout of the visual scene have an impact on the internal gain of the transformation. This gain can also be adapted by training under similar circumstances (Madelain and Krauzlis 2003). These results suggest that the brain uses both retinal and extra-retinal cues to maintain a permanent representation of target motion as needed for accurate pursuit in complex environments where moving objects can be transiently masked, shadowed, or illuminated. On-going work in Masson's group is investigating the circumstances under which one can force the motion integration stage to recompute 2D trajectory by solving the aperture problem on-line during steady-state pursuit, for instance by changing target shape or trajectory before it reappears. An alternative would be to mask only some informative parts of the visual object to probe how the integration of different motion signals can be quickly and efficiently recalibrated.

8.5 Tracking the Invisible Using Extra-Retinal Cues

The strong coupling between visual motion processing and smooth ocular pursuit has been taken as evidence that tracking eye movements exclusively depend on physical factors such as retinal motion. However, tracking responses can be both triggered and controlled using internal signals related either to other action systems such as arm movements (e.g. Vercher et al. 1995) or perceptual processes (e.g. Stone et al. 2000; Krauzlis and Adler 2001). Steinbach (1976) originally suggested that human subjects pursue the perceptual rather than the retinal stimulus: subjects were able to track the hidden, invisible corner of a rectangle or the invisible hub of a rolling wheel. Using simple line-drawing objects, Stone and colleagues have been able to demonstrate that human subjects pursue the coherent motion of an object, which is partly occluded, rather than the mere retinal image motion (Beutter and Stone 2000; Stone et al. 2000; see also Hafed and Krauzlis, Chap. 13).

Humans as well as monkeys can track imaginary figures defined only by parafoveal cues (man: Wyatt et al. 1994; Ilg and Thier 1999). It is not the foveal enclosure as previously suggested by Wyatt et al. (1994), which is critical for the tracking of imaginary figures. Similar steady-state gain was observed if the monkeys tracked the hidden corner of a simple triangle. However, if only a single spot was tracked parafoveally (6° eccentricity, which is similar to the line-endings eccentricity in the partially occluded triangles illustrated in Fig. 8.1d), the gain was clearly lower (Ilg and Thier 1999). This finding provides further support to the notion that pursuit is directed towards the perceptual target instead of the retinal target (Steinbach 1976). Moreover, high gain pursuit eye movements can be achieved without visual stimulation of the central visual field.

Traditionally, the existence of extra-retinal signals in pursuit-related activity was documented by brief disappearances of the pursuit target (Newsome et al. 1988). Human observers describe this condition as if the pursuit target would briefly move into the screen like a toy train in a tunnel. However, the eye velocity drops immediately as a consequence of target disappearance. So the interpretation of sustained or changed neuronal activity during the absence of the target is not completely conclusive. This problem can be circumvented if an imaginary target is used as pursuit target (Fig. 8.7). As already mentioned, during pursuit of the imaginary target, there is no stimulation of the central visual field. The results of these experiments are quite clear: no activity was observed in the middle temporal area MT when the monkey tracked the imaginary target. Thus, the activity recorded from MT does not include extra-retinal signals; MT can be best described as a purely visual motion decoder. In contrast, in the medial superior temporal area (MST) and in the frontal eye field (FEF) neurons were found whose activities were not significantly different during pursuit of the real and imaginary target, respectively (Ilg and Thier 2003). This suggests that these neurons are not only driven by visual, retinal image motion signals, but also receive an extra-retinal input related to the ongoing eye movements. Additional experiments have shown that MST also receives head-movement related signals (Thier and Erickson 1992; Ilg et al. 2004). The integration of retinal image motion, eye and head movement related information provides the ground for representing the trajectory of a tracked target in an external frame of reference.

8.6 Reacting and Predicting: Two Independent Recurrent Loops?

By investigating both reflexive and voluntary tracking responses to simple motion stimuli, we have been able to dissect out the temporal dynamics of motion integration (Fig. 8.8a). Studies on ocular following have shown that rapid tracking eye movements can be initiated by a linear combination of local 1D motion measurements. Such averaging is restricted to local motion within the plane of fixation which is also the plane where the object of interest is located, thanks to the binocular

Fig. 8.7 Neuronal and behavioral responses to motion of real and illusory targets. When a real target (see Fig. 8.1d) is tracked, the observed pursuit-related activity can be sensory or motor in origin. It is impossible to reveal the nature of this activity. However, if the pursuit-related activity is also present when the monkey is pursuing the imaginary target (see Fig. 8.1d), the visual origin can be excluded. Eye (*black and red*) and target (*green*) position during pursuit of a real (**a**, **b**, **e**) and an imaginary (**b**, **d**, **f**) target are shown. Black represents the preferred direction of the recorded neuron, red gives the non-preferred direction. The neuronal activity is shown as raster display and spike density functions (width of Gaussian kernel 20 ms). Time zero is set to the onset of target motion. (**a**) and (**b**) details the response recorded from the middle temporal area (MT), (**c**) and (**d**) from the medial superior temporal area (MST), and E and F the frontal eye field (FEF). Note that the neuron recorded from MT did not respond during pursuit of the imaginary target. The activity of the MST and FEF neuron during pursuit of the real and imaginary target in preferred direction was not significantly different. (*see Color Plates*)

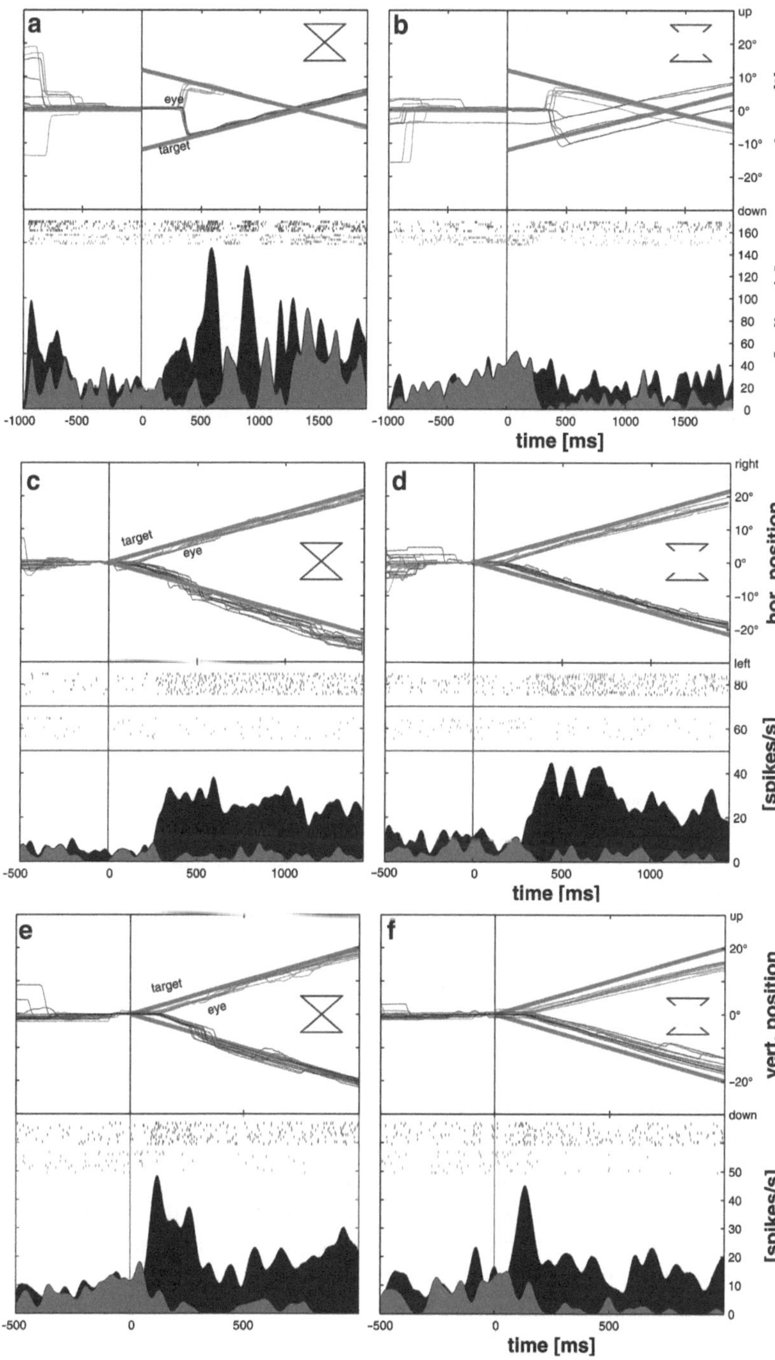

Fig. 8.7 (continued)

disparity tuning of early motion detectors (Busettini et al. 1996; Masson et al. 2001). The output of this linear integration is a coarse estimate of the object motion that will be progressively refined over time by integrating non-ambiguous local motion measurements such as 2D cues. One key idea is that the temporal dynamics comes from both the latency difference between each type of motion detection mechanism (see also Wilson et al. 1992) and the recurrent processing between motion detection and integration stages (Bayerl and Neumann 2004). The output of this recurrent processing is a visual representation of object 2D translation. Such mid-level representation can be under the influence of higher-level integration rules and other extra-retinal signals such as eye velocity memory or pursuit prediction to form a stable representation, which can be maintained despite large, transient fluctuations in the retinal image.

Accurate control of movements implies a rapid react to sudden appearance or change in the trajectory of the object of interest. However, under some circumstances, predictive signals regarding the upcoming target trajectory can play a major role in reducing the tracking error and ignoring spurious changes of the image. We have shown that, at pursuit initiation, retinal and extra-retinal information do not interact for solving the aperture problem and therefore computing the global motion of the newly presented stimulus (Montagnini et al. 2006). By contrast, the fact that correction of steady-state pursuit eye velocity at target reappearance (after a brief blinking) is not biased, suggests that a stable representation of global motion is now achieved, mainly from the interactions of retinal and extraretinal signals (Masson and Stone 2002). A similar combination of these signals may play a role in pursuing partly occluded objects (see Fig.8.1d; Ilg and Thier 2003). These results call for further experimental work in order to understand the precise interactions between retinal and extraretinal inflows in the context of complex visual targets. One intense field of research is to clarify the neural substrate of these interactions. The cortical pathway for pursuit is often seen as a cascade of cortical areas projecting onto subcortical targets. In brief, visual motion input for pursuit arises in area MT from V1 direction-selective cells (Movshon and Newsome 1996) and is transmitted via area MST to the frontal pursuit area (FPA) in the arcuate sulcus (FEF) and the SEF (see Krauzlis 2004 for a review). Interestingly, all three of these areas provide outputs to subcortical nuclei of the neural circuit for pursuit (Boussaoud et al. 1990). These areas are also most often recurrently connected, forming a cascade of cortico-cortical loops (Zeki and Shipp 1988). Here, we would like to suggest that the cortical pathway for pursuit might be viewed as a system articulated around two recurrent loops, with the MT-MST connection playing the role of a gear: the V1-MT loop is involved in retinal motion integration while a MST-FEF/SEF loop is involved in predicting target motion. That MT and MST are *functionally* connected mainly in a feedforward manner would ensure that the reacting and predicting mechanism are not fully interconnected. Thus, prediction about future motion cannot be propagated back to the earliest stage of motion processing, protecting the brain's ability to react to a sudden change in the object motion such as a change in trajectory during occlusion or in front of a visual obstacle.

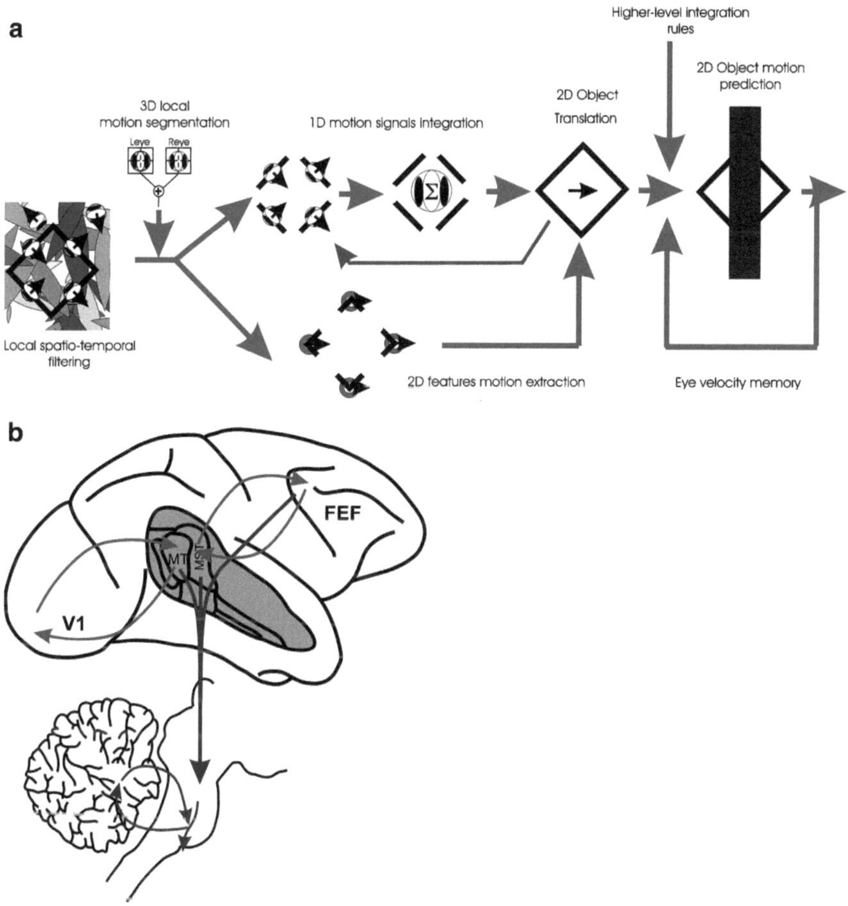

Fig. 8.8 Object motion representation as two cascaded recurrent loops. (**a**) Schematic representation of motion integration. See text for explanation. Modified from Masson (2004). (**b**) One hypothesis is that pursuit involves two recurrent loops with area MST playing the role of gear. A V1-MT recurrent loop can process retinal motion information dynamically. Such loop is critically dependent upon a constant visual input. On the contrary, a second recurrent loop involving MST-FEF maintains a stable representation of object motion despite transient changes in the images, using extra-retinal information. It is also responsible for anticipatory pursuit using internal representation of target motion. However, the fact that the two loops are not interconnected maintains the ability of the visual system to react to newly presented objects.

8.6.1 *What Are the Key Arguments for Such Two Independent Recurrent Mechanisms?*

We have seen above that inferring global motion direction from local motion is highly dynamical and that the neural solution of the aperture problem takes over ~100 ms to be fully completed (Pack and Born 2001). Several models have proposed a recurrent architecture to solve the motion integration problem (Koechlin et al. 1999; Bayerl and Neumann 2004; see Chap. 15 by Grossberg). In the model presented by Bayerl and Neumann (2004), the output of the motion integration stage is fed back on the local motion detection mechanisms and helps to amplify localized feature signals such as corners and line-endings and therefore to disambiguate the initial flow field estimation. Because motion detection and integration stages work at different spatial scales, diffusion of information in space and time helps to reconstruct the global motion signal over several computational cycles. This approach is similar to the dynamical Bayesian model that we have proposed above to render the time course of pursuit eye movements. This recurrent information processing might be implemented in the dense network of feedforward and feedback connections existing between area V1 and MT (see Bullier 2001; Sillito et al. 2006; see also Chap. 6). Thus, we propose that the V1-MT recurrent loop solves the aperture problem based on retinal information. Dynamics of MT neuron responses (Pack and Born 2001; Osborne et al. 2004; Smith et al. 2005) might reflect this iterative computation. Since MT neurons output convey sufficient information with only a few spikes (Osborne et al. 2004), these dynamics will be reflected all along their projection targets at cortical and subcortical levels implementing the sensorimotor transformation (Kawano 1999 for a review).

The pioneering work of Newsome et al. (1988) demonstrated that extinguishing retinal input for a short period of time during smooth pursuit eye movement silenced MT neurons for that duration. On the contrary, MST neurons showed a sustained activity during target blanking. This sustained activity in the absence of visual inputs was taken as evidence for an extra-retinal input to area MST, but not area MT. Ilg and Thier (2003) have shown that similar sustained activity was seen when the receptive field of MST neurons were centered onto the missing part of an imaginary target (see above). Again MT neurons were silenced in these conditions. Altogether, these results suggest that both retinal and extra-retinal signals are used to reconstruct target velocity independent of retinal input. As a consequence, at this stage, target motion representation becomes immune to transient and massive changes in the retinal image such as during occlusion (see Stone et al. 2000 for a similar idea in the context of perception and action coupling). Interestingly, evidence for extra-retinal signals related to target motion, independent of retinal image, are seen in both parietal and frontal pursuit-related neurons (Fukushima et al. 2002; Missal and Heinen 2004). Moreover, as described above, Ilg and Thier (2003) found similar responses properties in partly occluded target in both MST and FEF neurons during pursuit in macaques. Lastly, these neuronal activities related to pursuit are often seen during anticipatory pursuit, suggesting that the internal signal about incoming target motion might be generated within these populations of both FEF and MST neurons. Ilg (2003) suggested that anticipatory activity seen in area MST might be propagated backward from frontal areas, unveiling the role of the recurrent MST-FEF/SEF loop in target velocity representation and motion prediction. However, the absence of similar anticipatory responses in area MT as well as the fact that MT neurons reflect retinal image motion are both in favor of our hypothesis that such predictive target velocity signal is not propagated back to visual motion processing stages earlier than MST. Thus the link between areas MT and MST does act as a real gear which transmits information in one direction only. A critical test of this hypothesis would be to test how MT neurons will react to a transient blink of ambiguous motion stimuli such as used by Pack and Born (2001). If we were correct, direction selectivity of MT neurons shall be biased again by target orientation at reappearance. Moreover, tracking the time course of visual responses of neurons along the cortical pathways for pursuit using simple, line-drawing objects may provide us with large information about the dynamics of this representative specimen of sensorimotor transformation.

8.7 Conclusions

Early works on smooth pursuit eye movements were conducted by restricting visual motion to its simplest expression: a single dot over a dark background (see Lisberger et al. 1987; Ilg 2002). On the other extreme, several studies have shown that steady-state pursuit can be based on perceptual rather than on retinal input (Steinbach 1976; see Chap. 13) although the precise mechanisms by which perception can drive or influence eye movements are still unclear. In the same vein, cognitive influences on tracking performance has been highlighted by several groups (e.g. Kowler and Steinman 1979a, b). This opponency had led to radical different viewpoints on the oculomotor system, favoring either the "control system approach" ignoring perceptual and cognitive factors (Robinson 1986; Lisberger et al. 1987) or a more "ecletic approach" (Steinman 1986) open to these factors but less well defined in terms of control mechanisms. Recently, several attempts have been made by different groups to reconcile these two approaches, albeit a unified theoretical framework might still be missing (e.g. Madclain and Krauzlis 2003; De Hemptinne et al. 2006; Barnes and Schmidt 2002).

Using simple stimuli related to some fundamental aspects of biological motion processing, we have presented here the work of several groups who have highlighted key aspects of visual motion in the context of smooth eye movements. In particular, investigating the temporal dynamics of these processes appears to be essential in unveiling the underlying neural mechanisms. Going from a coarse estimate of retinal motion to a stable representation of 2D object motion trajectory in less than 300 ms unveils a coarse-to-fine strategy in controlling our actions. Put in more conventional terms for motor control, we suggest that smooth pursuit is an exquisite example of a general strategy that builds over time, a complex response on the root of reflexes. "*Launch first. Adjustments will follow*" might be the sensorimotor counterpart of the coarse-to-fine dynamics observed in sensory systems (e.g. Menz and Freeman 2003; Pack and Born 2001; Ringach et al. 1997). In the future, investigations concerning how these dynamics are affected under different contexts shall open the door to new, exciting findings.

8.8 Supplementary Materials (CD-ROM)

Movie 1 Pursing a single spot (file « *8_M1_Blobtrace.avi* ») Relative position of the eye and a single spot, moving rightward at 10° per second. Each position is the average over a 40 ms time window.

Movie 2 Pursing an upright single line (file « *8_M2_VertLinetrace.avi* ») Relative position of the eye and a single upright line target, moving rightward at 10° per second. Each position is the average over a 40 ms time window.

Movie 3 Pursing a tilted (CCW) line (file « *8_M3_TiltedLineCCWtrace.avi* ») Relative position of the eye and a tilted (–45°) line, moving rightward at 10° per second. Each position is the average over a 40 ms time window. Initial bias in eye velocity can be seen, driving the eye away from the center of the line.

Movie 4 Pursing a tilted (CCW) line (file « *8_M4_TiltedLineCWtrace.avi* ») Relative position of the eye and a tilted (+45°) line, moving rightward at 10° per second. Each position is the average over a 40 ms time window. Initial bias in eye velocity can be seen, driving the eye away from the center of the line.

Acknowledgments GM is supported by the CNRS, the Agence Nationale de la Recherche and the European Union (FACETS, FP6-2004-IST-FETPI-15879). AM is supported by a Marie Curie Intra-European Fellowship (GEMME, IEF-025213). We thank Lee Stone, Pascal Mamassian and Philippe Lefèvre for fruitful discussions about the ideas presented in the Chapter.

References

Adelson EH, Movshon JA (1982) Phenomenal coherence of moving visual patterns. Nature 300:523–525

Albright TD (1984) Direction and orientation selectivity of neurons in area MT of the macaque. J Neurophysiol 52:1106–1130

Albright TD, Desimone R (1987) Local precision of visuotopic organization in the middle temporal area (MT) of macaque. Exp Brain Res 65:582–592

Anderson B, Moore J (1979) Optimal filtering. Prentice Hall, Englewood Cliffs, NJ

Barnes G (1993) Visual-vestibular interactions in the control of head and eye movement: the role of visual feedback and predictive mechanism. Prog Neurobiol 41:435–472

Barnes G, Schmidt AM (2002) Sequence learning in human ocular ocular smooth pursuit. Exp Brain Res 144:322–335

Barthélemy FV, Perrinet LU, Castet E, Masson GS (2008) Dynamics of distributed 1D and 2D motion representations for short-latency ocular following. Vis Res 48:501–522

Bayerl P, Neumann H (2004) Disambiguating visual motion through contextual feedback modulation. Neural Comput 16:2041–2066

Bayerl P, Neumann H (2007) A fast biologically inspired algorithm for recurrent motion estimation. IEEE Trans Pattern Anal Mach Intell 29:246–260

Becker W, Fuchs AF (1985) Prediction in the oculomotor system: smooth pursuit during transient disappearance of a visual target. Exp Brain Res 57:562–575

Berzhanskaya J, Grossberg S, Mingolla E (2007) Laminar cortical dynamics of visual form and motion interactions during coherent object motion perception. Spatial Vis 20:337–395

Beutter BR, Stone LS (2000) Motion coherence affects human perception and pursuit similarly. Vis Neurosci 17:139–153

Born RT, Pack CC, Zhao R (2002) Integration of motion cues for the initiation of smooth pursuit eye movements. Prog Brain Res 140:225–237
Born RT, Pack CC, Ponce CR, Yi S (2006) Temporal evolution of 2-dimensional direction signals used to guide eye movements. J Neurophysiol 95:284–300
Boussaoud D, Ungerleider LG, Desimone R (1990) Pathways for motion analysis: cortical connections of the medial superior temporal and fundus of the superior temporal visual areas in the macaque. J Comp Neurol 15:462–495
Braddick OJ (1993) Segmentation versus integration in visual motion processing. Trends Neurosci 16:263–268
Bullier J (2001) Integrated model of visual processing. Brain Res Rev 36:96–107
Busettini C, Masson GS, Miles FA (1996) A role for binocular stereo cues in the rapid visual stabilization of the eyes. Nature 380:342–345
Carl J, Gellman RS (1987) Human smooth pursuit: stimulus dependant responses. J Neurophysiol 57:1446–1463
Castet E, Wuerger S (1997) Perception of moving lines: interactions between local perpendicular signals and 2D motion signals. Vis Res 37:705–720
Castet E, Lorenceau J, Shiffrar M, Bonnet M (1993) Perceived speed of moving lines depends on orientation, lenght, speed and luminance. Vis Res 33:1921–1936
Castet E, Charton V, Dufour A (1999) The extrinsic/intrinsic classification of two dimensional motion signals with barber-pole stimuli. Vis Res 39:915–932
Chang T, Grossberg S, Mingolla E (1998) Neural dynamics of motion integration and speed discrimination. Vis Res 38:2769–2786
Churchland MM, Chou I-H, Lisberger SG (2003) Evidence for object permanence in the smooth-pursuit eye movements of monkeys. J Neurophysiol 90:2205–2218
de Hemptinne C, Lefèvre P, Missal M (2006) Influence of cognitive expectation on the initiation of anticipatory and visual pursuit eye movements in the rhesus monkey. J Neurophysiol 95:3770–3782
Denève S (2004). Bayesian inference with recurrent spiking networks. In: Lawrence KS, Weiss Y, Bottou L (eds) Advances in neural information processing system. MIT Press, Cambridge, MA, pp 353–360
Denève S, Duhamel J-R, Pouget A (2007) Optimal sensorimotor integration in recurrent cortical networks: A neural implementation of Kalman filters. J Neurosci 27:5744–5756
Dow BM, Snyder AZ, Vautin RG, Bauer R (1981) Magnification factor and receptive field size in foveal striate cortex. Exp Brain Res 44:213–228
Fennema CL, Thompson WB (1979) Velocity determination in scenes containing several moving images. Comput Graph Image Process 9:301–315
Freyberg S, Ilg U (2008) Anticipatory smooth-pursuit eye movements in man and monkey. Exp Brain Res 186(2):203–214
Fukushima K, Yamatobe T, Shinmei Y, Fukushima J (2002) Predictive responses of periarcuate pursuit neurons to visual target motion. Exp Brain Res 145:104–120
Hafed ZM, Krauzlis RJ (2006) Ongoing eye movements contrain visual perception. Nat Neurosci 9:1449–1457
Hawken MJ, Gegenfurtner KR (2001) Pursuit eye movements to second-order motion targets. J Opt Soc Am A 18:2282–2296
Heinen SJ, Watamaniuk SN (1998) Spatial integration is human smooth pursuit. Vis Res 38:3785–3794
Heinen SJ, Badler TB, Ting W (2005) Timing and velocity randomization similarly affect anticipatory pursuit. J Vis 5:493–503
Hubel D, Wiesel TN (1965) Receptive fields and functional architecture in two striate visual areas (18 & 19) of the cat. J Neurophysiol 28:229–289
Hubel DH, Wiesel TN (1974) Uniformity of monkey striate cortex: a parallel relationship between field size, scatter and magnification factors. J Comp Neurol 158:295–305
Hueng X, Albright TD, Stoner GR (2007) Adaptive surround modulation in cortical area MT. Neuron 53:761–770

Ilg UJ (2002) Smooth pursuit eye movements: from low-level to high-level vision. Prog Brain Res 140:279–298
Ilg UJ (2003) Visual-tracking neurons in area MST are activated during anticipatory pursuit eye movements. NeuroReport 14:2219–2223
Ilg UJ, Thier P (1999) Eye movements of rhesus monkeys directed towards imaginary targets. Vis Res 39:2143–2150
Ilg UJ, Thier P (2003) Visual tracking neurons in primate area MST are activated by smooth-pursuit eye movements of an "imaginary" target. J Neurophysiol 90:1489–1502
Ilg UJ, Schumann S, Thier P (2004) Posterior parietal cortex neurons encode target motion in world-centered coordinates. Neuron 43:145–151
Kalman RE (1960) A new approach to linear filtering and prediction theory. Trans. ASME J. Basic Eng. 82:35–35
Kawano K (1999) Ocular tracking: behavior and neurophysiology. Curr. Opin. Neurobiol. 9(4): 467–473
Kettner RE, Leung H-C, Petersen BW (1996) Predictive smooth pursuit and complex two-dimensional trajectories in monkey: component interactions. Exp Brain Res 108:221–235
Kodaka Y, Kawano K (2003) Preparatory modulation of the gain of visuo-motor transmission for smooth pursuit in monkeys. Exp Brain Res 149:391–394
Koechlin E, Anton J-L, Burnod Y (1999) Bayesian inference in populations of cortical neurons: a model of motion integration and segmentation. Biol Cybern 80:25–44
Kowler E, Steinman R (1979a) The effect of expectations on slow oculomotor control II. Single target displacements. Vis Res 19:619–632
Kowler E, Steinman R (1979b) The effect of expectations on slow oculomotor control II. Single target displacements. Vis Res 19:633–646
Kowler E, Martins AJ, Pavel M (1985) The effect of expectations on slow oculomotor control IV. Anticipatory smooth eye movements depend on prior target motions. Vis Res 24:197–210
Krauzlis RJ (2004) Recasting the smooth pursuit eye movement system. J Neurophysiol 91:591–603
Krauzlis RJ, Adler SA (2001) Effects of directional expectations on motion perception and pursuit eye movements. Vis Neurosci 18:365–376
Krauzlis RJ, Stone LS (1999) Tracking with the mind's eye. Trends Neurosci 22:544–550
Lennie P, Movshon JA (2005) Coding of color and form in the geniculostriate visual pathway. J Opt Soc of Am A 22:2013–2033
Lindner A, Ilg U (2000) Initiation of smooth –pursuit eye movements to first-order and second-order motion stimuli. Exp Brain Res 133:450–456
Lisberger SG, Movshon JA (1999) Visual motion analysis for pursuit eye movements in area MT of macaque monkeys. J Neurosci 19:2224–2246
Lisberger SG, Morris EJ, Tyschen L (1987) Visual motion processing and sensory-motor integration for smooth pursuit eye movements. Annu Review Neurosci 10:97–129
Löffler G, Orbach HS (1999) Computing feature motion without feature detectors: a model for terminator motion without end-stopped cells. Vis Res 39:859–871
Lorenceau J, Alais D (2001) Form constraints in motion binding. Nat Neurosci 4:745–751
Lorenceau J, Shiffrar M, Wells N, Castet E (1993) Different motion sensitive units are involved in recovering the direction of moving lines. Vis Res 33:1207–1217
MacAvoy MG, Gottlieb JP, Bruce CJ (1991) Smooth-pursuit eye movement representation in the primate frontal eye field. Cereb Cortex 1:95–102
Madelain L, Krauzlis RJ (2003) Effect of learning on smooth pursuit during transient disappearance of a visual target. J Neurophysiol 90:972–982
Madelain L, Krauzlis RJ, Wallman J (2005) Spatial deployment of attention influences both saccadic and pursuit tracking. Vis Res 45:2685–2703
Majaj NJ, Carandini M, Movshon JA (2007) Motion integration by neurons in macaque MT is local, not global. J Neurosci 27:366–370
Masson GS (2004) From 1D to 2D via 3D: surface motion segmentation for gaze stabilisation in primates. J Physiol (Paris) 98:35–52

Masson GS, Castet E (2002) Parallel motion processing for the initiation of short-latency ocular following in humans. J Neurosci 22(12):5149–5163
Masson GS, Stone LS (2002) From following edges to pursuing objects. J Neurophysiol 88:2869–2873
Masson GS, Rybarczyk Y, Castet E, Mestre DR (2000) Temporal dynamics of motion integration for the initiation of tracking eye movements at ultra-short latencies. Vis Neurosci 17:753–767
Masson GS, Busettini C, Yang D-S, Miles FA (2001) Short-latency ocular following in humans: sensitivity to binocular disparity. Vis Res 41:3371–3387
Masson GS, Yang D-S, Miles FA (2002) Reversed short-latency ocular following. Vis Res 42:2081–2087
Menz MD, Freeman RD (2003) Stereoscopic depth processing in the visual cortex: a coarse-to-fine mechanism. Nat Neurosci 6:59–65
Miles FA (1998) The neural processing of 3-D visual information: evidence from eye movements. Eur J Neurosci 10:811–822
Miles FA, Kawano K, Optican LM (1986) Short-latency ocular following responses of monkey I. Dependence on temporospatial properties of visual input. J Neurophysiol 56:1321–1354
Mingolla E (2003) Neural models of motion integration and segmentation. Neural Network 16:939–945
Missal M, Heinen SJ (2004) Supplementary eye fields stimulation facilitates anticipatory pursuit. J Neurophysiol 92:1257–1262
Montagnini A, Spering M, Masson GS (2006) Predicting 2D target velocity cannot help 2D motion integration for smooth pursuit initiation. J Neurophysiol 96:3545–3550
Montagnini A, Mamassian P, Perrinet L, Castet E, Masson GS (2007) Bayesian modeling of dynamic motion integration. J Physiol (Paris) 101:64–77
Movshon JA, Newsome WT (1996) Visual response properties of striate cortical neurons projecting to area MT in macaque monkeys. J Neurosci 16(23):7733–7741
Newsome WT, Wurtz RH, Komatsu H (1988) Relation of cortical areas MT and MST to pursuit eye movements II. Differentiation of retinal and extraretinal inputs. J Neurophysiol 60:604–620
Nowlan SJ, Sejnowski TJ (1995) A selection model for motion processing in area MT of primates. Journal of Neuroscience 15:1195–1214
Osborne LC, Bialek W, Lisberger SG (2004) Time course of information about motion direction in visual area MT. J Neurosci 24:3210–3222
Osborne LC, Lisberger SG, Bialek W (2005) A sensory source for motor variation. Nature 437:412–416
Pack CC, Born RT (2001) Temporal dynamics of a neural solution to the aperture problem in visual area MT. Nature 409:1040–1042
Pack CC, Livingstone MS, Duffy KR, Born RT (2003) End-stopping and the aperture problem: two-dimensional motion signals in macaque V1. Neuron 39:671–680
Pack CC, Gartland AJ, Born RT (2004) Integration of contour and terminator signals in visual area MT of alert macaque. J Neurosci 31:3268–3280
Pack CC, Hunter JN, Born RJ (2005) Contrast dependence of suppressive influence in cortical area MT of alert macaque. J Neurophysiol 93:1809–1815
Pola J, Wyatt HJ (1985) Active and passive smooth eye movements: effects of stimulus size and location. Vis Res 25:1063–1076
Priebe NJ, Lisberger SG (2004) Estimating target speed from the population response in visual area MT. J Neurosci 24:1907–1916
Priebe NJ, Churchland MM, Lisberger SG (2001) Reconstruction of target speed for the guidance of eye movements. J Neurosci 21:3196–3206
Rao P (2004) Bayesian computation in recurrent neural circuits. Neural Comput 16:1–38
Rashbass C (1961) The relationship between saccadic and tracking eye movements. J Physiol (London) 159:326–338
Ringach DL, Hawken MJ, Shapley R (1997) Dynamics of orientation tuning in macaque V1. Nature 387:281–284

Robinson DA (1986) The systems approach to the oculomotor system. Vis Res 26:91–99
Sceniak MP, Ringach DL, Hawken MJ, Shapley R (1999) Contrast's effect on spatial summation in macaque V1 neurons. Nat Neurosci 2:733–739
Sheliga B, Chen KJ, Fitgibbon EJ, Miles FA (2005) Initial ocular following in humans: a response to first-order motion energy. Vis Res 45:3307–3321
Sillito AM, Cudeiro J, Jones HE (2006) Always returning: feedback and sensory processing in visual cortex and thalamus. Trends Neurosci 29:307–316
Simoncelli EP, Adelson EH, Heeger DJ (1991) Probability distributions of optical flow. In: Proceedings of the Conference on Computer Vision and Pattern Recognition, pp 310–315
Smith MA, Majaj NJ, Movshon JA (2005) Dynamics of motion signaling by neurons in macaque monkeys. Nat Neurosci 8:220–228
Steinbach LM (1976) Pursuing the perceptual rather than the retinal stimulus. Science 16:1371–1376
Steinman RM (1986) The need for an eclectic, rather than systems, approach to the study of the primate oculomotor system. Vis Res 26:101–112
Stocker AA, Simoncelli EP (2006) Noise characteristics and prior expectation in human visual speed perception. Nat Neurosci 9(4):578–585
Stone LS, Beutter BR, Lorenceau J (2000) Visual motion integration for perception and pursuit. Perception 29:771–787
Tabata H, Miura K, Kawano K (2005) Anticipatory gain modulation in preparation for smooth pursuit eye movements. J Cogn Neurosci 17:1962–1968
Tabata H, Miura K, Taki K, Matuura K, Kawano K (2006) Preparatory gain modulation of visuomotor transmission for smooth pursuit eye movements in moneys. J Neurophysiol 96:3051–3063
Thier P, Erickson RG (1992) Response to visual-tracking neurons from cortical area MST-l to visual, eye and head motion. Eur J Neurosci 4(6):539–553
Tlapale E, Masson GS, Viéville T, Kornprobst P (2007). Model of motion field diffusion controlled by form cues. In: Perception. 26th ECVP Abstracts, 36; Supplement, 215
Vercher JL, Quaccia D, Gauthier GM (1995) Oculo-manual coordination control: respective role of visual and non-visual information in ocular tracking of self-moved targets. Exp Brain Res 103:311–322
Verghese P, Stone LS (1995) Combining speed information across space. Vis Res 35:2811–2823
Wallace J, Stone LS, Masson GS (2005) Object motion computation for the initiation of smooth pursuit eye movements in humans. J Neurophysiol 93:2279–2293
Wallach H (1935). Uber Visuell Wahrgenommene Bewegungrichtung. Psychologishe Forschung 20:325–380. Translated by Wuerger S, Shapley B (1996) Perception 11:1317–1367
Weiss Y, Fleet D (2002). Velocity likelihood in biological and machine vision. In: Rao RPN, Olshausen BA, Lewicki MS (eds) Probabilistic models of the brain: perception and neural function. MIT Press, Cambridge, MA, pp 77–96
Weiss Y, Simoncelli EP, Adelson EH (2002) Motion illusions as optimal percepts. Nat Neurosci 5:598–604
Wilmer JB, Nakayama K (2007) Two distinct visual motion mechanisms for smooth pursuit: evidence from individual differences. Neuron 54:987–1000
Wilson HR, Ferrera VP, Yo C (1992) A psychophysically motivated model for two-dimensional motion perception. Vis Neurosci 9:79–97
Wyatt HJ, Pola J, Fortune B, Posner M (1994) Smooth pursuit eye movements with imaginary targets defined by extrafoveal cues. Vis Res 34:803–820
Yo C, Wilson HR (1992) Perceived direction of moving two-dimensional patterns that depends on duration, contrast and eccentricity. Vis Res 32:135–147
Zeki S, Shipp S (1988) The functional logic of cortical connections. Nature 335:311–317

Chapter 9
Interactions Between Perception and Smooth Pursuit Eye Movements

Ziad M. Hafed and Richard J. Krauzlis

Abstract When we see a moving object in our environment, our visual system analyzes and interprets this object's trajectory. This motion analysis is very often then used to generate a smooth pursuit eye movement that tracks the object in order to stabilize its image on the retina. Smooth pursuit generation therefore requires visual motion signals. However, by stabilizing the retinal image of a moving object, smooth pursuit necessarily causes motion of the retinal images of stationary objects. Since such induced retinal motion does not dramatically affect pursuit – and more importantly, does not cause stationary objects to be perceived as moving – it is clear that depending on the context, similar sensory inputs can either dramatically influence both perception and pursuit or not at all. Recent research has suggested that such interaction between motion processing and smooth pursuit constitutes the core of a more general interaction between high-level perceptual integration and eye movements. Specifically, drive signals that move the eyes need not be purely retinal in nature, but can also be integrated based on disparate retinal motion directions. Moreover, ongoing motor commands themselves are used not only to cancel motion percepts of stable objects, but also to perceptually disambiguate spatial relationships between sensory features of visual objects. These results suggest that bidirectional interactions between perception and action allow maximum flexibility in generating coherent percepts of our environment – and associated actions on this environment.

9.1 Introduction

Eye movements serve vision by compensating for the non-uniform photoreceptor distribution in our retina (Zeki 1993; Farah 2000; Sekuler and Blake 1994; Carpenter 1988). There are two types of voluntary eye movements that we employ

Z.M. Hafed (✉)
Systems Neurobiology Laboratories, The Salk Institute for Biological Studies, 10010North Torrey Pines Road, La Jolla, CA, 92037, USA
e-mail: zhafed@salk.edu

U.J. Ilg and G.S. Masson (eds.), *Dynamics of Visual Motion Processing: Neuronal, Behavioral, and Computational Approaches*,
DOI 10.1007/978-1-4419-0781-3_9, © Springer Science+Business Media, LLC 2010

in our everyday life: saccades, which are rapid flicks of the eye that re-orient the center of gaze towards objects of interest, and smooth pursuit, which is a smooth rotation of the eye with a speed and direction that closely match those of a moving object (Krauzlis 2005). Both of these types of eye movements are under the influence of sensory inputs from the retina. For example, spots of light that abruptly appear in the periphery are extremely effective in driving saccades towards them (Biscaldi et al. 1996; Reulen 1984; Posner 1980; Schall and Thompson 1999), and motion energy in images is known to drive smooth pursuit eye movements (Rashbass 1961; Westheimer 1954; Robinson 1965; Lisberger et al. 1987). However, it is this efficacy with which sensory inputs can influence eye movements that implies a richer interaction between the two in our visual system. Consider for example the case of smooth pursuit eye movements. When an object in our visual environment moves, the input motion signal from the retina is used to drive the eye as it tracks the object (Rashbass 1961; Westheimer 1954; Robinson 1965; Lisberger et al. 1987). However, such tracking causes another input motion signal from the retina: that of the stationary objects that the eye is not tracking (Fig. 9.1). Clearly, this additional motion signal is problematic both from the perspective of perception, since the stationary background should not be interpreted as moving during smooth pursuit, as well as from the perspective of motor control, since smooth pursuit velocity should be immune to the additional motion signal. Thus, the dynamics of visual motion processing in the brain involve interactions with the smooth pursuit system.

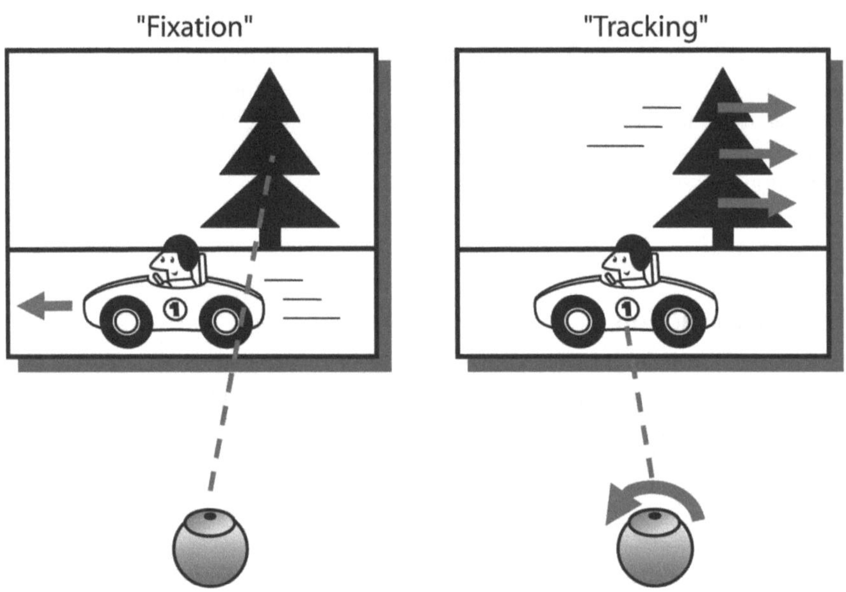

Fig. 9.1 Retinal motion can arise from object motion with a stationary eye (*left*) or eye motion with a stationary object (*right*)

In this chapter, we demonstrate that these interactions between motion processing and smooth pursuit eye movements form the basis for a more general interaction between visual perception and eye movements. After all, the ultimate goal of the visual system is not to analyze particular visual features, such as motion, color, or edge orientation, as much as it is to integrate and interpret various sources of information in order to facilitate interaction with our environment. Such integration is expected to manifest itself in two main forms when it comes to its effects on smooth pursuit eye movements. First, smooth pursuit initiation should be subject to cognitive influences, such as attention and intention, that are known to affect decision making and other voluntary behaviors, including saccades (Schall and Thompson 1999; Goldberg et al. 2006; Andersen and Buneo 2002; Colby and Goldberg 1999; Carello and Krauzlis 2003). Second, the drive signal for ongoing smooth pursuit has to account for the ambiguities that are very common in every natural visual scene that is scanned by eye movements. For example, occlusion can fragment the retinal image of a single object into disparate components, each of which providing a potentially erroneous cue for the smooth pursuit system (Lorenceau and Shiffrar 1992; Beutter and Stone 2000).

Another point that we make in this chapter is that the interaction between visual perception and eye movements is bidirectional – perception not only provides a key input to eye movements, but also depends on the feedback of information about these movements. Correctly perceiving the real-world motion – or lack thereof – of objects during smooth pursuit (as described above) is an example of the use of such feedback information. If the visual system were to also use this information in a more general sense (i.e. not restricted to motion perception, but also to form and object perception (Hafed and Krauzlis 2006)), then this would imply an additional mechanism for supporting visual scene analysis.

The picture that emerges based on the experiments reviewed in this chapter is that smooth pursuit eye movements form a closed action-perception loop that is similar in several respects to well-studied loops between saccades and visual perception (Sommer and Wurtz 2006; Sommer and Wurtz 2002; Ross and Ma-Wyatt 2004; Ross et al. 2001) or saccades and visual attention (Rizzolatti et al. 1987; Moore and Fallah 2004; Moore and Armstrong 2003). These observations open the door for future investigations of the interactions between smooth pursuit and the perception of coherent motions/objects, in turn allowing us to learn more about vision in general and motion processing in particular.

9.2 Sensory Processing for Smooth Pursuit Initiation

Smooth pursuit eye movements are predominantly driven by visual motion (Rashbass 1961; Berryhill et al. 2006). For example, trying to track an imagined object moving across a blank background gives rise to a series of saccadic eye movements along the imagined object trajectory with little or no smooth rotation of the eye during the inter-saccadic intervals. Therefore, the most basic interaction between smooth pursuit and visual perception involves the readout of a retinal motion signal, from the putative motion perception areas (MT and MST) (Lisberger et al. 1987; Komatsu and Wurtz 1988a, b; Newsome et al. 1988; Krekelberg and Albright 2005; Saito et al. 1986; Graziano et al. 1994; Duffy and Wurtz 1991; Born and Bradley 2005; Albright and Stoner 1995), in order to drive an eye movement output. However, unlike the optokinetic and ocular following reflexes that also respond to retinal motion (Ilg 1997), smooth pursuit eye movements are voluntary behaviors. This implies that smooth pursuit eye movements can be influenced by high-level sensory and cognitive processes in two main ways. First, given a particular perceptual evaluation of the input sensory evidence, smooth pursuit relies on selecting, through the aid of attention (Kastner and Ungerleider 2000), a moving target and deciding to track it. Second, for a given selected target, the metrics of smooth pursuit initiation (and subsequent pursuit stages, as we describe in the next section) continuously reflect the evolving perceptual interpretation of target motion, rather than its physical motion. This latter point suggests that pursuit initiation and motion perception share common neuronal resources in the brain (Krauzlis and Stone 1999). We elaborate on these two ideas in what follows.

Initiating smooth pursuit eye movements involves a break from a state of gaze fixation, much like initiating saccades. This conceptual similarity between smooth pursuit and saccades suggests that smooth pursuit initiation may be under similar high-level sensory and cognitive influences as saccades, for which there is an extensive literature on their links with perception. There are several lines of evidence to support this. For example, saccadic reaction times are strongly modulated by whether or not subjects are primed to release fixation (Saslow 1967; Reuter-Lorenz et al. 1991; Fischer and Weber 1993), and it is now known that a similar release from fixation takes place for smooth pursuit initiation. This is evidenced from experiments that replicate the well-known "gap effect" (Saslow 1967; Reuter-Lorenz et al. 1991; Fischer and Weber 1993) for smooth pursuit eye movements (Tam and Ono 1994; Merrison and Carpenter 1995; Krauzlis and Miles 1996a, b; Knox 1996; Kimmig et al. 2002). In those experiments, subjects fixate a stationary spot for a short period of time, after which the spot disappears and another one appears and moves in such a way as to induce a smooth pursuit eye movement. If a short temporal gap is introduced between the time of disappearance of the fixation spot and the time of appearance of the pursuit target, the latencies of smooth pursuit initiation are consistently reduced (Tam and Ono 1994; Merrison and Carpenter 1995; Krauzlis and Miles 1996a, b; Knox 1996; Kimmig et al. 2002), much like in the case of saccades (Saslow 1967; Reuter-Lorenz et al. 1991; Fischer and Weber 1993). Thus, smooth pursuit initiation is subject to the sensory – and potentially cognitive – influences associated with the "gap effect," and it is interesting to note that the neuronal substrates for this effect in smooth pursuit overlap those for saccades. For example, it has been shown that neurons in the superior colliculus (SC), a structure implicated in encoding the locations of salient targets for voluntary eye movements (Krauzlis et al. 2004), become more active if either a saccade or smooth pursuit target is preceded by a short temporal gap after fixation point offset than if it is not (Krauzlis 2003; Krauzlis et al. 2002).

Voluntary behaviors also involve selecting from multiple alternatives. Consider for example the scenario in which two spots of light appear concurrently at two different spatial locations and move in two different directions. In this scenario, the smooth pursuit system has to select one of these two targets as the eye movement is being initiated. Such selection benefits from prior information that is not necessarily sensory in nature. Specifically, if one of the two targets is cued before both of them appear, then the smooth pursuit system is more efficient in making the decision to track one of them, as evidenced by shortened smooth pursuit initiation times (Adler et al. 2002). Moreover, cueing the location of the target to be selected is often more effective than cueing its direction of motion or color, indicating that the selection process for smooth pursuit initiation is under the influence of spatial attention, much like saccades (Posner 1980; Schall and Thompson 1999; Goldberg et al. 2006; Colby and Goldberg 1999; Adler et al. 2002; Sheliga et al. 1994, 1997; Remington 1980; Findlay and Gilchrist 2001; Deubel and Schneider 1996). In fact, this similarity between smooth pursuit and saccades again has overlapping neuronal substrates. Specifically, sub-threshold electrical microstimulation in a spatial structure like the superior colliculus (SC) affects target selection for both smooth pursuit and saccades, with the effect of microstimulating a certain target location being similar to that of prior cueing at that location (Carello and Krauzlis 2003). So, as in the case of the release from gaze fixation, smooth pursuit initiation involves a target selection process that is similar to that for saccades and which is therefore under the influence of high-level processes, such as spatial attention, that also serve visual perception.

What about the readout of motion signals associated with these targets? As mentioned above, smooth pursuit relies on a sensory motion signal. In this example, the two targets are associated with two different spatial locations and two different motion signals. Thus, different populations of MT neurons are expected to be activated by these targets. Specifically, since MT neurons are retinotopically organized and exhibit speed and direction tuning when it comes to input retinal motion signals (Lisberger et al. 1987; Komatsu and Wurtz 1988a, b; Newsome et al. 1988; Krekelberg and Albright 2005; Born and Bradley 2005; Albright and Stoner 1995), our two targets are expected to activate two distinct groups of MT neurons. If smooth pursuit initiation involved readout of the population of active neurons in area MT, then the initial smooth pursuit response to two suddenly appearing targets might reflect an average of the speeds and directions encoded by the neurons representing these two targets. This is in fact the case: when two targets simultaneously appear and move, the eye tends to track neither of them and instead rotates in a direction that is the vector-average of the directions of motion of the two targets (Lisberger and Ferrera 1997). It should be noted, however, that this "vector averaging" behavior during smooth pursuit initiation does not necessarily imply that smooth pursuit cannot differentially select from among multiple motion signals. For example, spatial pre-cueing affects the initial smooth pursuit response (Adler et al. 2002). There is also evidence that a motion cue can bias the initial smooth pursuit motion trajectory towards that of the cued motion (Garbutt and Lisberger 2006), as if attention to a particular motion direction, and the subsequent enhancement of the response of a subset of MT neurons (Treue and Martinez-Trujillo 1999; Maunsell and Treue 2006), similarly influences the motor response to this direction.

The case of "vector averaging" by smooth pursuit initiation illustrates a form of dissociation between the motor output and the individual motion signals associated with the sensory input. This dissociation can also arise not because the readout mechanism for smooth pursuit pools across multiple sensory signals, but because pursuit accesses the *perceptual interpretation* of these signals, which could differ from their physical values. How might one uncover evidence that smooth pursuit is influenced by perception in this manner? One possibility is to correlate smooth pursuit performance with performance in visual discrimination tasks. Consider for example a simple pursuit task in which a stationary fixated spot of light starts to move smoothly with a constant speed and direction. Subjects typically start to track this spot with a combination of smooth pursuit and catch-up saccades that compensate for any imperfections in their smooth pursuit output (Rashbass 1961). In addition, subjects also vary significantly in the overall fidelity of their tracking. A recent study (Wilmer and Nakayama 2007) exploited this variability to study how smooth pursuit initiation relies on sensory motion perception, be it low-level (Nakayama 1985) or high-level (Seiffert and Cavanagh 1998). Specifically, each subject that was tested with the simple tracking task just described was also tested separately for his/her ability to perceive visual motion while fixating (i.e. without the need to transform the perceived motion into motor commands). Across many subjects, the authors found that an individual's perceptual ability in interpreting visual motion is correlated with his/her quality of smooth pursuit initiation. This suggests that the same neuronal resources that underlie the perception of motion are also necessary for driving the different components of the initial smooth pursuit response to target motion.

A more direct correlation between smooth pursuit initiation and perception can be achieved through the design of psychophysical tasks that require both an eye movement response as well as a fine perceptual discrimination of motion direction at the same time. In one study (Stone and Krauzlis 2003), subjects followed a moving spot whose trajectory deviated ever so slightly from either the vertical or horizontal axis, and the subjects' task was twofold: to track this trajectory as well as possible, and to make a perceptual judgment on whether this trajectory was to the right or left of pure vertical or whether it was above or below pure horizontal. Besides co-varying together on average, smooth pursuit direction and the perceptual judgments of target direction were also found to co-vary on a trial-by-trial basis, even when performance was near chance (Stone and Krauzlis 2003). This suggests a shared neuronal signal for supporting the perception of target motion as well as the generation of smooth pursuit. Such shared neuronal processing can also be inferred when the object to be pursued is an oriented bar. Due to the now well-known "aperture problem" (Marr and Ullman 1981), when such an oriented bar starts to move in a certain direction, the initial momentary perceived direction of motion is orthogonal to the bar's orientation and not the true motion direction (Yo and Wilson 1992; Lorenceau et al. 1993). This initial perceived direction has a neuronal correlate in MT neurons (Pack and Born 2001) as well as a direct behavioral correlate in the direction of smooth pursuit initiation (Pack and Born 2001; Born et al. 2006).

In summary, we have briefly described how smooth pursuit initiation is dependent on cognitive factors, such as attention, that also support and reflect the state of visual perception. These observations are reminiscent of findings made previously for the other type of voluntary eye movement which our visual system employs, that of saccades (Goldberg et al. 2006; Andersen and Buneo 2002; Colby and Goldberg 1999; Rizzolatti et al. 1987; Moore and Fallah 2004; Moore and Armstrong 2003; Kastner and Ungerleider 2000; Saslow 1967; Reuter-Lorenz et al. 1991; Fischer and Weber 1993). We believe that these findings will pave the way for future research on the interaction between smooth pursuit and visual perception. Specifically, the intention to generate a voluntary eye movement should give rise to several processes that prime the visual system for handling the retinal consequences of this movement. There is abundant evidence on how saccade preparation can influence perception in order to guarantee perceptual stability in the face of the upcoming eye movement (Sommer and Wurtz 2002, 2006; Ross et al. 2001) (also see Sect. 9.4 below). However, similar evidence for smooth pursuit initiation is still somewhat lacking, providing an important opportunity for future research. We have also described how the initial smooth pursuit response can be under the influence of the perceptual interpretation of sensory inputs. We elaborate on this point in the next section by describing the variety of ways in which smooth pursuit can be driven.

9.3 Perceptual and Cognitive Influence on Smooth Pursuit

Much of the work that we have referred to so far involves smooth pursuit of a small moving spot. With such a spot, the "perceived object" being tracked is identical to its retinal image. However, one of the reasons that vision is such a complicated process is that perceived objects can be dramatically different from the patterns of light that they project onto the retina. For example, cast shadows and occlusion fragment the retinal images of single objects into disparate components. As a result, processes of grouping, integration, and interpretation – or perception – are ultimately what constitute the sense of vision. It therefore comes as no surprise that smooth pursuit eye movements are strongly influenced by such processes.

One of the first lines of evidence that smooth pursuit eye movements can follow a perceived object that only has a sparse retinal image representation was provided by Steinbach in 1976 (Steinbach 1976). By starting with the observation that two points of light moving in a cycloidal pattern can be perceived as a rolling wagon wheel with two light sources on its rim (Fig. 9.2a and Supplementary Movie 1), Steinbach showed that subjects could easily track the center of the perceived wheel (Steinbach 1976). Unlike the two points of light, this center was not a visual stimulus on the retina, and it also moved with a trajectory that was decidedly different from the trajectories of the two lights. Moreover, Steinbach showed that the smooth tracking eye movement that his subjects generated was dependent on the percept of a rolling wheel, because he also asked subjects to track the same wheel but with only one of the two lights on its rim turned on. In this condition, perceiving a rolling wheel is difficult, and this difficulty was reflected in the quality of his subjects' eye movements (Steinbach 1976).

Related to Steinbach's results is evidence that smooth pursuit eye movements can also benefit from the perceptual grouping of different visual stimuli and the subsequent use of such grouping to modify the eye movement output. For example, it is possible to direct smooth pursuit eye movements towards the inferred, but invisible, midpoint between two visual stimuli that share a common motion trajectory (Fig. 9.2b and Supplementary Movie 2) (Wyatt et al. 1994; Ilg and Thier 1999) (also see Hafed and Krauzlis 2005; Hafed and Krauzlis 2008; Hafed et al. 2008). More recently, Beutter and Stone (2000) provided a clear and quantitative correlation between perceptual integration and smooth pursuit eye movements. Specifically, these authors used a class of partially-occluded line figures to ask whether smooth pursuit eye movements can follow the perceived object rather than the retinal stimulus. In their study, Beutter and Stone (2000) started with the visual scene depicted in Fig. 9.3a (left). In this scene, an outline diamond translates in front of a black background but behind a gray occluder containing two vertical apertures (Lorenceau and Shiffrar 1992; Beutter and Stone 2000; Lorenceau and Alais 2001). Importantly, the trajectory of the diamond and the geometry of the occluder are designed such that the vertices of the diamond are never revealed by the two vertical apertures. In other words, the diamond is always fragmented into four separate line segments, and the line segments individually appear to move in a purely vertical direction even though the diamond itself moves diagonally (Supplementary Movie 3) (Lorenceau and Shiffrar 1992).

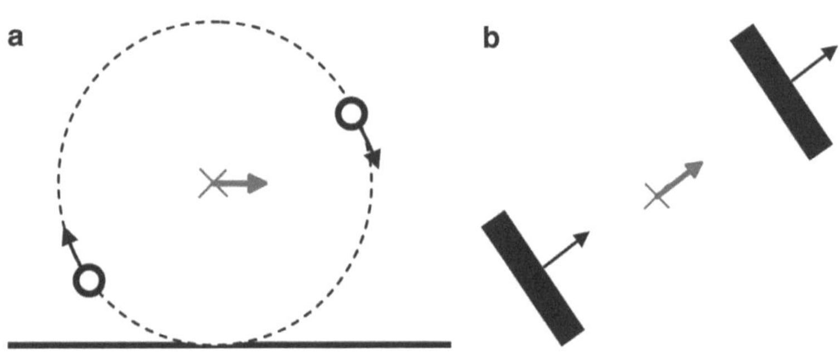

Fig. 9.2 Pursuit of the perceived stimulus. (**a**) If you view a rolling wagon wheel in complete darkness, the wheel and its motion can be perceived if two light sources are placed as shown on its rim. Such a percept can drive smooth pursuit eye movements to follow the invisible hub (*center*) of the wheel (Steinbach 1976). See Supplementary Movie 1. (**b**) Smooth pursuit can follow the 'imagined' center between two visual stimuli that are grouped together through common motion (Wyatt et al. 1994; Ilg and Thier 1999; Hafed and Krauzlis 2005; Hafed and Krauzlis, 2008; Hafed et al. 2008). See Supplementary Movie 2

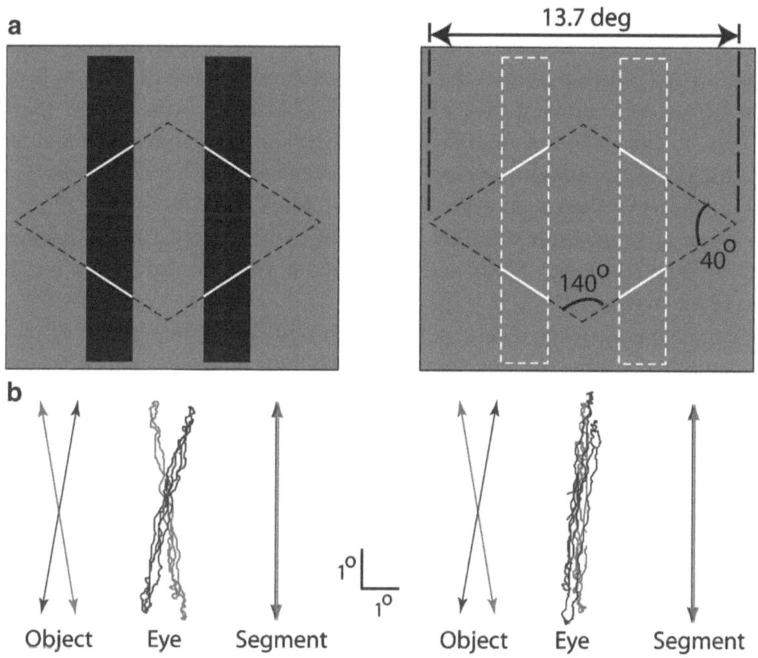

Fig. 9.3 Perceiving a rigid form allows smooth pursuit to track it correctly. (**a**) The scene on the left consists of a partially occluded outline of a diamond. The diamond translates in either a positive or a negative diagonal direction, such that its vertices are never visible though the two vertical apertures. Because of the apertures, each of the visible segments of the diamond moves in a purely vertical fashion (Lorenceau and Shiffrar 1992). The scene on the right is identical, but now the apertures reveal a background of identical luminance as the occluder. In this case, occlusion cues disappear, and it is difficult to perceive a rigid diamond (Lorenceau and Shiffrar 1992). See Supplementary Movies 3 and 4. (**b**) Eye movements track the motion of the diamond (*left*) when it is perceived correctly (**a**, *Left*) and the motion of the segments (*right*) when no percept of the diamond is possible (**a**, *right*). In both conditions, the physical motion of the visible line segments is identical, but the percept is different. Adapted from Krauzlis and Stone (1999) (*see Color Plates*)

One reason why this scene is an ideal stimulus to study the influence of perception on smooth pursuit eye movements is that previous work (Lorenceau and Shiffrar 1992) has shown that, despite the fragmentation of the diamond by occlusion, our visual system has no difficulty in interpreting the scene correctly and giving rise to a strong percept of a rigid object. In addition, the stimulus is designed such that it gives rise to retinal image motion which is different from the underlying object motion even during steady state smooth pursuit. Thus, if smooth pursuit reflects the object trajectory, which it does (Fig. 9.3b, left) (Beutter and Stone 2000), it can be concluded that the percept of a diamond is sufficient to drive the eye to track it.

To further demonstrate the validity of their result, Beutter and Stone (2000) exploited another advantage of the scene shown in Fig. 9.3. Specifically, if the occluded diamond now translates in front of a background whose luminance is identical to that of the occluder (Fig. 9.3a, right), the strong occlusion cue in the original condition disappears, and the visual system has a harder time grouping the four visible line segments into a single rigid diamond (Supplementary Movie 4) (Lorenceau and Shiffrar 1992) – the percept now becomes one of disjoint groups of line segments each moving in a purely vertical fashion, even though there was absolutely no change to any motion signals in the stimulus (Lorenceau and Shiffrar 1992). Interestingly, smooth pursuit eye movements now become more vertical, reflecting the new altered percept. Therefore, when contextual cues allow it, the visual system can integrate different sources of motion energy into a resultant trajectory (Beutter and Stone 2000), and this resultant trajectory can then drive smooth pursuit eye movements.

We can also experience a percept of motion without the integration of physical, albeit spatially disparate, motion signals. For example, under certain circumstances, a spot of light that jumps abruptly from one location to another gives rise to an "apparent motion" percept from the initial location to the final one (Nakayama 1985; Ramachandran and Anstis 1985). Such an apparent motion percept is sufficient to drive smooth pursuit eye movements (van der Steen et al. 1983; Lamontagne et al. 2002), and it therefore serves as a useful tool to study the relationship between perception and smooth pursuit eye movements even further. For example, since smooth pursuit reflects the perceptual interpretation of visual stimuli (Beutter and Stone 2000), one might ask: what happens to smooth pursuit when the perceptual interpretation of a stimulus changes? Madelain and Krauzlis (2003) investigated this question by using an apparent motion stimulus to drive smooth pursuit eye movements. More interestingly, they employed a perceptually bistable stimulus, which allowed them to investigate the temporal profile with which smooth pursuit follows perceptual reversals.

As Fig. 9.4a, b demonstrates, Madelain and Krauzlis (2003) used as a starting point the well-known Kanizsa square illusion (Kanizsa 1976). In this illusion, a contour of a square is perceptually completed based on four (partly-complete) circular discs around its four corners (Fig. 9.4a). If a row of such illusory squares is used, apparent motion (of an illusory square) along this row can be induced if the horizontal position of this row is continuously alternated between two values (Fig. 9.4b). In other words, a strong apparent motion percept can arise with this stimulus through the display of a continuously looping two-frame movie, in which the second frame is a horizontally shifted version of the first one (Supplementary Movie 5). Critically, if the second frame of the movie is shifted such that each new illusory square position is at the midpoint between two squares from the previous frame, then a percept of rightward apparent motion is equally likely as a percept of leftward apparent motion, resulting in bistability. Madelain and Krauzlis (2003) used this stimulus to first demonstrate that subjects could easily perceive rightward or leftward apparent motion and voluntarily switch its perceived direction. They then showed that subjects could voluntarily track along the perceived apparent motion direction. Interestingly, when they experienced a perceptual reversal during smooth tracking, subjects were often able to reverse their tracking direction smoothly and without the need to interrupt the ongoing eye movement with either fixation or saccades (Madelain and Krauzlis 2003) (Fig. 9.4c).

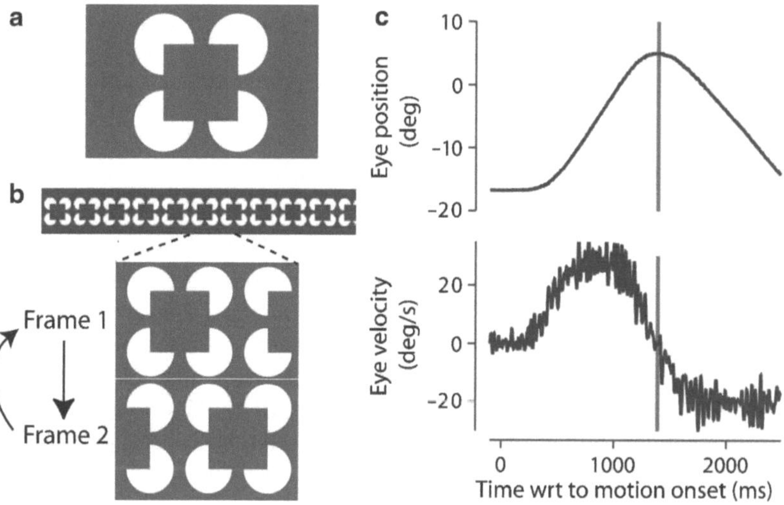

Fig. 9.4 Pursuit of bistable apparent motion of illusory objects. (**a**) Four partially complete circular discs can be placed in a way that creates an illusory percept of a square (Kanizsa 1976). (**b**) A row of such illuory squares can be used to create an apparent motion stimulus. With a two frame movie, either rightward or leftward apparent motion of an illusory square can be perceived, and the percept can be reversed at will (Madelain and Krauzlis 2003). See Supplementary Movie 5. (**c**) Sample eye position and velocity traces, showing a subject smoothly reversing his eye movement tracking direction in response to a reversal of his perceived apparent motion direction. Adapted from Madelain and Krauzlis (2003)

These observations allowed for fine quantification of the extent of temporal register between smooth pursuit and motion perception. Specifically, in addition to being instructed to track one direction of apparent motion and then voluntarily reverse this direction, subjects were also provided with a randomly occurring auditory signal during a trial. After trial end, the subjects had to indicate whether the auditory signal occurred before or after the perceptual reversal of apparent motion during the trial (irrespective of whether they thought the signal came before or after their actual motor reversal) (Madelain and Krauzlis 2003). Analysis of the subjects' behavioral and eye movement responses revealed that the motor reversals in the subjects' tracking started almost synchronously with the perceptual reversals, with the motor reversal lagging the perceptual reversal by ~50 ms on average (Madelain and Krauzlis 2003). Interestingly, when subjects tracked a *real* square that reversed its motion direction, the motor reversal lagged the perceptual reversal by a longer interval (~100 ms). So, smooth pursuit eye movements can provide a real-time estimate of the state of motion perception, and are better synchronized with perceptual state than when they are driven by explicit changes in the visual motion stimulus (Madelain and Krauzlis 2003). This close temporal synchrony may also be explained in part by an additional mechanism: that the perceptual reversal itself is facilitated by the intention to reverse tracking direction. While there is no explicit experiment demonstrating this, the next section describes some more recent findings that suggest that this is a distinct possibility (Hafed and Krauzlis 2006).

9.4 Closing the Action-Perception Cycle

Thus far, we have described how high-level sensory and cognitive factors such as attention and perception influence both the decision to initiate smooth pursuit eye movements as well as the trajectories of these movements. However, and as mentioned in the introduction, smooth pursuit eye movements alter the input sensory stream of information from the retina (Fig. 9.1). Thus, in addition to driving eye movements (Beutter and Stone 2000; Steinbach 1976; Madelain and Krauzlis 2003), perception itself has to be immune to the retinal consequences of these movements. The question of how the brain achieves perceptual stability despite eye movements has been debated for a very long time, and many theories have been proposed to answer it. One such theory suggests that vision is 'paused' during eye movements. For example, there exists evidence for the suppression of visual sensitivities to image displacements during saccades (Bridgeman et al. 1975), as well as evidence for large-scale change blindness in complex, natural scenes (Henderson and Hollingworth 2003; Rensink 2002; Rensink et al. 1997). However, in reality, vision is not entirely suppressed during eye movements. This is particularly true for smooth pursuit eye movements, which can span several seconds but clearly do not suppress vision completely. More elaborate theories of perceptual stability involve mechanisms such as template formation and comparison around an eye movement (Currie et al. 2000; Deubel et al. 1998; McConkie and Currie 1996), learning of the sensorimotor contingencies associated with an eye movement (O'Regan and Noe 2001), and updating of spatial representations before and during an eye movement (Duhamel et al. 1992; Umeno and Goldberg 1997; Walker et al. 1995). All of these mechanisms require that the brain be actively engaged in compensating for the retinal consequences of eye movements.

We have recently tested for a new interaction between visual perception and eye movements: that the motor commands for eye movements provide information that can constrain visual perception (Hafed and Krauzlis 2006). Specifically, if the brain already has access to efference copy information about an eye movement in order to maintain perceptual stability of the world (Sommer and Wurtz 2002, 2006; Duhamel et al. 1992; Umeno and Goldberg 1997; Walker et al. 1995), then it is possible that this information is also used to resolve the perceptual ambiguities that are often inherent in retinal images. If efference copy information were used in this manner, then ongoing smooth pursuit eye movements may have a significant effect on how we segment and analyze visual scenes. We showed that this is true by designing a set of experiments in which we kept the retinal input constant and varied the eye movement command being generated in the brain (Hafed and Krauzlis 2006).

In our most basic manipulation, our subjects viewed an outline chevron that translated along a circular trajectory behind an occluder having two vertical apertures

(Fig. 9.5a). This stimulus is very similar to that of Fig. 9.3, in that the motions of the visible object segments are ambiguous. In fact, previous research has shown that under fixation, this scene typically results in the perception of two unconnected groups of lines translating vertically within the apertures (Lorenceau and Alais 2001). Our aim was to show that information about ongoing eye movement commands can disambiguate this scene and give rise to a percept of a rigid chevron, even if the retinal input is unaltered. Specifically, we exploited the fact that, with the head fixed, retinal motion is detersmined by eye motion in the orbit as well as by object motion in space. That is, if $\dot{\mathbf{r}}(t)$ is the retinal velocity of the chevron, $\dot{\mathbf{e}}(t)$ is the velocity of the eye in the orbit and $\dot{\mathbf{o}}(t)$ is the world-centered velocity of the same chevron, then $\dot{\mathbf{r}}(t)$ is given by

$$\dot{\mathbf{r}}\,(t) = \dot{\mathbf{o}}(t) - \dot{\mathbf{e}}(t) \tag{9.1}$$

Therefore, different combinations of $\dot{\mathbf{o}}(t)$ and $\dot{\mathbf{e}}(t)$ can give rise to the same $\dot{\mathbf{r}}(t)$. We initially compared two combinations of $\dot{\mathbf{o}}(t)$ and $\dot{\mathbf{e}}(t)$: one with $\dot{\mathbf{e}}(t)$ being zero and $\dot{\mathbf{o}}(t)$ being circular motion (our baseline fixation condition), and the other with

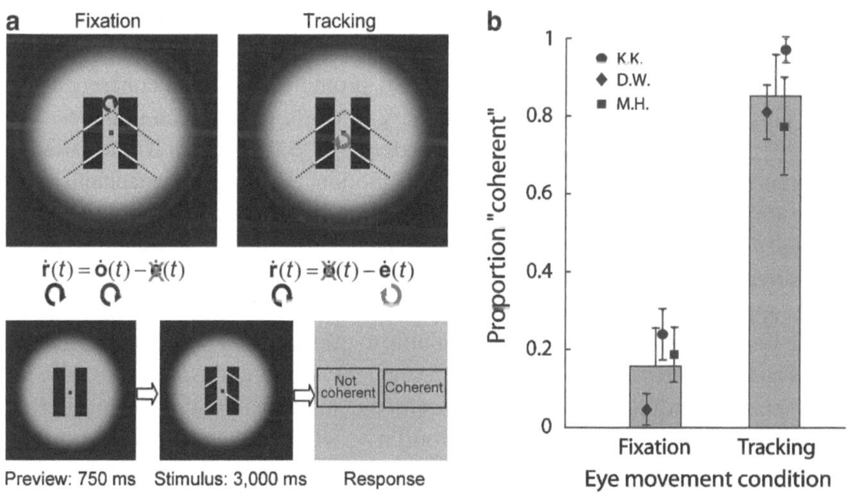

Fig. 9.5 Action for perception in the smooth pursuit system. (**a**) A scene similar to that of Fig. 9.3 was used except that the object of interest was an outline chevron. The same retinal motions of the chevron across different eye movement conditions were achieved through the use of (9.1) (*Top*). In fixation, the chevron translated in a circular trajectory while the eye remained stationary. In tracking, the chevron was translated to keep the retinal stimulus unchanged (Hafed and Krauzlis 2006). An initial preview period of 750 ms was enforced before the chevron appeared to ensure steady-state fixation or pursuit, and therefore ensure similarity of retinal events across the eye movements conditions (*Bottom*) (Hafed and Krauzlis 2006). (**b**) Subjects reported seeing a coherent chevron much more frequently during tracking than during fixation, suggesting that information about their ongoing eye movements was used to disambiguate their input retinal stimuli. See Supplementary movies 6 and 7. Other experiments exploited (9.1) further in order to test different combinations of eye and object motion. Adapted from Hafed and Krauzlis (2006)

$\dot{\mathbf{o}}(t)$ being zero and $\dot{\mathbf{e}}(t)$ being circular smooth pursuit (our tracking condition). With the phase of $\dot{\mathbf{e}}(t)$ in the tracking condition being 180° relative to that of $\dot{\mathbf{o}}(t)$ in the fixation condition, the retinal motion of the chevron was the same in both conditions. Such motion was accounted for entirely by motion of the chevron in space in the fixation condition but by ongoing eye movements in the tracking condition (Fig. 9.5a and Supplementary Movies 6 and 7).

We then asked our subjects to report on the perceptual coherence of the chevron – that is, on whether a rigid chevron was perceived or not. What we found was that the chevron was not easily perceived during fixation, consistent with previous results (Lorenceau and Alais 2001). However, and with the same retinal motion, the chevron was very easily perceived when it was viewed during an ongoing eye movement (Fig. 9.5b). Thus, information about the ongoing eye movement was sufficient to disambiguate the retinal stimulus.

The basic idea of using (9.1) above to match the retinal stimulus across different eye movement conditions can also allow for further exploration of how ongoing eye movements influence visual perception (Hafed and Krauzlis 2006). Consider, for example, the case in which the eye tracks along a horizontal axis and the chevron translates along a vertical one. In this case, circular motion of the chevron in retinal coordinates can still be achieved, just like in the case of the basic fixation and tracking conditions described above (Fig. 9.5), but with an orthogonal decomposition of eye and object motion. When we performed this decomposition, we were able to demonstrate that eye movements promote perceptual coherence even if the object being perceived is not stationary in world-centered coordinates. This rules out a "stable world assumption" (Wexler et al. 2001), which is often used by the brain's perceptual system during self-motion, as the sole mechanism for explaining these results. In addition, by having the eye track along one of the cardinal directions, we could now compare the benefits of the particular eye movement direction relative to the direction of inherent motion ambiguity in the stimulus. As mentioned earlier, the vertical apertures of Fig. 9.5a specifically force the visible line segments of the chevron stimulus to translate in a purely vertical direction. Thus, the horizontal motion of each segment is ambiguous. Interestingly, it is information regarding an ongoing *horizontal* eye movement that is the most effective in disambiguating the stimulus. Similarly, when the apertures are horizontal (imagine transposing the entire scene of Fig. 9.5a by 90°), information about an ongoing *vertical* eye movement is the most effective in disambiguating the stimulus (Hafed and Krauzlis 2006).

In all, these results suggest that the presence of eye movement information in the brain provides an aid to perception. As we have alluded to at the beginning of this section, such an aid to perception may represent an efficient use of computational resources in the ongoing process of vision; that is, this process continues unimpeded even during the periods in which the visual system is faced with eye movement contingent changes in retinal inputs. The fact that extra-retinal information about ongoing eye movements interacts with vision in this manner is plausible and garners support from other lines of evidence in the literature. Specifically, there is now convincing neurophysiological and brain imaging support for the use of corollary discharge information in the brain in order to maintain spatial stability across eye movements (Sommer and Wurtz 2002, 2006; Duhamel et al. 1992; Umeno and Goldberg 1997; Walker et al. 1995). Such information gives rise to dynamic, and often *predictive*, remapping of spatially organized visual and movement maps across eye movements in purely sensory (Nakamura and Colby 2000, 2002), sensorimotor (Duhamel et al. 1992; Umeno and Goldberg 1997), and motor (Walker et al. 1995) areas in the brain. Therefore, the same areas of the brain that perform processing geared towards perceptual integration and feature binding also have access to information about upcoming and ongoing eye movements.

In terms of smooth pursuit eye movements, there is also evidence for cancellation of pursuit-induced retinal image motion (Lindner et al. 2001) and for reduced sensitivity to retinal image smear during pursuit (Bedell et al. 2004; Bedell and Lott 1996) – similar to what happens for retinal image smear resulting from saccadic eye movements (Burr et al. 1999). Such a reduction is often greatest for global image motion opposite in direction to pursuit (Lindner et al. 2001), which again suggests that extra-retinal information is used by the visual system in such a way as to relieve perception of possibly irrelevant visual cues. Other results that support the notion of eye movements promoting visual perception emerge out of work that is mainly concerned with understanding the quality of corollary discharge signals in the brain and how these signals may be used in conjunction with retinal information in order to support vision. For example, such work has led to interesting findings on the pop-out of coherent motion during smooth pursuit (Greenlee et al. 2002) and on the influence of pursuit on depth perception (Naji and Freeman 2004). In addition, there is evidence that in the perception of object velocity, retinal image motion and eye velocity may interact in humans in a non-trivial [i.e. not simple superposition *à la* (9.1)] manner (Goltz et al. 2003). Evidence more related to the experiments of Fig. 9.5 is that oscillatory pursuit eye movements sometimes cause synchronization of reversal patterns in a necker cube stimulus (Scotto 1991), which may explain the near perfect temporal synchrony of perceptual and pursuit reversals in Madelain's and Krauzlis' (2003) stimuli mentioned earlier (Fig. 9.4). Similar effects on perceptual reversal patterns with a necker cube stimulus have also been found for saccadic eye movements (Ross and Ma-Wyatt 2004).

All of the above leads to an important question concerning the issue of closing the action-perception cycle: perception is known to guide eye movements, so how can internal neuronal processes concerned with generating and monitoring ongoing eye movements then influence perception? We believe that such an interaction between guidance of eye movements by perception and promotion of perception by eye movements has analogies in, for example, the interactions that are hypothesized to exist between visual attention and eye movements (Hamker 2005). Recent research on the neural substrates of attention has revealed attentional effects throughout the visual cortex, to the extent that it is now believed that the role of early visual areas is geared more towards highlighting behaviorally relevant stimuli than merely providing a sensory reconstruction of retinal inputs (Maunsell 1995; Reynolds and Chelazzi 2004; Treue 2001). While this observation may seem intuitive in retrospect, it suggests that attention is crucial for effective visual perception.

The fact that spatial attention is extremely tightly linked to eye movements (Rizzolatti et al. 1987; Sheliga et al. 1994, 1997; Corbetta et al. 1998; Hafed and Clark 2002; Clark et al. 2007; Kustov and Robinson 1996; Nobre et al. 2000) (see also Sect. 9.2 above), even possibly controlled by eye movement centers in the brain (Moore and Fallah 2004; Moore and Armstrong 2003; Hamker 2005; Bisley and Goldberg 2003; Cavanaugh and Wurtz 2004; Ignashchenkova et al. 2004; Muller et al. 2005; Wardak et al. 2004), thus suggests that vision may essentially be centered around eye movements. This hypothesis is gradually gaining support from anatomical and physiological studies. For example, anatomical studies of cortico-cortical and thalamo-cortical connections in the brain lend themselves to viewing sensory processes as 'monitors' of ongoing motor programs (Guillery and Sherman 2002). So eye movement information is and should be widespread in the brain.

In the light of this, it is interesting that there is evidence that global percepts also influence the early retinotopic areas of the brain (Leopold and Logothetis 1996; Ress and Heeger 2003). Such percepts and their reversals again probably arise in, and are controlled by, the so-called 'output' stages of the brain (Leopold and Logothetis 1999). It is thus plausible to argue that the role of these stages in the control of visual attention (Moore and Fallah 2004; Moore and Armstrong 2003; Hamker 2005; Bisley and Goldberg 2003; Cavanaugh and Wurtz 2004; Ignashchenkova et al. 2004; Muller et al. 2005; Wardak et al. 2004) and the link between visual attention and eye movements mean that there has to be an equally important link between perception and eye movements. A causal link between eye movement execution and the shifting of spatial attention (Moore and Fallah 2004; Moore and Armstrong 2003; Hamker 2005; Bisley and Goldberg 2003; Cavanaugh and Wurtz 2004; Ignashchenkova et al. 2004; Muller et al. 2005; Wardak et al. 2004) may also be paralleled by a causal link between eye movement execution and perception.

9.5 Concluding Remarks

In this chapter, we have reviewed some of the known interactions between smooth pursuit eye movements and visual perception. Specifically, we have shown that in addition to being driven by direct sensory inputs, smooth pursuit can also be driven and modified based on 'cognitively processed' inputs. Moreover, we have shown that smooth pursuit itself can influence the sensory interpretation of visual inputs, in a manner that is more sophisticated than simply altering the pattern of retinal stimulation. All of these results indicate that perception and action interact in a more complicated fashion than previously thought (Fig. 9.6). As we have mentioned above, we believe that such interaction is necessary for our visual system to provide us with a continuous, uninterrupted analysis of our visual environment. We also believe that this interaction provides our community with an interesting set of opportunities and challenges for future research.

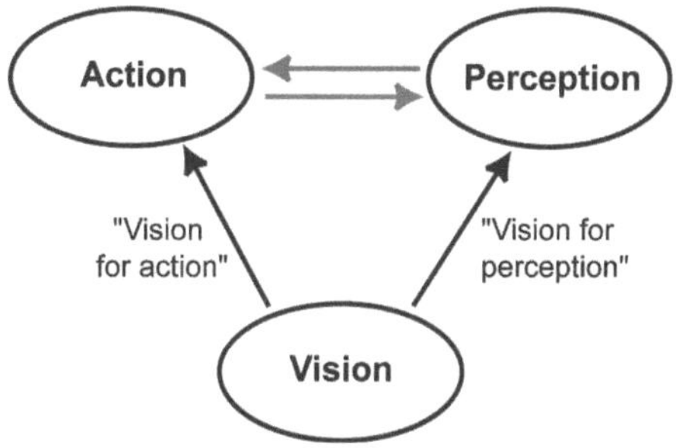

Fig. 9.6 Perception-for-action and action-for-perception in the visual system. The highly schematic figure shows information flow from the primary visual cortex (vision) along the dorsal (action) and ventral (perception) visual processing streams in the brain (Milner and Goodale 1995). Just like with other motor outputs, recent research on smooth pursuit eye movements has shown that the two visual processing streams interact in a bidirectional manner . This allows the brain maximum flexibility in sensing and reacting to the visual environment

One implication of the results described in this chapter is that smooth pursuit eye movements can serve as an overt measure of various covert processes that take place in the brain. For example, smooth pursuit can reflect the influences of spatial (Tam and Ono 1994; Merrison and Carpenter 1995; Krauzlis and Miles 1996a, b; Knox 1996; Kimmig et al. 2002; Krauzlis et al. 2004; Adler et al. 2002) and feature-based (Garbutt and Lisberger 2006) attention, as well as the current state of motion perception (Beutter and Stone 2000; Steinbach 1976; Medelain and Krauzlis 2003). Coupled with the fact that saccades, as well as involuntary eye movements like microsaccades (Hafed and Clark 2002; Clark et al. 2007), also reflect the influences of visual and non-visual cognitive events, this means that eye movements in general constitute an important window to the state of the visual system (and the brain in general). The use of eye movements to study motion processing, in particular, and visual perception, in general, therefore seems imperative in our future research. In addition, this use in populations with neurodevelopmental disorders, such as attention-deficit hyperactivity disorder (ADHD) (Sweeney et al. 2004), can shed light both on the neuronal basis of these disorders as well as on how the visual system functions in its normal state.

However, with the above opportunity provided by eye movements comes the challenge of knowing that ongoing eye movements themselves influence the state of visual perception (Hafed and Krauzlis 2006). Thus, understanding the intricate details of the action-perception cycle of eye movements requires a greater understanding of the neural mechanisms through which potential or actual eye movement commands contribute to sensory processing. This can be achieved through psychophysics, although neurophysiological studies involving controlled perturbation of the normal system through stimulation or inactivation of neuronal populations seem to be an attractive candidate. With such perturbations, one can test various models of how information about impending or ongoing eye movements is used in the brain (see for example Moore and Fallah 2004; Moore and Amstrong 2003, for studies of this kind with saccades).

Finally, while they are probably the most important for vision, eye movements are not the only type of action that has implications on how we perceive and subsequently interact with our environment. Eye movements are very often coordinated with head and body movements, and it would be interesting to investigate how issues of coordination and control of saccades and smooth pursuit eye movements by sensory (and other inputs) generalize to the control of head and body movements.

9.6 Supplementary Materials (CD-ROM)

Movie 1 Cycling dots (see Fig. 9.2) (file «*9_M1_Cyclingdots.mov*») This movie shows an animation of the stimulus described in Fig. 9.2a, and demonstrates that a percept of a rolling wheel is possible.

Movie 2 Pursuing 2 bars (see Fig. 9.2) (file «*9_M2_Pursuit2bars.mov*») Supplementary movie 2 shows an example of the stimulus described in Fig. 9.2b. In this example, the eye position (green crosshair) of a monkey subject from our laboratory is shown as this monkey tracked an invisible point between peripheral stimuli that shared common motion (Hafed and Krauzlis 2005; Hafed and Krauzlis, submitted). The movie runs at half the actual frame rate of the experiment.

Movies 3-4 Moving diamond with visible or invisible occluders (files «*9_M3_Occludeddiamond.mov*», «*9_M4_Invisibleoccluders.mov*»). Supplementary movies 3 and 4 show an occluded diamond with high and low contrast occluders, respectively. In the former, a percept of a rigid diamond is possible, and the motion of the diamond along a diagonal axis can be inferred. In the latter, such a percept is not possible, and the visible diamond segments appear to move in a purely vertical direction.

Movie 5 Pursuing multistable apparent motion (file «*9_M5_Packman.gif*») Supplementary movie 5 shows a movie of the apparent motion stimulus of Madelain and Krauzlis (2003) demonstrating its directional bistability. The reader will notice that it is easy to perceive rightward or leftward motion and to track this motion with his/her eyes.

Movies 6-7 Passive and active chevron motion (see Fig. 9.5) (files «*9_M6_Chevron1.mov*», «*9_M7_Activechevron.mov*»). Supplementary movies 6 and 7 show the basic fixation and tracking conditions respectively of the stimulus shown in Fig. 9.5,. In the fixation condition, a percept of a rigid chevron is very difficult (the whole chevron is shown for a portion of the movie to demonstrate the contrast in percepts when the chevron is partially occluded). However, when the same retinal stimulus is viewed during an ongoing eye movement, the chevron is easily perceived.

Acknowledgment This work was supported by NIH Grant EY-12212 from the National Eye Institute.

References

Adler SA, Bala J, Krauzlis RJ (2002) Primacy of spatial information in guiding target selection for pursuit and saccades. J Vis 2:627–644

Albright TD, Stoner GR (1995) Visual motion perception. Proc Natl Acad Sci USA 92:2433–2440
Andersen RA, Buneo CA (2002) Intentional maps in posterior parietal cortex. Annu Rev Neurosci 25:189–220
Bedell HE, Chung STL, Patel SS (2004) Attenuation of perceived motion smear during vergence and pursuit tracking. Vis Res 44:895–902
Bedell HE, Lott LA (1996) Suppression of motion-produced smear during smooth pursuit eye movements. Curr Biol 6:1032–1034
Berryhill ME, Chiu T, Hughes HC (2006) Smooth pursuit of nonvisual motion. J Neurophysiol 96:461–465
Beutter BR, Stone LS (2000) Motion coherence affects human perception and pursuit similarly. Vis Neurosci 17:139–153
Biscaldi M, Fischer B, Stuhr V (1996) Human express saccade makers are impaired at suppressing visually evoked saccades. J Neurophysiol 76:199–214
Bisley JW, Goldberg ME (2003) Neuronal activity in the lateral intraparietal area and spatial attention. Science 299:81–86
Born RT, Bradley DC (2005) Structure and function of visual area MT. Annu Rev Neurosci 28:157–189
Born RT, Pack CC, Ponce CR, Yi S (2006) Temporal evolution of 2-dimensional direction signals used to guide eye movements. J Neurophysiol 95:284–300
Bridgeman B, Hendry D, Stark L (1975) Failure to detect displacement of the visual world during saccadic eye movements. Vis Res 15:719–722
Burr DC, Morgan MJ, Morrone MC (1999) Saccadic suppression precedes visual motion analysis. Curr Biol 9:1207–1209
Carello CD, Krauzlis RJ (2003) Manipulating intent: evidence for a causal role of the superior colliculus in target selection. Neuron 43:575–583
Carpenter RHS (1988) Movements of the eyes, 2nd edn. Pion, London
Cavanaugh J, Wurtz RH (2004) Subcortical modulation of attention counters change blindness. J Neurosci 24:11236–11243
Clark JJ, Hafed ZM, Jie L (2007) Attention and Action. In: Harris LR, Jenkin MRM (eds) Computational vision in neural and machine systems. Cambridge University Press, Cambridge, pp 129–148
Colby CL, Goldberg ME (1999) Space and attention in parietal cortex. Annu Rev Neurosci 22:319–349
Corbetta M, Akbudak E, Conturo TE, Snyder AZ, Ollinger JM, Drury HA, Linenweber MR, Petersen SE, Raichle ME, Van Essen DC, Shulman GL (1998) A common network of functional areas for attention and eye movements. Neuron 21:761–773
Currie CB, McConkie GW, Carlson-Radvansky LA, Irwin DE (2000) The role of the saccade target object in the perception of a visually stable world. Percept Pscyhophys 62:673–683
Deubel H, Bridgeman B, Schneider WX (1998) Immediate post-saccadic information mediates space constancy. Vis Res 38:3147–3159
Deubel H, Schneider WX (1996) Saccade target selection and object recognition: Evidence for a common attentional mechanism. Vis Res 36:1827–1837
Duffy CJ, Wurtz RH (1991) Sensitivity of MST neurons to optic flow stimuli I. A continuum of response selectivity to large-field stimuli. J Neurophysiol 65:1329–1345
Duhamel JR, Colby CL, Goldberg ME (1992) The updating of the representation of visual space in parietal cortex by intended eye movements. Science 255:90–92
Farah MJ (2000) The cognitive neuroscience of vision. Blackwell Publishers, Oxford
Findlay JM, Gilchrist ID (2001) Visual attention: the active vision perspective. In: Jenkins M, Harris L (eds) Vision and attention. Springer Verlag, Berlin, pp 85–105
Fischer B, Weber H (1993) Express saccades and visual attention. Behav Brain Sci 16:553–610
Garbutt S, Lisberger SG (2006) Directional cuing of target choice in human smooth pursuit eye movements. J Neurosci 26:12479–12486
Goldberg ME, Bisley JW, Powell KD, Gottlieb J (2006) Saccades, salience and attention: the role of the lateral intraparietal area in visual behavior. Prog Brain Res 155:157–175
Goltz HC, DeSouza JFX, Menon RS, Tweed DB, Vilis T (2003) Interaction of retinal image and eye velocity in motion perception. Neuron 39:569–576

Graziano MS, Andersen RA, Snowden RJ (1994) Tuning of MST neurons to spiral motions. J Neurosci 14:54–67
Greenlee MW, Schira MM, Kimmig H (2002) Coherent motion pops out during smooth pursuit. NeuroReport 13:1313–1316
Guillery RW, Sherman SM (2002) The thalamus as a monitor of motor outputs. Phil Trans Roy Soc Lond B 357:1809–1821
Hafed ZM, Clark JJ (2002) Microsaccades as an overt measure of covert attention shifts. Vis Res 42:2533–2545
Hafed ZM, Krauzlis RJ (2005) Goal representations dominate superior colliculus activity during parafoveal pursuit (Abstract No. 475.14). Society for Neuroscience Annual Meeting, Washington, D.C
Hafed ZM, Krauzlis RJ (2006) Ongoing eye movements constrain visual perception. Nat Neurosci 9:1449–1457
Hafed ZM, Krauzlis RJ. (2008) Goal representations dominate superior colliculus activity during extrafoveal tracking. J Neurosci 28:9426–9439
Hafed ZM, Goffart L, Krauzlis RJ (2008) Superior colliculus inactivation causes stable offsets in eye position during tracking. J Neurosci 28:8124–8137
Hamker FH (2005) The reentry hypothesis: the putative interaction of the frontal eye field, ventrolateral prefrontal cortex, and areas V4, IT for attention and eye movement. Cereb Cortex 15:431–437
Henderson JM, Hollingworth A (2003) Global transsaccadic change blindness during scene perception. Psychol Sci 14:493–497
Ignashchenkova A, Dicke PW, Haarmeier T, Thier P (2004) Neuron-specific contribution of the superior colliculus to overt and covert shifts of attention. Nat Neurosci 7:56–64
Ilg UJ (1997) Slow eye movements. Prog Neurobiol 53:293–329
Ilg UJ, Thier P (1999) Eye movements of rhesus monkeys directed towards imaginary targets. Vis Res 39:2143–2150
Kanizsa G (1976) Subjective contours. Sci Am 234:48–52
Kastner S, Ungerleider LG (2000) Mechanisms of visual attention in the human cortex. Annu Rev Neurosci 23:315–341
Kimmig H, Biscaldi M, Mutter J, Doerr JP, Fischer B (2002) The initiation of smooth pursuit eye movements and saccades in normal subjects and in "express-saccade makers". Exp Brain Res 144:373–384
Knox PC (1996) The effect of the gap paradigm on the latency of human smooth pursuit of eye movement. NeuroReport 7:3027–3030
Komatsu H, Wurtz RH (1988a) Relation of cortical areas MT and MST to pursuit eye movements I. Localization and visual properties of neurons. J Neurophysiol 60:580–603
Komatsu H, Wurtz RH (1988b) Relation of cortical areas MT and MST to pursuit eye movements III. Interaction with full-field visual stimulation. J Neurophysiol 60:621–644
Krauzlis RJ (2003) Neuronal activity in the rostral superior colliculus related to the initiation of pursuit and saccadic eye movements. J Neurosci 23:4333–4344
Krauzlis RJ (2005) The control of voluntary eye movements: New perspectives. Neuroscientist 11:124–137
Krauzlis RJ, Dill N, Kornylo K (2002) Activity in the primate rostral superior colliculus during the "gap effect" for pursuit and saccades. Ann New York Acad Sci 956:409–413
Krauzlis RJ, Liston D, Carello C (2004) Target selection and the superior colliculus: goals, choices and hypotheses. Vis Res 44:1445–1451
Krauzlis RJ, Miles FA (1996a) Decreases in the latency of smooth pursuit and saccadic eye movements produced by the "gap paradigm" in the monkey. Vis Res 36:1973–1985
Krauzlis RJ, Miles FA (1996b) Release of fixation for pursuit and saccades in humans: evidence for shared inputs acting on different neural substrates. J Neurophysiol 76:2822–2833
Krauzlis RJ, Stone LS (1999) Tracking with the mind's eye. Trends Neurosci 22:544–550
Krekelberg B, Albright TD (2005) Motion mechanisms in macaque MT. J Neurophysiol 93:2908–2921

Kustov AA, Robinson DL (1996) Shared neural control of attentional shifts and eye movements. Nature 384:74–77
Lamontagne C, Gosselin F, Pivik T (2002) Sigma smooth pursuit eye tracking: Constant k values revisited. Exp Brain Res 143:130–132
Leopold DA, Logothetis NK (1996) Activity changes in early visual cortex reflect monkeys' percepts during binocular rivalry. Nature 379:549–553
Leopold DA, Logothetis NK (1999) Multistable phenomena: changing views in perception. Trends Cognit Sci 3:254–264
Lindner A, Schwarz U, Ilg UJ (2001) Cancellation of self-induced retinal image motion during smooth pursuit eye movements. Vis Res 41:1685–1694
Lisberger SG, Ferrera VP (1997) Vector averaging for smooth pursuit eye movements initiated by two moving targets in monkeys. J Neurosci 17:7490–7502
Lisberger SG, Morris EJ, Tychsen L (1987) Visual motion processing and sensory-motor integration for smooth pursuit eye movements. Annu Rev Neurosci 10:97–129
Lorenceau J, Alais D (2001) Form constraints in motion binding. Nat Neurosci 4:745–751
Lorenceau J, Shiffrar M (1992) The influence of terminators on motion integration across space. Vis Res 32:263–273
Lorenceau J, Shiffrar M, Wells N, Castet E (1993) Different motion sensitive units are involved in recovering the direction of moving lines. Vis Res 33:1207–1217
Madelain L, Krauzlis RJ (2003) Pursuit of the ineffable: perceptual and motor reversals during the tracking of apparent motion. J Vis 3:642–653
Marr D, Ullman S (1981) Directional selectivity and its use in early visual processing. Proc Roy Soc Lond B 211:151–180
Maunsell JHR (1995) The brain's visual world: representation of visual targets in cerebral cortex. Science 270:764–769
Maunsell JHR, Treue S (2006) Feature-based attention in visual cortex. Trends Neurosci 29:317–322
McConkie GW, Currie CB (1996) Visual stability across saccades while viewing complex pictures. J Exp Psychol Human Percept Perform 22:563–581
Merrison AFA, Carpenter RHS (1995) Express smooth pursuit. Vis Res 35:1459–1462
Milner AD, Goodale MA (1995) The visual brain in action. Oxford University Press, Oxford
Moore T, Armstrong KM (2003) Selective gating of visual signals by microstimulation of frontal cortex. Nature 421:370–373
Moore T, Fallah M (2004) Microstimulation of the frontal eye field and its effects on covert spatial attention. J Neurophysiol 91:152–162
Muller JR, Philiastides MG, Newsome WT (2005) Microstimulation of the superior colliculus focuses attention without moving the eyes. Proc Natl Acad Sci USA 102:524–529
Naji JJ, Freeman TCA (2004) Perceiving depth order during pursuit eye movement. Vis Res 44:3025–3034
Nakamura K, Colby CL (2000) Visual, saccade-related, and cognitive activation of single neurons in monkey extrastriate area V3A. J Neurophysiol 84:677–692
Nakamura K, Colby CL (2002) Updating of the visual representation in monkey striate and extrastriate cortex during saccades. Proc Natl Acad Sci USA 99:4026–4031
Nakayama K (1985) Biological image motion processing: a review. Vis Res 25:625–660
Newsome WT, Wurtz RH, Komatsu H (1988) Relation of cortical areas MT and MST to pursuit eye movements II. Differentiation of retinal from extraretinal inputs. J Neurophysiol 60:604–620
Nobre AC, Gitelman DR, Dias EC, Mesulam MM (2000) Covert visual spatial orienting and saccades: overlapping neural systems. Neuroimage 11:210–216
O'Regan JK, Noe A (2001) A sensorimotor account of vision and visual consciousness. Behav Brain Sci 24:939–1011
Pack CC, Born RT (2001) Temporal dynamics of a neural solution to the aperture problem in visual area MT of macaque brain. Nature 409:1040–1042
Posner MI (1980) Orienting of attention. Quart J Exp Psychol 32:3–25
Ramachandran VS, Anstis SM (1985) Perceptual organization in multistable apparent motion. Perception 14:135–143

Rashbass C (1961) The relationship between saccadic and smooth tracking eye movements. J Physiol (London) 159:326–338
Remington RW (1980) Attention and saccadic eye movements. J Exp Psychol Human Percept Perform 6:726–744
Rensink RA, O'Regan JK, Clark JJ (1997) To see or not to see: the need for attention to perceive changes in scenes. Psychol Sci 8:368–373
Rensink RA (2002) Change detection. Annu Rev Psychol 53:245–277
Ress D, Heeger DJ (2003) Neuronal correlates of perception in early visual cortex. Nat Neurosci 6:414–420
Reulen JP (1984) Latency of visually evoked saccadic eye movements I. Saccadic latency and the facilitation model. Biol Cybern 50:251–262
Reuter-Lorenz PA, Hughes HC, Fendrich R (1991) The reduction of saccadic latency by prior offset of the fixation point: an analysis of the gap effect. Percept Psychophys 49:167–175
Reynolds JH, Chelazzi L (2004) Attentional modulation of visual processing. Annu Rev Neurosci 27:611–647
Rizzolatti G, Riggio L, Dascola I, Umilta C (1987) Reorienting attention across the horizontal and vertical meridians: Evidence in favor of a premotor theory of attention. Neuropsychologica 25:31–40
Robinson DA (1965) The mechanics of human smooth pursuit eye movement. J Physiol (London) 180:569–591
Ross J, Ma-Wyatt A (2004) Saccades actively maintain perceptual continuity. Nat Neurosci 7:65–69
Ross J, Morrone MC, Goldberg ME, Burr DC (2001) Changes in visual perception at the time of saccades. Trends Neurosci 24:113–121
Saito H, Yukie M, Tanaka K, Hikosaka K, Fukada Y, Iwai E (1986) Integration of direction signals of image motion in the superior temporal sulcus of the macaque monkey. J Neurosci 6:145–157
Saslow MG (1967) Effects of components of displacement step stimuli upon latency for saccadic eye movements. J Opt Soc Am 57:1024–1029
Schall JD, Thompson KG (1999) Neural selection and control of visually guided eye movements. Annu Rev Neurosci 22:241–259
Scotto M (1991) Smooth periodic eye movements can entrain perceptual alternation. Percept Mot Skills 73:835–843
Seiffert AE, Cavanagh P (1998) Position displacement, not velocity, is the cue to motion detection of second-order stimuli. Vis Res 38:3569–3582
Sekuler R, Blake R (1994) Perception, 3rd edn. McGraw-Hill, New York
Sheliga BM, Craighero L, Riggio L, Rizzolatti G (1997) Effects of spatial attention on directional manual and ocular responses. Exp Brain Res 114:339–351
Sheliga BM, Riggio L, Rizzolatti G (1994) Orienting of attention and eye movements. Exp Brain Res 98:507–522
Sommer MA, Wurtz RH (2002) A pathway in primate brain for internal monitoring of movements. Science 296:1480–1482
Sommer MA, Wurtz RH (2006) Influence of the thalamus on spatial visual processing in frontal cortex. Nature 444:374–377
Steinbach MJ (1976) Pursuing the perceptual rather than the retinal stimulus. Vis Res 16:1371–1376
Stone LS, Krauzlis RJ (2003) Shared motion signals for human perceptual decisions and oculomotor actions. J Vis 3:725–736
Sweeney JA, Takarae Y, Macmillan C, Luna B, Minshew NJ (2004) Eye movements in neurodevelopmental disorders. Curr Opin Neurol 17:37–42
Tam WJ, Ono H (1994) Fixation disengagement and eye-movement latency. Percept Psychophys 56:251–260
Treue S (2001) Neural correlates of attention in primate visual cortex. Trends Neurosci 24:295–300
Treue S, Martinez-Trujillo JC (1999) Feature-based attention influences motion processing gain in macaque visual cortex. Nature 399:575–579

Umeno MM, Goldberg ME (1997) Spatial processing in the monkey frontal eye field I. Predictive visual responses. J Neurophysiol 78:1373–1383
van der Steen J, Tamminga EP, Collewijn H (1983) A comparison of oculomotor pursuit of a target in circular real, beta or sigma motion. Vis Res 23:1655–1661
Walker MF, Fitzgibbon EJ, Goldberg ME (1995) Neurons in the monkey superior colliculus predict the visual result of impending saccadic eye movements. J Neurophysiol 73:1988–2003
Wardak C, Olivier E, Duhamel JR (2004) A deficit in covert attention after parietal cortex inactivation in the monkey. Neuron 42:501–508
Westheimer G (1954) Eye movement responses to a horizontally moving visual stimulus. AMA Arch Ophthalmol 52:932–943
Wexler M, Panerai F, Lamouret I, Droulez J (2001) Self-motion and the perception of stationary objects. Nature 409:85–88
Wilmer JB, Nakayama K (2007) Two distinct visual motion mechanisms for smooth pursuit: evidence from individual differences. Neuron 54:987–1000
Wyatt HJ, Pola J, Fortune B, Posner M (1994) Smooth pursuit eye movements with imaginary targets defined by extrafoveal cues. Vis Res 34:803–820
Yo C, Wilson HR (1992) Perceived direction of moving two-dimensional patterns depends on duration, contrast and eccentricity. Vis Res 32:135–147
Zeki S (1993) A vision of the brain. Blackwell Scientific Publications, Oxford

Chapter 10
Perception of Intra-saccadic Motion

Eric Castet

Abstract A typical saccadic eye movement lasts about 40 ms. During this short period of time, the image of the stationary world around us rapidly moves on the retina with a complex accelerating and decelerating profile. The reason why this 40 ms retinal motion flow does not elicit motion perception in everyday life is an issue that has received considerable interest. The present chapter first presents a brief history of the main ideas and experiments bearing on this issue since the seventies. Some key experimental paradigms and results in psychophysics are then described in detail. Finally, some suggestions for future investigations, both psychophysical and physiological, are made. A major goal of the chapter is to pinpoint some fundamental confusions that are often encountered in the literature. It is hoped that understanding these confusions will help identify more clearly the theoretical points – among which the role of temporal masking – on which scientists strongly disagree.

10.1 Introduction

The stationary world around us does not appear to move during saccadic eye movements. Early authors already wondered why we are not aware of the activity elicited during the short saccadic period (about 40 ms) in which the image of the world does move on the retina (Dodge 1900, 1905; Holt 1903). This intra-saccadic issue should not be confused with another one usually referred to as the trans-saccadic fusion issue (Deubel et al. 2002). In the latter case, the problem is to understand why the

E. Castet (✉)
Dynamics of Visual Perception and Action, Institut de Neurosciences Cognitives de la Méditerranée, CNRS and Université de la Méditerranée, 31 Chemin Joseph Aiguier, 13402, Marseille, France
e-mail: eric.castet@incm.cnrs-mrs.fr

U.J. Ilg and G.S. Masson (eds.), *Dynamics of Visual Motion Processing: Neuronal, Behavioral, and Computational Approaches*,
DOI 10.1007/978-1-4419-0781-3_10, © Springer Science+Business Media, LLC 2010

2-frame shift in position occurring between the pre- and post-saccadic images does not usually elicit any displacement percept.

To explain why the world does not appear to move during each saccade, two extreme theories are proposed which are actually nonexclusive. The first theory postulates an active suppression process originating from central nervous structures and operating during the saccade in order to inhibit visual areas. In such a framework, "extra-retinal" signals, conceptually similar to an efference copy, are triggered by the oculo-motor command and sent to visual structures. The other general theory does not postulate any extra-retinal signal and relies on visual and/or retinal spatio-temporal processes such as the well-known temporal masking.

In the last two decades, a "preference" for the extra-retinal suppression theory seems to have emerged. More precisely, the idea that the motion-processing system, and thus motion perception, is actively and selectively suppressed during saccades can be often found in the literature. This is illustrated below by a few exemplary citations. "*There is now good evidence that perception of motion is strongly suppressed during saccades (rapid shifts of gaze), presumably to blunt the disturbing sense of motion that saccades would otherwise elicit.*" (Burr et al. 1999). "*During fast saccadic eye movements, visual perception is suppressed. This saccadic suppression prevents erroneous and distracting motion percepts resulting from saccade induced retinal slip.*" (Georg and Lappe 2007). "*[...] this fits well with the idea that saccadic suppression reflects the visual system's attempt to ignore the retinal image motion induced by saccades.*" (Kleiser et al. 2004). "*The purpose of the saccadic suppression of motion may be to block out unreliable motion signals that would be produced by a saccade*" (Shioiri and Cavanagh 1989). This preference is also found in several recent reviews (Burr and Morrone 2004; Ross et al. 1996, 2001).

The goal of the present chapter is to offer a more balanced view of the issue if only because it has been shown that intra-saccadic motion perception can be easily elicited in humans (Castet et al. 2002; Castet and Masson 2000). There is much confusion in the literature that might explain the tendency to systematically assert that motion processing and motion perception are suppressed during saccades. Notably, the expression "saccadic suppression" is extensively used as though it were a unique process relying on a homogenous set of experiments. In contrast, I will attempt to show that there are different classes of experimental effects that might actually reflect totally different visual processes. Another concern is related to the general problem of consciousness, which is strongly debated in visual neurosciences. When we do not consciously perceive a retinal event which lasts about 50 ms, does it mean that this event has to be erased in early visual areas, or does it mean that this brief period is "filled-in" by anterior and posterior retinal events? The answer to this question is crucial to make correct predictions regarding the neural processes leading to the intra-saccadic "blindness" in normal viewing.

The first section of the chapter outlines the evolution of the ideas since the seventies without describing in detail the experimental effects. This is to help understand why some ideas seem to have become predominant while overlooking some key results available in the early literature. Then, a few key experimental effects, and their possible interpretations, will be described without pretending to be exhaustive. I will rely mainly on psychophysical studies, as Chap. 10 by Mike Ibbotson will be devoted to physiological work on the intra-saccadic perception issue.

10.2 A Brief History of the Concepts

10.2.1 Up to 1982

In the seventies, it was believed that saccadic speeds were too fast for the visual system to resolve and caused therefore a blurring or smearing of the visual scene. In this context, two important studies showed that a form of **temporal masking** was the main factor preventing us from perceiving the smearing or "grey-out" induced by each saccade (Campbell and Wurtz 1978; Matin et al. 1972). The principle of Campbell and Wurtz's experiments, which extended those of Matin et al. was simple (Fig. 10.1). When the experimental room was illuminated only during the time of saccades, observers perceived the scene as being smeared or greyed out. By grey-out, a decrease in the apparent contrast of the image was meant. However, as the duration of the light was extended beyond the end of the eye movement, the amount of smearing became progressively less. Only 40 ms of post-saccadic illumination of the room was sufficient to restore a sharp percept of the scene. In this case, the authors insisted that subjects did not perceive a smeared image followed by a sharp image but instead reported a single sharp percept. It was thus the presence of a post-saccadic image of the scene which made it possible to avoid the perception of the brief intra-saccadic grey-out (the effect of a pre-saccadic image was shown to be as efficient as a post-saccadic image). The authors referred to this temporal masking mechanism as a "saccadic omission" process (instead of saccadic suppression) in order to emphasize their main theoretical point: the basic process needed by the visual system to prevent percepts induced by the intra-saccadic stimulations cannot rely on a suppression (or on a dampening) process. If it were the case, the temporal flow of our perception would be constantly interrupted by a dark (or a dimmer) brief percept whenever we make a saccade. What is needed is a mechanism that preserves the perceptual continuity between the pre- and post-saccadic images, so that the brief period corresponding to the intra-saccadic stimulation does not entail any conscious percept at all. I will call this conceptual requirement the saccadic "**temporal filling-in**" issue.

A few years later, the seminal study of Burr and Ross (1982) was published and turned out to have far-reaching and lasting consequences. This paper first started by noting that previous work had always assumed that the human visual system cannot resolve objects moving at high speeds. The authors decided to test this commonly held assumption by measuring the contrast threshold at which direction discrimination of very fast movements was possible – observers' eyes were static. Their striking result was that the use of low spatial frequency gratings (or wide bars) as stimuli allowed observers to perceive motion at incredibly high speeds (even higher than usual saccadic speeds). Moreover, peak contrast sensitivity was identical at all speeds up to 800°/s and corresponded to a temporal frequency of about 10 Hz.

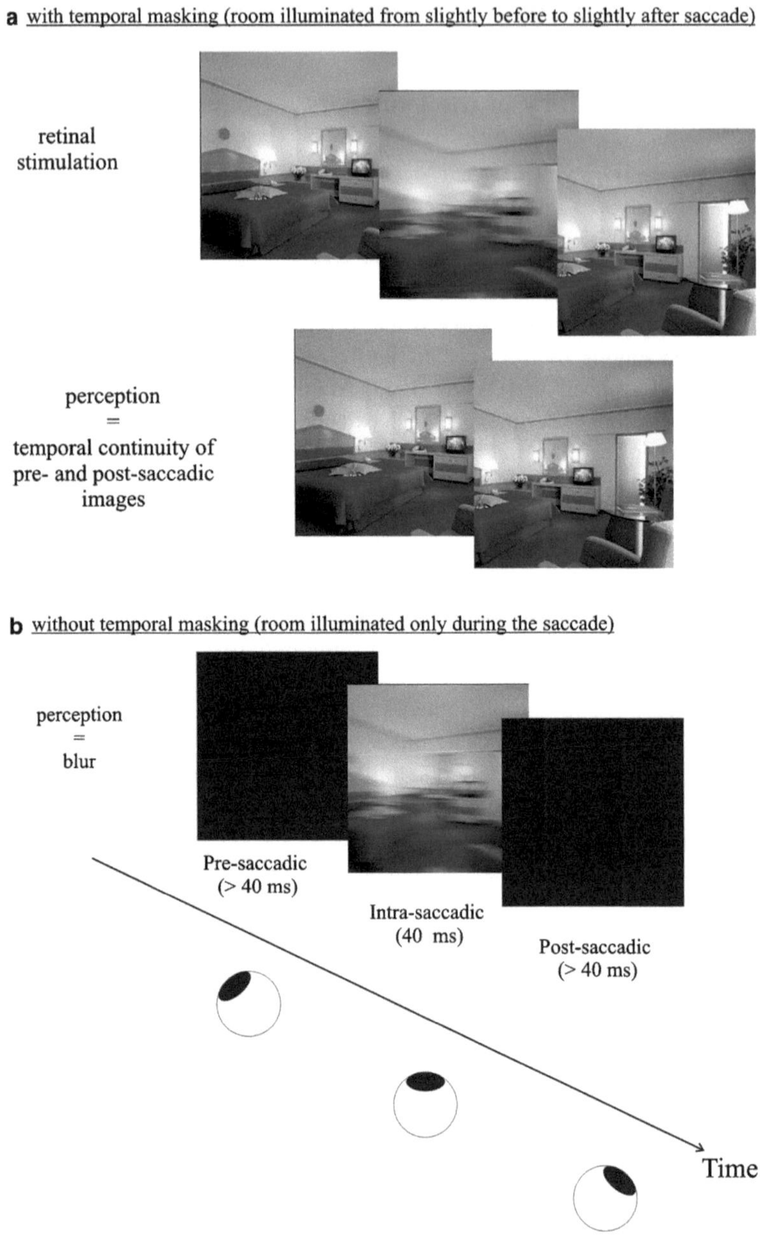

Fig. 10.1 Schematic representation of Campbell and Wurtz's (1978) results. (**a**) Intra-saccadic blur perception (or *grey-out*) is temporally masked by the pre- and post-saccadic images, thus preserving temporal continuity. (**b**) In the absence of temporal masking, intra-saccadic blur is clearly perceived

These amazing results led the authors to wonder why observers were not "startled during a saccade by the intrusion of low frequency components onto the scene?" To answer this question, they proposed for the first time "that during saccades motion sensitivity is dampened, precisely to avoid the disturbing consequences of saccadic image motion which would follow if it were left intact".

This motion sensitivity damping hypothesis was made more explicit in a paper which was published the same year (Burr et al. 1982). The paradigm and results of this study will be discussed later in order to focus on concepts in the present section. The authors' key idea was that the **contrast sensitivity** of motion mechanisms was selectively depressed during a saccade so that the visual system registered no motion despite the rapid movement of the image across the retina. Therefore, this theory will be referred to hereafter as a "**motion contrast-sensitivity reduction**" hypothesis. The essential point to be made here is that the introduction of a sensitivity reduction hypothesis by Burr et al. (1982) ignored the saccadic temporal filling-in problem induced by any such theory as pointed out by Campbell and Wurtz (1978): reducing contrast sensitivity of intra-saccadic motion signals cannot prevent us from perceiving them. Indeed, a 40 ms period of reduced contrast would not go unnoticed and we should perceive a dim motion of the world every time we make a saccade.

10.2.2 From 1982 to 1999

We can try to imagine what could have been, after 1982, a logical scientific agenda aimed at accommodating the results presented in the preceding section. The work of Campbell and Wurtz (1978) could have been fruitfully pursued in the following way. While these authors showed that temporal masking was able to induce an omission of intra-saccadic smearing, the question as to whether temporal masking was also able to avoid the perception of intra-saccadic motion signals was left open. This question was actually not even mentioned as it was believed at that time that high-speed motion signals could not be resolved by the visual system. Another reason to ignore this question was that observers never reported motion during intra-saccadic illumination of the room but only a grey-out (this was probably because of the absence of low spatial frequency components in the room). However, this issue should have been investigated after 1982 as soon as Burr and Ross (1982) had shown the possibility for the stationary eye to perceive the motion of gratings moving at saccadic speeds. In practical terms, Campbell and Wurtz's (1978) study should have been performed again by displaying low spatial frequency gratings in the room. As will be shown in the next section, it was not until 2002 that this issue was studied.

In addition, the two studies by Burr and colleagues left the following questions unanswered. In the experiments with a stationary eye (Burr and Ross 1982), the stimuli were moving at a constant speed and for an unlimited duration (terminated when the contrast adjustment was performed). This retinal stimulation is however very different from that induced by a saccade made over a stationary grating: in the latter case, for a typical saccade, speed increases from 0°/s to 300°/s within 20 ms and returns to 0°/s in another 20 ms. It could thus be argued that low level motion detectors cannot be activated by this intra-saccadic retinal flow either because of the too short duration or because of the acceleration/deceleration profile. Moreover, provided that such a simulated intra-saccadic velocity profile is susceptible to induce motion perception, the effect of temporal masking on this velocity profile could then be investigated. To my knowledge, these experiments have never been performed. Finally, the postulated "motion contrast-sensitivity reduction" process should have been explicitly considered as a backup mechanism whose probable role was to supplement the main temporal masking process. If the issue had been stated in these terms, experiments would have tried to quantitatively disentangle the respective roles of the two processes.

Surprisingly, none of the lines of research suggested above was ever followed. Instead, it seems that the debate on intra-saccadic perception evolved in a biased way. The "motion contrast-sensitivity reduction" hypothesis became so popular that Campbell and Wurtz's results and ideas were totally overlooked. This resulted in the common belief that we do not perceive intra-saccadic motion because we are motion-blind during saccades. The clear emergence of such an extreme claim probably arose after the publication of a very influential paper (Burr et al. 1994). The novel suggestion introduced in this paper was that the suppression of the motion system was actually a depression of the whole magno-cellular system. It was also proposed that this depression took place in the LGN and was triggered by a central signal associated with the oculo-motor command.

10.2.3 From 2000

In a context where most publications took for granted that we cannot perceive motion during saccades, psychophysical studies showed that it is actually quite easy to perceive intra-saccadic motion as long as the retinal stimulation is optimized for the motion-sensitive system (Castet and Masson 2000; Garcia-Perez and Peli 2001). Moreover, with principles similar to those used by Campbell and Wurtz (1978), it was suggested that temporal masking was a powerful factor allowing the visual system to omit intra-saccadic motion signals, or more precisely to temporally fill-in the period of intra-saccadic stimulation (Castet et al. 2002).

Altogether, these studies helped provide a more global picture of the intra-saccadic perception issue. A brief image displayed during a saccade elicits a percept that depends on its spatial frequency content, namely smearing with high spatial frequencies and motion (against the saccade) with low spatial frequencies. If the image extends by a few dozens of ms either before or after the saccade, thus eliciting a form of temporal masking, the image is perceived as static. Therefore, temporal masking seems to be a homogeneous and parsimonious process used by the visual system to prevent us from being startled either by smearing or by motion of the scene whenever we make saccades. There is thus no a priori reason to postulate an additional damping process, except if the latter is considered as a backup mechanism whose function is to facilitate temporal masking.

The start of the new millennium also triggered a promising line of research in the form of several exciting physiological studies that focused on intra-saccadic MT activity. These studies are described in Chap. 10 by Ibbotson.

10.3 Some Key Effects and Their Interpretation

The goal of the preceding section was to clarify the evolution of the ideas bearing on intra-saccadic motion perception. I tried to show that the main disagreements among vision scientists mainly rely on whether and how they synthesize the key experimental effects that are described in detail in the current section. I'll start with evidence that low spatial frequency gratings can elicit motion perception during saccades. This finding, although discovered recently, sounds as a good starting point in order to understand the general issue of intra-saccadic motion perception.

10.3.1 The Trailing Effect: Intra-Saccadic Motion Perception in the Direction of the Saccade

There are many examples in the history of science showing how common sense and experience are sometimes very influential in the development of theories. One famous example is the theory of spontaneous generation, which held for centuries that some living organisms are generated by decaying organic substances, like maggots spontaneously appearing in meat. While Sir Thomas Browne started to question this theory in the seventeenth century, his contemporary, Alexander Ross, wrote: "*To question this (i.e., spontaneous generation) is to question reason, sense and experience*". Common experience might also bias the theories bearing on the issue of intra-saccadic motion perception. The idea that intra-saccadic motion signals have to be actively suppressed is probably attractive because we indeed never perceive motion during saccades.

It is therefore important to consider the following finding, which contradicts our common experience (Castet and Masson 2000). Two important results were reported in this study: a) it can be quite easy to perceive motion during a saccade and b) this motion percept seems to rely on low-level motion detectors. We designed the following experimental paradigm. A low spatial frequency grating (say about 0.17 cpd) is continuously moving at very high speed (360°/s) on a CRT monitor (Fig. 10.2a). This grating is invisible when viewed with static eyes as its temporal frequency (60 Hz) is above critical fusion frequency (Fig. 10.2b). A crucial requirement in this paradigm is to have a high refresh rate monitor (160 Hz here) so that the grating temporal frequency can be set below the monitor's Nyquist

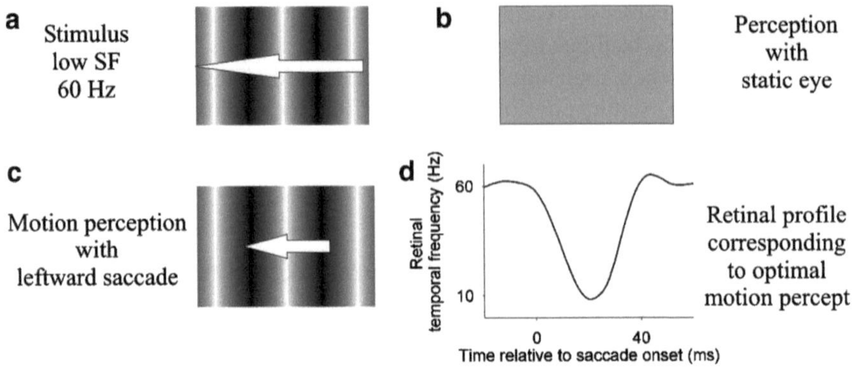

Fig. 10.2 Intra-saccadic motion perception: basic principle underlying the 'trailing eye effect' (Castet and Masson 2000). A grating moving at high speed is "invisible" with static eyes. A clear motion percept is induced when a saccade is made in the grating's direction if peak saccadic velocity is sligthly below grating's speed

frequency. If a saccade is made in the direction of the moving grating – for instance, horizontal saccade with a vertical grating – observers report three types of percepts, which depend on saccade amplitude. With small amplitudes (2°), observers still perceive a gray screen. With large amplitudes of about 12°, the grating appears as if it had been statically flashed on the screen. With medium amplitudes (about 6°), conspicuous motion perception of the vertical grating bars is reported in the saccade direction (Fig. 10.2c). For all amplitudes tested, the direction of the grating on the retina is always in the direction of the saccade (i.e. the speed of the eye is always smaller than (or equal to) the grating's speed). It must be noted that one advantage of this paradigm is the absence of any visible luminance contrast in the image before and after the saccade, thus avoiding any potential influence of temporal masking.

We interpreted the effect of saccade amplitude in terms of retinal temporal frequency elicited around the peak velocity time. With small amplitudes, the retinal temporal frequency at the peak is still very high so that the grating is still above fusion frequency, and hence invisible. With the large amplitudes tested, the saccadic peak velocity reaches the grating's speed so that the grating is momentarily stabilized on the retina, and hence perceived as a static flash. With medium amplitudes, the retinal frequency of the grating is around 10–25 Hz, i.e. within an optimal range for the motion sensitivity system, thus explaining the compelling motion percept (Fig. 10.2d). For about 20 ms around the peak velocity time, the grating is thus moving on the retina with an average speed that is slightly higher than that of the eye. We have therefore dubbed this phenomenon the "trailing eye" effect.

Thus, motion perception occurs when the grating moves in the saccade direction with a retinal average speed which is likely to stimulate motion selective cells in area MT (Movshon and Newsome 1996). To confirm the involvement of low level motion detectors in this percept, another experiment was performed on the basis of the classical direction-specific adaptation paradigm (Levinson and Sekuler 1975). Paradigm was the same except for the main following points. Each trial was preceded by an adaptation phase (eyes static with a 12 Hz grating). After adaptation, the high-speed grating (test) was presented either in the same or opposite direction with respect to the adaptation grating. Contrast sensitivity for the test grating was assessed with adaptive staircase procedures. Results showed that sensitivity was higher when adaptation direction did not coincide with test direction. This direction-specific adaptation is therefore evidence that direction-selective detectors underlie the intra-saccadic motion percept reported in our paradigm (Castet and Masson 2000).

The "trailing eye" paradigm is very convenient as a demonstration tool. Anyone, having access to a high refresh rate monitor (optimally 160–200 Hz), can use it in order to experience intra-saccadic motion perception without the necessity to measure eye movements. It usually takes only one or a few saccades before any observer spontaneously reports compelling intra-saccadic motion provided that saccades are of the appropriate amplitude. The apparent contrast of the grating is high so that the percept is really conspicuous. Hundreds of naïve observers have systematically perceived the "trailing eye effect" in our laboratory over the last years. They were unambiguously startled by the sudden appearance of the moving bars every time they made a saccade.

To sum up, the trailing eye effect unambiguously shows that we can easily perceive the motion of low spatial frequency gratings presented during a saccade. One crucial aspect of the paradigm is the absence of any potential temporal masking due to the pre- and post-saccadic images. The effect occurs when the grating's retinal temporal frequency is around 15–25 Hz around the time of the eye peak velocity. It is therefore clear that motion processing, and thus the magno-cellular system, is functional during saccades. Incidentally, one can wonder whether the theory of intra-saccadic magno-cellular suppression would have gone so far, had the proponents of this theory had the possibility to experience the trailing eye phenomenon.

10.3.2 Temporal Masking

10.3.2.1 Temporal Masking with Static Eyes

First of all, it should be noted that vision scientists have been investigating temporal masking – when observers have their eyes motionless – in literally thousands of studies over the last decades (Bachmann 1994, p. 11). The basic phenomenon is that a brief target of about 40 ms, which is visible when presented alone, becomes less visible or even invisible when it is preceded or followed by spatially overlapping visual masks inducing respectively forward and backward masking (Bachmann 1994; Breitmeyer 1984; Breitmeyer and Ogmen 2000, 2006). For instance, temporal masking is classically used as an experimental tool in the famous masked priming paradigm: masked targets are shown to have measurable behavioral effects although they are not consciously perceived (Dehaene et al. 2001; Kinoshita and Lupker 2003).

It is crucial to emphasize the phenomenology of temporal masking. In conditions of slight masking, the target becomes less visible. This means that observers perceive a temporal succession of three entities: the forward mask, the target (although degraded and/or dimmed), and the backward mask. However, in conditions of stronger masking, the target is not consciously perceived at all: the temporal flow of conscious perception only contains the forward mask and the backward mask. I believe therefore that the latter condition of "absolute temporal masking" (i.e. when the target is invisible) should be considered and referred to as a phenomenon of "temporal filling-in".

The interesting aspect of this "temporal filling-in" process is that vision scientists in the field of static vision have no doubt about its huge efficiency. Nobody seems astonished to observe that a 40 ms duration target can be made invisible by temporal masks of brief durations. However, within the framework of intra-saccadic motion perception, many vision scientists seem to be reluctant to recognize the functional role of temporal masking. This reluctance is all the more puzzling that several authors have proposed that temporal masking might have evolved to solve the perceptual problems associated with saccadic eye movements (Bachmann 1994; Breitmeyer 1984; Breitmeyer and Ganz 1976). In this respect, it has often been noted that temporal masking – with static eyes – is optimal when the target duration is around 40–50 ms, a duration which is strikingly similar to the typical saccade duration.

10.3.2.2 Temporal Masking and Saccades

There is actually only one study suggesting that temporal masking renders intra-saccadic motion signals invisible (Castet et al. 2002). In this study, vertical gratings – static on the screen – of different durations (from 18 to 50 ms) were briefly displayed while observers made horizontal saccades of about 40 ms (Fig. 10.3) Across trials, only two percepts were reported: either motion opposite the saccade or stationarity of the grating. The first clear-cut result was that motion perception was systematically reported when the grating was displayed only during the saccade, i.e. when the pre- and post-saccadic images were gray. The second result was that the probability of reporting motion, for a constant duration of intra-saccadic stimulation, decreased when the duration of pre- or post-saccadic duration increased. As an illustration, a 40 ms grating appearing at the onset of a 40 ms saccade elicits motion perception. However, motion perception is dramatically reduced if the grating is longer (50 ms) and thus induces a 10 ms post-saccadic stimulation.

While this study clearly suggests the involvement of temporal masking, it cannot rule out that a backup process, in the form of a motion contrast-sensitivity reduction, is acting in parallel. Future work should carry out the same kind of experiments for different contrasts of the grating in order to assess the respective quantitative weights of temporal masking and contrast-sensitivity reduction.

Temporal masking is a convenient way of describing the influence of forward and backward masks on a target. However, studies with static eyes have shown that temporal masking is not a homogenous process. There are for instance notable differences between masking by light and masking by pattern, thus suggesting the involvement of

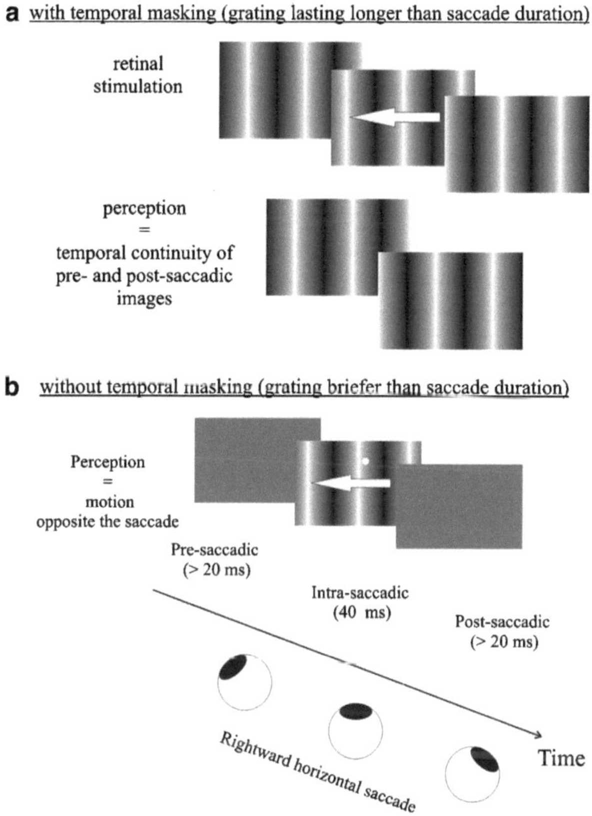

Fig. 10.3 Schematic representation of Castet et al.'s (2002) results. A grating, which is static on the screen, is displayed with different durations at different moments relative to saccade onset. (**a**) Intra-saccadic motion is temporally masked by the pre- and post-saccadic images. (**b**) Motion against saccade direction is perceived when a brief grating (i.e. shorter than saccade duration) is displayed during the saccade

different physiological mechanisms such as integration and/or interruption processes. Given the scarceness of data on the role of temporal masking on intra-saccadic processing, it seems currently premature to offer suggestions as to the exact nature of the processes involved in the temporal filling-in of the intra-saccadic period.

10.3.3 Contrast Sensitivity Reduction: Usually Referred to as "Saccadic Suppression"

10.3.3.1 The H-H Paradigm

In the seventies, the basic phenomenon of "saccadic suppression" had already been discovered and investigated in numerous studies (Matin 1974). It was reported that visual thresholds are elevated for stimuli such as flashes presented during or in close temporal proximity to saccades (from about 50 ms before to 50 ms after the saccade). This threshold elevation typically reached a maximum of 0.5 log unit of relative luminance to stimuli delivered in mid-saccade.

The main theoretical issue at that time was to explain the source of this threshold elevation. Four main possibilities were considered: a) central inhibition possibly associated with a corollary discharge, b) retinal smear, c) shear between the vitreous body and the retina, and d) visual masking. The potential influence of visual masking was often mentioned because of possible lateral masking interaction occurring while the eyes are moving. For instance, a given receptive field could be stimulated by the border of the screen at the beginning of a saccade: this initial stimulation would then be temporally integrated with the subsequent stimulation due to the target stimulus as the eye moves over the screen, thus constituting contour masking.

In this context, a very important study was performed (Volkmann et al. 1978). The experimental paradigm of this study is described here in detail as it has been used in many publications in the following decades. The novel and clever principle is the following: a horizontal grating is displayed while observers make horizontal saccades (Fig. 10.4) – this will therefore be referred to as the "**H-H paradigm**." The advantage of this experiment over previous ones is the minimization of retinal smear and contour masking. It is clear that retinal smear is absent when receptive fields are moving horizontally over a horizontal grating. Using gratings as stimuli has also the advantage of maintaining a constant mean luminance level for receptive fields moving over the stimulus, thus further minimizing contrast masking effects over time.

With this H-H paradigm, Volkmann et al. (1978) measured contrast sensitivity for gratings of different spatial frequencies lasting 10 ms. Measurements were made either at different moments relative to saccades or for a steadily fixating eye. Sensitivity measured with the stationary eye was compared to sensitivity measured during saccades. The clear-cut effect was that contrast sensitivity reduction was maximal at low spatial frequencies and was absent with higher spatial frequencies.

10.3.3.2 The "Central Origin" Interpretation of the "Saccadic Suppression" Effect

Volkmann et al. (1978) interpreted their results as evidence that the contrast-sensitivity reduction could not be due solely to contour masking or smearing factors and had therefore a significant central origin. This hypothesis has been pursued, along with an extensive use of the H-H paradigm, in the following decades. The "central origin" hypothesis usually relies on the three following well-established signatures of the "saccadic suppression" effect.

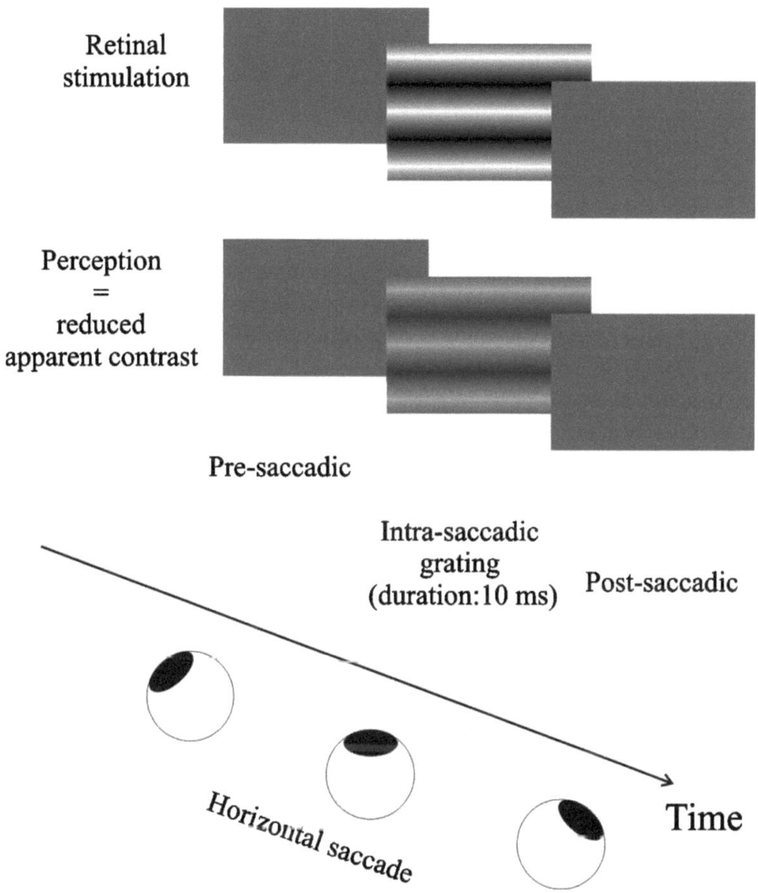

Fig. 10.4 Schematic principle of the paradigm designed by Volkman et al. (1978) and later used in many studies (ex. Burr et al. (1994)). The basic principle is that a brief horizontal grating is displayed during a horizontal saccade: we therefore call this paradigm the "H-H paradigm". Note that no retinal motion is induced in this kind of experiments. Contrast sensitivity measured in this paradigm is reduced when compared to sensitivity measured for a grating of same duration observed with static eyes

Time Course

In the seventies, the time course of the effect had already been investigated in several studies (Matin 1974). It was known that the reduction effect occurred not only for stimuli presented during the saccade but also for stimuli presented slightly before or after the saccade. The reduction observed for stimuli displayed before the saccade was mostly interpreted as evidence that a suppression signal of central origin, similar to an efference copy, reached the visual areas before the actual execution of the saccade. This time course was replicated by

Volkmann et al. (1978), who then emphasized that such result supports the "central origin" hypothesis.

Magno-Cellular Specificity

Initial evidence by Volkmann et al. (1978) that contrast sensitivity reduction observed during saccades is specific to low spatial frequencies has been replicated several times (Burr et al. 1982, 1994). In addition to this specificity, it was found, still using the H-H paradigm, that contrast sensitivity reduction observed during saccades at low spatial frequencies is only observed for achromatic gratings (Burr et al. 1982, 1994). When chromatic equiluminant gratings are used, intra-saccadic sensitivity is not reduced. This suggests that the effect is specific to the color-blind magno-cellular system. As this system is associated with motion processing, it was proposed that contrast-sensitivity of the magno-cellular system is actively decreased during saccades to avoid intra-saccadic motion perception.

This magno-specificity was also reported in a series of experiments which measured spectral-sensitivity functions during saccadic eye movement by the increment-threshold method (Sato and Uchikawa 1999; Uchikawa and Sato 1995). Increment thresholds for a brief monochromatic light of various wavelengths were measured either with static eyes or during saccades. The curve measured during saccades showed a dip in sensitivity for lights around 580 nm, a classic signature of color opponent mechanisms, thus suggesting that only the magno-cellular pathway had been affected by the saccade.

Early Site

There is converging evidence that contrast-sensitivity reduction occurs at a very early site within the visual hierarchy. This was first proposed on the basis of contrast masking experiments which suggested that reduction preceded the site of contrast masking, usually assumed to start in V1, and could thus occur as early as the lateral geniculate nucleus (Burr et al. 1994). The idea that the neural site of the contrast sensitivity reduction is very early, i.e. before V1, has been suggested in a few other studies (Burr et al. 1999; Thilo et al. 2004).

10.3.3.3 The "Retinal Origin" Interpretation of the "Saccadic Suppression" Effect

It seems currently accepted by many authors that saccadic contrast-sensitivity reduction has a central origin. As stated, a very convincing reason supporting this claim is that smearing and masking factors are indeed minimized in the commonly used H-H paradigm. Contour masking, for instance, only occurs at the borders of the stimulus (which is large) and should have a very minor influence.

More generally, it is quite clear that temporal masking cannot have any major role in this effect as the test stimulus is displayed only during the saccade and is thus unable to elicit any pre- or post-saccadic stimulation.

However, it has never been proved that an active process causes this contrast-sensitivity reduction or that it has a central origin. This effect could actually be related to activity occurring within the retina as has already been briefly outlined (Castet et al. 2001). **There is only one key assumption in our proposal: we assume that a brief decrease in the light adaptation level occurs within the retina during the saccade.**

For observers with static eyes, there is clear evidence that brief decrements of a homogenous background reduce the sensitivity to brief stimuli presented in close temporal vicinity (Poot et al. 1997; Schwartz and Godwin 1996). Figure 10.5 presents the results of Poot et al. (1997) in a schematic way: a 10 ms decrement of background luminance is preceded (or followed) by a 10 ms test pulse. The threshold values measured for this test are shown in the bottom part of the Fig. 10.5: these values are clearly elevated when compared to thresholds measured without any decrement (horizontal dotted line). Moreover, the time scale of the effect is very similar to that found for the temporal evolution of the "saccadic suppression" phenomenon. Note that the sensitivity is reduced even when the test is displayed before the background decrement. In the light adaptation literature, it has been known since Crawford's work that thresholds can start to increase before the physical

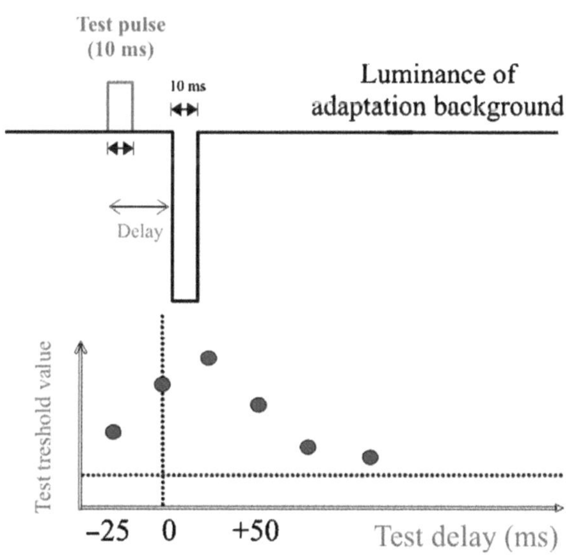

Fig. 10.5 Illustration of Poot et al.'s (1997) results. A test pulse is displayed at different times relative to a brief decrement in the luminance of the adaptation background. Sensitivity to the test pulse is decreased even when the test is displayed before the background luminance. The dynamics of this effect is very similar to that obtained in the H-H paradigm for a grating displayed at different moments relative to saccade

onset of a conditioning field (Crawford 1947; Pokorny et al. 2003). These effects, and many others, have recently been modeled within a retinal model of temporal processing of light input – one important characteristic of this model of early visual processing is the presence of a contrast gain control process (Snippe et al. 2000).

Thus, if an event equivalent to a brief decrement of background luminance occurred during saccades, the basic "saccadic suppression" effect would be expected, i.e. gratings briefly displayed (as in the H-H paradigm) around the time of saccades would show reduced luminance contrast sensitivity. Most importantly, this alternative interpretation is able to account for the three key signatures of the "saccadic suppression" effect as described below.

Time Course

As already stated, the time course of the effect of luminance decrements on test pulse sensitivity with respect to the decrement's onset (static eye experiment) is very similar to the time course of "saccadic suppression" with respect to saccade onset. Interestingly, sensitivity can be reduced even for a test pulse displayed before the luminance decrement of the background. Reduced contrast sensitivity occurring before saccade onset is usually taken as evidence that an extra-retinal suppression signal is sent to visual structures before the saccade occurs. However, this claim is not necessary any longer if we assume that the reduction in sensitivity results from the backward influence in time of an intra-saccadic retinal decrement. More generally, visual processing of an event occurring at time t must take into account events occurring later on because temporal integration within a relatively large window (up to 100 ms) is a common feature of low level processing.

Magnocellular Pathway Specificity

How does our interpretation account for the well-established magno-specific loss of sensitivity? The rapid influence of background luminance decrements, which induce a temporal luminance contrast, can be accounted for in terms of contrast gain control (Hood 1998; Snippe et al. 2000). The latter process is a clear feature of ganglion cells, which is present in the magnocellular pathway of the monkey but absent in the parvocellular stream (Benardete and Kaplan 1997, 1999; Lee et al. 1994). It is thus likely that the temporal contrast created by rapid changes in the adaptation level activates a process of contrast gain control, which exclusively affects the magnocellular neurons. This would therefore explain why intra-saccadic contrast sensitivity reduction is specific to the magnocellular system.

Early Site

As we propose that contrast sensitivity reduction occurs within the retina, it is obvious that our interpretation is consistent with results suggesting that the phenomenon takes place before the primary visual cortex (see “early site” section above). This idea is consistent with an old and overlooked work suggesting that “saccadic suppression” is of retinal origin. It was indeed shown that contrast sensitivity is reduced for flashes presented before and during passive saccades elicited by tapping the eyeball near the outer canthus (Richards 1968). In this case, any extra-retinal influence can clearly be excluded. To my knowledge, this crucial study has never been considered wrong in its design or in its methodology, but it has often been ignored.

Possible Causes of an Intra-Saccadic Decrease in the Retinal Light Adaptation Level?

It seems that several factors could account for a brief decrease in the retinal light adaptation level at the time of saccades. Actually, any factor, whether neural, optical, biophysical, or mechanical, which would eventually reduce the response of ganglion cells during saccades, could explain the intra-saccadic contrast sensitivity reduction.

Richards was the first to explicitly test this hypothesis when he measured the Stiles-Crawford effect using intra-saccadic visual stimuli (Richards 1969). The peak of the Stiles-Crawford effect he found suggested that the shearing forces occurring between the vitreous body and the retina as a result of intrasaccadic acceleration induced a bending of the photoreceptors. These intra-saccadic shearing forces are so strong that they are thought to be one major cause of retinal detachment (David et al. 1997, 1998).

We have previously suggested that the intra-saccadic tilt of the photoreceptors might induce an intra-saccadic decrease in the retinal light adaptation level (Castet et al. 2001). However, I emphasize now that this decrease in retinal light adaptation level might be induced in an additive way by other factors taking place within the retina. It must be reminded that less than 10% of incoming light is absorbed by photoreceptors thus showing the crucial importance of physiological optics to explain how photoreceptors respond to light (Baylor 1987). It is very likely that all cellular layers within the retina and especially those close to the vitreous gel (i.e. not only the photoreceptors) are differentially displaced by the intra-saccadic accelerations. It seems plausible that these mechanical displacements occurring between the inner and outer retinal layers are significantly blocking the trajectories of photons before the latter activate the photoreceptors (in primates, light must travel through the whole retina before reaching the photoreceptors). This would in itself be sufficient to entail an intra-saccadic reduction in the light adaptation level. In addition, it is possible that intra-saccadic shearing forces elicit mechanical pressures on the retinal cells, thus inducing detrimental biophysical effects. For instance, there is clear evidence that membranes of mammalian retinal cells contain mechano-gated K+ channels (Maingret et al. 1999). The opening of these channels induced by mechanical stretch during saccades would hyperpolarize the cells and thus entail reduced global activity. This hypothesis could also explain the old observation that sensitivity reduction also occurs with electrically induced visual phosphenes in darkness (Riggs et al. 1974).

10.3.3.4 Summary

The contrast sensitivity reduction – usually called "saccadic suppression" – observed for brief stimuli displayed during a saccade is a well-established result. However, the cause of this intra-saccadic threshold elevation still remains an open question.

10.3.4 Saccadic Suppression of Image Displacement

The issue of intra-saccadic perception should not be confused with another issue usually referred to as the trans-saccadic integration problem (Deubel et al. 2002). As any saccade changes the line of sight, there is always a mismatch between the pre-saccadic and the post-saccadic retinal images of the world (each one lasting at least 150–200 ms), i.e. the two images do not spatially coincide. Within this framework, the brief retinal motion flow induced by the saccade (for about 40 ms) is conceptually ignored and only the succession of the two fixation images is considered. When the eyes are static, such a two-frame displacement elicits a phenomenon known as "apparent motion" (Anstis 1970). However, despite the jump (shift) occurring on the retina between two successive fixations, this phenomenon is not reported and perceptual stability of the world is maintained. Many studies have been performed on the trans-saccadic fusion issue and many controversies subsist (Bridgeman et al. 1994).

Conceptually, it is clear that trans-saccadic integration (which could be called between-fixation integration) and intra-saccadic perception are independent issues. The first one is a problem of correspondence between two spatially-shifted static images of the world (before and after the saccade). The second one concerns the brief motion flow that is present only during the saccade (i.e. the world continuously moves on the retina during the saccade). Unfortunately, many confusing links between the two issues are often found in the literature. For instance, references concerning the trans-saccadic integration issue are commonly cited in the context of the intra-saccadic motion perception issue. It is also often asserted that results pertaining to the two different issues are actually providing evidence for a homogeneous process known as "saccadic suppression". Acknowledging this common confusion is not new. For instance, Bridgeman et al. (1994, p. 255) already noted: "*Many investigators have emphasized the need to distinguish between the problem of the stable position of the visual world despite eye movements and the problem of why no movement is seen with saccadic eye movements. Nevertheless, there is an irresistible tendency in handbooks and textbooks to combine the two …*".

One origin of the confusion is clearly semantic and related to a well-established effect originally quantified in 1975 and dubbed "Saccadic Suppression of Image

Displacement – SSID" in 1976 (Bridgeman et al. 1975; Stark et al. 1976). The basic experimental paradigm (which has led to many variants) to measure SSID is the following. A target (which may be either a dot or an extended image) is displayed at a certain location well before a saccade occurs. During the saccade, this target is shifted to another location in the smallest amount of time possible (e.g. within one frame when using a CRT monitor). The task of the observer is to report whether the change in location (i.e. displacement) has been perceived. The principal finding is that the displacement threshold is much higher in the saccadic condition than in a fixation condition. Several studies have reported that this rise in displacement threshold has a magnitude of up to one-third the size of the saccade (Bridgeman et al. 1975, Ilg and Hoffmann 1993; Stark et al. 1976).

Since 1975, Bridgeman and collaborators have always emphasized that the SSID effect has to be interpreted within the theoretical framework of the trans-saccadic issue. Their main point is that SSID unambiguously shows that trans-saccadic perceptual stability cannot be accounted for by a vectorial cancellation theory as initially proposed by Helmholtz (1866/1924). In such a theory, the expected saccadic displacement vector (an efference copy) would be subtracted from the displacement vector measured by the retinal displacement between the pre- and post-saccadic images. Perceptual stability would result from the zero sum of the efference copy signal and the trans-saccadic retinal displacement. A cancellation process would thus imply that changing target's location, as in Bridgeman's experiments, even by a small amount, should be detected. However, the SSID effect shows that this is not the case. It is beyond the scope of the present chapter to comment on the different interpretations and controversies concerning this effect (Bridgeman et al. 1994; Currie et al. 2000, Deubel et al. 1998, 2002; Deubel and Schneider 1996). The only point to remember here is that the SSID paradigm has been elaborated to tackle the trans-saccadic issue and has therefore not much to say about intra-saccadic perception.

Why then is there still a confusing tendency to combine the two issues? It seems that the confusion arises also because of a methodological constraint imposed on the SSID paradigm: it is usually convenient to detect saccades online so that the experimenter can trigger the change in position during the saccade[1] (with a 100 Hz CRT monitor, this change occurs within 10 ms). The fact that we don't perceive this transient event leads some authors to think that the SSID effect provides evidence for central intra-saccadic suppression. However, this conclusion again ignores the well-established role of temporal masking, or more generally of temporal integration processes, as described in previous sections. The SSID paradigm indeed involves forward- and backward masking as the target is present long before

[1]It should noted that this intra-saccadic manipulation is not a necessity. Ideally, these experiments should be carried out in the following way : the pre-saccadic target should be extinguished just before saccade onset and then displayed again right after saccade offset with a spatial shift. This would actually be the cleanest way of investigating the trans-saccadic integration issue, but it is currently impossible to perform this manipulation online because of technical limitations.

saccade onset and long after saccade offset. Considering temporal masking – by the pre- and post-saccadic images – is therefore sufficient to understand why the transient event induced by the small jump is not perceived (Campbell and Wurtz 1978; Matin 1974).

In summary, the SSID effect and its variants offer a powerful tool to investigate the trans-saccadic integration issue. This issue concerns the ability of the visual system to encode spatial location from one fixation to the next. This is for instance clearly illustrated by the following citation: "*Saccadic suppression of displacement is of interest because of its implications for the processing of information about egocentric spatial location*" across saccades (Deubel et al. 1996, p. 992). The semantic confusion between "Saccadic suppression of displacement – SSID" and the issue of intra-saccadic perception should therefore be avoided.

10.3.5 Physiological Studies

In the last decade, a few physiological studies have been performed in relation with the intra-saccadic motion perception issue. As this literature is reviewed in Chap. 10, the goal of the present section is to offer a few comments on the principles and rationales underlying some inescapable studies. Hopefully, these comments will help provide a few guidelines for the design of future physiological work.

There is now clear electrophysiological evidence that neurons in the Middle Temporal cortex (MT), an area devoted to motion processing, can respond to the intra-saccadic motion flow. It was first shown that MT neurons are transiently activated by the visual flow induced by fixational saccades in a directionally selective way (Bair and O'Keefe 1998). A more recent study showed that many directionally selective cells in MT and MST (Medial superior Temporal – another area linked to motion processing) are stimulated by the intra-saccadic motion flow induced by voluntary saccades (Thiele et al. 2002). This was confirmed in very recent studies (Ibbotson et al. 2007; Price et al. 2005). As psychophysical studies have shown that intra-saccadic motion perception is possible and thus suggested the functional integrity of the magno-cellular system (Castet et al. 2002, Castet and Masson 2000), it is reassuring to find that many MT direction-selective cells are responsive during saccades.

An important suggestion of recent work is that the process preventing us from perceiving intra-saccadic motion would take place in MT thanks to intra-saccadic modulations of direction selectivity (Ibbotson et al. 2007; Thiele et al. 2002). The key paradigm used in these studies relies on a comparison between active conditions (retinal stimulation is induced by a saccade) and passive conditions (retinal stimulation is induced on a static eye by a simulated saccadic visual flow). The results show complex differences between the two conditions and notably an attenuation of spiking activity in the active condition, thus suggesting an extra-retinal inhibiting influence. While this active/

passive paradigm opens the way to a promising line of research, the interpretation of the results in terms of extra-retinal suppression of motion perception is still unsatisfactory because it does not account for the clear intra-saccadic motion percepts reported in humans (Castet et al. 2002; Castet and Masson 2000). In other words, if there are extra-retinal influences operating in MT around the time of saccades, they might be related to visuo-motor processes that are not related to conscious motion perception.

More generally, it seems that there is still a long way to go before a clear link between intra-saccadic motion perception and physiological studies can be established. Here are a few of the points that should be taken into account in future work. First of all and most importantly, physiological studies have not yet provided evidence for a correlation between intra-saccadic motion perception and MT activity. This could be achieved in the future by having monkeys perform a perceptual judgment on motion direction while recording MT activity. This comment leads to a second limitation of extant studies: the visual stimulation, which is used in these studies, is not optimized to elicit strong high-speed motion signals. Intra-saccadic motion perception requires low spatial frequency gratings: this is indeed necessary in order to get sufficiently low intra-saccadic temporal frequencies within the range of the magno-cellular pathway. Finally, intra-saccadic motion perception only occurs in the absence of temporal masking (Castet et al. 2002; Castet and Masson 2000). In other words, the stimulus should be displayed only during the saccade, thus leaving a grey background both before and after the saccades.[2] Thanks to this careful control of the visual stimulation, optimal motion responses could be elicited. In a second step, a direct test of the effect of temporal masking on this motion response could be investigated by presenting gratings lasting slightly longer than saccade duration as in Castet et al. (2002). The latter experiment could assess whether the masking effect of pre- and post-saccadic images is operating before or after the MT level.

10.4 Suggestions for Future Studies

It is first suggested to use the expression "saccadic suppression" more advisedly in future studies in order to avoid the current confusions permeating the field. Notably, it should be borne in mind that "saccadic suppression" usually refers to a contrast sensitivity reduction observed with a static retinal stimulus as massively studied using

[2]The absence of a control for temporal masking effects is particularly annoying as illustrated by a finding that was emphasized in Thiele et al. (2002). The authors described a small subset of neurons that seemed to reverse their direction-selectivity only in the active condition. However, we already noted that this reversal response was much too late (it peaked 150 ms after saccade onset) to be interpreted as the result of an anticipatory suppressive extra-retinal influence (Castet et al. 2002). Moreover, Price et al. (2005) found no evidence for this reversal in direction tuning when analysing responses within 25–75 ms after saccade onset.

the H-H paradigm (see Sect. 10.3.3.1) In addition, I wonder whether it is productive to look for neural signatures of this effect in extra-striate areas. As it is agreed on by everyone that this effect occurs at an early site within the visual hierarchy (i.e. before V1), it should not be surprising to find correlates of this response in V1 and in other low-level cortical visual areas (as for instance in, Kleiser et al. 2004). As I believe that this effect has a retinal origin (see Sect. 10.3.3.3), I suggest instead to record the activity of retinal ganglion cells in response to the H-H paradigm. Finally, I have presented several arguments, both theoretical and experimental, suggesting that results obtained with the H-H paradigm (or its variants) should not be interpreted any longer as evidence that the main process preventing us from perceiving intra-saccadic motion results from a contrast-sensitivity reduction mechanism.

A more relevant line of research stems from electrophysiological studies investigating how MT direction-selective neurons respond to retinal motion either with passive or active saccades. However, as already suggested, it would be crucial to explicitly address the following questions:

1. *Is the retinal stimulation optimized to activate low-level motion detectors in response to high-speed motion?*
2. *Is there a link between motion perception and the measured response?*
3. *What is the role of temporal masking or of temporal integration processes?*
4. *What is the link between intra-saccadic motion signals induced by "normal" saccades, as discussed in the present chapter, and micro-saccades (Martinez-Conde et al.* 2004*)?*

Finally, it might be fruitful to investigate whether the intra-saccadic motion signals, which are clearly present in area MT, can serve a functional purpose even if they are not perceived consciously out of the laboratory. Indeed, it might be argued that these motion flow signals, rather than being useless, could be used for instance by oculo-motor processes.

10.5 Conclusions

During a typical saccade, lasting about 40 ms, the image of the stationary world moves on our retina against the saccade direction. This 40 ms retinal stimulation does not elicit, in normal viewing, a conscious motion percept. This is puzzling because observers with static eyes can perceive conspicuous motion when low spatial frequency components – which are present in natural images (Field 1987) – are moving at saccadic speeds Burr & Ross (1982). Understanding the reasons of this intra-saccadic motion blindness constitutes the intra-saccadic motion perception issue.

This issue is of considerable interest as it provides a way of studying the interaction between purely visual processes and central processes linked to oculo-motor programming. Many recent textbooks, papers, and reviews give the impression that the main factor involved in this issue is a central "saccadic suppression". This expression is used as if it referred to a well-characterized process, whereas it is only a convenient way of encompassing several disparate and probably not-related perceptual and physiological effects. In this chapter, I've tried to describe some of these key effects in an attempt to offer a clarification of the misleading terminology (given the wealth of studies including "saccadic suppression" as a keyword, it was impossible to be exhaustive).

In addition, I've tried to offer a more balanced view of the intra-saccadic motion perception issue by recalling a few overlooked points. I have emphasized the crucial conceptual point initially made by Campbell and Wurtz (1978): reducing – or suppressing – the contrast of the 40 ms intra-saccadic stimulation cannot be the functional process used to solve the intra-saccadic motion perception issue. Otherwise, the apparent contrast of the world around us would briefly diminish during each of our saccades. In contrast, temporal masking by both pre- and post-saccadic images is a process which is able to perceptually "fill-in" the intra-saccadic period and thus to induce a temporally continuous flow. According to this temporal integration analysis, the intra-saccadic stimulation enters into the visual stream of processing and is temporally masked at some, yet unknown, stage of the visual hierarchy. The validity of this analysis is confirmed by psychophysical studies showing that conspicuous intra-saccadic motion perception occurs when first, the retinal stimulation is optimized for the magno-cellular system and second, pre- and post-saccadic maskings are absent. When temporal masking is present, the percept is determined by the static retinal image provided by the extra-saccadic stimulation (Castet et al. 2002; Castet and Masson 2000). If there were a central process whose function was to prevent saccade-induced motion signals from entering conscious perception, the intra saccadic motion percepts reported in these studies would be impossible.

Acknowledgments I wish to thank Frédéric Chavane for his helpful comments concerning the possible physiological factors influencing retinal activity during saccades.

References

Anstis SM (1970) Phi movement as a subtraction process. Vision Res 10:1411–1430
Bachmann T (1994) Psychophysiology of visual masking. Nova Science Publishers, New York
Bair W, O'Keefe LP (1998) The influence of fixational eye movements on the response of neurons in area MT of the macaque. Visual Neurosci 15:779–786
Baylor DA (1987) Photoreceptor signals and vision. Investig Ophthalmol Visual Sci 28:34–49
Benardete EA, Kaplan E (1997) The receptive field of the primate P retinal ganglion cell, I: Linear dynamics. Visual Neurosci 14:169–185
Benardete EA, Kaplan E (1999) The dynamics of primate M retinal ganglion cells. Visual Neurosci 16:355–368
Breitmeyer BG (1984) Visual masking: an integrative approach. Oxford University Press, New York

Breitmeyer BG, Ganz L (1976) Implications of sustained and transient channels for theories of visual pattern masking, saccadic suppression, and information processing. Psychol Rev 83:1–36

Breitmeyer BG, Ogmen H (2000) Recent models and findings in visual backward masking: a comparison, review, and update. Perception Psychophys 62:1572–1595

Breitmeyer BG, Ögmen H (2006) Visual masking: Time slices through conscious and unconscious vision. Oxford University Press, New York

Bridgeman B, Hendry D, Stark (1975) Failure to detect displacement of the visual world during saccadic eye movements. Vision Res 15:719–722

Bridgeman B, Van der Heijden AHC, Velichkovsky BM (1994) A theory of visual stability across saccadic eye movements. Behav Brain Sci 17:247–292

Burr D, Morrone MC (2004) Visual perception during saccades. In: Chalupa LM, Werner JS (eds) The visual neurosciences, vol 2. MIT Press, Cambridge, Massachusetts, pp 1391–1401

Burr DC, Holt J, Johnstone JR, Ross J (1982) Selective depression of motion sensitivity during saccades. J Physiol (London) 333:1–15

Burr, DC, & Ross, J (1982) Contrast sensitivity at high velocities. Vision Res 22:479–484

Burr DC, Morrone MC, Ross J (1994) Selective suppression of the magnocellular `visual pathway during saccadic eye movements. Nature 371:511–513

Burr DC, Morgan MJ, Morrone MC (1999) Saccadic suppression precedes visual motion analysis. Curr Biol 9:1207–1209

Campbell FW, Wurtz RH (1978) Saccadic omission: why we do not see a grey-out during a saccadic eye movement. Vision Res 18:1297–1303

Castet E, Masson GS (2000) Motion perception during saccadic eye movements. Nat Neurosci 3:177–183

Castet E, Jeanjean S, Masson GS (2001) 'Saccadic suppression': no need for an active extra-retinal mechanism. Trends Neurosci 24:316–317

Castet E, Jeanjean S, Masson GS (2002) Motion perception of saccade-induced retinal translation. Proc Natl Acad Sci USA 99:15159–15163

Crawford BH (1947) Visual adaptation in relation to brief conditioning stimuli. Proc Roy Soc Lond B 134:283–302

Currie CB, McConkie GW, Carlson-Radvansky LA, Irwin DE (2000) The role of the saccade target object in the perception of a visually stable world. Perception Psychophys 62:673–683

David T, Smye S, Jame T, Dabbs T (1997) Time-dependent stress and displacement of the eye wall tissue of the human eye. Med Eng Phys 19:131–139

David T, Smye S, Dabbs T, James T (1998) A model for the fluid motion of vitreous humour of the human eye during saccadic movement. Phys Med Biol 43:1385–1399

Dehaene S, Naccache L, Cohen L, Le Bihan D, Mangin JF, Poline JB, Riviere D (2001) Cerebral mechanisms of word masking and unconscious repetition priming. Nat Neurosci 4:752–758

Deubel H, Schneider WX (1996) Saccade target selection and object recognition: evidence for a common attentional mechanism. Vision Res 36:1827–1837

Deubel H, Bridgeman B, Schneider WX (1998) Immediate post-saccadic information mediates space constancy. Vision Res 38:3147–3159

Deubel H, Schneider WX, Bridgeman B (2002) Transsaccadic memory of position and form. Prog Brain Res 140:165–180

Deubel H, Schneider WX, Bridgeman B (1996) Postsaccadic target blanking prevents saccadic suppression of image displacement. Vision Res 36:985–996

Dodge R (1900) Visual perception during eye movements. Psychol Rev 7:454–465

Dodge R (1905) The illusion of clear vision during eye movements. Psychol Bull 2:193–199

Field DJ (1987) Relations between the statistics of natural images and the response properties of cortical cells. J Opt Soc Am A 4:2379–2394

Garcia-Perez MA, Peli E (2001) Intrasaccadic perception. J Neurosci 21:7313–7322

Georg K, Lappe M (2007) Spatio-temporal contingency of saccade-induced chronostasis. Exp Brain Res 180(3):535–539

Helmholtz Hv (1866/1924) Helmholtz's treatise on physiological optics. The Optical Society of America, Electronic edition (2001). University of Pennsylvania. URL: http//psych.upenn.edu/backuslab/helmholtz.
Holt EB (1903) Eye-movement and central anaesthesia I. The problem of anaesthesia during eye-movement. Psychol Monogr 4:3–46
Hood DC (1998) Lower-level visual processing and models of light adaptation. Annu Rev Psychol 49:503–535
Ibbotson MR, Price NS, Crowder NA, Ono S, Mustari MJ (2007) Enhanced motion sensitivity follows saccadic suppression in the superior temporal sulcus of the macaque cortex. Cereb Cortex 17:1129–1138
Ilg UJ, Hoffmann KP (1993) Motion perception during saccades. Vision Res 33:211–220
Kinoshita S, Lupker S (2003) Masked priming: the state of the art. Psychology Press, New York
Kleiser R, Seitz RJ, Krekelberg B (2004) Neural correlates of saccadic suppression in humans. Curr Biol 14:386–390
Lee BB, Pokorny J, Smith VC, Kremers J (1994) Responses to pulses and sinusoids in macaque ganglion cells. Vision Res 34:3081–3096
Levinson E, Sekuler R (1975) The independence of channels in human vision selective for direction of movement. J Physiol (London) 250:347–366
Maingret F, Fosset M, Lesage F, Lazdunski M, Honore E (1999) TRAAK is a mammalian neuronal mechano-gated K+ channel. J Biol Chem 274:1381–1387
Martinez-Conde S, Macknik SL, Hubel DH (2004) The role of fixational eye movements in visual perception. Nat Rev Neurosci 5:229–240
Matin E (1974) Saccadic suppression: a review and an analysis. Psychol Bull 81:899–917
Matin E, Clymer AB, Matin L (1972) Metacontrast and saccadic suppression. Science 178:179–182
Movshon JA, Newsome WT (1996) Visual response properties of striate cortical neurons projecting to area MT in macaque monkeys. J Neurosci 16:7733–7741
Pokorny J, Sun VC, Smith VC (2003) Temporal dynamics of early light adaptation. J Vision 3:423–431
Poot L, Snippe HP, van Hateren JH (1997) Dynamics of adaptation at high luminances: adaptation is faster after luminance decrements than after luminance increments. J Opt Soc Am A 14:2499–2508
Price NS, Ibbotson MR, Ono S, Mustari MJ (2005) Rapid processing of retinal slip during saccades in macaque area MT. J Neurophysiol 94:235–246
Richards W (1968) Visual suppression during passive eye movement. J Opt Soc Am 58:1159–1160
Richards W (1969) Saccadic suppression. J Opt Soc Am 59:617–623
Riggs LA, Merton PA, Morton HB (1974) Suppression of visual phosphenes during saccadic eye movements. Vision Res 14:997–1011
Ross J, Burr D, Morrone C (1996) Suppression of the magnocellular pathway during saccades. Behav Brain Res 80:1–8
Ross J, Morrone MC, Goldberg ME, Burr DC (2001) Changes in visual perception at the time of saccades. Trends Neurosci 24:113–121
Sato M, Uchikawa K (1999) Increment-threshold spectral sensitivity during saccadic eye movements in uniform visual field. Vision Res 39:3951–3959
Schwartz SH, Godwin LD (1996) Masking of the achromatic system: implications for saccadic suppression. Vision Res 36:1551–1559
Shioiri S, Cavanagh P (1989) Saccadic suppression of low-level motion. Vision Res 29:915–928
Snippe HP, Poot L, van Hateren JH (2000) A temporal model for early vision that explains detection thresholds for light pulses on flickering backgrounds. Visual Neurosci 17:449–462
Stark L, Kong R, Schwartz S, Hendry D (1976) Saccadic suppression of image displacement. Vision Res 16:1185–1187

Thiele A, Henning P, Kubischik M, Hoffmann KP (2002) Neural mechanisms of saccadic suppression. Science 295:2460–2462
Thilo KV, Santoro L, Walsh V, Blakemore C (2004) The site of saccadic suppression. Nat Neurosci 7:13–14
Uchikawa K, Sato M (1995) Saccadic suppression of achromatic and chromatic responses measured by increment-threshold spectral sensitivity. J Opt Soc Am A 12:661–666
Volkmann FC, Riggs LA, White KD, Moore RK (1978) Contrast sensitivity during saccadic eye movements. Vision Res 18:1193–1199

Chapter 11
Intrasaccadic Motion: Neural Evidence for Saccadic Suppression and Postsaccadic Enhancement

Michael R. Ibbotson

Abstract Primates have relatively small foveas within their retinas. There are sufficient photoreceptors only in the fovea to allow high spatial resolution vision. Thus, to obtain high-resolution images in the entire visual field, there is a need to point the fovea at targets of interest, and this is achieved using saccadic eye movements. Humans make around three saccades per second. How is smooth, uninterrupted visual perception maintained in the face of the frequent image displacements generated by saccades? It has been known for many years that visual perception is modified before, during, and after saccades and, more recently, evidence has accumulated showing how neural activity is also modulated. This review will describe some of the recent electrophysiological work that shows how neural activity in the visual brain changes at the time of saccades. The review is divided into three sections. The first section describes theories relating to the saccadic suppression and some of the neural evidence to support and refute those theories. The second section describes the recently discovered alterations to the timing of visual responses in the nervous system. It turns out that the latencies of visual responses are reduced at the time of saccades, providing possible explanations for the changes in the time perception that have been observed. The third section describes how neural responsiveness is increased after saccades. Recent findings have shown that when visual stimuli are presented within the receptive fields of visual neurons soon after saccades, responses are larger than that occurring in the absence of saccades. This enhanced responsiveness persists for approximately the same duration as typical intersaccadic periods. Thus, visual responsiveness is normally enhanced between saccades. Conversely, during long periods of fixation visual sensitivity is relatively low. These latter findings suggest that the motor and visual systems work closely together to maximize the sensitivity and efficiency of the visual system.

M.R. Ibbotson (✉)
Visual Sciences, Research School of Biological Sciences, Australian National University, Canberra, Australia
e-mail: Michael.Ibbotson@anu.edu.au

U.J. Ilg and G.S. Masson (eds.), *Dynamics of Visual Motion Processing: Neuronal, Behavioral, and Computational Approaches*,
DOI 10.1007/978-1-4419-0781-3_11,

11.1 Saccadic Suppression

Primates, including humans, use saccadic eye movements to change their direction of gaze several times per second. Saccades have a central role in vision, which has led to the term "active vision" (Findlay and Gilchrist 2003). In active vision, saccades are used to keep the high-acuity foveal regions of the eyes centered on objects of interest. As a result, a wide-field high-resolution image can be built by fitting all the pieces of the visual jigsaw together, each piece derived from a foveal snap-shot (e.g., Miller and Bockisch 1997). Large areas of the brain are devoted to coordinating gaze direction with the ongoing perception of the visual environment, such as the visual and parietal cortices, the frontal eye fields (FEF), and numerous subcortical regions (e.g., the superior colliculus) (for a review, Findlay and Gilchrist 2003). While saccades offer a major advantage to the primate visual system, they also cause several major problems (Ross et al. 2001a). With respect to the content of this chapter, the most important problem associated with saccades is that every time a saccade occurs the visual world is swept at high speed across the retina.

Despite frequent exposure to rapid, whole-field image motion produced by saccades, the motion is rarely perceived (Ross et al. 2001a). The inability to perceive the image motion suggests that either (1) the motion is too fast to be detected by biological motion detectors or that (2) some mechanism suppresses the motion signals. It is simple to disprove the first concept because external saccade-like image motion produces a powerful sense of movement (Ross et al. 2001a). Taking the second viewpoint, Holt (1903) went as far as to suggest that there might be a central anesthesia during saccades. However, several investigators have shown that visual perception is not completely suppressed during saccades (Latour 1962; Krauskopf et al. 1966; Zuber and Stark 1966; Riggs et al. 1974) and that some direction sensitivity is retained (Ilg and Hoffmann 1993; Castet and Masson 2000). The term saccadic suppression is commonly used to describe the loss of visual sensitivity that occurs during saccades, but this term may be too strong. While the term "saccadic suppression" will be used throughout this chapter, it is worth noting that "selective visual attenuation" is a more accurate description.

Recordings from motion sensitive neurons throughout the visual pathways of many species reveal that strong responses are generated by saccade-like image movements in the absence of saccades, thus providing direct evidence that the visual system is able to detect very rapid image motion (e.g., Price and Ibbotson 2001; Price et al. 2005). In support of the visual attenuation model (rather than complete suppression), electrophysiology shows that many neurons in various areas of the brain retain some responsiveness to the retinal slip (image motion) during real saccades, e.g., in the lateral geniculate nucleus (LGN) (Ramcharan et al. 2001; Reppas et al. 2002); early visual cortical areas such as V1/V2 (Wurtz 1969; Toyama et al. 1984; Battaglini et al. 1986); ventral stream regions such as V4 (Tolias et al. 2001), and motion-specialized cortical regions in the dorsal stream such as the middle temporal (MT) and medial superior temporal (MST) areas (Thiele et al. 2002; Price et al. 2005; Ibbotson et al. 2006, 2007). Given that the machinery to detect rapid image motion is present and that these signals flow through to perception in the absence of saccades, what mechanisms lead to the attenuation of motion perception during saccades?

Two theories will be discussed here. The first suggests that saccades cause the photoreceptors in the retina to bend away from the optical axis (and therefore, the pupil) of the eye due to mechanical shearing forces (Richards 1969; Castet and Masson 2000; Castet et al. 2001). As the photoreceptors act as wave-guides (Snyder and Pask 1973), any movement away from the optical axis will generate a reduction in the visual sensitivity (Stiles and Crawford 1933). As explained below, this could lead to a cascade of visual effects that reveal themselves perceptually as suppression. The second theory suggests some form of internally generated or "active extra retinal" suppression of the visual motion processing pathways (e.g., Holt 1903; for review: Ross et al. 2001a). The first part of this chapter discusses both the theories briefly.

11.1.1 Discussion of the Castet et al. (2001) Theory

The eyes move at very high speeds during saccades (peaks of 100–500°/s: Carpenter 1988). The rapid acceleration generated by these eye movements (initially around $400°s^2$) may cause shearing forces near the vitreous-scleral boundary that tilt the photoreceptors relative to the optical axis (Richards 1969; Castet et al. 2001). Richards investigated if photoreceptors tilted during saccades and estimated that they did tilt by around 2°. Light that passes longitudinally through human foveal cones, which are relatively long and thin photoreceptors, is more efficient than light that strikes the receptors at an angle (Stiles and Crawford 1933; Snyder and Pask 1973). Thus, Castet and colleagues suggest that "the intrasaccadic tilt of the receptors should produce an overall decrease in luminance." However, a fall in luminance by itself is unlikely to be the only source of saccadic attenuation of visual sensitivity because the retina primarily extracts contrast information, rather than luminance, at most stages beyond the photoreceptors (Kufler 1953; Kaplan and Shapley 1986). Thus, Castet and colleagues further suggest that the luminance decrease leads to a sudden change in the visual adaptation level (Rushton 1965), which in turn leads to an immediate and short-lived reduction in sensitivity (Snippe et al. 2000). It would be expected that a strong effect related to this mechanism should occur soon after the saccade-start when the acceleration, and the perceptual suppression close to the start of saccades, is maximum (Diamond et al. 2000). Of course, during a saccade a point is reached where the acceleration is zero (Carpenter 1988). Subsequently (allowing for inertia), it might be expected that the saccadic suppression would be released. Then, at the end of the saccade deceleration a peak is reached, suggesting that the saccadic suppression should reveal itself strongly just after the saccade (again giving time for inertia). Richards (1969) did show that photoreceptor tilting persisted for a short time after saccades. For Castet et al.'s theory to be supported, we would expect to see evidence that the saccadic suppression has a triphasic time course. That is, strong soon after the saccade, weak in the middle, and strong again soon afterwards. Investigations into the time course of suppression reveal only a monophasic suppression that peaks at saccade onset and does not show a second peak associated with eye deceleration (Fig. 11.1a, Diamond et al. 2000). It could be argued that while the tilting of the photoreceptors would be triphasic, the adaptation process could have a longer time constant. Nonetheless, if this is true strong suppression should be evident for some time after the saccade and this does not appear to be the case (Diamond et al. 2000). In fact, as described in Sect. 11.3 of this chapter, there is substantial postsaccadic enhancement of neural activity in the immediate wake of saccades (Ibbotson et al. 2007).

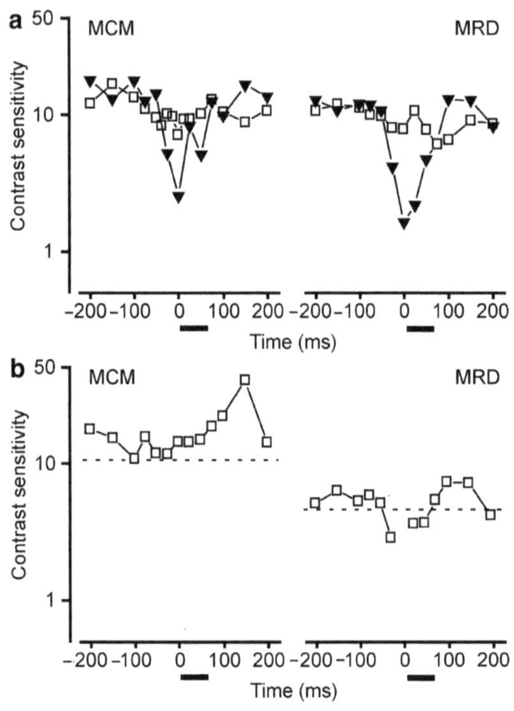

Fig. 11.1 (**a**) For two subjects (MCM and MRD), contrast sensitivity of luminance-modulated gratings during saccades (*solid triangles*) and during motion in which the subject fixates straight ahead but the image is moved in a saccade-like fashion (*open squares*). The subjects had to identify the brightness polarity of the midline of a flashed grating. Contrast sensitivity during saccades drops by a log unit and is minimal at saccade onset. There are minimal changes in the contrast sensitivity caused by the movement of the image in the absence of saccades. (**b**) Contrast sensitivity during saccades for gratings modulated in color (red-green) at equiluminance (e.g., isoluminant stimuli). Dotted lines represent contrast sensitivity without saccades. There is little sign of saccadic suppression, but there is an indication of postsaccadic enhancement (see Sect. 11.3). The black bars below the abscissa in all graphs indicate the duration of the motion. Redrawn from Figs. 4 and 7 in Diamond et al. (2000), with permission from the Journal of Neuroscience.

11.1.2 Discussion of the Extra-Retinal Model

11.1.2.1 Perceptual Evidence

Studies of visual perception during saccades reveal that the sensitivity to luminance-modulated gratings is greatly attenuated, particularly at low spatial frequencies (Burr et al. 1982; Shioiri and Cavanagh 1989). For example, contrast sensitivity is reduced by an order of magnitude during saccades (Fig. 11.1a; Diamond et al. 2000). For a luminance-modulated stimulus, the texture is provided by differences in brightness in the different regions (e.g., black and white bars). Investigation of changes in contrast sensitivity using isoluminant stimuli reveal a quite different pattern (Diamond et al. 2000). For an isoluminant stimulus, the pattern texture is provided by differences in color (e.g., red and green bars) but the luminance of the differently colored areas is equal. It turns out that the contrast sensitivity for isoluminant stimuli is not reduced during saccades (Fig. 11.1b; Diamond et al. 2000). On the path between retina and cortex, the visual system largely segregates the processing of luminancemodulated stimuli and color into two routes, the so-called magnocellular and parvocellular pathways, respectively (e.g., Kaplan and Shapley 1986; Kaplan et al. 1990; Schiller et al. 1990). From this work it is apparent that isoluminant stimuli are processed primarily by the parvocellular pathway. Putting this physical segregation of visual pathways together with the observation that isoluminant stimuli are not suppressed has led to the theory that active extra-retinal mechanisms specifically suppress magnocellular neurons (Burr et al. 1994; Ross et al. 2001a). In their alternate theory, Castet et al. (2001) have suggested that this specificity for the magnocellular pathway could be explained by the differing adaptive properties of the two visual streams. They presume that the rapid changes in adaptation state in the magnocellular pathway could saturate the responses of magnocellular neurons. Therefore, a perceptual effect that reveals itself as reduced contrast sensitivity would be most prominent for luminance-modulated stimuli.

Based on the evidence so far, the extra-retinal model and Castet et al.'s theories both have merit. However, without doubt the most compelling evidence for an extra-retinal mechanism to explain the saccadic suppression arises from the timing of the effect. In early studies, several authors noted that maximal saccadic suppression occurred just prior to saccades (e.g., Latour 1962; Zuber and Stark 1966). Under highly controlled conditions, Diamond et al. (2000) investigated the time course of the saccadic suppression by measuring the contrast sensitivity before, during, and after saccades. They found that the contrast sensitivity was significantly reduced >50 ms before the saccade-onset when using luminance-modulated stimuli but not at any time when using isoluminant stimuli (Fig. 11.1). It is very difficult to reconcile this data with a theory that depends initially on the rapid movement of the eye during the saccade (Richards 1969). It is not possible for a physical movement of photoreceptors to occur before the saccade begins. Therefore, an effect that starts with the tilting of the photoreceptors struggles to explain the reduction in the contrast sensitivity that occurs before the eye starts to move.

Several other observations argue against Castet et al.'s explanation for saccadic suppression (Ross et al. 2001b). It is established that the Stiles-Crawford effect occurs primarily in cone photoreceptors and is not present in rod photoreceptors (Stiles and Crawford 1933). Saccadic suppression has been demonstrated at very low light intensities (scotopic conditions), which would only activate rods (Diamond et al. 2000; Zuber and Stark 1966; Burr et al. 1982). Furthermore, saccadic suppression is weak or absent for isoluminant stimuli, while it is very powerful for luminance-modulated patterns. As it is the responses of cones that are primarily attenuated by the Stiles-Crawford effect, it might be expected that the sensitivity to isoluminant stimuli would be altered in some way during saccades, regardless of differing adaptation mechanisms. Ross et al. (2001b) also point out that the photoreceptors would have to bend by an estimated 12.5° to create the reduction in sensitivity that is observed in human perception, while the evidence points towards significantly lower tilt angles (around 2°: Richards 1969).

11.1.2.2 Physiological Evidence

Further to the perceptual observations, Bremmer et al. (2002) recorded neurons in several regions of the parietal cortex of alert, behaving monkeys (e.g., MT; MST; and the ventral and lateral intraparietal areas: VIP and LIP). They presented a flashed bar inside the receptive field of each neuron before, during, and after saccades. They reported in abstract form that visual responses in MT, MST, and VIP, but not LIP were suppressed before and during the saccades with a time course similar to that observed perceptually in humans (Fig. 11.1a; Diamond et al. 2000). Thus, neurons in the parietal cortex of alert monkeys exhibit saccadic suppression prior to saccades, at least for flashed stimuli.

Thiele et al. (2002) used a paradigm in which they compared the responses generated by MT/MST neurons of alert macaque monkeys in two conditions. In the first condition monkeys made 10° saccades across a textured background (active case). While in the second condition, the monkeys fixated a central target and the background image was moved in such a way as to replicate the image motion during saccades (passive case). It was found that responses in the active case were generally smaller than responses in the passive condition, and the directionality of the cells often changed. These observations were extended in later studies (Price et al. 2005; Ibbotson et al. 2007). The later investigations used stimuli with much higher luminance and contrast, and found that the effects described by Thiele and colleagues were even stronger. Moreover, the use of high-contrast stimuli revealed another important result: response latencies in the active condition were significantly shorter than in the passive condition (for details, see Sect. 11.2).

Figure 11.2 summarizes the findings from the above investigations for an example MST neuron (Ibbotson, unpublished observations). The figure shows the arrival time of every spike in the form of raster plots and the spike density functions (SDFs) derived from those plots. In the passive case (upper SDF row), motion in the preferred and anti-preferred directions generated strong responses. There is a slight bias towards

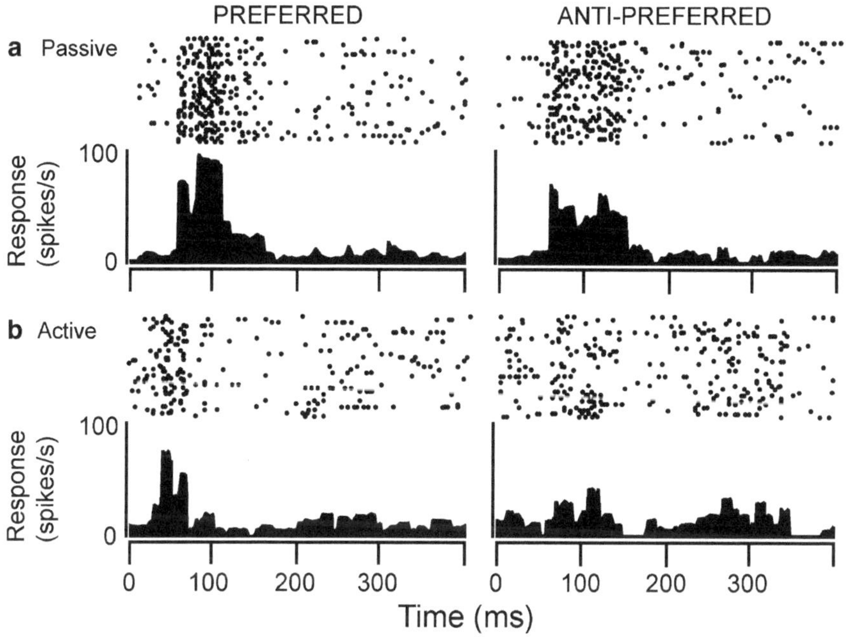

Fig. 11.2 (**a**) The top row shows raster plots of spike arrival times when a wide-field textured stimulus was moved in either the preferred or anti-preferred directions relative to a cell's preferred tuning (passive stimulation). In this case the monkey fixated a central target. Preferred direction motion produced a slightly higher spike frequency, but both directions generated strong responses: i.e., the cell was only weakly direction-selective. Below the raster plot is the spike density function derived from the spikes shown. Using the same format, (**b**) shows responses generated when the monkey made saccades back-and-forth between fixation targets (active stimulation). The time course and speed of motion on the retina was as similar as possible between the active and passive conditions. It was apparent that the responses in the active case had shorter latencies, lower amplitudes for both motion directions and that the responses became more direction-selective. For example, in the active case, the response in the anti-preferred direction was far smaller than for the preferred direction. Zero on the x-axis shows the time of stimulus onset (**a**) or the time of saccade onset (**b**). Saccades had amplitudes of 10° and durations of approximately 30 ms. The reproduced saccade-like movement of the image in (**a**) also lasted 30 ms

higher frequencies for the preferred direction, but the cell is only weakly direction-selective. In the active case (lower SDF row), a saccade moving such that the retinal slip was in the cell's preferred direction produced a strong excitation at very short latency. However, a saccade in the opposite direction generated virtually no response. A direct comparison between the passive and active cases for the same direction reveals clearly that active motion responses have lower spike frequencies than passive responses, suggesting saccadic suppression. Also, very clear is the fact that in one active direction the response is so strongly suppressed that virtually no activity is evident. This differential reduction in spiking frequency leads to a clear change in the directional tuning of the cell (Thiele et al. 2002). That is, the cell goes from being only weakly directional in the passive case to being highly direction-selective.

To show the suppression effect at the population level, the response amplitudes of 54 cells in the passive and active conditions are shown in Fig. 11.3 for saccades in both directions along the preferred motion axes of the cells. If the responses in the active and passive cases were the same, all the data points in Fig. 11.3 would fit on the dashed line of equality. In the majority of cases, the response amplitudes fall below the line of equality showing that active responses are smaller than passive responses. For many cells, responses become inhibited below the cell's spontaneous firing rate in the active case while the responses were clearly excitatory in the passive condition. It is also clear from this population data that regardless of the direction of the saccade (or saccade-like image motion), the response amplitudes during the active case were smaller in virtually all cells. Thus, the direction of the saccade appears to play little part in the effect. That is, all saccade directions lead to saccadic suppression of MT/MST neurons, even though the cells are clearly direction-selective.

How do these results support the concept of an extra-retinal mechanism for saccadic suppression? First, the fact that the directionality of the cells changes indicates that suppression is selective. That is, the system is wired up in such a way that the suppression maximally influences responses in certain motion directions. This selective wiring might have important functional consequences

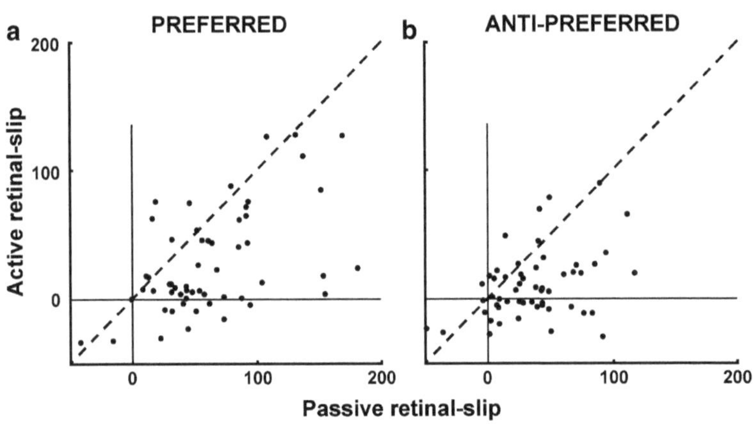

Fig. 11.3 Scatter plots showing saccadic suppression in 54 MT/MST neurons. All graphs plot active versus passive response amplitudes. The left-hand graph compares the two conditions for retinal motion in each cell's preferred direction (defined based on the passive case). The right column compares active and passive cases for anti-preferred retinal motion. All response amplitudes were the response minus the mean spontaneous rate for the cell. The thin vertical and horizontal lines show a normalized spontaneous rate for each cell. Thus, points that fall to the left or beneath the thin lines represent cells that are inhibited by motion. It is evident that the great majority of cells have response amplitudes that are smaller in the active case (points below the diagonal lines of equality). In a large proportion of the cells active motion actually suppresses the responses below the spontaneous rate (e.g., points below the thin horizontal line). It is evident that suppression occurred during saccades in both directions along the preferred motion axis. The values on the axes were the response spike rate minus the mean spontaneous activity for each cell

(Thiele et al. 2002). A generalized suppression, arising from tilting of the photoreceptors, could only influence all motion directions. Clearly, there is some suppression for all saccade directions but it has a greater impact on the motion processing network for certain directions. Secondly, the other important observation made during these studies was that the latencies of the responses generated in the active condition were significantly shorter than in the passive case (e.g., Fig. 11.2). For 62 MT neurons, average response latencies during active and passive stimulation were 30 ± 5 (SD) ms and 67 ± 15 ms, respectively (Price et al. 2005). The difference was significant for all cells (*t*-test, $p < 0.01$). For 42 MST neurons (Ibbotson et al. 2007), 35 showed response latencies that were significantly shorter in the active case (*t*-test, $p < 0.01$). The mean latencies were 38 ± 11 ms (active) and 69 ± 18 ms (passive), respectively. Seven MST cells showed no saccade-related changes in response latency (active: 79 ± 18 ms; passive: 76 ± 14 ms). It is interesting that the shorter latencies during saccades occur even though the response amplitudes are decreased due to saccadic suppression. It is common in visual neurons that when response amplitudes decrease (e.g., due to decreased stimulus contrast), response latencies increase (e.g., Ibbotson et al. 1994). Thus, a simple explanation of latency changes based on the biophysics of the cells from which the recordings were obtained cannot explain the reduced latencies. Rather, the data suggest a complex mechanism that occurs over many serial processing stages.

The shortened response latencies argue for an active mechanism that in some way primes the visual system before the saccade occurs. The latencies described above appear very short, leading to questions about whether they could be visual in origin. However, very short latencies for responses to visual stimuli (30–40 ms) have been observed previously in cortical neurons, even in the absence of saccades (e.g., V1: Maunsell and Gibson 1992; Mazer et al. 2002; MT/MST: Petersen et al. 1985; Raiguel et al. 1989; 1999; Kawano 1999). Moreover, very short-latency (50 ms) reflexive eye movements have been measured in the immediate wake of saccades in macaque monkeys (Miles et al. 1986; Kawano and Miles 1986). As these eye movements require full transmission through the visual and motor systems before the eye can start to move, the neural circuits that control them must have the capacity for very short latencies, at least when image motion occurs close to a saccade (see Sect. 11.3).

11.2 Conclusion

Combining all the observations above, we believe that an active extra-retinal mechanism is operating to minimize latencies and suppress the neural activity around the time of saccades. However, we cannot completely discount the theory that some perceptual consequence related to photoreceptor tilting, the associated Stiles-Crawford effect, and visual adaptation is also influencing the perception in the later stages of each saccade.

11.3 Perceptual Consequences of Reduced Visual Latencies During Saccades

As described above, changes to visual processing and thus to perception occur around the time of saccades (Ross et al. 2001a). What has been made clear is that the saccadic suppression is not absolute, so some stimuli are perceived. While this is the case, there is no doubt that perception is not normal: the perception of space is compressed (Ross et al. 2001a), as is the perception of time (Morrone et al. 2005). Two time-related perceptual changes result from this altered processing: time compression before and during saccades is followed by slight postsaccadic time expansion (Morrone et al. 2005). More extensive time expansions after saccades have been reported but they are not specific to saccades (Yarrow et al. 2001). Morrone et al. (2005) presented successive flashed visual stimuli to subjects and found four important results. First, they found that the inter-stimulus interval was underestimated if the first flash was presented slightly before or during a saccade. Secondly, similar time compressions did not occur for audible clicks, so the phenomenon appears to be restricted to the visual system. As an example of the magnitude of the visual effect, subjects underestimated a 110 ms interval by around 50–60 ms. Thirdly, and perhaps most remarkably, they found that there was an inversion of time for critical inter-stimulus intervals that occurred at specific times prior to the saccade-start. That is, time appeared to run backwards. The time inversion was observed by asking subjects to report the temporal order of bars flashed in different spatial locations (i.e., did the upper or lower bar appear first?). Observers consistently reported the second flash as occurring first for inter-flash intervals of 20–75 ms if the first flash occurred just before the saccade-start. Fourthly, they made the rather counterintuitive observation that the precision of time estimations was increased during saccades, despite the fact that time perception was no longer veridical.

As outlined in previous sections (Sect. 11.1.2.2), we have shown that neurons in visual areas of primate parietal cortex reduce their response latencies to visual stimulation at the time of saccades (Price et al. 2005; Ibbotson et al. 2007). I will explain below how these observations may provide a neural explanation for the perception of perisaccadic time compression and postsaccadic time expansion (Ibbotson et al. 2006). We compared the responses to image motion generated when monkeys made saccades (active case) with responses to the same image motion profile when we moved the stimulus (passive case) (Price et al. 2005; Ibbotson et al. 2007). These results show clearly that the latencies of responses in the active case were significantly shorter than in the passive case. These results also suggest a simple explanation for the perceptual observations of time compression (Ibbotson et al. 2006; Burr and Morrone 2006). Let us take the following example. If two flashed stimuli are presented 100 ms apart, the neurons will respond to both flashes with the same latency (say 65 ms) so that the inter-response interval remains 100 ms (Fig. 11.4; red trace). If the second flash is presented at saccade onset, the response latency to the first flash will be 65 ms but that to the second flash will be

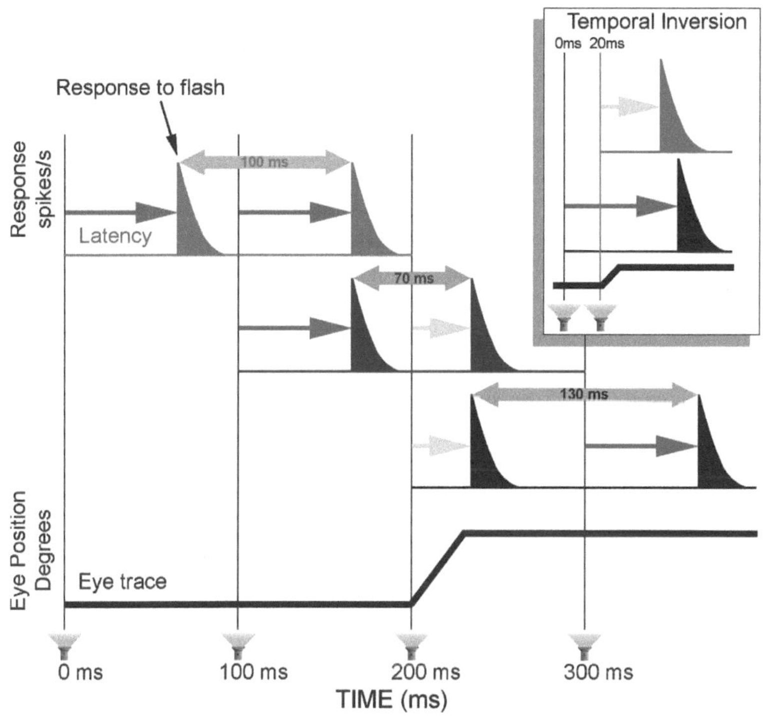

Fig. 11.4 Theoretical flashed stimuli are presented at 100 ms intervals (vertical black lines). A saccade begins at the onset of the third flash (deviation in eye trace). The cell responds to each flash with a burst of action potentials (filled response profiles). In the theoretical drawings, response amplitudes are not suppressed close to saccades, but it would be expected that responses before and during saccades would be smaller in reality. The latency to the response in the theoretical example is 65 ms (*horizontal gray arrows*) except for the response to the flash at saccade onset where the latency is 35 ms (*horizontal yellow arrows*). For the three flash intervals shown the perceived inter-flash intervals are 100 ms (control; red responses), 70 ms (peri-saccadic time compression; blue responses) and 130 ms (postsaccadic time expansion; black responses). The inset panel shows how time inversion could occur for two flashed stimuli with an inter-stimulus interval of 20 ms. In this case, the response to the second flash (red) arrives before the response to the first flash (black). Figure reproduced from original artwork with permission from Current Biology (*see Color Plates*)

shorter (say 35 ms). Thus, the inter-response interval will be reduced to 70 ms (Fig. 11.4; blue trace). Alternatively, if the first flash is presented at the saccade onset and the second after the saccade, the inter-response interval will expand to 130 ms (Fig. 11.4; black trace). How do we explain the time reversal? If the inter-flash interval was 20 ms and the second flash occurred at the saccade onset, the response to the first flash would arrive 65 ms after the first flash while the response to the second flash would arrive 55 ms after the first flash (20 ms interval plus 35 ms latency). Consequently, there would be a reversal in the temporal order of flash-response arrivals. That is, the response to the second flash will arrive 10 ms before the response to the first flash (Fig. 11.4; inset).

The theory outlined above assumes that a downstream clock remains unaffected by saccades and that initial flashes do not alter the processing of subsequent flashes (Ibbotson et al. 2006). Neurons in area LIP of the parietal cortex appear to keep track of elapsed time between behaviorally important events (Leon and Shadlen 2003). It is not known if the time coding in the LIP cells is influenced by saccades but it is known that LIP neurons do not exhibit saccadic suppression (Bremmer et al. 2002). The data from the study by Ibbotson et al. (2007) reveal a small number of MST neurons that do not have saccade-related reductions in latency. This result suggests that the time-coding properties of some neurons in the parietal cortex are not influenced by saccadic eye movements.

Morrone et al. (2005) showed that the temporal precision was improved at the time of saccades. Thus, while the perception of time is clearly incorrect in absolute terms, the repeatability of the measurement is improved. As shown above, the standard deviations of the mean latencies for MT/MST neurons in the active case are smaller than those for the passive case. The reduction in variance was significant (F ratio test, $p << 0.01$). These data indicate an increase in the precision of response timing during saccades, which could account for the peri-saccadic perceptual improvement in temporal precision. This observation makes good physiological sense. If a mechanism reduces the visual latency, it is only possible that the latency can be compressed before the effect saturates. A visual signal needs to be generated in the photoreceptors and then transmitted to the parietal cortex, all of which takes time. If the mechanism has pushed the latency to its lowest possible value, it is likely that the latencies will be highly reliable from repetition-to-repetition within a cell and that the neuron-neuron variability will be reduced. Highly repeatable timing events in the nervous system will inevitably reveal themselves as precise time measurements perceptually, even if the absolute time measure is incorrect.

Is there a functional role for time compression either perceptually or neurally? We suggest that the visual nervous system and the associated eye movement control areas have evolved a mechanism by which all delays in transmission time imposed by additional processing stages are removed during and after saccades. Thus, just at the time when there is maximal change in the visual scene due to gaze shifts, the system is able to operate at maximum speed. It is likely that this increase in speed is accompanied by a loss in processing capacity but we presume that the compensation for these losses is that important signals can reach the cortex quickly. Only further research will establish what information might be lost. From a perceptual perspective, the compression of time might have a role in perceptually removing the blanking periods (gray-outs) that are induced by saccadic suppression. Simply put: if the perception of time during a saccade is compressed, there is less opportunity for it to be noticed. Previous studies have shown that the perception of time is expanded after saccades: the so-called chronostasis (Yarrow et al. 2001). This post-saccadic time expansion is thought to compensate for the time lost during the blanking period associated with a saccade and in this way might be linked to the time compression during saccades. However, there is controversy related to the link between chronostasis and time changes around the times of saccades (Morrone et al. 2005). Specifically, chronostasis is found in other sensory and motor domains and thus does not appear to be a purely saccade-related phenomenon (Park et al. 2003; Yarrow and Rothwell 2003). In summary, the observation of time compression and its neural basis have only very recently appeared in the literature. Future research will help to clarify if these time warping phenomena have any functional advantage.

11.4 Postsaccadic Enhancement

In macaque monkeys, the sudden movement of large-field textured patterns generates reflexive tracking eye movements known as ocular following. If this sudden movement occurs immediately after a saccade, ocular following is generated with very short latencies: 50–60 ms from saccade-end (Kawano and Miles 1986). Similar short-latency postsaccadic ocular following responses are observed in humans with latencies of ~70 ms (Gellman et al. 1990; Masson and Castet 2002). In macaques, the initial eye speed of ocular following decays in an exponential fashion after saccade-end with a time constant of 60 ms (Kawano and Miles 1986). Consequently, immediately after a saccade the initial eye speed generated by stimulus motion can be 3–4 times larger than if the same stimulus is presented 300 ms after a saccade. These data show that the initial ocular following speeds are enhanced by prior saccades. Moreover, the initial speed of the ocular following depends to a large extent on the retinal slip that occurs during the saccade (Kawano and Miles 1986). That is, the mechanism may have partial visual origins.

Short-latency visuomotor responses to moving patterns imply that a neural motion signal is rapidly calculated and transferred to the extraocular muscles to generate tracking eye movements. To investigate this pathway, Kawano and colleagues (Kawano et al. 1992, 1994, 1997; Kawano 1999) recorded from several likely sites on the visuomotor pathway while alert monkeys were presented with moving patterns 50 ms after saccade-end. These sites included the cortical area MST, the brain-stem dorsolateral pontine nucleus (DLPN), and the ventral paraflocculus (VPFL). They revealed that direction-selective neurons in MST had a mean response latency of 47 ms, with many as short as 40 ms. These responses preceded the initial change in eye speed of ocular following by an average of 8–9 ms. The latency distributions of the neurons in these brain areas suggests information flows from MST through the DLPN to the VPFL. Lesion and anatomical studies confirm the involvement of all three areas in the pathway generating short-latency ocular following (for review, Kawano 1999).

Behavioral findings show that the initial ocular following speeds are enhanced in the wake of saccades and it appears that MST has an important role in driving those eye movements. However, it was not until recently that studies have investigated whether the response amplitudes of MST (and MT) neurons are enhanced after saccades. Ibbotson et al. (2007) compared the response magnitudes of MT/MST neurons soon after saccades with responses to the same stimulus with no prior eye movements. They revealed that MT/MST neurons produced significantly larger responses to image motion in the 150 ms after a saccade than they did under no-saccade conditions (Fig. 11.5). Moreover, it was shown that the latencies of the MT/MST responses were much shorter immediately after the saccade than they were under control conditions. For example, the optimal visual stimulation of all MT/MST cells with moving patterns 50 ms after a saccade produced latencies of 48 ± 10 ms, which compares to 69 ± 18 ms without a prior saccade. Overall, the data described above provide a neural correlate of the postsaccadic enhancement of eye speed and very short-latency ocular following responses observed by Miles et al. (1986).

The postsaccadic enhancement of neural activity shown in MT and MST correlates well with behavioral observations of ocular following responses (Ibbotson et al. 2007). However, evidence for a direct link between the neural activity and behavior is still lacking. Nonetheless, the enhanced motion sensitivity after saccades presumably has some role in preventing the eye drift after saccades (glissades) and in all behaviors where rapid tracking is required soon after changing the direction of gaze (Kawano and Miles 1986). For example, when making a saccade to a moving football, it is essential to start tracking the football as rapidly as possible after the saccade has directed the fovea towards it. Moreover, one would expect the eye tracking system to operate with high gain once the saccade is over, thus providing a functional explanation for postsaccadic enhancement of ocular following. The present results potentially have broad implications outside the field of eye movement control because MT/MST neurons are known to have a role in motion perception (MT: Newsome and Pare 1988; MST: Celebrini and Newsome 1994, 1995). It would, therefore, be interesting to look for modulations of motion perception in the 500 ms period following saccades. We predict that substantial reductions in motion detection thresholds (increased sensitivity) should be found in the first 200 ms following a saccade. Clues that contrast sensitivity might increase in the wake of saccades are available (Burr et al. 1994; Diamond et al. 2000). In those studies there is evidence of increased color discrimination and contrast sensitivity, as well as weak indications of increases in luminance discrimination in the first 200 ms after saccade-end. If the perception of motion is enhanced in some way after saccades it may have a role in ensuring maximal sensitivity to moving objects that appeared while the eye was in motion and the visual system was undergoing

Fig. 11.5 (continued) respond to the image motion with latencies slightly shorter (10 ms) than the first eye movements. (**a**) Eye speed profiles generated by motion either 50 ms (*thick line*) or 500 ms (*thin line*) after saccade-end (80 repeats). The gray area shows the difference between the two eye speed profiles. The postsaccadic background motion was 40°/s. Using the same line conventions, spike density functions are shown for single MT (**b**) and MST (**c**) neurons stimulated 50 or 500 ms after a saccade. It is clear that eye speed during ocular following is transiently enhanced when the image motion occurs soon after a saccade (Kawano and Miles 1986). Moreover, the spike frequencies generated by MT and MST neurons are enhanced in the same period after saccades. Note that both the MT and MST neuron produce a small burst of spikes in response to the movement of the image generated by the saccade (saccade response). Figure reproduced from original artwork with permission from Cerebral Cortex

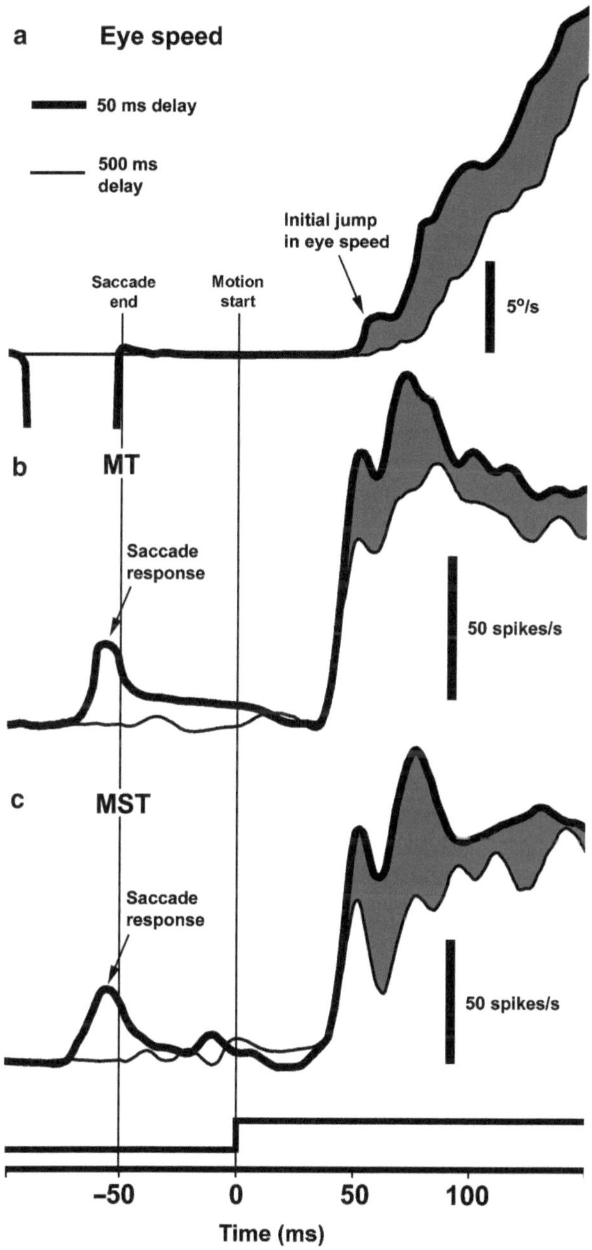

Fig. 11.5 Eye speed and spike rate profiles from a macaque monkey performing the short-latency ocular following paradigm (from Ibbotson et al. 2007). Monkeys make saccades across a textured background to a central target. Once the eye arrives at the target there is a 50 ms (or 500 ms) pause, then the textured background moves in a recorded cells preferred direction. This sudden image motion generates a short-latency ocular following response in which the eyes track the scene (Miles et al. 1986). Simultaneous recordings from MT or MST areas reveal that many cells

saccadic suppression. There may also be a function in recovering from the "temporary visual impairment" generated by saccadic suppression.

11.5 Final Conclusions

The results described in this chapter show that while great neural effort is put into visual-motor coordination of eye movements (Findlay and Gilchrist 2003), there is also substantial neural processing devoted to suppressing neural activity before and during saccades, while transiently enhancing the activity after saccades. The reasons for the suppression and enhancement appear to be related to the need to retain a stable visual percept despite continuous saccadic intrusions. It is reasonable to suggest that the suppression acts to attenuate perception of rapid image motion before and during saccades. Conversely, it can be suggested that the postsaccadic enhancement has a role in boosting sensitivity at the exact moment after a saccade where the system is trying to re-establish a stable percept.

At the beginning of the chapter two theories were reviewed, both suggesting mechanisms by which the visual sensitivity might be reduced during saccades. It is perfectly plausible that both theories are correct. Active extra-retinal suppression must occur because neural responses and visual perception are greatly attenuated prior to saccades. The suppression before the saccades largely excludes the possibility that the reduced sensitivity arises from the eye movement itself. However, once the eye has moved there is no existing evidence that completely excludes an additional suppressive effect that originates from a tilting of the photoreceptors.

If we accept from the evidence outlined above that there is an active extra-retinal mechanism that causes the saccadic suppression, where do the signals arise? To account for the presaccadic suppression observed perceptually and the reduced latencies we observe in MT/MST, it is likely that an efference copy of eye movement commands must be involved. This could arise in areas such as the FEF, where significant activity occurs hundreds of milliseconds before saccades (Bruce and Goldberg 1985). We presume that this input would be inhibitory. Additional active saccadic suppression could arise later during the saccade from short-latency subcortical neurons in the pretectum that are sensitive to rapid image motion and project to the LGN (Schmidt 1996; Price and Ibbotson 2001). Since the thalamic projections from the pretectum are putatively GABAergic in primates they could provide a peri- or postsaccadic inhibition of activity.

It remains to be resolved where the saccade-related modulation of activity first acts. Anatomical (Cucchiaro et al. 1993; Schmidt 1996), physiological (Lee and Malpeli 1998; Reppas et al. 2002), and stimulation studies (Thilo et al. 2004) all point towards modulation acting as early as the LGN. It is then possible that the saccadic modulation of cortical activity, such as that reported here in MST, is inherited from the LGN. Alternatively, it is possible that modulation may act simultaneously at a range of cortical and subcortical sites. Given the strength of anatomical connections between MT/MST and motor planning regions such as the FEF (Leichnetz 1989), it is possible that the effects observed in MT/MST arise due to a combination of direct and inherited saccadic modulation.

Acknowledgments Some of the work presented here was conducted using funds from the Australian Research Council Centre of Excellence in Vision Science (CE0561903) and two NIH grants (EY06069; RR0165). Thanks go to Professor Mustari, Drs Cloherty, Price, Crowder, Hietanen and Ono and to Tracey Broznya, Katia Peixoto and Anthony Gazy for help with experiments and animal care.

References

Battaglini PP, Galletti C, Aicardi G, Squatrito S, Maioli MG (1986) Effect of fast moving stimuli and saccadic eye movements on cell activity in visual areas V1 and V2 of behaving monkeys. Arch Ital Biol 124:111–119

Bremmer F, Kubischik M, Hoffmann KP, Krekelberg B (2002) Neural dynamics of saccadic suppression; Program No. 57.2. Abstract Viewer/Itinerary Planner. Society for Neuroscience, Washington, DC. Online

Bruce CJ, Goldberg ME (1985) Primate frontal eye fields I. Single neurons discharging before saccades. J Neurophysiol 53:603–635

Burr DC, Morrone MC (2006) Perception: transient disruptions to neural space-time. Curr Biol 16:R847–R849

Burr DC, Holt J, Johnstone JR, Ross J (1982) Selective depression of motion sensitivity during saccades. J Physiol (London) 333:1–15

Burr DC, Morrone MC, Ross J (1994) Selective suppression of the magnocellular visual pathway during saccadic eye movements. Nature 371:511–513

Carpenter RHS (1988) Movements of the eyes, 2nd edn. Pion Press, London

Castet E, Masson GS (2000) Motion perception during saccadic eye movements. Nat Neurosci 3:177–183

Castet E, Jeanjean S, Masson GS (2001) 'Saccadic suppression'-no need for an active extra-retinal mechanism. Trends Neurosci 24:316–317

Celebrini S, Newsome WT (1994) Neuronal and psychophysical sensitivity to motion signals in extrastriate area MST of the macaque monkey. J Neurosci 14:4109–4124

Celebrini S, Newsome WT (1995) Microstimulation of extrastriate area MST influences performance in a direction discrimination task. J Neurophysiol 73:437–448

Cucchiaro JB, Uhlrich DJ, Sherman SM (1993) Ultrastructure of synapses from the pretectum in the A-laminae of the cat's lateral geniculate nucleus. J Comp Neurol 334:618–630

Diamond MR, Ross J, Morrone MC (2000) Extraretinal control of saccadic suppression. J Neurosci 20:3449–3455

Findlay JM, Gilchrist ID (2003) Active vision: the psychology of looking and seeing. Oxford, Oxford University Press

Gellman RS, Carl JR, Miles FA (1990) Short latency ocular-following responses in man. Vis Neurosci 5:107–122

Holt EB (1903) Eye movements and central anaesthesia. Psychol Rev 4:3–45

Ibbotson MR, Mark RF, Maddess T (1994) Spatiotemporal response properties of direction-selective neurons in the nucleus of the optic tract and dorsal terminal nucleus of the wallaby, *Macropus eugenii*. J Neurophysiol 72:2927–2943

Ibbotson MR, Crowder NA, Price NSC (2006) Neural basis of time changes around the time of saccades. Curr Biol 16:R834–R836

Ibbotson MR, Price NSC, Crowder NA, Ono S, Mustari MJ (2007) Enhanced motion sensitivity follows saccadic suppression in the superior temporal sulcus of the macaque cortex. Cereb Cortex 17:1129–1138

Ilg UJ, Hoffmann KP (1993) Motion perception during saccades. Vis Res 33:211–220
Kaplan E, Shapley RM (1986) The primate retina contains two types of ganglion cells, with high and low contrast sensitivity. Proc Natl Acad Sci U S A 83:2755–2757
Kaplan E, Lee BB, Shapley RM (1990) New views of primate retinal function. Prog Retin Res 9:273–336
Kawano K (1999) Ocular tracking: behavior and neurophysiology. Curr Opin Neurobiol 9:467–473
Kawano K, Miles FA (1986) Short-latency ocular following responses of monkey II. Dependence on a prior saccadic eye movement. J Neurophysiol 56:1355–1380
Kawano K, Shidara M, Yamane S (1992) Neural activity in dorsolateral pontine nucleus of alert monkey during ocular following responses. J Neurophysiol 67:680–703
Kawano K, Shidara M, Watanabe Y, Yamane S (1994) Neural activity in cortical area MST of alert monkey during ocular following responses. J Neurophysiol 71:2305–2324
Kawano K, Inoue Y, Takemura A, Kitama T, Miles F (1997) A cortically mediated visual stabilization mechanism with ultra-short latency in primates. In: Thier P, Karnath H-O (eds) Parietal lobe contributions to orientation in 3D space. Springer, Heidelberg, pp 185–199
Krauskopf J, Graf V, Gaarder K (1966) Lack of inhibition during involuntary saccades. Am J Psychol 79:73–81
Kufler SW (1953) Discharge patterns and functional organization of mammalian retina. J Neurophysiol 16:37–68
Latour PL (1962) Visual threshold during eye movements. Vis Res 2:261–262
Lee D, Malpeli JG (1998) Effects of saccades on the activity of neurons in the cat lateral geniculate nucleus. J Neurophysiol 79:922–936
Leichnetz GR (1989) Inferior frontal eye field projections to the pursuit-related dorsolateral pontine nucleus and middle temporal area (MT) in the monkey. Vis Neurosci 3:171–180
Leon MI, Shadlen MN (2003) Representation of time by neurons in the posterior parietal cortex of the macaque. Neuron 38:317–327
Masson GS, Castet E (2002) Parallel motion processing for the initiation of short-latency ocular following in humans. J Neurosci 22:5149–5163
Maunsell JH, Gibson JR (1992) Visual response latencies in striate cortex of the macaque monkey. J Neurophysiol 68:1332–1344
Mazer JA, Vinje WE, McDermott J, Schiller PH, Gallant JL (2002) Spatial frequency and orientation tuning dynamics in area V1. Proc Natl Acad Sci U S A 99:1645–1650
Miles FA, Kawano K, Optican LM (1986) Short-latency ocular following responses of monkey I. Dependence on temporospatial properties of visual input. J Neurophysiol 56:1321–1354
Miller JM, Bockisch C (1997) Visual perception – where are the things we see? Nature 386:550–551
Morrone MC, Ross J, Burr DC (2005) Saccadic eye movements cause compression of time as well as space. Nat Neurosci 8:950–954
Newsome WT, Pare EB (1988) A selective impairment of motion perception following lesions of the middle temporal visual area (MT). J Neurosci 8:2201–2211
Park J, Schlag-Rey M, Schlag J (2003) Voluntary action expands perceived duration of its sensory consequence. Exp Brain Res 149:527–529
Petersen SE, Baker JF, Allman JM (1985) Direction-specific adaptation in area MT of the owl monkey. Brain Res 346:146–150
Price NSC, Ibbotson MR (2001) Pretectal neurons optimised for the detection of saccade-like movements of the visual image. J Neurophysiol 85:1512–1521
Price NSC, Ibbotson MR, Ono S, Mustari MJ (2005) Rapid processing of retinal slip during saccades in macaque area MT. J Neurophysiol 94:235–246
Raiguel SE, Lagae L, Gulyas B, Orban GA (1989) Response latencies of visual cells in macaque areas V1, V2 and V5. Brain Res 493:155–159
Raiguel SE, Xiao DK, Marcar VL, Orban GA (1999) Response latency of macaque area MT/V5 neurons and its relationship to stimulus parameters. J Neurophysiol 82:1944–1956
Ramcharan EJ, Gnadt JW, Sherman SM (2001) The effects of saccadic eye movements on the activity of geniculate relay neurons in the monkey. Vis Neurosci 18:253–258

Reppas JB, Usrey WM, Reid RC (2002) Saccadic eye movements modulate visual responses in the lateral geniculate nucleus. Neuron 35:961–974
Richards W (1969) Saccadic suppression. J Opt Soc Am 59:617–623
Riggs LA, Merton PA, Morton HB (1974) Suppression of visual phosphenes during saccadic eye movements. Vis Res 14:997–1011
Ross J, Morrone MC, Goldberg ME, Burr DC (2001a) Changes in visual perception at the time of saccades. Trends Neurosci 24:113–121
Ross J, Morrone MC, Goldberg ME, Burr DC (2001b) Response: 'saccadic suppression' – no need for an active extra-retinal mechanism. Trends Neurosci 24:317–318
Rushton WA (1965) Visual adaptation. Proc R Soc Lond B 16:20–46
Schiller PH, Logothetis NK, Charles ER (1990) Functions of the colour-opponent and broad-band channels of the visual system. Nature 343:68–70
Schmidt M (1996) Neurons in the cat pretectum that project to the dorsal lateral geniculate nucleus are activated during saccades. J Neurophysiol 76:2907–2918
Shioiri S, Cavanagh P (1989) Saccadic suppression of low-level motion. Vis Res 29:915–928
Snippe HP, Poot L, van Hateren JH (2000) A temporal model for early vision that explains detection thresholds for light pulses on flickering backgrounds. Vis Neurosci 17:449–462
Snyder AW, Pask C (1973) The stiles-crawford effect-explanation and consequences. Vis Res 13:1115–1137
Stiles WS, Crawford BS (1933) The luminous efficency of rays entering the eye pupil at different points. Proc R Soc Lond B 112:428–450
Thiele A, Henning P, Kubischik M, Hoffmann KP (2002) Neural mechanisms of saccadic suppression. Science 295:2460–2462
Thilo KV, Santoro L, Walsh V, Blakemore C (2004) The site of saccadic suppression. Nat Neurosci 7:13–14
Tolias AS, Moore T, Smirnakis SM, Tehovnik EJ, Siapas AG, Schiller PH (2001) Eye movements modulate visual receptive fields of V4 neurons. Neuron 29:757–767
Toyama K, Kimura M, Komatsu Y (1984) Activity of the striate cortex cells during saccadic eye movements of the alert cat. Neurosci Res 1:207–222
Wurtz RH (1969) Comparison of effects of eye movements and stimulus movements on striate cortex neurons of the monkey. J Neurophysiol 32:987–994
Yarrow K, Rothwell JC (2003) Manual chronostasis: tactile perception precedes physical contact. Curr Biol 13:1134–1139
Yarrow K, Haggard P, Heal R, Brown P, Rothwell JC (2001) Illusory perceptions of space and time preserve cross-saccadic perceptual continuity. Nature 414:302–305
Zuber BL, Stark L (1966) Saccadic suppression: elevation of visual threshold associated with saccadic eye movements. Exp Neurol 16:65–79

Part III
Modeling Dynamic Processing

Chapter 12
Maximizing Causal Information of Natural Scenes in Motion

Dawei W. Dong

Abstract Natural scenes contain a huge amount of information if counted by spatial pixels and temporal frames. However, most of the information is redundant because the pixels and the frames are highly correlated. The optical flow, generated by motions of the objects and the observer, contributes significantly to the statistical regularity of such spatiotemporal correlations. The visual system of an animal such as the human is highly adapted to this statistical regularity such that the visual sensitivity follows the same contours as the spatiotemporal correlations of natural scenes in motion, in particular, along two axes: space and motion instead of space and time. Furthermore, vision is an active process, during which eye movements interact with visual scenes and select images that arrive on the retina: pursuits and fixations on objects significantly alter the image velocity distributions on the fovea and the periphery, which lead to the dependence of the visual sensitivity on the retinal eccentricity; saccades between objects change the natural scene statistics dynamically, which lead to the dependence of the visual sensitivity on the time relative to saccades. All of these can be accounted for by the proposed ecological theory that the visual system maximizes the causal information of the natural visual input.

12.1 Introduction: An Approach Based on Information Theory

Animals live in a dynamical world. Everything changes and fluctuates. It seems that nothing is certain but the passage of time. Yet, the physical world, however dynamic and probabilistic, does have some causal connections from its past to its future, governed by the law of physics. A fundamental function of the brain is to afford an

D.W. Dong
Center for Complex Systems and Brain Sciences, Florida Atlantic University, Boca Raton, FL, 33431
e-mail: dawei@ccs.fau.edu

U.J. Ilg and G.S. Masson (eds.), *Dynamics of Visual Motion Processing: Neuronal, Behavioral, and Computational Approaches*,
DOI 10.1007/978-1-4419-0781-3_12,

animal some ability of knowing the future from the past, i.e., to acquire the causal information.

Through the sensors which are essential for the survival of an animal species, an individual who knows more about the future has an obvious evolutionary advantage over those who know less. Due to the natural selection, it is conceivable that the neural system which processes the information from those sensors tends to maximize the causal information — and equivalently, to minimize the non-causal information — for given physiological constraints on the processing capacity.

The hypothesis of this paper/chapter is that maximizing causal information is an organization principle of the visual system, in particular, in dynamically processing the motion information. To apply this principle, we first gain a good understanding of the statistical properties of the natural visual input to the brain. We then reveal how the visual processing depends on the statistics of the natural visual input. We apply this approach in two cases: first, to the stationary case in which we relate the average natural scene statistics to the average visual sensitivities; second, to the non-stationary case in which we reveal how the statistics of the natural visual input and the visual processing depend on scenes and eye movements.

12.2 Statistics of Natural Scenes: Scaling to Motion

In recent years, quantitative measurements have been conducted extensively on the statistical properties of the visual input — in particular, the images "natural" for the human visual system — in many aspects, such as color (Webster and Mollon 1997; Parraga et al. 1998; Ruderman et al. 1998; Tailor et al. 2000), stereo (Li and Atick 1994; Hoyer and Hyvarinen 2000), space (Burton and Moorhead 1987; Field 1987, 1994; Ruderman and Bialek 1994; Olshausen and Field 1996), and space-time (Dong and Atick 1995a, VanHateren and Ruderman 1998). The discovery most relevant to this paper is that the Fourier transform of the spatiotemporal correlation matrix, or the power spectrum, of natural scenes, is a non-separable function of spatial and temporal frequencies and exhibits an interesting scaling behavior:

$$R(f, w) \sim f^{-m-1} P(w/f)$$

where f and w are the spatial and the temporal frequencies, $-m$ is the exponent of the spatial power spectrum $\sim f^{-m}$, and $P(v)$ is the probability density distribution function of the optical flow velocity v (Dong 2001a).

To see the scaling behavior, the measured power spectrum is plotted in Figure 1 (top) as a function of f for different w/f ratios. The curves are just a horizontal shift from each other and all of them follow a straight line in the log-log plot. In fact, when multiplying the spectrum by $m + 1$ power of f, i.e., when plotting $f^{m+1} R(f, w)$ as a function of w/f, all curves coincide very well, as shown in Figure 1 (bottom).

It has been shown earlier that such spatiotemporal correlation can be derived from the first principles, under the assumption that the motions of objects relative

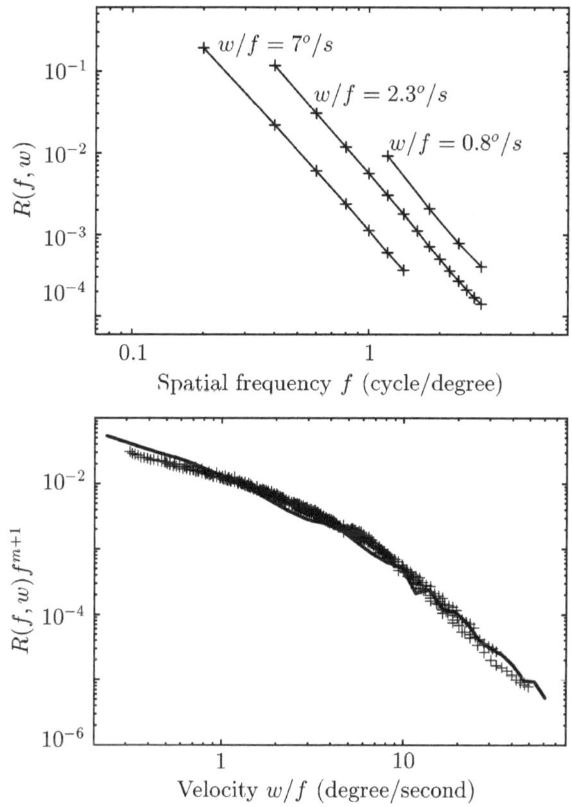

Fig. 12.1 Scaling behavior of spatiotemporal power spectrum. **Top:** the power spectrum is plotted for three velocities — ratios of temporal and spatial frequencies — (0.8, 2.3, 7) degree/second. The curves have the same shape ~ f^{-m-1}. **Bottom:** the power spectrum is replotted as a function of w/f after multiplication by f^{m+1}, with $m = 2.3$. All the data points fall on a single curve. The solid curve is the probability density distribution of the optical flow velocity (adapted from Dong 2001a)

to the observer follow a certain velocity distributions (Dong and Atick 1995a). Supporting this physical explanation is the clear agreement between the scaled power spectrum f^{m+1} $R(f, w)$ and the measured optical flow velocity distribution (Dong 2001a) shown together in Fig. 1 (bottom).

Of course, this physical scenario has correlations higher than the pairwise second order correlation (or equivalently the spatiotemporal power spectrum). Then, what is the reason to study the second order corrlation? The second order correlation matrix, which should be thought of as a constraint on the joint probability distribution, is the simplest quantity that captures the statistical dependency in space and time of natural time-varying images. Furthermore, we believe that the second order correlation is the quantity that neurons the early stages of the visual system are able to evaluate and take advantage of in recoding the visual input (see Section 3 and Section 4).

This constraint suggests that the spatial and temporal correlations arise predominantly from motions and in general, the spatial and temporal parts of such correlations are not separable and hence visual processing cannot be fully characterized in space independently of time. The scaling behavior suggests that a natural way to separate the correlations is in space and motion — just as Newtonian (Hamiltonian) physical systems should be. More interestingly, it suggests that a better way to examine spatiotemporal tuning data from visual systems is to plot the data not as a function of f and w separately but as a function of f and w/f — if the visual processing is indeed adapted to the natural scene statistics. In this representation we expect that neural responses will exhibit more universal behavior.

12.3 Scaling of Human Visual Sensitivity

We will show in the next section the theoretical connection between the spatiotemporal power spectrum of natural scenes and the response properties of visual systems. But before showing the theory, we will first show the scaling behavior of the visual sensitivity, which is strikingly similar to the scaling behavior of the natural scenes.

Figure 2 (top) shows the human visual sensitivity to spatiotemporal gratings, determined by the threshold modulation amplitude I_t. Here, instead of the usual way of plotting along spatial and temporal frequencies, the curves are plotted along the spatial frequency and the velocity, i.e., the ratio of the temporal and the spatial frequencies (Kelly 1979). Clearly, the curves for different velocities have similar shapes. In particular, the rising slopes of the curves in the log-log plot are half of $m + 1$, which ensure that the natural power spectrum with a slope of $-(m + 1)$ (for a constant f/w ratio, see Figure 1, top) is whitened. It was shown that the sensitivity curves approximately coincide with each other when shifted by a factor of $P(f/w)$ according to the statistics of natural scenes (Dong 2001a). This scaling behavior in human visual sensitivity shows the predominant role of image motion in visual processing as suggested by the natural scene statistics shown in Section 2.

Figure 2 (bottom) shows a non-stationary aspect of visual processing: the human visual sensitivity adapting to a simple scene statistics, I_0, the mean light level. At a given temporal frequency which is relatively low, the amplitude sensitivity is inversely proportional to I_0. Furthermore, the rising slopes of the curves in the log-log plot are approximately 1, which decorrelate (whiten) the natural scenes with the velocity distribution shown in Figure 1 (bottom) which has the asymptotic slope near -2 for reasonably high velocities. Because the sensitivity is inversely proportional to the mean light, the decorrelation is maintained at all mean light levels for relatively low spatial and temporal frequencies shown in Figure 2.

Next, let's see how we can make a theoretical connection between the scaling to motion in the statistics of natural scenes and the scaling of visual response properties shown in human visual sensitivity.

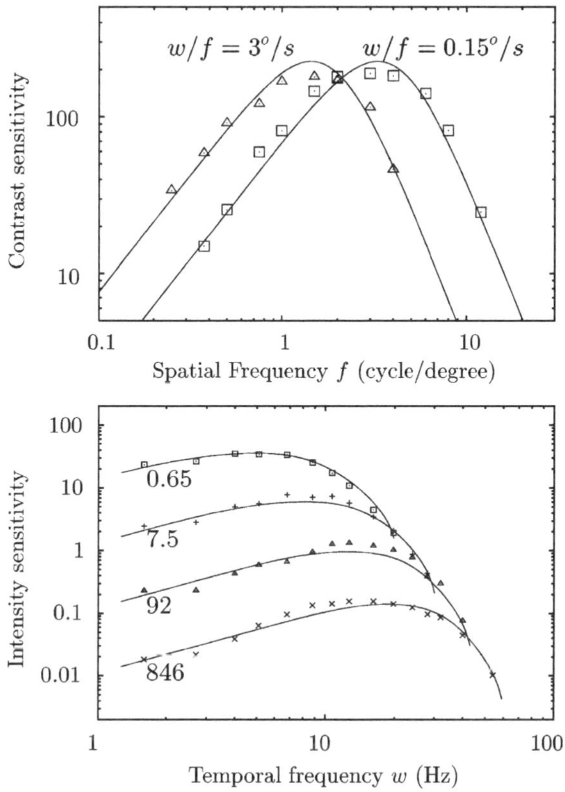

Fig. 12.2 Top: the spatiotemporal visual sensitivity. The amplitude sensitivity data ($1/I_t$, in the unit of trolands^{-1}, Kelly 1979) are plotted as a function of spatial frequency for two *w/f* ratios — 0.15, 3 degree/second (at a fixed mean light level of 300 troland). The curves at low spatial frequencies have the positive slopes which exactly decorrelates the natural image power spectrum. **Bottom:** the temporal visual sensitivity adapting to the mean light level. The amplitude sensitivity data ($1/I_t$, in the unit of trolands^{-1}, Kelly 1961) are plotted for four different mean light levels — 0.65, 7.1, 77, 850 trolands (at a very low spatial frequency). The curves at low temporal frequencies have the positive slopes which exactly decorrelates the natural image power spectrum. The solid lines are the theoretical prediction

12.4 Maximizing Causal Information

It is worth to review some earlier approaches based on the information theory (Shannon and Weaver 1949). The brain has to process an enormous amount of information from the sensory input. It is reasonable to expect that neurons in the sensory pathways developed to take advantage of certain regularities and statistical properties in the sensory input to build more efficient representations of the world (Attneave 1954; Barlow 1961; Srinivasan et al. 1982; Linsker 1988; Atick and Redlich 1990). There has been a vast amount of evidence showing the connection between the properties of natural stimuli and the properties of the sensory systems, and it is quite clear that better characterization of the properties of natural signals leads to deeper understanding of neural functions (see references in Atick 1992; Dong and Atick 1995b; Simoncelli and Olshausen 2001; Zhaoping 2002).

One fundamental assumption common to most of the earlier works is that one needs to know what information/noise is **before** applying a theory (see Linsker 1989, Atick and Redlich 1990, VanHateren 1992). However, in this paper, we propose a theory which will discover what information/noise is **when** applied to the visual system. We will describe the theory briefly on the conceptual level without going into mathematical details.

At any given moment of time, let's divide the visual input S into two parts: S_- and S_+ for the past (occurred) and the future (incoming) visual input, respectively. Because of the causality, the visual system can only process and represent its input based on S_-. Our hypothesis is that the goal of the visual processing is to transform S_- into an internal representation O such that O contains the maximum amount of information about S_+ (let's denote this information as $I(O, S_+)$). This requirement can be easily fulfilled if there is no limitation to the maximum capacity of the internal representation: just simply let $O = S_-$, so $I(O, S_+) = I(S_-, S_+)$ which contains all the information about the future from the past. However, this representation is redundant and inefficient since S_- has the spatiotemporal correlations as shown in Section 2. Given a limitation to the capacity (let's denote it as $C(O)$), one can build a better representation $O = K(S_-)$ by choosing a transfer function K which maximizes $I(O, S_+)$ while minimizing $C(O)$ or keeping $C(O)$ constant. In this representation, O will be more independent (decorrelated).

Both $I(O, S_+)$ and $C(O)$ are well defined in terms of the information theory for a given statistical ensemble of visual input S and a given transfer function K. In this chapter, we will restrict ourselves to the class of linear transfer functions and to the second order statistics.

If natural scenes had been changing randomly over time, the power spectrum would be flat in time and the causal information $I(O, S_+)$ would be zero, i.e., everything would be noise, i.e., one could not predict the future from the past. However, natural scenes are dominated by objects which move according to Newtonian (Hamiltonian) dynamics, and as a result, the temporal power spectrum is not flat — in fact, dominated by a characteristic velocity distribution, which give rise to the predictability.

Using the natural scene power spectrum R as shown in Section 2, one can derive the optimal linear transfer function (filter) K in the Fourier domain for a given capacity C: $K(f, w) = K(R, C)$.*[1] Since the power spectrum is the product of the spatial power $f^{-(m+1)}$ and the velocity distribution $P(w/f)$, for a given $w/f = v$, the transfer function K is the same in the form of K(fs-())$K\left(f_s^{-(m+1)}, C\right)$ — in which f_s is the spatial frequency f scaled by a factor of $P(v)^{-\frac{1}{m+1}}$. This explains why both the power spectrum and the visual sensitivity have the same scaling behavior.

[1]This is true when the spatial and temporal frequencies f and w are low. When f and w are high, the optical transfer function will have an impact (see Atick 1992 and VanHateren 1992).

Furthermore, for low spatial frequencies, $K \sim f_s^{\frac{m+1}{2}}$, such that $|K|^2 R$ is constant, i.e., the output is whitened or decorrelated. The amplitude of the derived K is shown in Figure 2 as solid lines to compare with human visual sensitivity. One immediately sees the good agreement between the theoretical prediction and the experimental data.

The derived K not only covers different spatial and temporal frequencies but also holds for different background mean light intensities (Figure 2, bottom) where the sensitivity K is inversely proportional to the mean light level. The maximization determines the spatiotemporal receptive field as well. The resulting receptive field is similar to the minimum phase solution (Dong and Atick 1995b), and is shown in Figure 3. This filter does decorrelate the natural visual input on average.

12.5 Statistics of the Visual Input on the Retina

So far we did not take into account a very important aspect of vision. Vision is an active process. In the natural environment of animals such as humans, eye movements interact with visual scenes and select images that arrive on the retina. It is very important to study the statistical properties of such images on the retina, even for the averaged or stationary properties; since after all, those images are the ones which constitute the visual input to the brain. Furthermore, there are two important non-stationary aspects of the visual input: first, the eye movement itself is a state variable, it is important to investigate the dependency of the input statistics on eye movements; second, different scenes have different intrinsic properties and interact differently with eye movements, it is important to investigate the dependency of the visual input statistics on scenes.

The interest in investigating the properties of such visual input to the brain dates back to the early days when eye movements were first recorded as human subjects explored various natural images (Yarbus 1967). But the images used then were static scenes. Natural images also change over time due to the movements of objects in a scene. Pursuits of moving objects as well as saccades between them are essential to the active visual process. It was shown in simulation that smooth tracking of objects could significantly alter the image velocity distributions on the fovea and the periphery (Eckert and Buchsbaum 1993). It was also argued that small eye movements between saccades (drifts and micro-saccades during pursuits and fixations) are essential for processing visual input when objects are otherwise stationary on the retina (Hubel 1988) and for generating input correlations necessary to the development of visual systems (Rucci et al. 2000). Yet partially due to technical difficulties, direct experimental measurements of the statistical properties of the natural visual input to the brain, i.e., the natural time-varying images on the retina during free-viewing, have been wanting.

Recently, the statistics of natural time-varying images were studied with an account for eye movements (Dong 2001b). The recorded eye positions during free

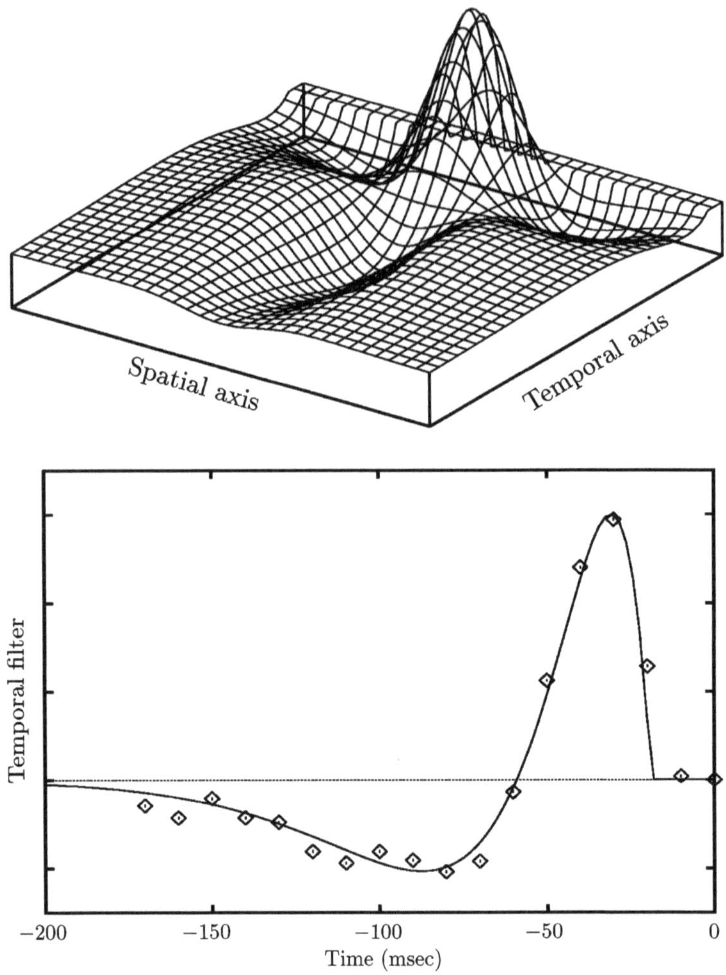

Fig. 12.3 **Top:** the predicted spatiotemporal receptive field. **Bottom:** the temporal filter $K(t)$. The data (diamond symbol) are calculated through the reverse correlation of the spike train and the stimuli at the center of the spatial receptive field of a cat LGN cell (Cai et al. 1997). The solid line is the theoretical prediction, sliced through the spatial center of the spatiotemporal receptive field on the top (note: there is 20 msec time delay from visual stimuli to the LGN)

viewing were used to derive the images on the retina and to serve as state variables. For several different visual scenes, each of which was a large segment of natural time-varying images, the velocity distribution and the spatiotemporal correlation function of the images on the retina were measured. It confirmed the earlier result that the correlation function exhibits a scaling behavior in space and time, which is related to the spatial correlation function and the velocity distribution. Based on this scaling behavior, a spatially decorrelated representation (called *contrast signal* in this paper, see Appendix) was derived, which completely separates the effects

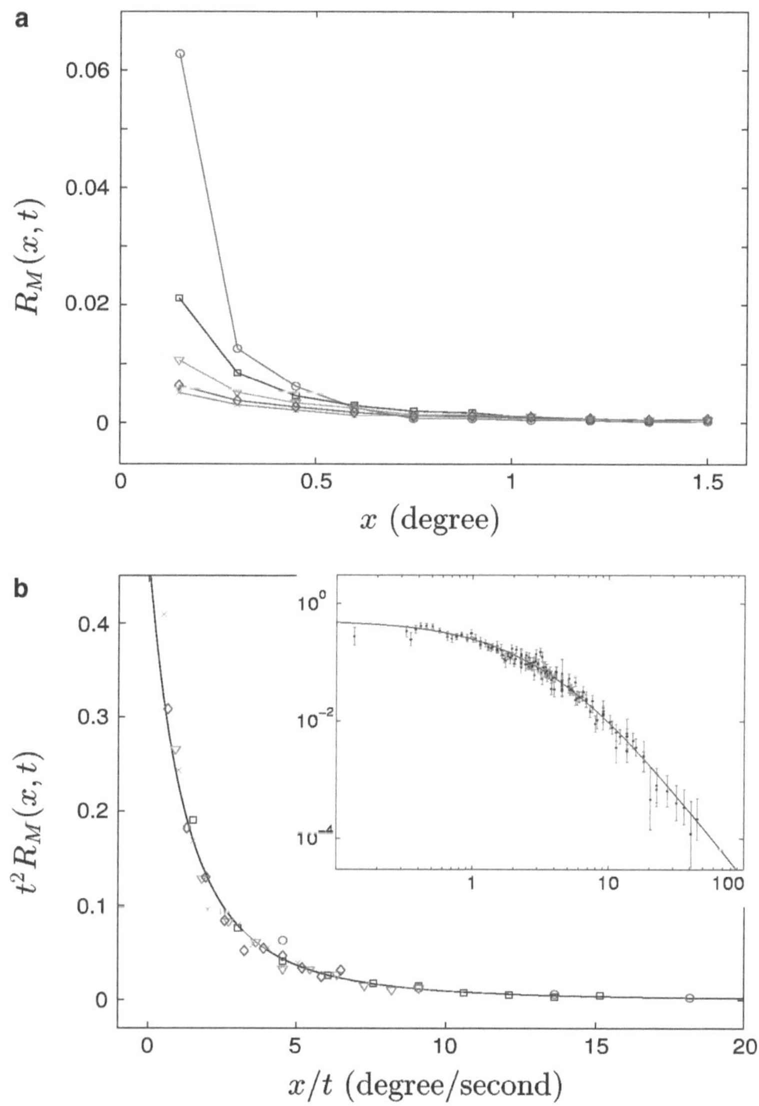

Fig. 12.4 Measured spatiotemporal correlation of the contrast signal $R_M(x,t)$. (**a**) the curves in circle, square, triangle, diamond, and cross, for $t = 33, 100, 167, 233$, and 300 msec, respectively. (**b**) all the data points in (A) re-plotted as a function of x/t after multiplication by t^2. Inset: $t^2 R_M(x,t)$ plotted in the log-log scale with standard error bars for all data points of $x < 1.5$ deg and $t < 300$ msec. Also plotted in (B) and the inset is the curve $P(x/t)$, while $P(v)$ is the best fit curve for the velocity distribution (*see Color Plates*)

of the spatial correlation function and the velocity distribution (see Fig. 4) and is particularly useful for studying non-stationary statistics. It was shown that both visual scenes and eye movements give rise to non-stationary statistics.

In the following sections we will summarize the non-stationary statistics and relate those to the non-stationary visual processing.

12.6 Non-Stationary Velocity Distribution on the Retina

During pursuits and fixations, the optical flow velocity distribution is different for different retinal eccentricities and for different visual scenes. The average optical flow velocity is lower on the fovea than on the periphery. On the fovea, more dynamic scenes have higher averages of optic flow velocity.

Figure 5 (top) shows the optical flow velocity distributions for two different scenes. Although both scenes have power law tails for high velocities, the slopes in the log-log plot are different and the average velocities are different. The more dynamic scene (picking fruits in a forest) has a longer tail for high velocities and lower probabilities at low velocities than the less dynamic scene (watching birds on a river bank). The average velocities are different by a factor more than four. The visual system has to be able to deal with a wide range of retinal image velocities due to the non-stationary velocity distribution of visual scenes.

At the bottom of Figure 5, the distributions of the average velocity for five scenes are shown for the fovea (the horizontal axis) and for 3 degrees away from the fovea (the vertical axis). This shows one more aspect of the non-stationary velocity distribution: the difference due to retinal eccentricity. For all five scenes, the average optical flow velocity is higher for more peripheral locations, indicated by the data points clustering above the diagonal line. This is a direct confirmation that the efficient coding requires that the visual sensitivity to higher velocities increases with increasing retinal eccentricities (Eckert and Buchsbaum 1993).

12.7 Non-Stationary Correlations of the Contrast Signal

The eye movements have two effects: the spatiotemporal correlation is greatly reduced by saccadic eye movements and the velocity distribution between saccades is more concentrated in the regime of lower velocities. These findings reveal some important aspects of how information is processed by the sensory systems with the active participation of the motor systems.

We use $R_1(x,t)$ to denote the correlations of two contrast signals with one saccade in between and $R_0(x, t)$ to denote the correlations of two contrast signals without any saccade in between. Figure 6 shows those two correlation functions (scaled by t^2 as before) of the contrast signal. From Figure 6, it is clear that

$$R_0(x,t) >> R_1(x, t) \quad \text{and} \quad R_1(x, t) \sim 0 \tag{1}$$

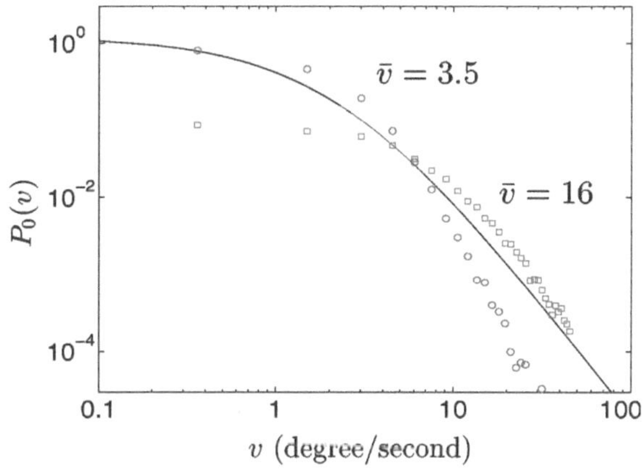

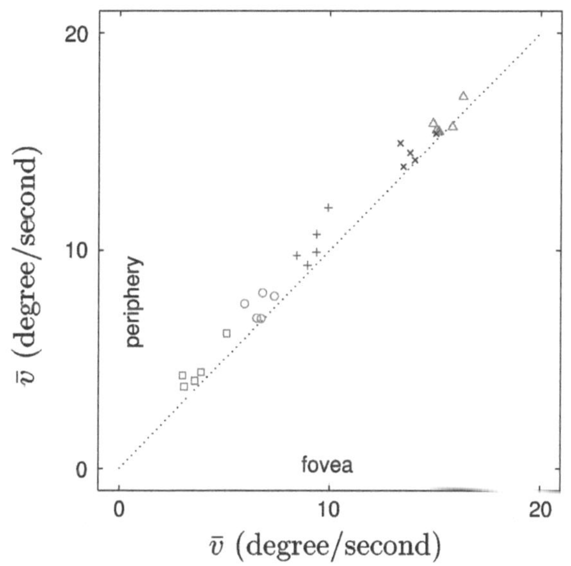

Fig. 12.5 **Top:** the measured probability density distribution of the retinal image velocity $P_0(v)$ for two different scenes (triangle: watching birds on a river bank, circle: picking fruits in a forest). Also plotted is $P_0(v) = \frac{v_0}{(v+v_0)^3}$ curve with the best fit v_0 = 2.3 deg/sec for the velocity distribution averaged over five scenes. Bottom: the averages of the retinal image velocity $\bar{v}$ of the fovea (horizontal) versus the periphery (vertical). Different symbols are for different scenes and different data points of the same symbol are for different observers. The average retinal image velocity at 3 degrees away from the fovea (the vertical axis) is clearly higher than on the fovea (the horizontal axis) for all scenes (*see Color Plates*)

for the range of x/t from zero to several degrees per second. Within this range, the contrast signals at different times are uncorrelated if there is a saccade occurring in between, but correlated if there is none.

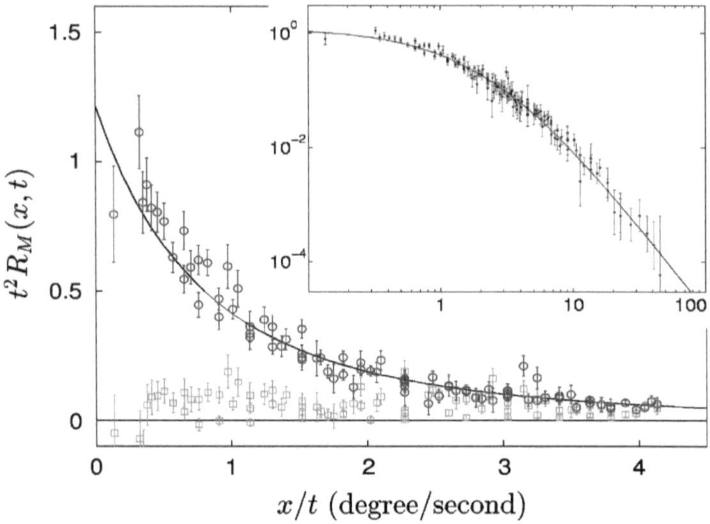

Fig. 12.6 Measured spatiotemporal correlation functions R_1 (with saccades) and R_0 (without saccades). The scaled correlation functions $t^2 R_1(x,t)$ and $t^2R_0(x,t)$ in square and circle, respectively, are shown for a small x/t range. Inset: $t^2R_0(x,t)$ is shown in the log-log plot with standard error bars for all data points of $x < 1.5$ deg and $t < 300$ msec. Also plotted in both is the curve $P_0(x/t)$, where $P_0(v)$ is the best fit curve for the velocity distribution without any saccade in between (*see Color Plates*)

The other important effect of eye movements is the concentration of $R_0(x,t)$ toward the region of lower x/t. This is shown in Figure 6 (inset) in a log-log plot, in comparison to Figure 4B (inset). Also plotted in both Figure 6 and its inset is the $P_0(x/t)$ curve, which has the same shape as $P(x/t)$ in Figure 4B with the asymptotic power law of an exponent 3, but has a smaller average velocity and thus a higher peak around zero.

It has been proposed that the first stage of visual processing (retina) removes the spatial correlation and that the retinal output is the contrast signal (Atick and Redlich 1990, 1992). It is clear that saccades remove the spatiotemporal correlations of the contrast signal. However, the contrast signal still has spatiotemporal correlations between saccades. It has been proposed that the remaining correlations are removed by the second stage of visual processing (LGN) (Dong and Atick 1995b). Dan et al. (1996) directly verified the decorrelation theory of the LGN. They stimulated paralyzed and anesthetized cats with natural movies. However, since the correlations are non-stationary for awake animals, the LGN processing also has to be non-stationary to maintain decorrelation and efficient coding (Dong 2001b; Truccolo and Dong 2001). Again, this can be derived from maximizing causal information but once more again, let's take a look at the experimental facts first.

12.8 Eye Movements and Saccadic Modulation in the LGN

Lee and Malpeli (1998) have recently studied the excitability of LGN neurons as the cat made saccadic eye movements under light and dark conditions. They found that a period of suppression began about 200 msec prior to the saccade, peaked about 100 msec prior to the saccade onset and quickly reversed, so that there was a peak of facilitation about 100 msec after the saccade end. As this function persisted in the dark, oculomotor signals must have manipulated LGN gain. Potential sources of these eye movement signals include sensory signals from muscle afferents in the brainstem (Donaldson and Dixon 1980; Lal and Friedlander 1990ab), cortico-geniculate input (Swadlow and Weyand 1987), and pretectal influence (Schmidt 1996). Other reports also indicate that eye movement signals, via corollary discharge or muscle afferents, reach the LGN (see the review by Buisseret 1995).

More than one experiment showed that excitability around the time of eye movement changes because extra-retinal influences appear to alter the gain of retino-geniculate processing during the peri-saccadic interval (Guido and Weyand 1995; Lee and Malpeli 1998; Ramcharan et al. 2001; Reppas et al. 2002). Such phenomenon contributes not only to saccadic suppression, an attenuation of the ability to detect certain stimuli before and during the time of eye movements, but also to saccadic disinhibition, an increased excitability of the LGN after the end of a saccade. Functionally, such suppression would minimize the "blurring" effects caused by the retina rapidly sweeping by a textured environment and such disinhibition would allow the free flow of visual information at the new gaze location (Singer 1977).

Guido and Weyand (1995), Lee and Malpeli (1998) and Ramcharan et al. (2001) have noted a relatively increased probability of high frequency "bursting" following saccadic eye movements. Since these bursts are linked to the low-threshold calcium conductance, a period of hyperpolarization must precede the burst. Our recent theoretical study showed that the hyperpolarization — even with a small amount of burst spikes — can change the temporal filtering of the LGN to help improve efficiency (Truccolo and Dong 2001). From the efficient coding point of view, a change of temporal filtering properties according to eye movement can help the visual system to get more information (see Section 9).

12.9 Maximizing Causal Information of Non-Stationary Statistics

Although in Section 4 we apply the efficient coding theory to stationary statistics, there is no reason to believe that the theory should not be applicable to non-stationary statistics. In fact, the visual system has short-term adaptations, on the order of tens of seconds, which acts to reduce dependencies between neurons (Atick et al. 1993; Dong 1994; Webster and Miyahara 1997; Carandini and Ferster 1998; Dragoi et al. 2000). Furthermore, individual neurons adapt to changes in visual input statistics on very short time scales, on the order to seconds (Smirnakis et al. 1997; Muller et al. 1999; Brenner et al. 2000).

If the statistical properties of the visual input for animals are non-stationary, a dynamic coding is needed for the decorrelation. A dynamic coding — receptive field — adapting to non-stationary statistics, such as those generated by the scene changes and the eye movements, will be predicted from the efficient coding theory, which has been recently formulated for such fast changing, non-stationary input statistics (Dong 2001b). In this paper, the general efficient coding principles are applied to make predictions about the specific dynamic neural processing related to saccadic eye movements (but see also Section 4 for different mean light levels).

To demonstrate the method in the non-stationary case, we will derive the non-stationary optimal linear transfer function (filter), which takes into account the timing of the eye movement. Since the eye-movement information is assumed to be available to LGN cells, the relative timing τ to eye movements is taken into K, specifically, the temporal filter will be different depending on the time to the nearest saccade (note: long before saccade or long after saccade, the temporal filter will be the same). For the demonstration purpose, we also make a simplified assumption that the LGN only does temporal filtering of its input from the retina. So our LGN temporal filter $K = K(t, \tau)$ relates the LGN input $S(t)$ to the LGN output $O(\tau)$ by

$$O(\tau_j) = \sum_{t_i \le \tau_j} K(t_i, \tau_j) S(t_i) \tag{2}$$

in which the summation is over t for $t \le \tau$ (the causality). It is important to understand the meaning of $K(t, \tau)$: it is the temporal filter of the LGN at time τ relative to the nearest saccade (note: $\tau < 0$ is before saccade, $\tau > 0$ after saccade). Because of the causality, $K(t, \tau) = 0$ for any $t > \tau$.

A solution is shown in Figure 7 (top). The introduction of reference time τ can achieve better coding on the time scale of intervals between saccades, which is several hundred milliseconds on average. This makes the predicted temporal filter very dynamic. During and right after a saccade, the predicted receptive-field behaves as a temporal low-pass filter; whereas between two saccades, the predicted receptive-field behaves as a temporal difference (band-pass) filter. The predicted filter during saccades has a smaller response to a stimulus at certain midrange temporal frequencies, as expected from saccadic suppression. However, it has bigger responses for stimuli at relatively low and relatively high temporal frequencies. As a result, the LGN output decorrelation is maintained all the time (Figure 7, bottom).

12.10 Concluding Remarks

The paper/chapter shows that the spatiotemporal statistics of the visual input to the brain arise from the coupling of two separate sources: the spatial statistics and the retinal image motion statistics. This indicates that the visual world is spatial scale invariant and objects at different distances move with certain velocity distributions.

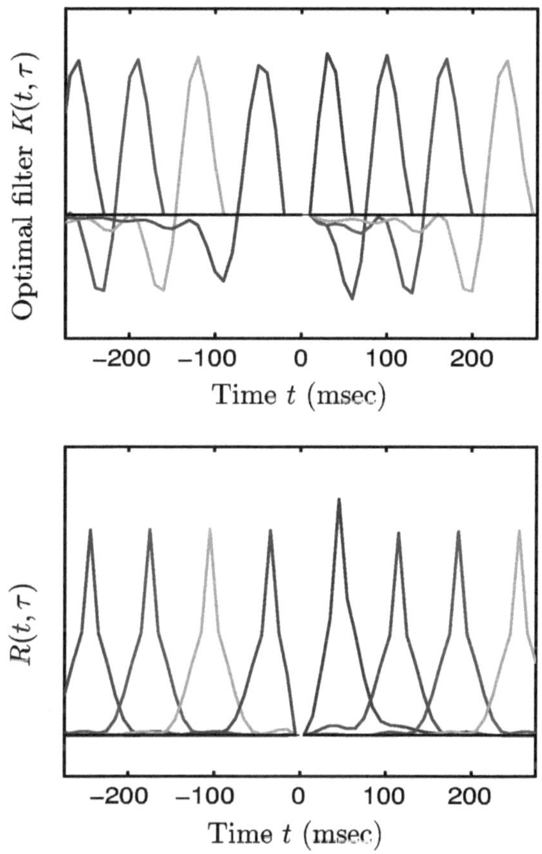

Fig. 12.7 The predicted optimal temporal filter $K(t,\tau)$ (top) and the corresponding output correlation $R(t,\tau)$ (bottom). The filter $K(t,\tau)$ at eight different τ are plotted, in which τ is the time relative to the saccade ($\tau = -245, -175, -105, -35, 35, 105, 175, 245$ msec from left to right). It is clear that the optimal temporal filter of the LGN depends on the time relative to saccades. The output correlations $R(t,\tau)$ are close to zero except near $t = \tau$ (*see Color Plates*)

Furthermore, the paper shows that the statistics are non-stationary and there are two major non-stationary effects, eye movements and visual scenes, both significantly affecting the retinal image motion statistics. This paper proposes that a fundamental principle of sensory coding is to maximize the causal information. Consequently, the visual processing is shaped according to the statistics of natural scenes in motion. These findings reveal how information is processed by the visual-sensory system with the active participation of the ocular-motor system, in an "optimal" way by maximizing causal information.

The main effect of the observed saccadic eye movements is to bring different parts of a visual scene into the fovea for processing. The fovea has the highest density of cones — the photoreceptors which operate under normal daylight conditions. The size of human fovea is 5.2 deg in diameter (1.7 deg for rod-free fovea) (Wandell 1995).

The average saccade amplitude during the natural viewing is 2.7 ± 1.7 deg — the right amount for bringing one part of a visual scene from right outside the fovea into the fovea. As shown in Figure 6, those saccades greatly reduce the spatiotemporal correlation of the visual input in the space-time range where the human fovea has the highest sensitivity: the optical flow velocity or x/t from 0.2 to 3 deg/sec and the spatial separation x from 0.1 to 0.5 deg (i.e. the spatial frequency from 5 to 1 cycle/deg) (Kelly 1979). Thus the saccades help to reduce the redundancy of the input, i.e., to increase the information flow through the fovea, during the natural viewing.

To code the natural visual input efficiently, the visual system needs to distinguish between the spatiotemporal changes generated by the motion of objects in a scene and the spatiotemporal changes generated by the motion of the observer — in particular saccadic eye movements. Since saccadic movements are predictable to the observer and do not carry information about the scene, visual responses containing information about saccadic movements are not efficient. Retinal responses generated by the model shown in Section 4 or the more realistic model using a spatio-temporal retinal filter derived from various experiments (Wehmeier et al. 1989) inevitably contain information about saccades and thus are less efficient than the proposed LGN model shown in Section 9. The latter changes according to saccades in order to be more efficient and thus contains less information about saccades. Recent experiments support this prediction (Dastjerdi et al. 2003; Dastjerdi 2007).

The other eye movements between saccades (drifts and micro-saccades during pursuits and fixations) do generate the retinal image statistics, especially the velocity distributions, which are suitable for the visual system to process. It is well known that images stabilized on the retina fade away after about a second (Riggs and Ratliff 1952; Ditchburn and Ginsborg 1952). It was hypothesized that eye movements between saccades help to change the retinal image characteristics to match the properties of the neurons in the early visual pathway to facilitate information processing, Hubel (1988). The measurements in this paper show that eye movements between saccades do increase (comparing Figure 4B and Figure 6) the input signal in the velocity range where the human fovea has the highest sensitivity: the retinal image velocity or x/t from 0.2 to 3 deg/sec (Kelly 1979). However, the measurements also show that for different scenes, the retinal image statistics are different, which leave the possibility that receptive fields dynamically adapt to different scenes to process the visual input better.

It is conceivable that with the theoretical consideration outlined in this chapter/paper, carefully designed experiments can improve our understanding of sensory systems enormously by using natural stimuli and taking into account the non-stationary aspects. After all, the dynamic nature of the brain makes it more powerful in information processing in the natural environment. In particular, for the visual system, natural time-varying images should be used to probe visual responses, and the eye movements of awake/roused animals should be recorded not only to register images on the retina but also to be counted as state variables. Doing so will truly help understand the dynamics of visual motion processing.

12.11 Appendix

12.11.1 Calculation of Spatiotemporal Correlation

In this paper/chapter, the following conventions are used: given light intensity $S(\mathbf{x}', t')$, the correlation between two points separated by spatiotemporal distance $\mathbf{x}$, t is

$$R(\mathbf{x},t,t') = \langle S(\mathbf{x}+\mathbf{x}',t+t')S(\mathbf{x}',t')\rangle_{\mathbf{x}'} \tag{3}$$

in which the $\langle\rangle_{\mathbf{x}'}$ denotes the average over the spatial location $\mathbf{x}'$. For an ergodic system, the expectation value of $S(\mathbf{x} + \mathbf{x}', t + t')S(\mathbf{x}', t')$ is independent of $\mathbf{x}'$, t', but in general, it is a function of the retinal eccentricity and relative timing to the saccades (see "Classification of eye movements"). In this paper/chapter, the focus is on the statistics near the fovea, so it is averaged over the $4^o \times 4^o$ region around the center of gaze.
When the dependence on the absolute timing is ignored

$$R(\mathbf{x},t) = \langle\langle S(\mathbf{x}+\mathbf{x}',t+t')S(\mathbf{x}',t')\rangle_{\mathbf{x}'}t\rangle \tag{4}$$

in which the second $\langle\rangle_{t'}$ is the average over t'. This equation is used to measure the average or stationary statistics.

To understand the effect of saccades on the retinal image statistics, the correlation is calculated around the saccades. In this case, the time t' of the signal $S(x', t')$ is measured by the advance time τ to the onset of next saccade at time s. So the time $t' = s - \tau$. Ignoring the dependence on the absolute timing but accounting for the dependence on the relative timing to saccade,

$$R(\mathbf{x},t,\tau) = \langle\langle S(\mathbf{x}+\mathbf{x}',t+t')S(\mathbf{x}',t')\rangle_{\mathbf{x}'}\rangle_{s'} \tag{5}$$

in which the second $\langle\rangle_s$ is the average over all saccades. This equation is used to measure the non-stationary statistics.

The light intensity $S(\mathbf{x}, t)$ in all the equations is mean-removed and normalized, such that the $R(0, 0) = 1$.[2] For the purposes of this paper/chapter the dependence on spatial orientation is ignored and all calculations are averaged over all orientations, thus only one spatial dimension x — the radial distance — is shown in all the figures. All the calculations are also averaged over viewers and scenes, unless mentioned otherwise. Also used in all calculations are the natural units of the measurements: 1 spatial unit is 0.15 deg, 1 temporal unit is 33 msec, and 1 velocity unit is 4.5 deg/sec. But for the convenience of the readers, the units of degree and second are used in all illustrations.

[2]The illumination does not affect the results in a wide range. see Conclusion and Discussion.

12.11. 2 *Calculation of Contrast Signal*

To further reveal the nature of the spatiotemporal correlation of the visual input, the retinal input is spatially whitened using the localized whitening filters (Atick and Redlich 1992; Atick et al. 1993; Penev and Atick 1996) calculated from the measured spatial correlation function as the following.

The spatial correlation function is calculated first by

$$R_s(\mathbf{x},\mathbf{x}') = \langle S(\mathbf{x},t)S(\mathbf{x}',t')\rangle t \tag{6}$$

in which the $\langle\rangle_t$ is the average over time. Obviously it is positive definite, so all the eigenfunctions $\{E_i(\mathbf{x})\}$ are real and all the eigenvalues $\{\lambda_i\}$ are positive. The localized whitening filters are

$$K_s(\mathbf{x},\mathbf{x}') = \sum_i E_i(\mathbf{x})\lambda_i^{-\frac{1}{2}} E_i(\mathbf{x}') \tag{7}$$

in which the summation is over all the eigenfunctions.

Those whitening filters are approximately shift invariant within two degrees of the gaze, and simulate the spatial process of the retina — of course, the real retina does more than spatial whitening (see references in Dong and Atick 1995b; Sterling 1998; Victor 1999). The filter output is

$$S_M(x,t) = \iint K_s(\mathbf{x},\mathbf{x}')S(\mathbf{x}',t)\,\mathrm{d}\mathbf{x}' \tag{8}$$

This is the spatially whitened visual input and is called the contrast signal in this paper/chapter. The spatiotemporal correlation of the contrast signal is calculated in the same way as of the light intensity signal in the previous subsection. To make a distinction, a subscript "*M*" is used for the contrast signal, e.g., $R_M(\mathbf{x}, t)$.

11. 3 *Quantification of retinal image motion*

The retinal image motion is characterized by the optical flow velocity of the visual input. It is another important statistical property of the visual input — as shown in the chapter, it is an integral part of the spatiotemporal correlation and the main contributor to the non-stationary aspects.

The velocity is determined by calculating the optical flow fields between two consecutive frames using the following method in two steps. For a small area ($1^o \times 1^o$) of image at the center of gaze in one frame, first find its translated one in the next frame, i.e., get the large movement in number of pixels (d_x, d_y), through minimizing the square-difference,

$$\left\langle (S(x+d_x, y+d_y, t+1) - S(x,y,t))^2 \right\rangle_{xy} \tag{9}$$

in which $\langle\rangle_{xy}$ is the average over the small area. Second, calculate the sub-pixel small movements (δ_x, δ_y) by minimizing another square-difference,

$$\left\langle\left(\delta_x \frac{\partial S}{\partial x} + \delta_y \frac{\partial S}{\partial y} - \frac{\partial S}{\partial t}\right)^2\right\rangle_{xy} \tag{10}$$

in which $\langle\rangle_{xy}$ is the average over the same area.[3] The overall velocity **v** has $v_x = d_x + \delta_x$ and $v_y = d_y + \delta_y$ for horizontal and vertical velocities, respectively. Since the light intensity $S(x, y, t)$ on the retina is used for the calculation, the velocity **v** is the velocity of the retinal image motion, i.e., the optical flow velocity on the retina — or an operational definition of it.

Assuming that the pixel values and their partial derivatives are not degenerate, $v_x = d_x + \delta_x$ and $v_y = d_y + \delta_y$ are well determined. For rigid-body image motion, they give the optical flow velocity. For non-rigid body motion, they give a definition of an "average" optical flow velocity. The dimension of the averaging area is chosen to be 1°, similar to the average spatial separation in calculating the correlation function. For $1^o \times 1^o$ area (400 pixels), it is very unlikely that the degenerate situation ever happens for natural time-varying images. It is straightforward to find the minimum error solutions (for similar methods, see Cafforio and Rocca 1975; Jain and Jain 1981; Horn and Schunk 1981).

But here is a note of caution. Although one never fails to find a minimum, it doesn't mean that there is no inaccuracy in calculating the optical flow velocity, e.g., in calculating the velocity component parallel to a high contrast edge. For non-rigid body motions, the "average" optical flow velocity obviously depends on the area size and location. What is used here is the average on the central part of the fovea. Furthermore, the optical flow velocity calculated through Equation (5) and (6) is by no means the same as the projected velocity of self-motion (Koenderink and VanDoorn 1987) or scene-motion (Verri and Poggio 1989).

12.11.4 *Classification of Eye Movements*

The information processing by the brain is dynamic, i.e., the brain actively seeks information. In the human visual system, this active and dynamic process is characterized by periods of fixations and pursuits interleaved by saccades, which change the center of gaze from one direction to another in a very short period of time, and thus change the visual input significantly. Therefore it is very interesting to quantitatively compare between saccades and other eye movements for their effects on the visual input.

---

[3]The three partial derivatives are estimated by 1/2(S(x+1, y, t)–S(x - 1, y,t)), 1/2(S(x, y + 1, t) - S(x, y – 1, t)), and $(S(x + d_x, y + d_y, t + 1) - S(x, y, t))$, respectively.

In this paper/chapter, saccade is defined as an eye movement of more than 0.5 deg within 20 msec. From the measurements during a typical free viewing of one of the natural scene movies, the average saccade amplitude is ~ 2.7 ± 1.7 deg, the average saccade duration is ~ 28 ± 8 msec, the average saccade speed is ~ 88 ± 29 deg/sec, and the average inter-saccade interval is ~ 490 ± 470 msec. For another subject viewing another movie, the corresponding numbers are 2.4 ± 1.5 deg, 26 ± 8 msec, 87 ± 27 deg/sec, and 520 ± 400 msec.

Acknowledgments The author wishes to thank Dr. Theodore Weyand for many interesting discussions about the LGN function and to thank Dr. Anna Kashina for her critical reading of the manuscript. This work was supported in part by FAU under the grant No. RIA-25, by NIMH under the grant No. MH019116, and by NSF under the grant No. PHY99-07949.

References

1.. Atick JJ (1992) Could information theory provide an ecological theory of sensory processing? *Network-Compu. Neural.* **3**, 213–251.
2. Atick JJ, Redlich AN (1990) Towards a theory of early visual processing. *Neural Comp.* **2**, 308–320.
3. Atick JJ, Redlich AN (1992) What does the retina know about natural scenes? *Neural Comp.* **4**, 196–210.
4. Atick JJ, Li Z, Redlich AN (1993) What does post-adaptation color appearance reveal about cortical color representation. *Vision Res.* **33**, 123–129.
5. Attneave F (1954) Some informational aspects of visual perception. *Psychol Rev* **61**, 183–193.
6. Barlow HB (1961) Possible principles underlying the transformation of sensory messages. *In: Sensory Communication* (Rosenblith WA, ed, MIT Press, Cambridge) 217–234.
7. Brenner N, Bialek W, VanSteveninck RR (2000) Adaptive Rescaling Maximizes Information Transmission. *Neuron* **26**, 695–702.
8. Buisseret P (1995) Influence of extraocular-muscle proprioception on vision. *Physiological Reviews* **75**, 323–338.
9. Burton GJ, Moorhead IR (1987) Color and spatial structure in natural scenes. *Applied Optics* **26**, 157–170.
10. Cafforio C, Rocca F (1975) Methods for measuring small displacements of television images. *IEEE Info. Theo.* **22**, 573–579.
11. Cai DQ, DeAngelis GC, Freeman RD (1997) Spatiotemporal receptive-field organization in the lateral geniculate-nucleus of cats and kittens. *J of Neurophysiol* **78**, 1045–1061.
12. Carandini M, Ferster D (1998) A tonic hyperpolarization underlying contrast adaptation in cat visual cortex. *Science* **276**, 949–952.
13. Dan Y, Atick JJ, Reid RC (1996) Efficient coding of natural scenes in the lateral geniculate nucleus: Experimental test of a computational theory. *J Neurosci* **16**, 3351–3362.
14. Dastjerdi M (2007) Efficient representation of natural visual input in the thalamus. *Ph D thesis, University Microfilms International, Ann Arbor, MI.*
15. Dastjerdi M, Weyand TG, Dong DW (2003) The spatiotemoral receptive field (STRF) properties of the lateral geniculate nucleus (LGN) in the awake cats during free-viewing natural time-varying images. *Society for Neuroscience Abstract* **29**, 229.3.
16. Ditchburn RW, Ginsborg BL (1952) Vision with a stabilized retinal image. *Nature* **170**, 36–37.
17. Donaldson IML, Dixon RA (1980) Excitation of units in the lateral geniculate and contiguous nuclei of the cat by stretch of extrinsic ocular muscles. *Exp. Brain Res.* **38**, 245–255.

18. Dong DW (1994) Associative decorrelation dynamics: a theory of self-organization and optimization in feedback networks. *In: Advances in Neural Information Processing Systems 7* (Tesauro G, Touretzky DS, Leen TK, eds, MIT Press, Cambridge, MA), 925–932.
19. Dong DW (2001a) Spatiotemporal inseparability of natural images and visual sensitivities. *In: Computational, neural & ecological constraints of visual motion processing* (Zanker JM, Zeil J, eds, Springer Verlag, Berlin Heidelberg New York), 371–380.
20. Dong DW (2001b) Effects of eye movements on information processing: Visual input statistics during free-viewing natural time-varying images. *Invest ophth & Vis Sci* **42**, S3347.
21. Dong DW, Atick JJ (1995a) Statistics of natural time-varying images. *Network-Compu. Neural.* **6**, 345–358.
22. Dong DW, Atick JJ (1995b) Temporal decorrelation: a theory of lagged and nonlagged responses in the lateral geniculate nucleus. *Network-Compu. Neural.* **6**, 159–178.
23. Dragoi V, Sharma J, Sur M (2000) Adaptation-induced plasticity of orientation tuning in adult visual cortex. *Neuron* **28**, 287–298.
24. Eckert MP, Buchsbaum G (1993) Efficient coding of natural time varying images in the early visual system. *Phil. Trans. R. Soc. Lond. B* **339**, 385–395.
25. Field DJ (1987) Relations between the statistics of natural images and the response properties of cortical cells. *J. Opt. Soc. Am. A* **4**, 2379–2394.
26. Field DJ (1994) What is the goal of sensory coding. *Neural Computation* **6**, 559–601.
27. Guido W, Weyand TG (1995) Burst responses in thalamic relay cells of the awake, behaving cat. *J. Neurophysiol.* **74**, 1782–1786.
28. Horn BKP, Schunk BG (1981) Determining optical flow. *Artificial Intelligence* **17**, 185–203.
29. Hoyer PO, Hyvarinen A (2000) Independent component analysis applied to feature extraction from colour and stereo images. *Network-Compu. Neural.* **3**, 191–210.
30. Hubel DH (1988) Eye, Brain, and Vision. (Freeman and Company, New York).
31. Jain JR, Jain AK (1981) Displacement measurement and its application in interframe image coding. *IEEE T. Commun.* **29**, 1799–1808.
32. Kelly DH (1961) Visual responses to time-dependent stimuli .1. amplitude sensitivity measurements. *Journal of the Optical Society of America* **51**, 422–429.
33. Kelly DH (1979) Motion and vision. II. Stabilized spatio-temporal threshold surface. *J. Opt. Soc. Am.* **69**, 1340–1349.
34. Koenderink JJ, VanDoorn AJ (1987) Facts on Optical Flow. *Bio. Cybern.* **56**, 247–254.
35. Lee D, Malpeli JG (1998) Effects of saccades on the activity of neurons in the cat lateral geniculate nucleus. *J. Neurophysiol* **79**, 922–936.
36. Lal R, Friedlander MJ (1990a) Effect of eye position changes on retinogeniculate transmission in the cat. *J. Neurophysiol.* **63**, 502–522.
37. Lal R, Friedlander MJ (1990b) Effect of passive eye movement on retinogeniculate transmission in the cat. *J. Neurophysiol.* **63**, 523–538.
38. Li Z, Atick JJ (1994) Efficient stereo coding in the multiscale representation. *Network-Compu. Neural.* **5**, 1–18.
39. Linsker R (1988) Self-organization in a perceptual network. *Computer* **21**, 105–117.
40. Linsker R (1989) An application of the principle of maximum information preservation to linear systems. *In: Advances in Neural Information Processing Systems 1* (Touretzky DS, ed, Morgan Kaufman, San Mateo, CA), 186–194.
41. Muller JR, Metha AB, Krauskopf J, Lennie P (1999) Rapid Adaptation in Visual Cortex to the Structure of Images. *Science* **285**, 1405–1408.
42. Olshausen BA, Field DJ (1996) Emergence of simple-cell receptive-field properties by learning a sparse code for natural images. *Nature* **381**, 607–609.
43. Parraga CA, Brelstaff G, Troscianko T (1998) Color and luminance information in natural scenes. *J Opt Soc Am* **15**, 563–569.
44. Penev PS, Atick JJ (1996) Local feature analysis: A general statistical theory for object representation. *Network-Compu. Neural.* **7**, 477–500.
45. Ramcharan EJ, Gnadt JW, Sherman SM (2001) The effects of saccadic eye movements on the activity of geniculate relay neurons in the monkey. *Visual Neuroscience* **18**, 253–258.

46. Reppas JB, Usrey WM, Reid RC (2002) Saccadic eye movements modulate visual responses in the lateral geniculate nucleus. *Neuron* **35**, 961–974.
47. Riggs LA, Ratliff F (1952) The effects of counteracting the normal movements of the eye. *J. Opt. Soc. Am.* **42**, 872–873.
48. Rucci M, Edelman GM, Wray J (2000) Modeling LGN responses during free-viewing: A possible role of microscopic eye movements in the refinement of cortical orientation selectivity. *J Neurosci* **20**, 4708–4720.
49. Ruderman DL, Bialek W (1994) Statistics of natural images: scaling in the woods. *Phy. Rev. Let.* **73(6)**, 814–817.
50. Ruderman DL, Cronin TW, Chiao CC (1998) Statistics of cone responses to natural images: implications for visual coding. *J Opt Soc Am* **15**, 2036–2045.
51. Schmidt M (1996) Neurons in the cat pretectum that project to the dorsal lateral geniculate nucleus are activated during saccades. *J. Neurophysiol.* **76**, 2907–2918.
52. Shannon, CE, Weaver W (1949) A mathematical theory of communication. *Univ. of Illinois Press, Urbana.*
53. Simoncelli EP, Olshausen BA (2001) Natural image statistics and neural representation. *Ann. Rev. Neurosci.* **24**, 1193–1216.
54. Singer W (1977) Control of thalamic transmission by corticofugal and ascending reticular pathways in the visual system. *Phys. Reviews* **57**, 386–420.
55. Smirnakis SM, Berry MJ, Warland DK, Bialek W, Meister M (1997) Adaptation of retinal processing to image contrast and spatial scale. *Nature* **386**, 69–73.
56. Srinivasan MV, Laughlin SB, Dubs A (1982) Predictive coding - a fresh view of inhibition in the retina. *Proc. R. Soc. B* **216**, 427–459.
57. Sterling P (1998) Retina. *In: The Synaptic Organization of the Brain* (Shepherd GM, ed, Oxford, New York) 205–253.
58. Swadlow HA, Weyand TG (1987) Corticogeniculate neurons, corticotectal neurons, and suspected interneurons in visual cortex of awake rabbits: Receptive field properties, axonal properties, and effects of EEG arousal. *J. Neurophysiol.* **57**, 977–1001.
59. Tailor DR, Finkel LH, Buchsbaum G (2000) Color-opponent receptive fields derived from independent component analysis of natural images. *Vision Research* **40**, 2671–2676.
60. Truccolo WA, Dong DW (2001) Dynamic temporal decorrelation: information theoretic and biophysical model of the functional role of lateral geniculate nucleus. *Neurocomputing* **38–40**, 993–1001.
61. VanHateren JH (1992) Theoretical predictions of spatiotemporal receptive fields of fly LMCs, and experimental validation. *J. Comp. Physiol. A* **171**, 157–170.
62. VanHateren JH, Ruderman DL (1998) Independent component analysis of natural image sequences yields spatio-temporal filters similar to simple cells in primary visual cortex. *P Roy Soc Lond B Bio* **265**, 2315–2320.
63. Verri A, Poggi T (1989) Motion Field and Optical Flow: qualitative properties. *IEEE T. Pat. Ana. Mac. Intel.* **11**, 490–498.
64. Victor JD (1999) Temporal aspects of neural coding in the retina and lateral geniculate. *Network-Comp. Neural.* **10**, R1–R66
65. Wandell BA (1995) Foundations of Vision (Sinauer, Sunderland).
66. Webster MA, Mollon JD (1997) Adaptation and the color statistics of natural images. *Vision Res.* **37**, 3283–3298.
67. Webster MA, Miyahara E (1997) Contrast adaptation and the spatial structure of natural images. *J Opt Soc Am A* **14**, 2355–2366.
68. Wehmeier U, Dong DW, Koch C, VanEssen DC (1989) Modeling the mammalian visual system. *In: Methods in Neuronal Modeling: from Synapses to Networks* (Koch C, Segev I, eds, MIT Press, Cambridge), 335–360.
69. Yarbus AL (1967) Eye Movements and Vision (Plenum, New York).
70. Zhaoping L (2002) Optimal Sensory Encoding. *In: The Handbook of Brain Theory and Neural Networks (Second Edition)* (Arbib MA, ed, MIT Press, Cambridge, MA), 815–819.

Chapter 13
Neural Models of Motion Integration, Segmentation, and Probabilistic Decision-Making

Stephen Grossberg

Abstract What brain mechanisms carry out motion integration and segmentation processes that compute unambiguous global motion percepts from ambiguous local motion signals? Consider, for example, a deer running at variable speeds behind forest cover. The forest cover is an occluder that creates apertures through which fragments of the deer's motion signals are intermittently experienced. The brain coherently groups these fragments into a trackable percept of the deer and its trajectory. Form and motion processes are needed to accomplish this using feedforward and feedback interactions both within and across cortical processing streams. All the cortical areas V1, V2, MT, and MST are involved in these interactions. Figure-ground processes in the form stream through V2, such as the separation of occluding boundaries of the forest cover from boundaries of the deer, select the motion signals which determine global object motion percepts in the motion stream through MT. Sparse, but unambiguous, feature tracking signals are amplified before they propagate across position and are integrated with far more numerous ambiguous motion signals. Figure-ground and integration processes together determine the global percept. A neural model predicts the processing stages that embody these form and motion interactions. Model concepts and data are summarized about motion grouping across apertures in response to a wide variety of displays, and probabilistic decision making in parietal cortex in response to random dot displays.

S. Grossberg (✉)
Department of Cognitive and Neural Systems, Center for Adaptive Systems, Center of Excellence for Learning in Education, Science and Technology, Boston University, Boston, MA, USA
e-mail: steve@cns.bu.edu

U.J. Ilg and G.S. Masson (eds.), *Dynamics of Visual Motion Processing: Neuronal, Behavioral, and Computational Approaches*,
DOI 10.1007/978-1-4419-0781-3_13, © Springer Science+Business Media, LLC 2010

13.1 Introduction: The Interdependence of Motion Integration and Segmentation

13.1.1 Aperture Problem and Feature Tracking Signals

Visual motion perception solves the two complementary problems of *motion integration* and of *motion segmentation.* The former joins nearby motion signals into a single object, while the latter keeps them separate as belonging to different objects. Wallach (1935; translated by Wuerger et al. 1996) first showed that the motion of a featureless line seen behind a circular aperture is perceptually ambiguous: No matter what may be the real direction of motion, the perceived direction is perpendicular to the orientation of the line. This phenomenon was called the *aperture problem* by Marr and Ullman (1981). The aperture problem is faced by any localized neural motion sensor, such as a neuron in the early visual pathway, which responds to a moving local contour through an aperture-like receptive field. Only when the contour within an aperture contains features, such as line terminators, object corners, or high contrast blobs or dots, can a local motion detector accurately measure the direction and velocity of motion (Shimojo et al. 1989).

These problems become most challenging when an object moves behind multiple occluders. Although the various object parts are then segmented by occluders, the visual system can often integrate these parts into a percept of coherent object motion that crosses the occluders. Studying conditions such as these under which the visual system can and cannot accomplish correct segmentation and integration provides important cues to the processes that are used by the visual system to create object motion percepts during normal viewing conditions.

To solve the interlinked problems of motion integration and segmentation, the visual system uses the relatively few unambiguous motion signals arising from image features, called *feature tracking* signals, to select the ambiguous motion signals that are consistent with them, while suppressing the more numerous ambiguous signals that are inconsistent with them. In addition, the visual system uses contextual interactions to compute a consistent motion direction and velocity from arrays of ambiguous motion signals when the scene does not include any unambiguous feature tracking signals. A particular challenge is to explain how motion percepts can change from ones of integration to segmentation in response to small changes in object or contextual cues.

This chapter summarizes efforts to develop a neural model of the cortical form and motion processes that clarify how such motion integration and segmentation processes occur (Fig. 13.1). This 3D FORMOTION model has been progressively developed over the years to explain and predict an ever-broadening set of data about motion perception; (e.g., Baloch and Grossberg 1997; Berzhanskaya et al. 2007; Baloch et al. 1998; Chey et al. 1997, 1998; Grossberg et al. 2001; Grossberg and Rudd 1989, 1992). Comparisons with related models are found in these archival articles.

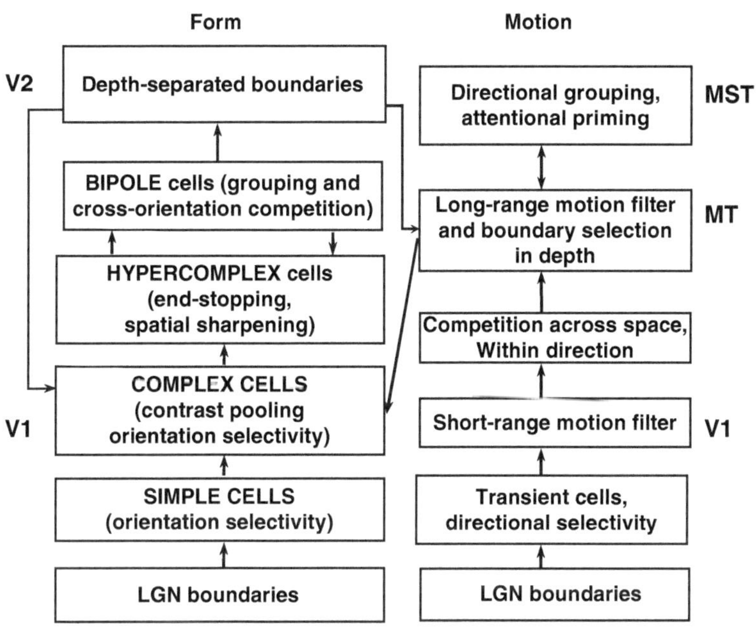

Fig. 13.1 The 3D FORMOTION model processing stages. See text for details [Reprinted with permission from Berzhanskaya et al. (2007).]

13.1.2 Neurophysiological Support for Predicted Aperture Problem Solution

In addition to model explanations of known data, the model has predicted data that were subsequently reported. In particular, Chey et al. (1997) explained how feature tracking estimates can gradually propagate across space to capture consistent motion directional signals, while suppressing inconsistent ones, in cortical area MT. Such motion capture was predicted to be a key step in solving the aperture problem. Pack and Born (2001) reported neurophysiological data that directly support this prediction. As simulated in the model, MT neurons initially respond primarily to the component of motion perpendicular to a contour's orientation, but over a period of approximately 60 ms the responses gradually shift to encode the true stimulus direction, regardless of orientation. Pack and Born also collected data which support the concept that motion signals are used for target tracking; that is, the initial velocity of pursuit eye movements deviates in a direction perpendicular to local contour orientation, suggesting that the earliest neural responses influence the oculomotor response.

Many psychophysical data also illustrate how feature tracking signals can propagate gradually across space to capture consistent ambiguous signals. Castet et al. (1993) described a particularly clear illustration of this. Figure 13.2a summarizes their data. They considered the horizontal motion of both a vertical and a tilted line that

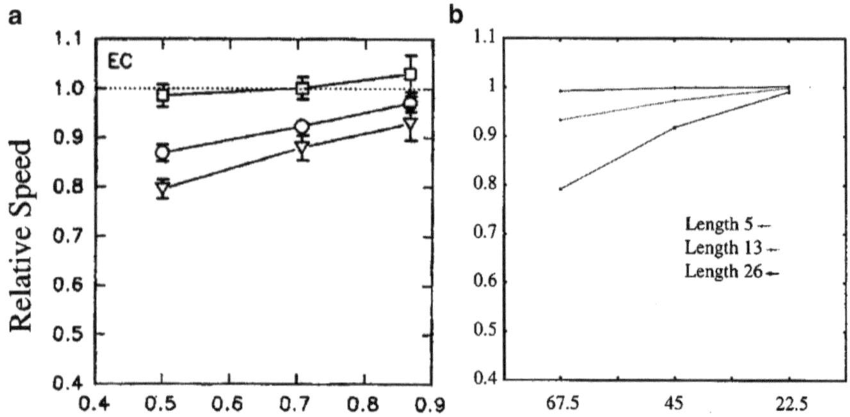

Fig. 13.2 Effects of line length and orientation on perceived speed of horizontally moving lines. Relative perceived speed for three different line orientations and lengths are shown as percentages of the perceived speed of a *vertical line* of the same length. (**a**) shows data from Castet et al. (p. 1925). Each data line corresponds to a different line length (0.21, 0.88, and 1.76°). The horizontal axis shows the ratio of the speed normal to the line's orientation relative to the actual translation speed. The three data points from left to right for each line length correspond to line angles of 60, 45, and 30° from vertical, respectively. The *horizontal dotted line* indicates a veridical speed perception; results below this *line* indicate a bias toward the perception of slower speeds. (**b**) shows simulation results, also for three lengths and orientations. In both cases perceived relative speed decreases with line length and angle from vertical. Simulated lines use slightly different orientations from those in the experiments so that the simulated input conforms to the Cartesian grid [Reprinted with permission from Chey et al. (1997).]

are moving at the same speed. Suppose that the unambiguous feature tracking signals at the line ends capture the ambiguous motion signals near the line middle. The preferred ambiguous motion direction and speed are normal to the line's orientation. In the case of the vertical line, the speed of the feature tracking signals at the line ends equals the preferred ambiguous speed near the line's middle. For the tilted line, however, the preferred ambiguous speed is less than the feature tracking speed. If the speed of the line is judged using a weighted average of feature signals and ambiguous signals, then the tilted line will be perceived to move slower than the vertical line.

To further test this idea, Castet et al. (1993) also showed that the ambiguous speeds have a greater effect as line length increases when the line is viewed for a brief duration. These additional data strongly support the idea that feature tracking signals at the line ends propagate inwards along the line to capture the ambiguous motion speed and direction there. As capture takes longer to complete when lines are longer, the ambiguous motion signals have a larger effect on longer lines. Chey et al. (1997) simulated these data, as shown in Fig. 13.2b.

In addition to simulating data of Castet et al. (1993) on how the perceived speeds of moving lines are affected by their length and angle, Chey et al. (1997) used similar model mechanisms to also simulate, among other percepts, how the barberpole illusion (Wallach 1976) is produced, how it can be affected by various configurational changes, and how plaid patterns move both coherently and incoherently. In addressing plaid pattern motion, the model provides explanations of when plaid patterns cohere or do not (Adelson and Movshon 1982; Kim and Wilson 1993; Lindsey and Todd 1996), how contrast affects the perceived speed and direction of moving plaids (Stone et al. 1990), and why the movement of so-called Type 2 patterns differs from those of Type 1 patterns (Ferrera and Wilson 1990, 1991; Yo and Wilson 1991). All of these data may be explained by an interaction of figure-ground separation mechanisms in the form system interacting with motion capture mechanisms in the motion stream.

13.1.3 Formotion Binding by Laminar Cortical Circuits

As the model name 3D FORMOTION suggests, it proposes how form and motion processes interact to generate coherent percepts of object motion in depth. Among the problems that the model analyses are the following form-motion (or *formotion*) binding issues, which go beyond the scope of other models: How do form-based 3D figure-ground separation mechanisms in cortical area V2 interact with directionally selective motion grouping mechanisms in cortical areas MT and MST to preferentially bind together some motion signals more easily than others? In cases where form-based figure-ground mechanisms are insufficient, how do motion and attentional cues from cortical area MT facilitate figure-ground separation within cortical area V2 via MT-to-VI-to-V2 feedback? These processes help to explain and simulate many motion data, including the way in which the global organization of the motion direction field in areas MT and MST can influence whether the percept of an object's form looks rigid or deformable through time.

The model also goes beyond earlier motion models by proposing how *laminar* cortical circuits realize these mechanisms (Fig. 13.3) These laminar circuits embody explicit predictions about the functional roles that are played by identified cells in the brain. The 3D FORMOTION model extends to the motion system laminar models of cortical circuits that have previously explained challenging perceptual and brain data about 3D form perception in cortical areas V1, V2, and V4 (e.g., Cao and Grossberg 2005; Grossberg 1999, 2003; Grossberg and Raizada 2000; Grossberg and Seitz 2003; Grossberg and Swaminathan 2004; Grossberg and Williamson 2001; Grossberg and Yazdanbaksh 2005), as well as about cognitive working memory, sequence learning, and variable-rate sequential performance (Grossberg and Pearson 2008).

13.1.4 Intrinsic and Extrinsic Terminators

A key issue concerns the assignment of motion to an object boundary when it moves relative to an occluder. How does the brain prevent motion integration across

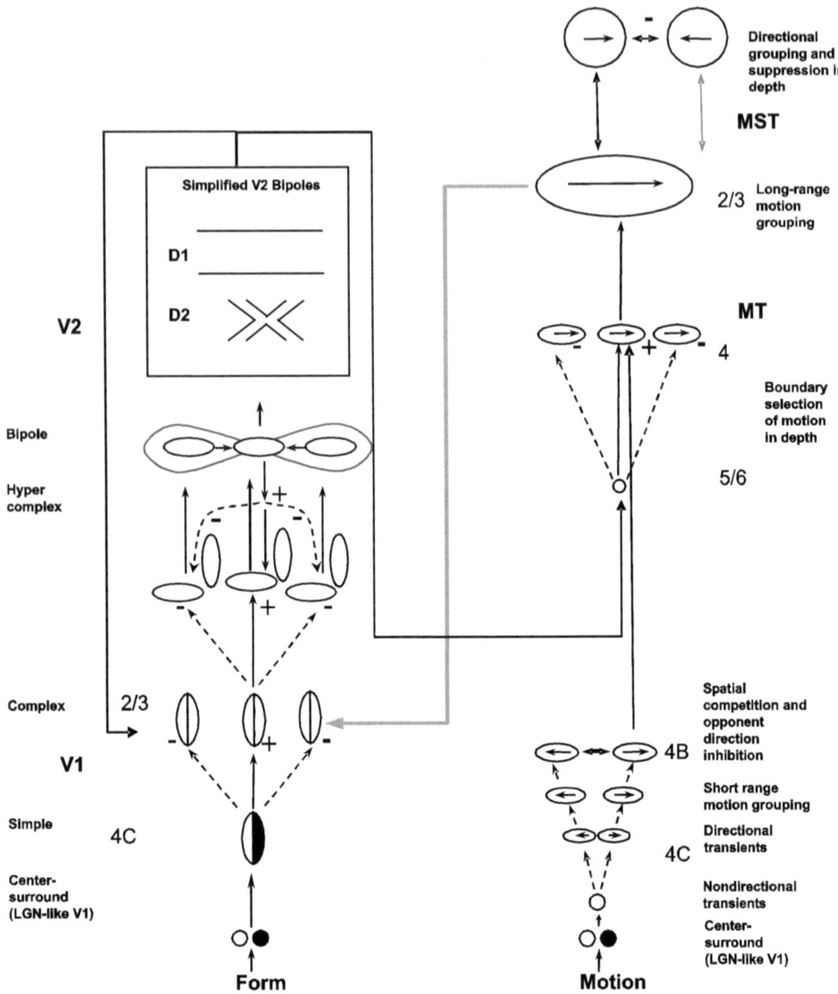

Fig. 13.3 Laminar circuits of 3D FORMOTION model. See text for details [Reprinted with permission from Berzhanskaya et al. (2007).]

both the occluder and its occluded objects? In the example in Fig. 13.4, motion of the left line end corresponds to the real motion of the line. The right line end is formed by the boundary between the line and a stationary occluder. Its motion provides little information about the motion of the line. Bregman (1981) and Kanizsa (1979), and more recently Nakayama et al. (1989), have discussed this problem. Nakayama et al. use the terminology of *intrinsic* and *extrinsic* terminators to distinguish the two cases. An intrinsic terminator belongs to the moving object, whereas an extrinsic one belongs to the occluder. Motion of intrinsic terminators is incorporated into an object's motion direction, whereas motion of extrinsic terminators

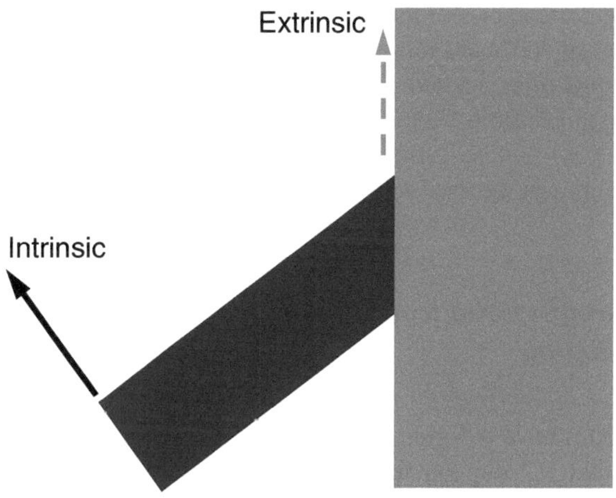

Fig. 13.4 Extrinsic and intrinsic terminators: The local motion of the intrinsic terminator on the *left* reflects the true object motion, while the local motion of the extrinsic terminator on the *right* follows the vertical outline of the occluder

is attenuated or eliminated (Duncan et al. 2000; Lidén and Mingolla 1998; Shimojo et al. 1989).

The FACADE model (Grossberg 1994, 1997; Kelly and Grossberg 2000) of 3D form vision and figure-ground separation proposed how boundaries in 3D scenes or 2D images are assigned to different objects in different depth planes, and thereby offered a mechanistic explanation of the properties of extrinsic and intrinsic terminators. The 3D FORMOTION model (Berzhanskaya et al. 2007; Grossberg et al. 2001) proposed how FACADE depth-selective figure-ground separation in cortical area V2, combined with depth-selective formotion interactions from area V2 to MT, enable intrinsic terminators to create strong motion signals on a moving object, while extrinsic terminators create weak ones. The model starts with motion signals in V1, where the separation in depth has not yet occurred, and predicts how V2-to-MT boundary signals can select V1-to-MT motion signals at the correct depths, while suppressing motion signals at the same visual locations but different depths.

13.1.5 Form and Motion Are Complementary: What Sort of Depth Does MT Compute?

The prediction that V2-to-MT signals can capture motion signals at a given depth reflects the hypothesis that the form and motion streams compute *complementary* properties (Grossberg 1991, 2000). The V1–V2 cortical stream, acting alone, is predicted to compute precise oriented depth estimates in the form of 3D boundary representations, but coarse directional motion signals. In contrast, the V1–MT cortical stream computes coarse depth estimates, but precise directional motion estimates.

Overcoming the deficiencies of the form and motion cortical streams in computing precise estimates of form-and-motion-in-depth is predicted to occur via V2-to-MT inter-stream interactions, called *formotion* interactions. These interactions use depth-selective signals from V2 to capture motion signals in MT to lie at the correct depths. In this way, precise form-and-motion-in-depth estimates are achieved in MT, which can, in turn, be used to generate good target tracking estimates.

13.1.6 Neurophysiological Support for Formotion Capture Prediction

Ponce et al. (2008) have reported neurophysiological data that are consistent with the prediction that V2 imparts finer disparity sensitivity onto MT: When V2 is cooled, depth selectivity, but not motion selectivity, is greatly impaired in MT. These data do not support the alternative view that fine depth estimates are computed directly in MT.

There are many psychophysical data that support this view of motion capture. Indeed, the V2-to-MT motion selection mechanism clarifies why we tend to perceive motion of visible objects and background features, but not of the intervening empty spaces between them. For example, consider an example of *induced motion* (Duncker 1929/1937) wherein a frame moving to the right caused a stationary dot within the frame to appear to move to the left. Motion signals must propagate throughout the interior of the frame in order to reach and influence the dot. Despite this global propagation, the homogeneous space between the frame and the dot does not seem to move. The 3D FORMOTION model predicts that this occurs because there are no boundaries between the frame and the dot whereby to capture a motion signal. More generally, the model proposes that the formotion interaction whereby V2 boundaries select compatible MT motion signals may be necessary for a conscious percept of motion to occur when such boundaries are active.

V2-to-MT formotion signals overcome one sort of uncertainty in cortical computation. Another sort of uncertainty is overcome by using MT-to-V1 feedback signals. These top-down modulatory signals can help to separate boundaries in V1 and V2 where they cross in feature-absent regions. Such feature-absent signals are illustrated, for example, by the chopstick illusion (Anstis 1990); see Fig. 13.5. Here, attention or internal noise signals can amplify motion signals of one chopstick more than that of the other via MST–MT interactions. This stronger chopstick can send its enhanced signals to V1 from MT. These enhanced signals, in turn, can use V1-to-V2 figure-ground separation mechanisms to separate the two chopsticks in depth, with the stronger boundary pulled nearer than the weaker one. The nearer boundary can now be completed by perceptual grouping mechanisms. In addition, FACADE mechanisms show how the intrinsic boundaries of the nearer chopstick can be detached from the farther chopstick, thereby enabling the farther chopstick boundaries to also be completed in depth behind those of the occluding chopstick.

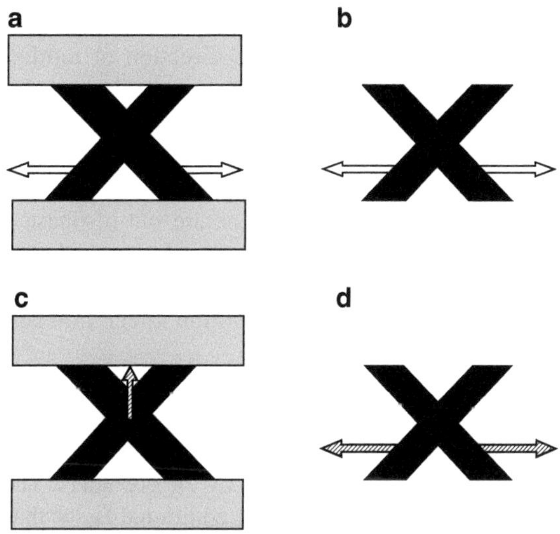

Fig. 13.5 Chopsticks illusion: Actual chopsticks motion (*clear arrows*, *top*) is equivalent in (**a**) and (**b**), with visible and invisible occluders, respectively. Visible occluders result in a coherent vertical motion percept (**c**, *hatched arrow*). Invisible occluders result in the percept of two chopsticks sliding in opposite directions (**d**) [Reprinted with permission from Berzhanskaya et al. (2007).]

As these boundaries are completed, they are injected back into MT from V2 to generate a final percept of two separated figures in depth.

Another factor that influences motion perception is adaptation. This can be accomplished by a process of transmitter habituation, inactivation, or depression. For example, motion signals at the positions of a static extrinsic terminator can adapt and therefore become weaker through time. Moving intrinsic terminators, on the other hand, generate strong motion signals. The adaptation process hereby simplifies the computation of intrinsic motion signals on a relatively short time scale.

On a longer time scale, bistable motion percepts can occur as a result of the interaction of cooperative–competitive model mechanisms with habituative mechanisms when multiple moving objects overlap (Figs. 13.1 and 13.3). For example, percepts of pairs or of moving plaids random dot patterns can alternate between at least two possible perceptual outcomes (Ferrera and Wilson 1987, 1990; Kim and Wilson 1993; Snowden et al. 1991; Stoner and Albright 1998; Stoner et al. 1990; Trueswell and Hayhoe 1993). One possible outcome is a transparent motion percept, where two gratings or two dot-filled planes slide one over the other in depth. Alternatively, if the directions of motions are compatible, then displays can produce a percept of coherent motion of a unified pattern, and no separation in depth occurs. Under prolonged viewing, the same display can perceptually alternate between coherent plaid motion and different motions separated in depth (Hupé and Rubin 2003).

Similar mechanisms can explain and simulate percepts of object shapes that are more complex than lines or dots. For example, Lorenceau and Alais (2001) studied different shapes moving in a circular-parallel motion behind occluders (Fig. 13.6). Observers had to determine the direction of motion, clockwise or counterclockwise. The percent of correct responses depended on the type of shape, and on the visibility of the occluders. In the case of a diamond (Fig. 13.6a), a single, coherent, circular motion of a partially occluded rectangular frame was easy to perceive across the apertures. In the case of an arrow (Fig. 13.6c), two objects with parallel sides were seen to generate out-of-phase vertical motion signals in adjacent apertures. Local motion signals were identical in both displays, and only their spatial arrangement differed. Alais and Lorenceau suggested that certain shapes (such as arrows) "veto" motion integration across the display, while others (such as diamond) allow it.

The 3D FORMOTION model explains the data without using a veto process. The model proposes that the motion grouping process uses anisotropic direction-sensitive receptive fields (see Fig. 13.3) that preferentially integrate motion signals within a given direction across gaps produced by the occluders. The explanation of Fig. 13.6d–f follows in a similar way, with the additional factor that the ends of the bars possess intrinsic terminators that can strongly influence the perceived motion direction of the individual bars.

Motion grouping also helps to explain percepts of rotational motion using the "gelatinous ellipses" display (Vallortigara et al. 1988; Weiss and Adelson 2000). When "thin" (high aspect ratio) and the "thick" (low aspect ratio) ellipses rotate around their centers, the perception of their shapes is strikingly different. The thin ellipse is perceived as a rigid rotating form, whereas the thick one is perceived as deforming non-rigidly through time. Here, the differences in 2D geometry result in differences of the spatiotemporal distribution of motion direction signals that are grouped together through time. When these motion signals are consistent with the coherent motion of a single object, then the motion grouping process within the model MT-MST processing stages (Fig. 13.1) generates a percept of a rigid rotation. When the motion field decomposes after grouping into multiple parts, with motion trajectories incompatible with a rigid form, a non-rigid percept is obtained. The ability of nearby "satellites" to convert the non-rigid percept into a rigid one can also be explained by motion grouping. In contrast, Weiss and Adelson (2000) proposed that such a percept can be explained via a global optimization process. We believe that motion grouping provides a biologically more plausible explanation.

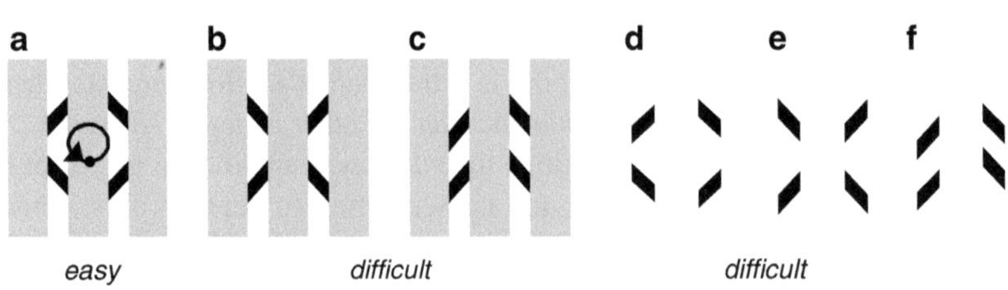

Fig. 13.6 Lorenceau–Alais displays: Visible (**a**–**b**) and invisible (**d**–**f**) occluder cases. See text for details

Data about probabilistic decision making in response to moving dot patterns will be discussed after the model is summarized.

13.2 3D FORMOTION Model

The 3D FORMOTION model (Figs. 13.1 and 13.3) comprises six key interactions involving the brain's form and motion systems. Because model processing stages are analogous to areas of the primate visual system, they are called by the corresponding anatomical names: (1) V1-to-MT filtering and cooperative–competitive processes set the stage for resolving the aperture problem by amplifying feature tracking signals and attenuating ambiguous motion signals so that the feature tracking signals have a chance to overwhelm numerically superior ambiguous motion signals. (2) 3D boundary representations, in which figures are separated from their backgrounds, are formed in cortical area V2. (3) These depth-selective V2 boundaries select motion signals at the appropriate boundary positions and depths in MT via V2-to-MT signals. (4) A spatially anisotropic motion grouping process propagates across perceptual space via MT-MST feedback to integrate veridical feature-tracking and ambiguous motion signals and thereby determines a global object motion percept. This is the motion capture process that solves the aperture problem. (5) MST–MT feedback can convey an attentional priming signal from higher brain areas that can influence the motion capture process, and have an influence via MT-to-V1 feedback in V1 and V2. (6) Motion signals in MT can disambiguate locally incomplete or ambiguous boundary signals in V2 via MT-to-V1-to-V2 feedback.

These interactions provide a functional explanation of many neurophysiological data. Table 13.1 summarizes the key anatomical connections and neuron properties that occur in the model, alongside selected references supporting those connections or functional properties. Table 13.1 also lists the model's key physiological predictions that remain to be tested. As illustrated in Figs. 13.1 and 13.3, these interactions are naturally understood as part of a *form processing stream* and a *motion processing stream*.

13.2.1 The Form Processing System

The model's form processing system comprises six stages that are displayed on the left sides of Figs. 13.1 and 13.3. Inputs are processed by distinct ON and OFF cell networks whose cells obey membrane, or shunting, equations while they undergo on-center off-surround and off-center on-surround network interactions, respectively, that are similar to those of LGN cells. These cells excite simple cells in cortical area V1 to register boundary orientations, followed by complex and hypercomplex stages that perform pooling across simple cells tuned to opposite contrast

Table 13.1 Functional projections and properties of model cell types and predictions [Reprinted with permission from Berzhanskaya et al. (2007).]

Connection/functional property	Selected references
Functional projections	
V1 4Ca to 4B	Yabuta et al. (2001) and Yabuta and Callaway (1998)
V1 to MT	Anderson et al. (1998), Rockland (2002), Sincich and Horton (2003), and Movshon and Newsome (1996)
V1 to V2	Rockland (1992) and Sincich and Horton (2002)
V2 to MT	Anderson and Martin (2002), Rockland (2002), Shipp and Zeki (1985), and DeYoe and Van Essen (1985)
MT to V1 feedback	Shipp and Zeki (1989), Callaway (1998), Movshon and Newsome (1996), and Hupé et al. (1998)
V2 to V1 feedback	Rockland and Pandya (1981) and Kennedy and Bullier (1985)
Properties	
V1 adaptation	Abbott et al. (1997), Chance et al. (1998), (rat), and Carandini and Ferster (1997), (cat)
V1(4ca) transient nondirectional cells	Livingstone and Hubel (1984)
V1 spatially offset inhibition	Livingstone (1998), Livingstone and Conway (2003), and Murthy and Humphrey (1999) (cat)
V2 figure-ground separation	Zhou et al. (2000) and Bakin et al. (2000)
MT figure-ground separation and disparity sensitivity	Bradley et al. (1998), Grunewald et al. (2002), and Palanca and DeAngelis (2003)
MT center-surround receptive fields	Bradley and Andersen (1998), Born (2000), and DeAngelis and Uka (2003)
Some MT receptive fields elongated in preferred direction of motion	Xiao et al. (1997)
Attentional modulation in MT	Treue and Maunsell (1999)
Predictions	
Short-range anisotropic filter in V1 (motion stream)	
Long-range anisotropic filter in MT (motion)[a]	
V2 to MT projection carries figure-ground completed-form-in-depth separation signal	
MT to V1 feedback carries figure-ground separation signal from motion to form stream MST to MT feedback helps solve aperture problem by selecting consistent motion directions	

[a]Although Xiao et al. (1997) found that some MT neurons have receptive fields that are elongated along the preferred direction of motion, there is no direct evidence that these neurons participate preferentially in motion grouping

polarities, divisive normalization that reduces the amplitude of multiple ambiguous orientations in a region, end-stopping that enhances activity at line-ends, and spatial sharpening. These cells input to the perceptual grouping circuit in layer 2/3 of V2. Here bipole cells receive signals via long-range horizontal interactions from approximately collinear cells whose orientation preferences lie along, or near, the collinear axis. These cells are indicated by the figure-8 shape in Fig. 13.3. They act like statistical "and" gates that permit grouping only when there is sufficient evidence from pairs or greater numbers of inducers on both sides of the cell body (Grossberg 1994; Grossberg and Mingolla 1985a, b). Grouping is followed by a stage of cross-orientation competition that reinforces boundary signals with greater support from neighboring boundaries while weakening spatially overlapping boundaries of non-preferred orientations. Boundaries are assigned into different depths, as follows:

13.2.1.1 Perceptual Grouping and Figure-Ground Separation of 3D Form

The FACADE boundary completion process includes separation of extrinsic vs. intrinsic boundaries in depth (Grossberg 1994, 1997; Kelly and Grossberg 2000) within the pale stripes of V2. One cue of occlusion in a 2D image is a T-junction, as illustrated in Fig. 13.4 where the moving black bar intersects the stationary gray rectangular occluder. The top of the T belongs to the occluding gray rectangle, while the stem belongs to the occluded black bar. Bipole long-range excitatory horizontal interactions can strengthen the boundary of the gray occluder where it intersects the black bar, while short-range competition (Fig. 13.3) weakens, or totally inhibits, the boundary of the black occluded bar where it touches the gray occluder. This *end gap* in the black boundary initiates the process of separating occluding and occluded boundaries. In other words, perceptual grouping properties are predicted to initiate the separation of figures from their backgrounds, without the use of explicit T-junction operators. This prediction has received support from psychophysical experiments (e.g., Dresp et al. 2002; Tse 2005). Such figure-ground separation enables the model to distinguish extrinsic from intrinsic terminators, and to thereby select motion signals at the correct depths.

The 3D FORMOTION model, to the present, has not simulated all stages of boundary and surface interaction that are predicted to be used in 3D figure-ground separation. These mechanisms are, however, fully simulated in Fang and Grossberg (2009) and Grossberg and Yazdanbaksh (2005) using laminar cortical V1 and V2 circuits, as well as in Kelly and Grossberg (2000) using non-laminar circuits. Instead, to reduce the simulation computational load, as soon as T-junctions were detected by the model dynamical equations, V2 boundaries were algorithmically assigned the depths that a complete figure-ground simulation would have assigned them. In particular, static occluders are assigned to the near depth (D1 in Fig. 13.3) and lines with extrinsic terminators are assigned to the far depth (D2 in Fig. 13.3). These V2 boundaries are used to provide both V2-to-MT motion selection signals and V2-to-V1 depth-biasing feedback. While V2-to-V1 feedback is orientation-specific, the V2-to-MT projection sums boundary signals over all orientations, just as motion signals do at MT (Albright 1984).

13.2.1.2 Motion Induction of Figure-Ground Separation

When form cues are not available to initiate figure-ground separation, motion cues may be able to do so via feedback projections from MT to V1 (Figs. 13.1 and 13.3). Such a feedback projection has been reported both anatomically and electrophysiologically (Bullier 2001; Jones et al. 2001; Movshon and Newsome 1996) and it can benefit from attentional biasing within MT/MST (Treue and Maunsell 1999). As explained above, this mechanism can help to separate chopsticks in depth (see Fig. 13.5b). Focusing spatial attention at one end of a chopstick can enhance that chopstick's direction of motion in MT and MST. Enhanced MT-to-V1 feedback can selectively strengthen the boundary signals of one chopstick in Fig. 13.5b enough to trigger its boundary completion via V1-to-V2 interactions, as well as figure-ground separation that assigned the occluded chopstick to a farther depth. Then, by closing the V2-to-MT loop, these two overlapping but depth-separated bars can support depth-selective motions by the chopsticks in opposite directions (Bradley et al. 1998; Grossberg et al. 2001).

13.2.2 The Motion Processing System

The model's motion processing stream consists of six stages that represent cell dynamics homologous to LGN, V1, MT, and MST (Figs. 13.1 and 13.3, right).

13.2.2.1 Level 1: Input from LGN

ON and OFF cell inputs from retina and LGN, which are lumped into a single processing stage, activate model V1 (Xu et al. 2002). These inputs are not depth-selective. In response to a 2D picture, this depth-selectivity will come from figure-ground separated V2 boundaries when they select consistent motion signals in MT.

The 3D FORMOTION model uses both ON and OFF input cells. For example, if a bright chopstick moves to the right on a dark background, ON cells respond to its leading edge, but OFF cells respond to its trailing edge. Likewise, when the chopstick reverses direction and starts to move to the left, its leading edge now activates ON cells and its trailing edge OFF cells. By differentially activating ON and OFF cells in different parts of this motion cycle, these cells have more time to recover from habituation, so that the system remains more sensitive to repetitive motion signals.

13.2.2.2 Level 2: Transient Cells

The second stage of the motion processing system consists of non-directional transient cells, inhibitory directional interneurons, and directional transient cells. The non-directional transient cells respond briefly to a change in the image luminance, irrespective of the direction of movement. Such cells respond well to moving boundaries and poorly to the static occluder because of the habituation of the process that activates the transient signal. Adaptation is known to occur at several stages in the visual system, including retinal Y cells (Enroth-Cugell and Robson 1966;

Hochstein and Shapley 1976a, b) and cells in V1 (Abbott et al. 1997; Carandini and Ferster 1997; Chance et al. 1998; Varela et al. 1997) and beyond.

The non-directional transient cells send signals to inhibitory directional interneurons and directional transient cells, and the inhibitory interneurons interact with each other and with the directional transient cells (Fig. 13.7). The directional inhibitory interneuronal interaction enables the directional transient cells to realize directional selectivity at a wide range of speeds (Grossberg et al. 2001). This predicted interaction is consistent with retinal data concerning how bipolar cells interact with inhibitory starburst amacrine cells and direction-selective ganglion cells, and how starburst cells interact with each other and with ganglion cells (Fried et al. 2002). The possible role of starburst cell inhibitory interneurons in ensuring directional selectivity at a wide range of speeds has not yet been tested.

A directionally selective neuron fires vigorously when a stimulus is moved through its receptive field in one direction (called the preferred direction), while motion in the reverse direction (called the null direction) evokes little response (Barlow and Levick 1965). Mechanisms of direction selectivity include asymmetric

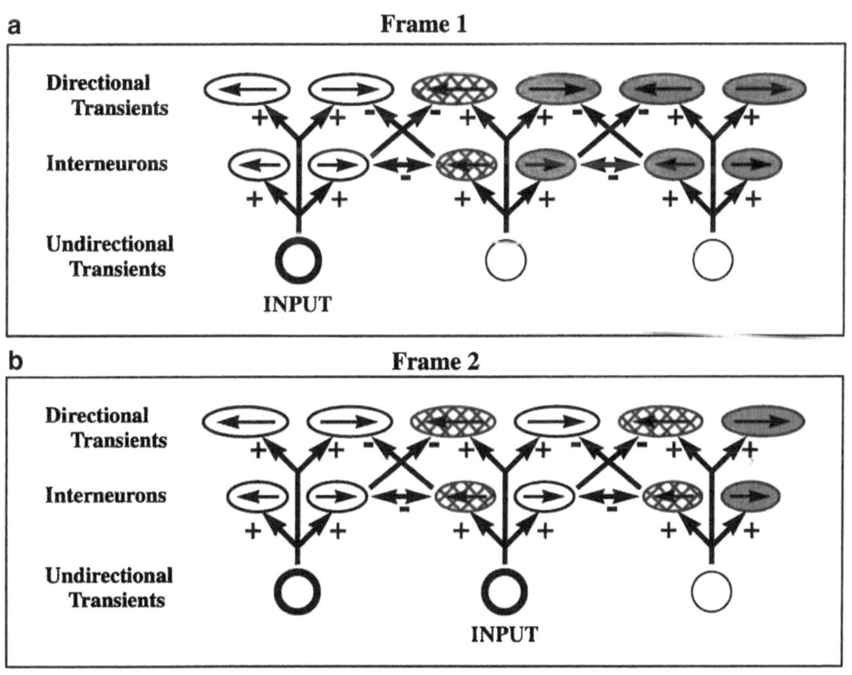

Fig. 13.7 Schematic diagram of a 1D implementation of the transient cell network showing the first two frames of the motion sequence. Thick *circles* represent active unidirectional transient cells while *thin circles* are inactive unidirectional transient cells. *Ovals* containing *arrows* represent directionally selective neurons. *Unfilled ovals* represent active cells, *cross-filled ovals* are inhibited cells, and *gray-filled ovals* depict inactive cells. Excitatory and inhibitory connections are labeled by "+" and "–" signs respectively [Reprinted with permission from Grossberg et al. (2001).]

inhibition along the preferred cell direction, notably an inhibitory veto of null-direction signals.

As noted above, after the transient cells adapt to a static boundary, boundary segments that belong to a static occluder – that is, extrinsic terminators – in the chopsticks display with visible occluders (Fig. 13.5a) produce weaker signals than those that belong to the continuously moving parts of the chopstick. On the other hand, in the invisible occluder chopsticks display (Fig. 13.5b), the horizontal motion signals at the chopstick ends will continually move, hence will be strong, and can thus significantly influence the conscious motion percept.

13.2.2.3 Level 3: Short-Range Filter

The short-range filter (Fig. 13.3) helps to selectively strengthen unambiguous feature tracking signals, relative to ambiguous motion signals. Cells in this filter accumulate evidence from directional transient cells of similar directional preference within a spatially anisotropic region that is oriented along the preferred direction of the cell; cf., Braddick (1980). Short-range filter cells amplify feature-tracking signals at unoccluded line endings, object corners, and other scenic features. The short-range spatial filter, followed by competitive selection, eliminates the need to solve the feature correspondence problem that various other models use (Reichardt 1961; van Santen and Sperling 1985).

13.2.2.4 Level 4: Spatial Competition and Opponent Direction Competition

Two kinds of competition further enhance the relative advantage of feature tracking signals. These competing cells are proposed to occur in layer 4B of V1 (Fig. 13.3, bottom-right). Spatial competition among like-directional cells of the same spatial scale further boosts the amplitude of feature tracking signals relative to those of ambiguous signals. This happens because feature tracking locations are often found at motion discontinuities, and thus get less inhibition than ambiguous motion signals that lie within an object's interior. Opponent-direction competition also occurs at this processing stage, with properties similar to those of V1 cells that may play this functional role (Rust et al. 2002).

Data of Pack et al. (2004) support properties of cells at this model stage. In their data, V1 cells exhibit suppression of responses to motion along visible occluders. Suppression occurs in the model because of the adaptation of transient inputs to static occluding boundaries. In addition, V1 cells in the middle of a grating, where ambiguous motion signals occur, respond more weakly than cells at the edge of the grating, where intrinsic terminators occur. Model spatial competition between motion signals explains this property through its properties of divisive normalization and endstopping. Together these properties amplify directionally unambiguous feature tracking signals at line ends relative to the strength of aperture-ambiguous signals along line interiors, which compete among themselves for normalized activity at their position.

13.2.2.5 Level 5: Long-Range Filter and Formotion Selection

The long-range filter pools together motion signals with the same, or similar, directional preference from moving features with different orientations, contrast polarities, and eyes. These motion signals may be carried from model layer 4B of V1 input to model area MT. Its cell targets have properties in the motion stream through MT that are homologous to those of complex cells in the form stream through V2.

Area MT also receives a projection from V2 (Anderson and Martin 2002; Rockland 1995). As described above, this V2-to-MT formotion projection is predicted to carry depth-specific figure-ground separated boundary signals. These V2 form boundaries selectively assign to different depths the motion signals coming into MT from layer 4B of V1.

Formotion selection is proposed to occur via a modulatory on-center, off-surround projection from V2 to layer 4 of MT. For example, in response to the chopsticks display with visible occluders (Fig. 13.5a), motion signals which lie along the visible occluder boundaries are selected in the near depth and are suppressed by the off-surround at other locations at that depth. The selected signals will be weak because the bottom-up input from V1 habituates along the selected occluder boundary positions. The V2 boundary signals that correspond to the moving boundaries select strong motion signals at the farther depth.

Boundary-gated signals from layer 4 of MT are proposed to input to the upper layers of MT (Fig. 13.3, top-right), where they activate a directionally-selective, spatially *anisotropic* long-range filter via long-range horizontal connections. The hypothesis that the long-range filter uses an anisotropic filter is consistent with data showing that approximately 30% of the cells in MT show a preferred direction of motion that is aligned with the main axis of their receptive fields (Xiao et al. 1997).

The predicted long-range filter cells in layer 2/3 of MT are proposed to play a role in motion grouping that is homologous to the role played by bipole cells in form grouping within layer 2/3 of the pale stripes of cortical area V2 (Grossberg 1999; Grossberg and Raizada 2000). As noted above, the anisotropic long-range motion filter allows motion signals to be selectively integrated across occluders in a manner that naturally explains the percepts generated by the Lorenceau–Alais displays of Fig. 13.6.

13.2.2.6 Level 6: Directional Grouping

The first five model stages can amplify feature tracking signals and assign motion signals to the correct depths. However, they do not explain how feature tracking signals propagate across space to select consistent motion directions from ambiguous motion directions and suppress inconsistent motion directions, all the while without distorting their speed estimates. They also cannot explain how motion integration can compute a vector average of ambiguous motion signals across space to determine the perceived motion direction when feature tracking signals are not present at that depth. The final stage of the model accomplishes this goal by using a motion grouping network that is interpreted to exist in ventral MST (MSTv), which is known to be important for target tracking (Berezovskii and Born 1999; Born and Tootell 1992; Eifuku and Wurtz 1998; Pack et al. 2001; Tanaka et al. 1993). We predict that this motion grouping network determines the coherent motion direction of discrete moving objects.

During motion grouping, cells that code the same, or similar, directions in MT send convergent inputs to cells in model MSTv via the motion grouping network. Within MSTv, directional competition at each position determines a winning motion direction. This winning directional cell then feeds back to its source cells in MT. This feedback selects activities of MT cells that code the winning direction, while suppressing activities of cells that code other directions. Using this broad feedback kernel, the motion grouping network enables feature tracking signals to select similar directions at nearby ambiguous motion positions, while suppressing other directions there. In other words, motion capture occurs and disambiguates ambiguous motion positions.

The next cycle of the feedback process allows these newly unambiguous motion directions to select consistent MSTv grouping cells at positions near them. As the grouping process cycles between MT and MSTv, the motion capture process propagates across space. Chey et al. (1997) and Grossberg et al. (2001) used this process to simulate data showing how the present model solves the aperture problem via a gradual process of motion capture, and Pack and Born (2001) provided supportive neurophysiological data by directly recording from MT cells, as noted above.

13.2.2.7 Ubiquitous Circuit Design for Selection, Attention, and Learning

Both the V2-to-MT and the MSTv-to-MT signals carry out their selection processes using modulatory on-center, off-surround interactions. The V2-to-MT signals select motion signals at the locations and depth of a moving boundary. The MSTv-to-MT signals select motion signals in the direction and depth of a motion grouping. Adaptive Resonance Theory predicted that such a modulatory on-center, off-surround network would be used to carry out attentive selection and modulation of adaptive tuning within all brain circuits wherein fast and stable learning of appropriate features is needed. In the V2-to-MT circuit, a formotion association is learned. In the MST-to-MT circuit, directional grouping cells are learned. Grossberg (2003) and Raizada and Grossberg (2003) review behavioral and neurobiological data that support this prediction in several brain systems. The Ponce et al. (2008) study supports the V2-to-MT prediction, but does not study how this association is learned. There do not seem to be any direct neurophysiological tests of the MSTv-to-MT prediction.

13.3 Temporal Dynamics of Decision-Making During Motion Perception

13.3.1 Motion Capture in Perceptual Decision-Making

The 3D FORMOTION model sheds new light on how the brain makes movement decisions, in particular saccadic eye movements, in response to probabilistically defined motion stimuli. It is well known that speed and accuracy of perceptual decisions covary with certainty in the input, and correlate with the rate of evidence accumulation in parietal and frontal cortical neurons. An enhancement of the 3D FORMOTION model with a parietal, indeed an LIP, directional movement processing stage that is gated by the basal ganglia (Fig. 13.8) is sufficient to explain many data of this kind (Grossberg and Pilly 2008; Pilly and Grossberg 2005, 2006).

In particular, this enhanced model can quantitatively simulate dynamic properties of decision-making in response to the types of ambiguous visual motion stimuli that have been studied in LIP neurophysiological recordings by Newsome, Shadlen, and colleagues. The most important circuits of this enhanced model already lie within the 3D FORMOTION model, because the rate of motion capture in the MT-MST grouping network covaries with the activation rate and amplitude of LIP cells that control a monkey's observable behavior in the experiment. The model hereby clarifies how brain circuits that solve the aperture problem, notably the circuits that realize motion capture, control properties of probabilistic decision making in real time. This is not surprising when one interprets the motion capture process as a resolution of ambiguity that selects the best consensus movement that is compatible with motion data.

13.3.2 Are the Brain's Decisions Bayesian?

These results are of particular interest because some scientists, including Newsome and Shadlen, have proposed that perception and decision-making can be described as Bayesian inference, which estimates the optimal interpretation of the stimulus given priors and likelihoods. However, Bayesian concepts do not provide a way to discover the neocortical mechanisms that make decisions. The present model explains data that Bayesian models have heretofore failed to explain, does so without an appeal to Bayesian inference, and, unlike other existing models of these data, generates perceptual representations in response to the experimental visual stimuli. The model quantitatively simulates the time course of LIP neuronal dynamics, as well as behavioral accuracy and reaction time properties, during both correct and error trials at different levels of input ambiguity in both fixed duration and reaction time tasks. Model MST computes the global direction of random dot motion stimuli as part of the motion capture process, while model LIP computes the perceptual decision that leads to a saccadic eye

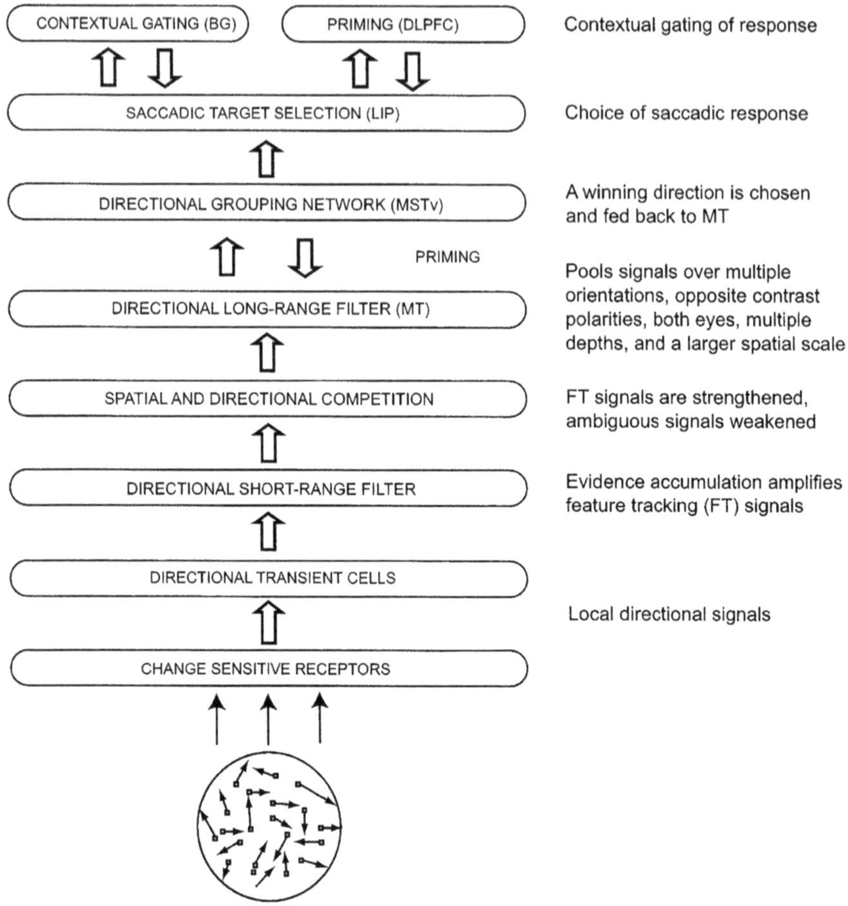

Fig. 13.8 Retina/LGN-V1-MT-MST-LIP-BG model processing stages. See text and Appendix for details. The random dot motion stimuli are preprocessed by the model Retina/LGN and processed by the model cortical V1–MT–MST stream. They contextually transform locally ambiguous motion signals into unambiguous global object motion signals with a rate, amplitude, and direction that covaries with the amount of dot coherence. These spatially distributed global motion signals then feed into model area LIP to generate appropriate directional saccadic eye movement commands, which are gated by the model basal ganglia [Reprinted with permission from Grossberg and Pilly (2008).]

movement. This self-organizing system thus trades accuracy against speed, and illustrates how cortical dynamics go beyond Bayesian concepts, while clarifying why probability theory ideas are initially so appealing.

Concerning the appeal of statistical, in particular Bayesian, concepts, it should be noted that the shunting on-center off-surround networks (Grossberg 1973, 1980) that occur ubiquitously in the brain, and also in the 3D FORMOTION model, tend to normalize the activities across a neural network. The spatially distributed pattern of these normalized activities may be viewed as a type of real-time probability distribution.

In addition, any filtering operation, such as that of the short-range and long-range filters, may be interpreted as a prior (namely, the current neural signal) multiplied by a conditional probability or likelihood (namely, the filter connection strength to the target cell). Likewise, a contrast-enhancing operation, such as that of the LIP recurrent on-center off-surround network that selects a winning direction from filter inputs, may be viewed as maximizing the posterior. These insights have been known in the neural modeling literature for a long time (Grossberg 1978). However, as Figs. 13.1, 13.3, and 13.8 illustrate, such local processes do not embody the computational intelligence of an entire neural system that has emerged through evolution to realize particular behavioral competences, such as motion perception and decision-making.

13.3.3 Two Movement Tasks

Newsome, Shadlen, and colleagues studied neural correlates of perceptual decision-making in macaques which were trained to discriminate motion direction. Random dot motion displays, covering a 5° diameter aperture centered at the fixation point, were used to control motion coherence; namely, the fraction of dots moving non-randomly in a particular direction from one frame to the next in each of three interleaved sequences. Varying motion coherence provided a quantitative way to control the ambiguity of directional information that the monkey used to make a saccadic eye movement to a peripheral choice target in the perceived motion direction, and thus the task difficulty.

Two kinds of tasks were employed, namely *fixed duration* (FD) and *reaction time* (RT) tasks. In the FD task (Roitman and Shadlen 2002; Shadlen and Newsome 2001), monkeys viewed the moving dots for a fixed duration of 1 s, and then made a saccade to the target in the judged direction after a variable delay. In the RT task (Roitman and Shadlen 2002), monkeys had theoretically unlimited viewing time, and were trained to report their decision as soon as the motion direction was perceived. The RT task allowed measurement of how long it took the monkey to make a decision, which was defined as the time from the onset of the motion until when the monkey initiated a saccade.

Neurophysiological recordings were done in LIP while the monkeys performed these tasks. The recorded neurons had receptive fields (RF) that encompassed just one target, and did not include the circular aperture in which the moving dots were displayed. Also, they were among those that showed sustained activity during the delay period of a memory-guided saccade task. Even though there is no motion stimulus within their classical receptive fields, these neurons still respond with directional-selectivity, probably because of extensive training on the task during which an association was learned (Bichot and Schall 1996). This property has also been observed for neurons in superior colliculus whose movement field contains just one target (Horwitz et al. 2004; Horwitz and Newsome 2001).

The recorded LIP neurons show visuo-motor responses. On correct trials during the decision-making period, more coherence in the favored direction causes faster LIP cell activation, on average, in both the tasks (Fig. 13.9), and also higher

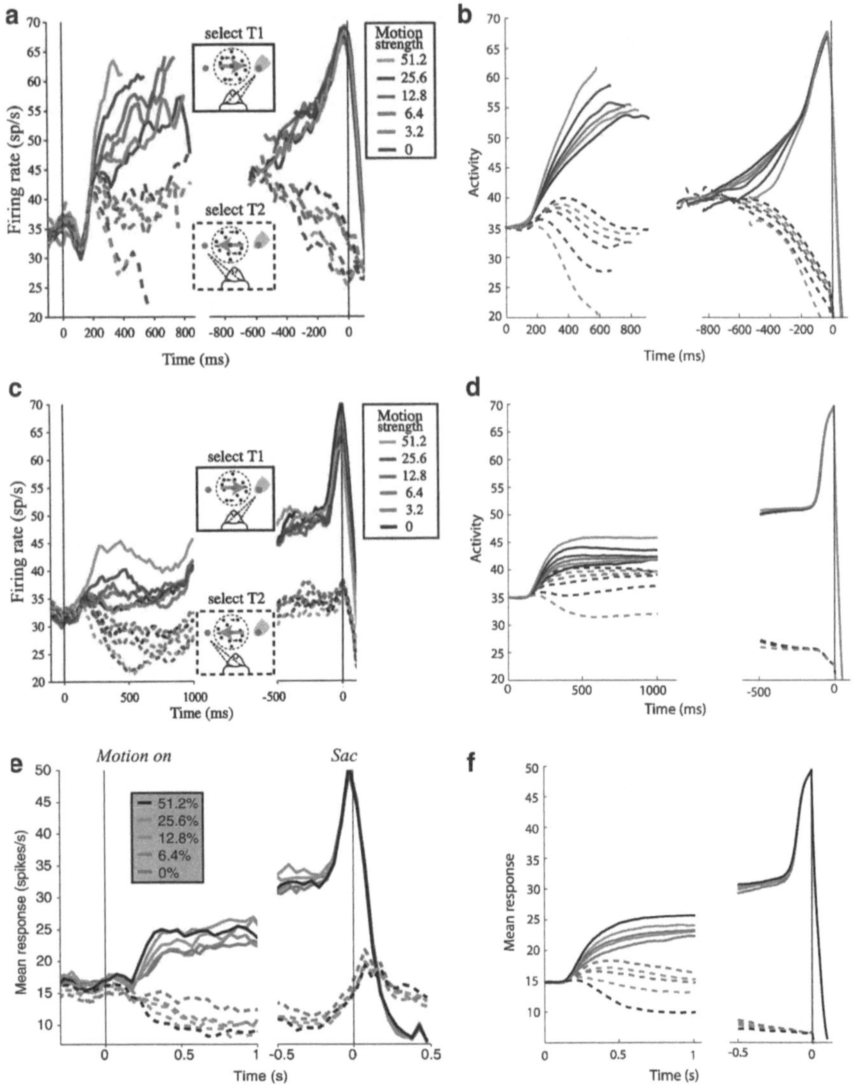

Fig. 13.9 Temporal dynamics of LIP neuronal responses during the fixed duration (FD) and reaction time (RT) tasks. (**a**) Average responses of a population of 54 LIP neurons among correct trials during the RT task (Roitman and Shadlen 2002). The *left part* of the plot is time-aligned to the motion onset, and includes activity only up to the median RT, and excludes any activity within 100 ms backward from saccade initiation (which corresponds to presaccadic enhancement). The *right part* of the plot is time-aligned to the saccade initiation, and excludes any activity within 200 ms forward from motion onset (which corresponds to initial transient dip and rise). (**b**) Model simulations replicate LIP cell recordings during the RT task. In both data and simulations for the RT task, the average responses were smoothed with a 60 ms running mean. (**c**) Average responses of a population of 38 LIP neurons among correct trials during the 2002 FD task (Roitman and Shadlen 2002), during both the motion viewing period (1 s) and a part (0.5 s) of the delay period before the saccade is made. (**d**) Model simulations mimic LIP cell recordings during the 2002 FD task.

maximal cell activation in the FD task (Fig. 13.9c–f). More coherence in the opposite direction causes faster cell inhibition in both the tasks, and also lower minimal cell activation in the FD task.

13.3.4 *Comparing Trackable Features with Coherently Moving Dots*

There are many details that need to be carefully discussed to quantitatively simulated data from this paradigm. These details should not, however, obscure the main point, which is that a clear mechanistic homolog exists between sparse feature tracking signals and sparse but coherent moving dots.

We have already discussed that the brain needs to ensure that a sparse set of unambiguous feature tracking motion signals can gradually capture a vastly greater number of ambiguous motion signals to determine the global direction and speed of object motion. In the case of random dot motion discrimination tasks, the signal dots at any coherence level produce unambiguous, though short-lived, motion signals. The model shows how the same mechanisms that help resolve the aperture problem can also enable a small number of coherently moving dots to capture the motion directions of a large number of unambiguous, but incoherently moving, dots. The intuitive idea is that the MT-MST feedback loop needs more time to capture the incoherent motion signals when there are more of them, and cannot achieve a high level of asymptotic response magnitude when more of them compete with the emerging winning direction. In other words, the effectiveness of the motion capture process depends on the input coherence. LIP then converts the inputs from MST into an eye movement command, and thereby enables the monkey to report its decision via a saccade.

13.3.5 *Experiments that Directly Probe Brain Design vs. Those that Do Not*

Another point worth noting is that a display of moving dots does not experience an aperture problem. All of the dots are capable, in principle, of generating unambiguous

Fig. 13.9 (continued) (**e**) Average responses of a population of 104 LIP neurons among correct trials during the 2001 FD task (Shadlen and Newsome 2001), during both the motion viewing period (1 s) and a part (0.5 s) of the delay period before the saccade is made. (**f**) Model simulations emulate LIP cell recordings during the 2001 FD task. In (**a–f**), *solid* and *dashed curves* correspond to trials in which the monkey correctly chose the right target (T1) and the left target (T2), respectively. Cell dynamics (rate of rise or decline, and response magnitude) reflect the incoming sensory ambiguity (note the different colors; the color code for the various coherence levels is shown in the corresponding data panels), and the perceptual decision (note the two line types). For 0% coherence, even though there is no correct choice per se, the average LIP response rose or declined depending on whether the monkey chose T1 or T2, respectively [Reprinted with permission from Grossberg and Pilly (2008).] (*see Color Plates*)

directional motion signals. However, the model's circuits reflect, I would argue, a brain design that has evolved to overcome the aperture problem. As a result, the brain can compute unambiguous object motion direction estimates in response to locally ambiguous motion signals. The brain can thereby successfully track important moving targets in the environment even under probabilistically defined environmental conditions.

One might argue that the best experiments are ones that most directly probe brain design. From this perspective, experiments with moving dots are not the best possible probes of a system that has evolved to solve the aperture problem. Of course, it is not possible to confidently design such experiments until one has a strong modeling hypothesis about what this design may be, and that can only be gleaned by a sustained theoretical analysis of many different kinds of parametric experimental data. The 3D FORMOTION model contributes to such an analysis, while also articulating key features of the brain's design for generating object motion percepts.

References

Abbott LF, Sen K, Varela JA, Nelson SB (1997) Synaptic depression and cortical gain control. Science 275:220–222

Adelson EH, Movshon JA (1982) Phenomenal coherence of moving visual patterns. Nature 300:523–525

Albright TD (1984) Direction and orientation selectivity of neurons in visual area MT of the macaque. J Neurophysiol 52:1106–1130

Anderson JC, Martin KAC (2002) Connection from cortical area V2 to MT in macaque monkey. J Comp Neurol 443:56–70

Anderson JC, Binzegger T, Martin KAC, Rockland KS (1998) The connection from cortical area V1 to V5: a light and electron microscopic study. J Neurosci 18:10525–10540

Anstis SM (1990) Imperceptible intersections: the chopstick illusion. In: Blake A, Troscianko T (eds) AI and the eye. Wiley, London, pp 105–117

Bakin JS, Nakayama K, Gilbert CD (2000) Visual responses in monkey areas V1 and V2 to three-dimensional surface configurations. J Neurosci 20:8188–8198

Baloch AA, Grossberg S (1997) A neural model of high-level motion processing: line motion and formotion dynamics. Vision Res 37:3037–3059

Baloch AA, Grossberg S, Mingolla E, Nogueira CAM (1998) A neural model of first-order and second-order motion perception and magnocellular dynamics. J Opt Soc Am A 16:953–978

Barlow HB, Levick WR (1965) The mechanism of directionally selective units in rabbit's retina. J Physiol (London) 178:477–504

Berezovskii V, Born RT (1999) Specificity of projections from wide-field and local motion-processing regions within the middle temporal visual area of the owl monkey. J Neurosci 20:1157–1169

Berzhanskaya J, Grossberg S, Mingolla E (2007) Laminar cortical dynamics of visual form and motion interactions during coherent object motion perception. Spat Vis 20:337–395

Bichot NP, Schall JD (1996) Thompson KG Visual feature selectivity in frontal eye fields induced by experience in mature macaques. Nature 381:697–699

Born RT (2000) Center-surround interactions in the middle temporal visual area of the owl monkey. J Neurophysiol 84:2658–2669

Born RT, Tootell RBH (1992) Segregation of global and local motion processing in macaque middle temporal cortex. Nature 357:497–499

Braddick OJ (1980) Low-level and high-level processes in apparent motion. Philos Trans R Soc Lond B Biol Sci 290:137–151

Bradley DC, Andersen RA (1998) Center-surround antagonism based on disparity in primate area MT. J Neurosci 18:552–565

Bradley DC, Chang GC, Andersen RA (1998) Encoding of three-dimensional structure-from-motion by primate area MT neurons. Nature 392:714–717

Bregman AL (1981) Asking the "what for" question in auditory perception. In: Kubovy M, Pomerantz JR (eds) *Perceptual organization*. Earlbaum Associate, Hillsdale, NJ, pp 99–118

Bullier J (2001) Integrated model of visual processing. Brain Res Brain Res Rev 36:96–107

Callaway EM (1998) Local circuits in primary visual cortex of the macaque monkey. Annu Rev Neurosci 21:47–74

Cao Y, Grossberg S (2005) A laminar cortical model of stereopsis and 3D surface perception: closure and da Vinci stereopsis. Spat Vis 18:515–578

Carandini M, Ferster D (1997) Visual adaptation hyperpolarizes cells of the cat striate cortex. Science 276:913–914

Castet E, Lorenceau J, Shiffrar M, Bonnet C (1993) Perceived speed of moving lines depends on orientation, length, speed and luminance. Vision Res 33:1921–1936

Chance FS, Nelson SB, Abbott LF (1998) Synaptic depression and the temporal response characteristics of V1 cells. J Neurosci 18:4785–4799

Chey J, Grossberg S, Mingolla E (1997) Neural dynamics of motion grouping: From aperture ambiguity to object speed and direction: OSA, 14:2570–2594

Chey J, Grossberg S, Mingolla E (1998) Neural dynamics of motion processing and speed discrimination. Vision Res 38:2769–2786

DeAngelis GC, Uka T (2003) Coding of horizontal disparity and velocity by MT neurons in the alert macaque. J Neurophysiol 89:1094–1111

DeYoe EA, Van Essen DC (1985) Segregation of efferent connections and receptive field properties in visual area V2 of the macaque. Nature 317:58–61

Dresp B, Durand S, Grossberg S (2002) Depth perception from pairs of overlapping cues in pictorial displays. Spat Vis 15:255–276

Duncan RO, Albright TD, Stoner GR (2000) Occlusion and the interpretation of visual motion: perceptual and neuronal effects of context. J Neurosci 20:5885–5897

Duncker K (1937) Induced motion. In: Ellis WE (ed) A sourcebook of Gestalt psychology. Routledge and Kegan Paul, London (Original work published in 1929)

Eifuku S, Wurtz RH (1998) Response to motion in extrastriate area MSTl: center-surround interactions. J Neurophysiol 80:282–296

Enroth-Cugell C, Robson J (1966) The contrast sensitivity of retinal ganglion cells of the cat. J Physiol (London) 187:517–552

Fang L, Grossberg S (2009) From stereogram to surface: how the brain sees the world in depth. Spat Vis 22(1):45–82

Ferrera VP, Wilson HR (1987) Direction specific masking and the analysis of motion in two dimensions. Vision Res 27:1783–1796

Ferrera VP, Wilson HR (1990) Perceived direction of moving two-dimensional patterns. Vision Res 30:273–387

Ferrera VP, Wilson HR (1991) Perceived speed of moving two-dimensional patterns. Vision Res 31:877–893

Fried SI, Münch TA, Werblin FS (2002) Mechanisms and circuitry underlying directional selectivity in the retina. Nature 420:411–414

Grossberg S (1973) Contour enhancement, short-term memory, and constancies in reverberating neural networks. Stud Appl Math 52:213–257

Grossberg S (1978) A theory of human memory: self-organization and performance of sensory-motor codes, maps, and plans. In: Rosen R, Snell F (eds) *Progress in theoretical biology*, Volume 5. Academic, New York, NY, pp 233–374

Grossberg S (1980) How does a brain build a cognitive code? Psychol Rev 87:1–51

Grossberg S (1991) Why do parallel cortical systems exist for the perception of static form and moving form? Percept Psychophys 49:117–141
Grossberg S (1994) 3-D vision and figure-ground separation by visual cortex. Percept Psychophys 55:48–121
Grossberg S (1997) Cortical dynamics of three-dimensional figure-ground perception of two-dimensional pictures. Psychol Rev 104:618–658
Grossberg S (1999) How does the cerebral cortex work? Learning, attention and grouping by the laminar circuits of visual cortex. Spat Vis 12:163–185
Grossberg S (2000) The complementary brain: unifying brain dynamics and modularity. Trends Cogn Sci 4:233–246
Grossberg S (2003) How does the cerebral cortex work? Development, learning, attention, and 3D vision by laminar circuits of visual cortex. Behav Cogn Neurosci Rev 2:47–76
Grossberg S, Mingolla E (1985a) Neural dynamics of form perception: boundary completion, illusory figures, and neon color spreading. Psychol Rev 92:173–211
Grossberg S, Mingolla E (1985b) Neural dynamics of perceptual grouping: textures, boundaries, and emergent segmentations. Percept Psychophys 38:141–171
Grossberg S, Pearson L (2008) Laminar cortical dynamics of cognitive and motor working memory, sequence learning and performance: towards a unified theory of how the cerebral cortex works. Psychol Rev 115(3):677–732
Grossberg S, Pilly P (2008) Neural dynamics of probabilistic decision making during motion perception in the visual cortex. Vision Res 48(12):1345–1373
Grossberg S, Raizada RD (2000) Contrast-sensitive perceptual grouping and object-based attention in the laminar circuits of primary visual cortex. Vision Res 40:1413–1432
Grossberg S, Rudd M (1989) A neural architecture for visual motion perception: group and element apparent motion. Neural Netw 2:421–450
Grossberg S, Rudd ME (1992) Cortical dynamics of visual motion perception: short-range and long-range apparent motion. Psychol Rev 99:78–121
Grossberg S, Seitz A (2003) Laminar development of receptive fields, maps and columns in visual cortex: the coordinating role of the subplate. Cereb Cortex 13:852–863
Grossberg S, Swaminathan G (2004) A laminar cortical model for 3D perception of slanted and curved surfaces and of 2D images: development, attention, and bistability. Vision Res 44:1147–1187
Grossberg S, Williamson JR (2001) A neural model of how horizontal and interlaminar connections of visual cortex develop into adult circuits that carry out perceptual grouping and learning. Cereb Cortex 11:37–58
Grossberg S, Yazdanbaksh A (2005) Laminar cortical dynamics of 3D surface perception: stratification, transparency, and neon color spreading. Vision Res 45:1725–1743
Grossberg S, Mingolla E, Viswanathan L (2001) Neural dynamics of motion integration and segmentation within and across apertures. Vision Res 41:2351–2553
Grunewald A, Bradley DC, Andersen RA (2002) Neural correlates of structure-from-motion perception in macaque V1 and MT. J Neurosci 22:6195–6207
Hochstein S, Shapley RM (1976a) Linear and nonlinear spatial subunits in Y cat retinal ganglion cells. J Physiol (London) 262:265–284
Hochstein S, Shapley RM (1976b) Quantitative analysis of retinal ganglion cell classifications. J Physiol (London) 262:237–264
Horwitz GD, Newsome WT (2001) Target selection for saccadic eye movements: direction-selective visual responses in the superior colliculus. J Neurophysiol 86:2527–2542
Horwitz GD, Batista AP, Newsome WT (2004) Direction-selective visual responses in macaque superior colliculus induced by behavioral training. Neurosci Lett 366:315–319
Hupé JM, Rubin N (2003) The dynamics of bi-stable alternation in ambiguous motion displays: a fresh look at plaids. Vision Res 43:531–548
Hupé JM, James AC, Payne BR, Lomber SG, Girard P, Bullier J (1998) Cortical feedback improves discrimination between figure and background by V1, V2 and V3 neurons. Nature 394:784–787

Jones HE, Grieve KL, Wang W, Sillito AM (2001) Surround suppression in primate V1. J Neurophysiol 86:2011–2028

Kanizsa G (1979) Organization in vision: essays on Gestalt perception. Praeger Press, New York, NY

Kelly FJ, Grossberg S (2000) Neural dynamics of 3-D surface perception: figure-ground separation and lightness perception. Percept Psychophys 62:1596–1619

Kennedy H, Bullier J (1985) A double-labeling investigation of the afferent connectivity to cortical areas V1 and V2 of the macaque monkey. J Neurosci 5:2815–2830

Kim J, Wilson HR (1993) Dependence of plaid motion coherence on component grating directions. Vision Res 33:2479–2489

Lidén L, Mingolla E (1998) Monocular occlusion cues alter the influence of terminator motion in the barber pole phenomenon. Vision Res 38:3883–3898

Lindsey DT, Todd JT (1996) On the relative contributions of motion energy and transparency to the perception of moving plaids. Vision Res 36:207–222

Livingstone MS (1998) Mechanisms of direction selectivity in macaque V1. Neuron 20:509–526

Livingstone MS, Conway BR (2003) Substructure of direction-selective receptive fields in macaque V1. J Neurophysiol 89:2743–2759

Livingstone MS, Hubel DH (1984) Anatomy and physiology of a color system in the primate visual cortex. J Neurosci 4:309–356

Lorenceau J, Alais D (2001) Form constraints in motion binding. Nat Neurosci 4:745–751

Marr D, Ullman S (1981) Directional selectivity and its use in early visual processing. Proc R Soc Lond B Biol Sci 211:151–180

Movshon JA, Newsome WT (1996) Visual response properties of striate cortical neurons projecting to area MT in macaque monkeys. J Neurosci 16:7733–7741

Murthy A, Humphrey AL (1999) Inhibitory contributions to spatiotemporal receptive-field structure and direction selectivity in simple cells of cat area 17. J Neurophysiol 81:1212–1224

Nakayama K, Shimojo S, Silverman GH (1989) Stereoscopic depth: its relation to image segmentation, grouping, and the recognition of occluded objects. Perception 18:55–68

Pack CC, Born RT (2001) Temporal dynamics of a neural solution to the aperture problem in visual area MT of macaque brain. Nature 409:1040–1042

Pack C, Grossberg S, Mingolla E (2001) A neural model of smooth pursuit control and motion perception by cortical area MST. J Cogn Neurosci 13:102–120

Pack CC, Gartland AJ, Born RT (2004) Integration of contour and terminator signals in visual area MT of alert macaque. J Neurosci 24:3268–3280

Palanca BJ, DeAngelis GC (2003) Macaque middle temporal neurons signal depth in the absence of motion. J Neurosci 23:7647–7658

Pilly PK, Grossberg S (2005) Brain without Bayes: temporal dynamics of decision-making in the laminar circuits of visual cortex. Society of Neuroscience Abstracts, 591.1

Pilly PK, Grossberg S (2006) Brain without Bayes: temporal dynamics of decision-making during form and motion perception by the laminar circuits of visual cortex. J Vis 6:886

Ponce CR, Lomber SG, Born RT (2008) Integrating motion and depth via parallel pathways. Nat Neurosci 11(2):216–223

Raizada RD, Grossberg S (2003) Towards a theory of the laminar architecture of cerebral cortex: computational clues from the visual system. Cereb Cortex 13:100–113

Reichardt W (1961) Autocorrelation, a principle for evaluation of sensory information by the central nervous system In W.A. Rosenblith (ed.), *Sensory communication* (p. 303). Wiley, New York, NY, pp 303–317

Rockland KS (1992) Laminar distribution of neurons projecting from area V1 to V2 in macaque and squirrel monkeys. Cereb Cortex 2:38–47

Rockland KS (1995) Morphology of individual axons projecting from area V2 to MT in the macaque. J Comp Neurol 355:15–26

Rockland KS (2002) Visual cortical organization at the single axon level: a beginning. Neurosci Res 42:155–166

Rockland KS, Pandya DN (1981) Cortical connections of the occipital lobe in the rhesus monkey: interconnections between areas 17, 18, 19 and the superior temporal sulcus. Brain Res 212:249–270

Roitman JD, Shadlen MN (2002) Response of neurons in the lateral intraparietal area during a combined visual discrimination reaction time task. J Neurosci 22:9475–9489

Rust NC, Majaj NJ, Simoncelli EP, Movshon JA (2002) Gain control in Macaque area MT is directionally selective. Society for Neuroscience Abstracts, Orlando, FL

Shadlen MN, Newsome WT (2001) Neural basis of a perceptual decision in the parietal cortex (area LIP) of the rhesus monkey. J Neurophysiol 86:1916–1936

Shimojo S, Silverman GH, Nakayama K (1989) Occlusion and the solution to the aperture problem for motion. Vision Res 29:619–626

Shipp S, Zeki S (1985) Segregation of pathways leading from area V2 to areas V4 and V5 of macaque monkey visual cortex. Nature 315:322–325

Shipp S, Zeki S (1989) The organization of connections between areas V5 and V1 in Macaque visual cortex. Eur J Neurosci 1:309–332

Sincich LC, Horton JC (2002) Divided by cytochrome oxidase: a map of the projections from V1 to V2 in macaques. Science 295:1734–1737

Sincich LC, Horton JC (2003) Independent projection streams from macaque striate cortex to the second visual area and middle temporal area. J Neurosci 23:5684–5692

Snowden RJ, Treue S, Erickson RG, Andersen RA (1991) The response of area MT and V1 neurons to transparent motion. J Neurosci 11:2768–2785

Stone LS, Watson AB, Mulligan JB (1990) Effect of contrast on the perceived direction of a moving plaid. Vision Res 30:1049–1067

Stoner GR, Albright TD (1998) Luminance contrast affects motion coherency in plaid patterns by acting as a depth-from-occlusion cue. Vision Res 38:387–401

Stoner GR, Albright TD, Ramachandran VS (1990) Transparency and coherence in human motion perception. Nature 344:153–155

Tanaka K, Sugita Y, Moriya M, Saito H (1993) Analysis of object motion in the ventral part of the medial superior temporal area of the macaque visual cortex. J Neurophys 69:128–142

Treue S, Maunsell JHR (1999) Effects of attention on the processing of motion in macaque middle temporal and medial superior temporal visual cortical areas. J Neurosci 19:7591–7602

Trueswell JC, Hayhoe MM (1993) Surface segmentation mechanisms and motion perception. Vision Res 33:313–323

Tse PU (2005) Voluntary attention modulates the brightness of overlapping transparent surfaces. Vision Res 45:1095–1098

Vallortigara G, Bressan P, Bertamini M (1988) Perceptual alternations in stereokinesis. Perception 17:4–31

van Santen JP, Sperling G (1985) Elaborated Reichardt detectors. J Opt Soc Am A 2:300–3321

Varela JA, Sen K, Gibson J, Fost J, Abbott LF, Nelson SB (1997) A quantitative description of short-term plasticity at excitatory synapses in Layer 2/3 of rat primary visual cortex. J Neurosci 17:7926–7940

Wallach H (1935) On the visually perceived direction of motion. Psychol Forsch 20:325–380

Wallach H (1976) *On perception.* Quadrangle, New York, NY

Weiss Y, Adelson EH (2000) Adventures with gelatinous ellipses – constraints on models of human motion analysis. Perception 29:543–566

Wuerger S, Shapley R, Rubin N (1996) 'On the visually perceived direction of motion' by Hans Wallach: 60 years later. Perception 25:1317–1367

Xiao DK, Raiguel S, Marcar V, Orban GA (1997) The spatial distribution of the antagonistic surround of MT/V5 neurons. Cereb Cortex 7:662–677

Xu X, Bonds AB, Casagrande VA (2002) Modeling receptive-field structure of koniocellular, magnocellular, and parvocellular LGN cells in the owl monkey (Aotus trivigatus). Visual Neurosci 19:703–711

Yabuta NH, Callaway EM (1998) Functional streams and local connections of layer 4C neurons in primary visual cortex of the macaque monkey. J Neurosci 18:9489–9499
Yabuta NH, Sawatari A, Callaway EM (2001) Two functional channels from primary visual cortex to dorsal visual cortical areas. Science 292:297–300
Yo C, Wilson HR (1991) Perceived speed of moving two-dimensional patterns depends on duration, contrast and eccentricity. Vision Res 32:877–893
Zhou H, Friedman HS, von der Heydt R (2000) Coding of border ownership in monkey visual cortex. J Neurosci 20:6594–6611

Chapter 14
Features in the Recognition of Emotions from Dynamic Bodily Expression

Claire L. Roether, Lars Omlor, and Martin A. Giese

Abstract Body movements can reveal important information about a person's emotional state. The visual system efficiently extracts subtle information about the emotional style of a movement, even from point-light stimuli. While much existing work has addressed the problem of style perception from a holistic perspective, we try to investigate which features are critical for the recognition of emotions from full-body movements. This work is inspired by the motor-control concept of "synergies," which define spatial components of movements that encompass only a limited set of degrees of freedom that are jointly controlled. We present an algorithm that learns a highly compact generative model for the joint-angle trajectories of emotional body movements. The model approximates movements by nonlinear superpositions of a small number of basis components. Applying sparse feature learning, we extracted from this representation the spatial components that are characteristic for happy, sad, fearful and angry movements. The extracted features for walking were highly consistent with emotion-specific features of gait, as described in the literature. We further show that this type of result is not restricted to locomotor movements. Compared to other techniques, the proposed algorithm requires significantly fewer basic components to accomplish the same level of accuracy. In addition, we show that feature learning based on such less compact representations does not result in easily interpretable local features. Based on the features extracted from the trajectory data, we studied how spatio-temporal components that convey information about emotional styles of body movements are integrated in visual perception. Using motion morphing to vary the information content of different components, we show that the integration of spatial features is slightly suboptimal compared to a Bayesian ideal observer. Besides, integration was worse for components that matched the components extracted from the movement trajectories.

M.A. Giese (✉)
Section for Computational Sensomotorics, Department of Cognitive Neurology, Hertie Institute for Clinical Brain Research & Center for Integrative Neuroscience, Frondsbergstr. 23, 72074 Tuebingen, Germany
e-mail: martin.giese@uni-tubebingen-de

U.J. Ilg and G.S. Masson (eds.), *Dynamics of Visual Motion Processing: Neuronal, Behavioral, and Computational Approaches*,
DOI 10.1007/978-1-4419-0781-3_14,

This result is inconsistent with the hypothesis that emotional body movements are recognized by a parallel internal simulation of the underlying motor behavior. Instead, it seems that the recognition of emotion from body movements is based on a purely visual process that is influenced by the distribution of attention.

14.1 Introduction

For humans as a highly social species, the reliable recognition of the emotional states of conspecifics is a central visual skill. Accordingly, communication about affect states is considered as a valuable evolutionary adaptation, an idea already proposed in Darwin's seminal work on emotion expression in different species (Darwin 2003). This adaptive value is obvious in cases where immediate survival is concerned, since for instance the sight of a frightened person usually implies a threatening situation nearby. However, the reading of affects from subtle signals, such as motion styles, can also be advantageous in many social situations, as shown by a recent study demonstrating greater success in negotiations in persons who are more adept at reading emotional expressions (Elfenbein et al. 2007).

Most research on the topic of emotion expression has focused on emotional faces. There is now general consensus that humans reliably recognize a number of emotional expressions that are characterized by different arrangements of facial features. In particular, there is a set of six emotions (anger, happiness, sadness, fear, disgust and surprise) that is recognized universally across many different cultures (Ekman 1992; Ekman and Friesen 1971; Izard 1977).

Emotions are complex states involving a multitude of physiological changes (Cacioppo et al. 2000). It is thus not surprising that the expression of emotions by humans is not restricted to the face: imagine watching the audience at a large sports event. Details of people's facial expression may not be visible. However, it seems easy to infer from the body movements whether a part of the crowd is emotionally positive or negative about current events on the field, even in the absence of auditory cues. In recent years, there has been growing interest in such possible associations between body movements and emotion attributions, or for short, emotional body expressions (Atkinson et al. 2004, 2007; Boone and Cunningham 1998; Clarke et al. 2005; de Gelder 2006; de Gelder and Hadjikhani 2006; de Meijer 1989, 1991; Ekman 1965; Ekman and Friesen 1967; Grezes et al. 2007; Hietanen et al. 2004; Montepare et al. 1987, 1999; Pollick et al. 2001, 2002; Sogon and Masutani 1989; Walk and Homan 1984; Wallbott 1998; Wallbott and Scherer 1986). Such studies involve the recording of emotionally expressive body movements by video or motion capture. Emotions are often evoked by specific scenarios. Such movements can be highly expressive, as demonstrated by the finding that human observers can classify them with accuracies that are significantly above chance level.

The moving human body can communicate emotions in many different ways. A fair number of studies have been performed to elucidate which features of the movement should be considered expressive. Multiple types of body movement can be used to transport emotional messages (e.g., Ekman and Friesen 1972; Friesen et al. 1979). One class of emotional movements is gestures from which an emotional state can be inferred. Examples are manipulators such as scratching oneself, or culturally shaped emblems as the eye-wink or the thumbs-up sign. But even when gestures are not considered, the style variations of one and the same movement can be sufficient to support the percept of an emotional state in the sender (e.g., Hietanen et al. 2004; Montepare et al. 1987; Pollick et al. 2001). It is this type of emotional body expression that we focus on.

Most studies in which the association between characteristics of the movement and the perception of emotion were investigated have taken a "holistic approach": observers rated specific features of the movement of the whole body, for example, the overall level of activity, the spatial extent of a movement and its jerkiness. Some characteristics are consistently found in such studies, for example, that angry movements tend to be large, fast and relatively jerky, whereas both fearful and sad movements are smaller and slower (Montepare et al. 1987; Sogon and Masutani 1989; Wallbott 1998). Certain body parts also appear particularly important for the expression of different emotions. An often-cited example is that the expression of sadness is characterized by a hanging head and hunched shoulders. Similarly, the movement of a single arm has been shown to be sufficient for above-chance recognition of emotional states (Pollick et al. 2001; Sawada et al. 2003). Parts of the human figure can thus serve as features for the perception of emotion from body movements.

Related work, in particular in computer graphics, has investigated how motion styles can be modeled and parameterized (Brand and Hertzmann 2000; Safonova 2004; Unuma 1995). Recent work has employed such techniques for the study of the visual processing of style variations for movements with different degrees of complexity, including emotional movement styles (Giese and Lappe 2002; Jordan et al. 2006; Mezger et al. 2005; Troje 2002). However, all of these studies have addressed the perception of emotional style in a holistic fashion, not specifically analyzing which spatio-temporal features are contributing to the perception of individual emotional styles.

In this chapter, we try to address the question which spatio-temporal features are important for the perception of emotional styles from body movements. For this purpose, we first provide an analysis of the motor behavior during the execution of emotional movements. We present a new unsupervised learning method for the approximation of full-body movements by highly compact generative models. Based on this compact representation, applying sparse feature learning, we extract the spatio-temporal features that are most characteristic for different emotional styles. On the basis of this analysis of the motor behavior, in a second step, we then investigate the visual perception of emotional body movements. We study how different spatio-temporal features are integrated during the formation of emotional-style judgments. More specifically, we compare the integration of different available cues in the visual system with an ideal-observer model that assumes feature integration by linear combination. Such models have been shown to be adequate for cue integration in vision and between different sensory modalities (Alais and Burr 2004; Ernst and Banks 2002; Hillis et al. 2004; Knill 2007).

In the following, we first briefly describe the data basis with emotional movements on which our theoretical and experimental studies were based (Sect. 14.2). We will then in Sect. 14.3 address the execution of emotional body movements, and will discuss the algorithm for the extraction of informative spatio-temporal features. Section 14.4 presents a set of psychophysical experiments that study how different emotion-specific spatio-temporal components are integrated in the visual perception of emotional body movements.

14.2 Database of Emotional Movements

Our experimental and theoretical studies were based on a database of emotional movements that was recorded with a motion-capture system. The database contained different types of movements (walking and forehand tennis swing) that were executed with five emotional styles (happy, sadness, anger, fear and neutral).

14.2.1 Actors

For the analysis of the movement features, the movements from 14 right-handed lay actors with several years of acting experience were recorded (five male, nine female, mean age 27 years 3 months). None of the actors had orthopedic or other problems could interfere with normal movement behavior.

14.2.2 Recording Procedure

During the recordings actors walked on a straight line for approximately five meters, first neutrally, and then expressing anger, happiness, sadness and fear in counterbalanced order. Each condition was repeated three times. To ensure spontaneous expression of emotions, the actors were instructed to imagine emotionally charged life events while they performed expressive gestures, vocalization and facial expression. The pure imagination of life events by itself had been shown to be an effective mood-induction procedure (Westermann et al. 1996). Once reaching the intended mood state, the actors started the trials of the recordings, during which the performed movements were prescribed, with the instruction to execute the movement with the adopted emotional affect. In addition, participants were instructed to avoid any extra gestures. Recordings and psychophysical experiments were performed with informed consent of participants. All experimental procedures had been approved by the responsible local ethics board of the University of Tübingen (Germany).

Movement trajectories were recorded using an eight-camera VICON 612 motion capture system (VICON, Oxford, UK). The system has a sampling frequency of

120 Hz and determines the three-dimensional positions of 41 reflective markers (1.25 cm diameter) with spatial error below 1.5 mm. The markers were attached to skin or tight clothing with double-sided adhesive tape, and commercial VICON software was used to reconstruct the three-dimensional marker positions, and to interpolate short missing parts of the trajectories.

14.2.3 Data Processing

For further data processing, a single gait cycle was selected from each trial, re-sampled with 100 time steps, and smoothed by spline interpolation. For computing the joint angles, the marker positions were approximated by a hierarchical kinematic body model with 17 joints (head, neck, spine, and right and left clavicle, shoulder, elbow, wrist, hip, knee and ankle) and coordinate systems attached to all rigid segments of the skeleton. With small deviations of the basis vectors from orthogonality corrected by singular-value decomposition and with the pelvis coordinate system serving as the beginning of the kinematic chain, the rotations between adjacent coordinate systems were characterized by Euler angles. Any jumps caused by ambiguities in the computation of Euler angles were removed by unwrapping, and differences between start and end points of the trajectories were corrected by spline interpolation between the five first and last frames of each trajectory. Trajectories were additionally smoothed by fitting with a third-order Fourier series.

14.3 Features in Motor Behavior of Emotional Expressions

As first step of our analysis, we investigated what differentiates body motion trajectories that were recorded from humans executing the same actions with different emotional styles. A naïve approach that just lists the many significant variations in quantities that characterize the kinematics and the dynamics between movements with different emotional styles seemed not very satisfying as a description of the critical emotion-specific spatio-temporal features. Instead, we tried to devise an algorithm that automatically extracts a small number of highly informative spatio-temporal features. The algorithm comprises two algorithmic steps: (1) learning of a highly compact model for the trajectories of emotional body movements applying a special method for blind source separation that is adopted to the statistical structure of the data; and (2) sparse feature learning in order to automatically select a small number of critical spatio-temporal features that account for the differences between trajectories expressing different emotional styles.

In the following, we first briefly describe the new algorithm for the learning of compact trajectory models and demonstrate its performance in comparison with other popular methods for dimension reduction for trajectory data. We then introduce the algorithm for sparse feature learning and demonstrate that it extracts meaningful spatio-temporal features for emotional gaits, which largely match previous data in the literature. In addition, we present results from other classes of movements for which no such previous data exist.

14.3.1 Learning of Generative Models for Body Movement

Since the seminal work of Bernstein in the 1920s it has been a classical idea in motor control that complex motor behavior might be based on the combination of simpler primitives, termed "synergies," that encompass only a limited set of degrees of freedom (Bernstein 1967). Classically, this concept has been proposed as possible solution for the "degrees of freedom problem," that is, the problem to devise efficient control algorithms for biological motor systems with large numbers of degrees of freedom. A variety of proposals have been made to simplify this problem by decomposing the control of complex movements into smaller components, like movement primitives (e.g., Flash and Hochner 2005) or basis vector fields (d'Avella and Bizzi 2005; Poggio and Bizzi 2004).

Based on this theoretical idea, several studies have applied dimension reduction methods for the extraction of basic components from measurements of motor behavior by unsupervised learning. Some of these studies have applied PCA or factor analysis to movement trajectories (Ivanenko et al. 2004; Santello and Soechting 1997). Others have analyzed EMG data applying non-negative matrix factorization, recovering a small number of components that are sufficient for the reconstruction of the muscle activities in the frog (d'Avella and Bizzi 2005). Similar approaches for dimension reduction have also been applied in computer graphics and computer vision for the synthesis and tracking of full-body movements from components learned from motion capture data (Safonova et al. 2004; Yacoob and Black 1999). For an accurate approximation of complex body movements typically a significant number of basis components (e.g., >8 principle components) were required. While such models with a large number of basic components provide excellent approximations of trajectory data, the model parameters are typically difficult to interpret since the variance of the data is distributed over a large number of model terms. For obtaining models with easily interpretable parameters it is thus critical to concentrate the variance onto a few highly informative terms using a model that is adjusted to the statistical properties of the data.

Our method for the learning of compact trajectory models is based on independent component analysis (ICA). The development of the algorithm started with the observation that, like PCA, ICA requires a quite significant number of source terms to achieve very accurate approximations of the trajectories of all degrees of freedom of full-body movements.

If, however, the same analysis is applied to the angle trajectories of individual joints, it turns out that a very small number of source terms (independent components) is sufficient to accomplish very accurate approximations (e.g., explaining

>96% of the variance with only three components). Even more interesting is the observation that the obtained source components for different joints often are extremely similar in shape, and mainly differ by phase or time-shifts. Figure 14.1 shows the first three ICA components extracted from the shoulder and elbow trajectories of a set of arm movements, consisting of right-handed throwing, golf swing and tennis swing.

After appropriate phase shifting, the components extracted from the elbow and shoulder joint are almost identical. A measure for the similarity that is invariant against phase shifts is given by the maximum of the correlation over all possible phase shifts. The values of this measure for two example movements are listed in Table 14.1.

This specific statistical property of the trajectory data motivates an approximation of the data using a generative model that deviates from the linear mixing model underlying normal PCA and ICA. These methods are defined by instantaneous mixtures of the form:

$$x_i(t) = \sum_{j=1}^{n} \alpha_{ij} s_j(t) \tag{14.1}$$

In this model the functions s_j identify source signals that are orthogonal or statistically independent. The trajectories result from these signals by weighted linear superposition, where the weights of the sources in the mixtures are defined

Fig. 14.1 First three ICA components extracted from the shoulder and elbow trajectories in arm movements before and after phase shifting. The duration of all movements was normalized, so the time axis indicates the percentage of the movement completed

Table 14.1 Maximum cross-correlations between source signals (over all phase shifts) extracted from the trajectories of three exemplary arm joints for walking (left) and non-periodic arm movements (throwing, golf swing and tennis swing) (right part of table)

	Walking			Arm movements		
Movements	Right shoulder	Right elbow	Right wrist	Right shoulder	Right elbow	Right wrist
Right shoulder	1	0.83	0.88	1	0.93	0.85
Right elbow	0.83	1	0.80	0.93	1	0.83
Right wrist	0.88	0.80	1	0.85	0.83	1

by the mixing weights α_{ij}. In this model phase shifts between different joints can only be accommodated for by introducing a sufficient number of source terms, defining a basis that is rich enough to approximate the same signal form with multiple time shifts. This model is thus not ideally suited for modeling variations of the coordination between multiple joints.

A model that takes the specific structure of the trajectory data better into account includes time shifts τ_{ij} for the individual source terms:

$$x_i(t) = \sum_{j=1}^{n} \alpha_{ij} s_j(t - \tau_{ij}) \tag{14.2}$$

Models with time shifts have been previously proposed for the analysis of EMG signals, assuming positivity of the approximated signals (d'Avella and Bizzi 2005). Here we present an algorithm that extends this approach for signals with arbitrary sign. In acoustics this kind of model is called *anechoic mixture*. Classical applications of anechoic mixtures arise in electrical engineering, for example, when signals from multiple antennas that are received asynchronously, or in acoustics, for example, when sound signals are recorded with multiple microphones resulting in different running times. Only few algorithms have been proposed for the solution of under-determined anechoic mixing problem. In this case, the number of sources exceeds the number of signals. Almost no work exists about over-determined anechoic mixing problems, where the number of source signals is smaller than the number of original signals. This case is most important for dimension reduction.

14.3.2 Algorithm for Blind Source Separation

The mathematical structure of the mixture (14.2) suggests that a solution of the demixing problem might be obtained exploiting the framework of time-frequency analysis. (Because of the time delays an analysis in frequency space using usual Fourier transformation would lead to complex mixtures of frequency-dependent phase terms.) Signal representations that reflect the properties of the signal with respect to frequency bands as well as with respect to time lie at the core of time-frequency analysis. In some sense time-frequency transformations are similar to a music chart that also presents time and frequency information simultaneously. A well-known example of such representations is the time-windowed Fourier transform. A variety of representations of this type have been proposed in signal processing and acoustics.

Our algorithm is based on a popular quadratic representation, the Wigner-Ville spectrum (WVS), which is particularly appealing due to its close connection to energy and correlation measures. The WVS of a random process $x(t)$ is defined as the partial Fourier transform of the symmetric autocorrelation function of x:

$$W_x(t,\omega) := \int E\left\{x\left(t+\frac{\tau}{2}\right)\overline{x\left(t-\frac{\tau}{2}\right)}\right\}e^{-2\pi i\omega\tau}\,d\tau \tag{14.3}$$

The WVS is basically a bivariate function defined over the time-frequency plane, which can loosely be interpreted as a time-frequency distribution of the mean energy of x.

Applying this integral transform (14.3) to the (14.1) results in the following relationship in time-frequency space:

$$W_{xi}(t,\omega) = \sum_j \alpha_{ij}^{\,2} W_{s_j}(t-\tau_{ij},\omega) \tag{14.4}$$

This equality holds true only if the sources are assumed to be statistically independent. As a two-dimensional representation of one-dimensional signals, (14.4) is redundant and can be solved by computing a set of projections onto lower dimensional spaces that specify the same information as the original problem (14.2). Computing projections that integrate over unbounded domains with respect to the time parameter t are particularly useful, since they eliminate the dependence of the unknown time shifts τ_{ij}.

A simple set of projections is obtained by computing the first- and zero-order moments of (14.4). These terms can be computed analytically resulting in the identities:

$$|Fx_i(\omega)|^2 = \sum_j \alpha_{ij}^{\,2}\, |Fs_j|^2 \tag{14.5}$$

$$|Fx_i(\omega)|^2 \frac{\partial}{\partial\omega}\arg(Fx_i(\omega)) = \sum_j \alpha_{ij}^{\,2}\, |Fs_j|^2 \left(\frac{\partial}{\partial\omega}\arg(Fs_j(\omega)) + \tau_{ij}\right) \tag{14.6}$$

In these equations F denotes the normal Fourier transform and arg the complex argument.

The two equations are solved iteratively by consecutive execution of the following two steps, until convergence is achieved:

1. Solve (14.5): This equation defines a normal linear (instantaneous) mixture problem with positivity constraints for the parameters. This problem is tractable with different published algorithm for ICA and non-negative matrix factorization (Hojen-Sorensen et al. 2002; Lee and Seung 1999).
2. Inserting the result obtained in step 1, (14.6) is solved numerically, determining the time delays and unknown phases of the Fourier transforms.

Further details about the algorithm and its application to other data sets can be found in Omlor and Giese (2007).

14.3.3 Approximation of Emotional Body Movements

We applied the developed algorithm, and a number of other unsupervised learning methods to our data. In particular, we compared popular methods for blind source separation (ICA and PCA) assuming linear instantaneous mixtures (without time delays) and Fourier analysis with our method.

Figure 14.2a shows a comparison of the tested methods for emotional gaits, plotting the explained variance against the number of source terms (components) included in the model.

Instead of evaluating approximation quality by the usual "explained variance," which is defined by the expression:

$$1-\left(\frac{\|\mathbf{D}-\mathbf{F}\|}{\|\mathbf{D}\|}\right)^2$$

where **D** signifies the original data matrix and **F** its approximation by the model and ||.|| indicates the Frobenius norm. Figure 14.2a shows a measure for approximation quality that is given by the expression:

$$1-\left(\frac{\|\mathbf{D}-\mathbf{F}\|}{\|\mathbf{D}\|}\right)$$

This measure has the advantage that it varies linearly with the residual-norm, opposed to the explained variance that is typically difficult to interpret for small residuals, since it is then very close to one.

Clearly, the method based on the mixture model (14.2) outperforms traditional PCA and ICA (Bell and Sejnowski 1995). To reach an accuracy level for which the approximation quality is higher than 90% PCA and ICA require at least five to six

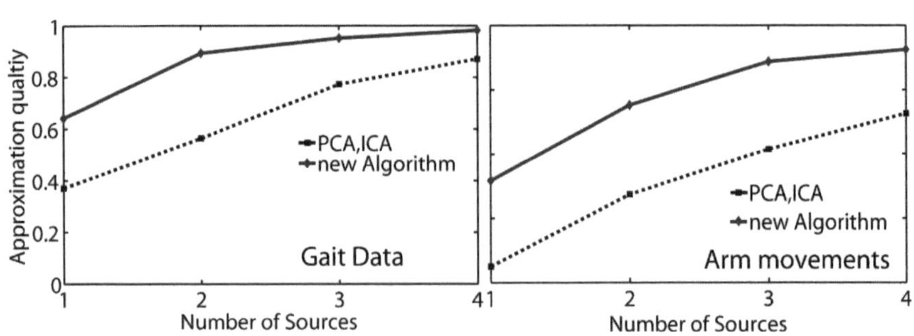

Fig. 14.2 Approximation quality as a function of the number of sources for traditional blind source separation algorithms (PCA/ICA) and our new algorithm

components, while the same level of accuracy is achieved with only two to three sources for the models based on (14.2).

Qualitatively the same result is obtained for the non-periodic arm movements (throwing, golf and tennis swing). As expected, the total approximation quality is lower than for walking, reflecting the higher variability of this dataset. An approximation quality of 90% is achieved using the mixture model (14.2) with four to five terms while normal PCA and ICA require more than seven terms to achieve the same level of accuracy. This shows that the proposed special structure of the generative model is beneficial not only for periodic movements.

14.3.4 Algorithm for the Learning of Spatio-Temporal Features

After training the proposed nonlinear mixture provides a structured accurate model for the trajectory data that can be used as basis for the learning of spatio-temporal features that are indicative for different emotional styles. In the following, we describe a simple algorithm for the learning of such features.

The model (14.2) parameterizes trajectories in terms of the source signals, the time delays and the mixing weights. We first applied this mixture model separately for trajectories with different emotions. Comparison of the estimated source function and delays revealed almost no differences between the different emotions. Delays varied between different joints, but almost not between different emotions. This motivated the estimation of a common source functions and delays over all emotions for the different actions. Within this parameterization differences between the emotions are encoded in the mixing weights α_{ij}. Concatenating all mixing weights into a single vector $\mathbf{a}$, one can approximate each style by a single vector. Assuming that $\mathbf{a}_0$ signifies the vector corresponding to the neutral version of the action, the style of the action corresponding to emotion j can be characterized by a vector $\mathbf{a}_j$. This movement can be characterized by its deviation from the neutral action according to the equation

$$\mathbf{a}_j = \mathbf{a}_0 + \mathbf{C}\mathbf{e}_j \tag{14.7}$$

where $\mathbf{e}_j$ signifies the corresponding unit vector. The columns of the matrix $\mathbf{C}$ specify the differences between emotional and neutral versions of the same action.

The idea of our algorithm for the learning of emotion-specific spatio-temporal features is to sparsify the matrix $\mathbf{C}$. This implies finding an approximate solution of (14.7) with a large number of zero entries of this matrix. A simple way for the estimation of such a solution is the minimization of an error function of the form:

$$L(\mathbf{C}) = \sum_j |\mathbf{a}_j - \mathbf{a}_0 - \mathbf{C}\mathbf{e}_j|^2 + \gamma \sum_{ij} |C_{ij}| \tag{14.8}$$

The second term measures the L_1 norm of the entries of the matrix **C** and punishes solutions with many small non-zero entries in this matrix. It is well-established that L_1 norm regularization terms result in such sparse solutions (Andrew 2004). The positive constant γ controls the influence of this term and was chosen to accomplish solutions with about 40% nonzero terms. The solution of the convex minimization problem (14.9) can be obtained by quadratic programming.

14.3.5 Application to Emotional Body Movements

The proposed algorithm for the automatic extraction of important spatio-temporal features was first applied to the trajectories of emotional gaits, because for this class of movements some data about important emotion-specific features were available in the literature, based on psychophysical rating studies (de Meijer 1989; Montepare et al. 1987; Wallbott 1998). Comparing the features extracted from the trajectories by our algorithm with this published data, we were able to verify whether our algorithm extracts features with biological relevance. In addition, we compared results from feature extraction for the anechoic mixture model described in Sect. 14.3.1 with other popular models for the approximation of movement trajectories.

Figure 14.3a shows the weight matrix obtained by the minimization of the error function (9) for different emotional gaits, as color-coded plot. Since only the weights corresponding to the source signal $s_1(t)$ resulted in non-zero elements of the matrix **C** we dropped the weight contribution of the other sources from the plot. The figure shows if the weights of the corresponding joints are increased (plain grayscale tile) or decreased (grayscale tile with white triangle) compared to neural walking. The signs in the figure represent a summary of results from the psychophysical literature (de Meijer 1989; Montepare et al. 1987; Wallbott 1998) and indicate the amplitude changes that human raters judged to be characteristic for different emotions (compared to neutral walking). The match between the features extracted automatically by our algorithm and the features from the literature is almost perfect. The only major deviation is that the algorithm detects a reduction of the knee angle amplitudes for fearful walking, a feature that we found also indicative in own psychophysical experiments. This implies that the proposed algorithm seems to extract features that are biologically meaningful, matching dominant features extracted by human raters.

Figure 14.3b shows the result obtained with the same feature extraction method applied to a representation with three sources extracted by PCA, a standard technique for dimension reduction in engineering that has been applied to gait trajectories by many groups. The number of significant features was matched between the two analyses. Opposed to our method there is no clear consistence between the features obtained from the PCA representation and the literature data. Also often the signs of the extracted joint angle changes are incorrect. Finally, Fig. 14.3c shows the results obtained by the same feature extraction algorithm if applied for a generative model that combines PCA and subsequent fitting of the weights by a

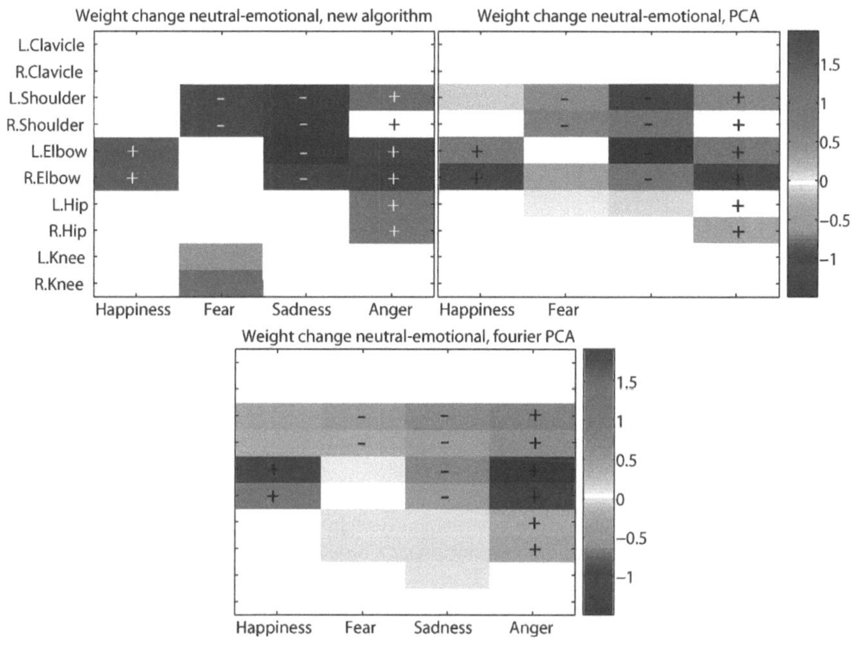

Fig. 14.3 Joint-specific changes of the contributions (weights) of the first source in emotional walking compared to neutral walking for three different generative models. Kinematic features that have been shown to be important for the perception of emotions from gait in psychophysical experiments are indicated by the plus and minus signs (details see text) (*see Color Plates*)

Fourier series (Fourier PCA). Modeling of gait styles by interpolation of Fourier series has been a classical approach in computer graphics (Unuma et al. 1995). Fourier PCA has been popularized in psychology as model for gait morphing and potential basis of explaining biological motion recognition in the brain (Troje 2002). For our set of gait data this technique required the introduction of eight source terms to accomplish an approximation quality with >95% explained variance. This is about two times more than the model defined by (14.9) (counting the two terms per frequency of the Fourier series as one to equate the number of free parameters). As illustrated in Fig. 14.3c, application of feature learning to this representation results in a profile that strongly deviates from the data in the literature.

Summarizing, these results indicate that application of a generative model that matches the intrinsic structure of the data results in more interpretable results for the extraction of the features that carry information about emotional style.

Subsequently, we applied the same feature extraction method to non-periodic emotional movements. For many classes of movements previous studies have reported a strong connection between expressiveness of the movement and simple kinematic features like speed or amplitude (Amaya et al. 1996; Pollick et al. 2001). The proposed feature extraction method supplements these studies as the shape of the trajectories is analyzed for emotion-specific content and its localization in

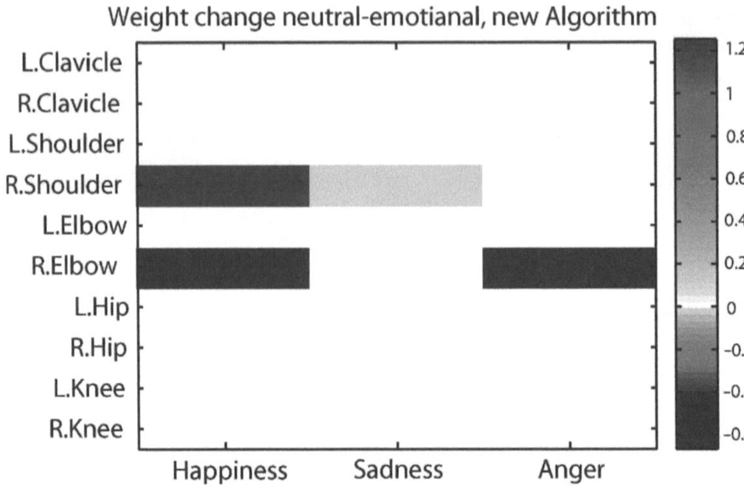

Fig. 14.4 Joint-specific change in the contribution (weight) of the first source in a right-handed tennis swing compared to the neutral movement (*see Color Plates*)

specific joints. In addition, the feature extraction assumed data with normalized movement time. It thus extracts more subtle emotion-specific features that go beyond the obvious and well-established result that, for example, slow movements tend to be rated as "sad" while fast movements are rather interpreted as "happy" or "angry."

An example from this ongoing analysis is shown in Fig. 14.4 that shows the extracted features for a right-handed tennis swing (only the right arm was moving). The feature analysis shows emotion-specific characteristic profiles of the amplitudes of the elbow and the shoulder joints. Consistent with observations for other movements in the literature, some emotion-specific expressive features are shared between emotions (Ekman and Friesen 1967), like the decrease of the elbow amplitude between anger and happiness.

14.4 Spatial Components in the Perception of Emotional Body Expressions

The results presented in the previous sections show that dynamic emotional body expressions can be characterized by characteristic spatio-temporal features. In addition, the features visually perceived as emotionally expressive largely overlap with emotion-specific features extracted from the motor behavior. This motivates the question how such spatio-temporal features are integrated in the visual perception of emotional body movements. More specifically, we studied whether the integration of such features shares properties with cue integration of other perceptual functions, which often can be modeled by linear cue integration models. In addition, we tried to study whether visual features that match the features extracted from the motor behavior of emotional expressions (Sect. 14.3.5) are particularly efficiently integrated in perception. This prediction was motivated by the popular hypothesis that perception of motor acts, and potentially also of emotions, might be based on an internal simulation of the underlying motor behavior (Gallese 2006; Wolpert et al. 2003). Recognition should thus be most accurate and sensitive if the structure of external stimuli matches the structure of such internal models as closely as possible.

For the investigation of these questions we devised a computational technique that permits precise control of the emotion-specific information in different spatio-temporal components of moving human figures presented as point-light stimuli. Human observers rated the expressiveness of emotional body movements that were generated with this method. The ratings were then compared to the behavior of a statistically optimal ideal-observer model in order to investigate to what extent feature integration is statistically optimal.

In the following, we first introduce the computational method for the variation of the information content of individual components (Sect. 14.4.1). We then discuss the design of the experiment (Sect. 14.4.2) and introduce a simple statistically optimal ideal-observer model (Sect. 14.4.3). The experimental results are discussed in Sect. 14.4.4.

14.4.1 Component-Based Motion Morphing

To vary the emotion-specific information in our stimuli we applied a technique for *motion morphing*. Deviating from the usual applications of such techniques, we morphed different spatial components, defined by groups of dots of a point-light figure, separately. This made it possible to specify, for example, strong emotion-specific information for the arm movement, but low emotion-specific information for the movement of the legs. All morphs were generated by linearly combining the trajectories of a prototypical emotional walk (angry, fearful, sad) with the trajectory of an emotionally neutral walk from the same actor.

Motion-morphing algorithms generate new trajectories by blending or interpolating between prototype movements with different style properties (Bruderlin and Williams 1995; Wiley and Hahn 1997). Such methods are highly suitable for the generation of stimuli for psychophysical experiments investigating the perception of movement style (Giese and Lappe 2002; Jordan et al. 2006; Troje 2002). We applied a morphing algorithm that creates new trajectories by linear combination of trajectories in space-time (Giese and Poggio 2000). Formally, the morphs can be characterized by the equation

$$\mathbf{x}_{\text{new}} = (1-m)\cdot\mathbf{x}_{\text{neutral}} + m\cdot\mathbf{x}_{\text{emot,k}} \tag{14.9}$$

where m represents a morphing parameter that determines the information about the emotion contained in the morph. The variables $\mathbf{x}_{\text{neutral}}$ and $\mathbf{x}_{\text{emot,k}}$ signify the trajectories of the neutral walk and of the walk with emotion k from the same actor. The multiplication signs signify linear combination in space-time, rather than the simple linear combination of the trajectory values time-point by time-point (Giese and Poggio 2000). All morphs were computed for the movements of only one actor in order to avoid artifacts caused by differences in the body geometry of different actors. By variation of the parameter m we generated the whole continuum between emotionally neutral and emotionally expressive walking (see Supplementary Movie 1).

We have previously shown that this morphing method produces natural-looking morphs even for different locomotion patterns (Giese and Lappe 2002), with perceived properties that interpolate between the ones of the prototypes. An additional study has shown that the metric of the morphing-parameter space for locomotion patterns closely matches the perceptual metric reconstructed by applying multi-dimensional scaling to human similarity judgments (Giese and Poggio 2003). The same method produces highly natural-looking morphs even for very complex movements, like karate techniques, and is suitable for applications in computer graphics (Mezger et al. 2005).

In order to vary the information content of different spatial components of point-light patterns separately, we applied the same algorithm to the trajectories of subgroups of dots. The parameters m_1 and m_2 referring to the morphing parameters of two different spatial components, each of them defined by a number of dots of the point-light stimulus, one can formally describe the resulting morph by the equations:

$$\begin{aligned}\mathbf{x}_{\text{new}}^{(1)} &= (1-m_1)\cdot\mathbf{x}_{\text{neutral}}^{(1)} + m_1\cdot\mathbf{x}_{\text{emot,k}}^{(1)}\\ \mathbf{x}_{\text{new}}^{(2)} &= (1-m_2)\cdot\mathbf{x}_{\text{neutral}}^{(2)} + m_2\cdot\mathbf{x}_{\text{emot,k}}^{(2)}\end{aligned} \tag{14.10}$$

The variables $\mathbf{x}_{\text{new}}^{(i)}$ signify the generated trajectories of the dots that belong to spatial component (i). Likewise, $\mathbf{x}_{\text{neutral}}^{(i)}$ and $\mathbf{x}_{\text{emot,k}}^{(i)}$ signify the trajectories of the corresponding prototypes. By varying the morphing parameters m_1 and m_2, the information content in the two spatial components can be gradually changed. This change can be applied to both components together and at the same level, that is, $m_1=m_2$. For this type of stimulus the choice $m_1=m_2=1$ defines a morph with full information content in both components, that is, corresponding to the emotional prototype, while $m_1=m_2=0$ specifies a neutral walk. Stimuli with information content only in the first component would correspond to parameter combinations with $m_1>0$ and $m_2=0$, while the combination $m_1=0$ and $m_2=1$ defines a stimulus with no information about emotion in the first component, but full information in the second.

14.4.2 Experimental Design

For studying how information is integrated across different spatial components we chose two different ways of dividing the dots into spatial components. The first division ("Upper–lower") was defined by the feature combinations found in the analysis of the motor patterns, which showed a strong right–left symmetry (Sect. 14.3.5). Comparing the changes relative to neutral walking, arms and legs emerged as separate spatial components that show emotion-specific changes. The walker was thus separated in the upper and lower half at the level of the pelvis (upper: head, arms and spine; lower: hips and legs, as shown in Fig. 14.5). With the second type of division (Right–left) we explicitly tried to violate the right–left symmetry observed in the analysis of the motor behavior. In this case, the components were defined by one arm and one leg from opposite sides of the body (the head was part of the component containing left arm and right leg).

The chosen spatial components always comprised at least one or more complete limbs of the point-light walker (see Supplementary Movies 2 and 3). This ensured a minimum violation of kinematic constraints, confirmed by our informal observation during debriefing that none of the observers reported the observation of strange-looking kinematic features or irregularities that would make it difficult to "imitate" the observed movement.

For comparison with the ideal-observer models we generated three different stimulus classes by variation of the morphing weights, for each of the two types of division (Fig. 14.5). For the first two classes information about emotion was present only in one of the spatial components (weight combinations with $m_1 \geq 0$, $m_2 = 0$ or $m_1 = 0$, $m_2 \geq 0$). We refer to this type of stimulus as "first-component" or "second-component," respectively. In particular, these two components for the "Upper–lower" division are referred to as "upper-body" and "lower-body," whereas the terms "left–right" and "right–left" denote the two component conditions of the "Right–left" component set. The two component conditions were used to determine the free parameters of the ideal-observer model. The third condition, which we refer to as "full-body," specified information about emotional style simultaneously for both spatial components ($m_1 = m_2 \geq 0$). The ratings of emotional expressiveness in this condition were predicted from the ideal-observer model and compared to the ratings measured with the full-body stimuli. Deviations of a subject's responses from the predicted statistically optimal ratings were indicative of suboptimal integration of the information provided by the two spatial components.

The prototype trajectories for the morphing were selected from the database (Sect. 14.2). A pilot experiment with 15 observers showed that the selected emotion prototypes were recognized at a minimum of 80% correct. The morph weights were adjusted for the individual emotions in order to achieve an optimal sampling of the response curves. The weights for the different stimulus classes are listed in Table 14.2. All stimuli were shown at the walking speed of neutral walking for the relevant actor.

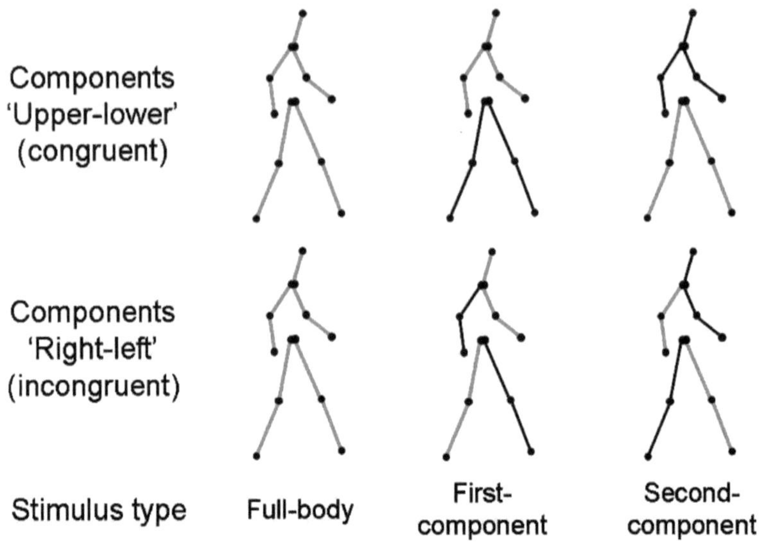

Fig. 14.5 Sets of spatial components used in the perception experiment. (Lines connecting the point-light walker's dots were not shown in the experiment.) *Light gray lines* indicate the two components that specified different amounts of emotion-specific information. *Black lines* denote parts of the figure moving as in neutral walking. *Top row*: *Upper–lower* components that are consistent with the features extracted from motor behavior. *Bottom row*: *Right–left* components consisting of an opposite arm and leg of the walker, violating the right–left symmetry observed in the motor behavior. The emotional style of the head movement is modulated together with the component containing the left arm and the right leg

Table 14.2 Morphing weights of the emotional prototypes for the different types of spatial components and different emotions. For the second component of the Upper–lower division different weights had to be chosen for sadness than for the other two emotions, to ensure an optimal sampling of the rating function, because the recognizability of sadness from the leg movements was smaller than for the other two emotions

	Upper–lower			Right–left
Full body	Component 1	Component 2		Component 1 and 2
All emotions	All emotions	Anger, Fear	Sadness	All emotions
0.05	0.05	0.1	0.1	0.05
0.10	0.10	0.2	0.2	0.10
0.15	0.15	0.3	0.3	0.15
0.20	0.20	0.4	0.5	0.20
0.25	0.25	0.5	0.6	0.25
0.30	0.30	0.6	0.8	0.30
0.50	0.50	0.8	1.0	0.50
0.80	0.80	1.0	1.3	0.80

The two sets of components were tested in separate experiments withnon-overlapping participants. The participants were students at the University of Tübingen, and they all had normal or corrected-to-normal vision. They were tested individually and were paid for their participation. For the Upper–lower components 11 participants (6 male, 5 female, mean age 23.6 years) and for the Right–left components 13 participants (5 male, 8 female, mean age 22.9 years) were included in the analysis.

Each of the two experiments consisted of three blocks, one for each of the three emotions anger, sadness and fear. The order of emotions counterbalanced across participants. In each block a total of 330 stimuli was shown, in random order: neutral walking was shown 90 times, and each of the morphed stimuli was repeated 10 times. On each trial one stimulus was shown, and the participant rated the intensity of expression of the target emotion stimulus on a seven-point scale (ranging from "not expressing the emotion" to "expressing the emotion very strongly"), responding by pressing the number keys 1 to 7. When a response key was pressed, a gray screen was shown for an inter-stimulus interval of 500 ms, followed by presentation of the next stimulus. The gray screen was also shown if the subject had not responded after 2.5 consecutively presented step cycles. Testing took place in a small, dimly lit room. Stimuli were displayed and participants' responses were recorded using the Psychophysics Toolbox (Brainard 1997) on a PowerBook G4 (60 Hz frame rate; 1280×854 pixel resolution), viewed from a distance of 50 cm.

Stimuli were presented as point-light walkers consisting of 13 dots, as shown in Fig. 14.5. The positions of these dots were computed from the morphed three-dimensional trajectories by parallel projection. We chose a profile view, the figure always facing to the observer's left. The walkers were moving as if on a treadmill, simulated by fixing the center of gravity of the figures to a constant point in space. The point-light stimuli consisted of black dots (diameter 0.47° of visual angle) on a uniform gray background. The overall figures subtended approximately 4 by 8.6° of visual angle.

14.4.3 Cue-Fusion Model

Many perceptual tasks require the integration of multiple sensory cues for making perceptual decisions. Such cues might arise from the same sensory modality, as in the context of depth-perception that integrates diverse cues such as shape and texture, motion, retinal disparity, or from different sensory modalities, as for the integration of haptic and visual estimates of for example, object size. The sensory estimate obtained in presence of multiple cues can often be well approximated by a linear combination of the estimates provided by the individual cues (Alais and Burr 2004; Knill 2007; Landy and Kojima 2001; Landy et al. 1995). Assuming normal distributions and independence for the individual cues, one can derive the statistically optimal estimator (by maximum likelihood estimation) resulting in a linear combination where the cues are weighted by their relative reliabilities (Alais and Burr 2004; Ernst and Banks 2002; Hillis et al. 2004; Knill 2003).

We applied the theoretical framework of such cue integration models to different spatial cues in the perception of emotional body expressions. We assumed that, as for the perception of objects that likely integrates information from different spatial parts or features (Harel et al. 2007; Logothetis et al. 1995), the recognition of emotions from body movements might integrate different spatio-temporal components.

The information content of individual spatial components was varied by the motion morphing technique described in Sect. 14.4.1. The morph parameters m_1 and m_2 defined thus the true information contents in the spatial components of the stimulus. Subjects rated the emotional expressiveness of each stimulus, defining the perceptual rating y. With the assumption that the emotional expressiveness ratings obtained from the individual cues are linearly related to the morph parameters m_i and normally distributed one can derive the model prediction for the rating (see Appendix):

$$\hat{y} = w_0 + w_1 m_1 + w_2 m_2 \tag{14.11}$$

Since it was difficult to obtain reliable ratings of emotional expressiveness from stimuli containing only one spatial component (especially for stimuli with emotional information restricted to lower-extremity movement) we did not try to estimate the reliability of the individual cue estimates directly. Instead we chose an approach where we directly fitted the model parameters w_i model (14.11) based on the first- and second-component stimulus sets, for which the emotion-specific information was restricted to one of the spatial components. We then used this model to predict the ratings of the subjects for the full-body stimuli, where emotional-style information was present in both spatial components (Sect. 14.4.2). The parameters w_i were estimated by linear regression. The prediction quality of the model was assessed by comparing the model prediction with a General Linear Models fitted directly to the predicted data (see Appendix for details).

14.4.4 Experimental Results

The results of the experiments, averaged across subjects, are shown in Fig. 14.6. As expected, the rated emotional intensity of the stimuli generally increased with increasing morphing level. This finding provides support that the morphing technique was effective in gradually varying the information content of the stimuli. In addition, the ratings vary almost linearly with the morph parameters, supporting the adequateness of the linearity assumption that was central for the derivation of the model (14.11).

For all emotions and both sets of components, the regression line for the full-body condition always had the steepest slope, indicating that the emotional information

Fig. 14.6 Results of the cue-integration experiment. Two types of spatial components were tested: *Upper–lower* (corresponding to components extracted from motor behavior) (**a**), and *Right–left* (**b**). Mean intensity ratings are shown as a function of the morph parameters m_1 and m_2 (linear weight). The first column shows the ratings measured with the full-body stimuli (*solid lines*) and the prediction from the first- and second-component conditions (*dashed line*). The other two columns show the ratings for the component conditions. Standard errors were not plotted because they were very small (<0.15)

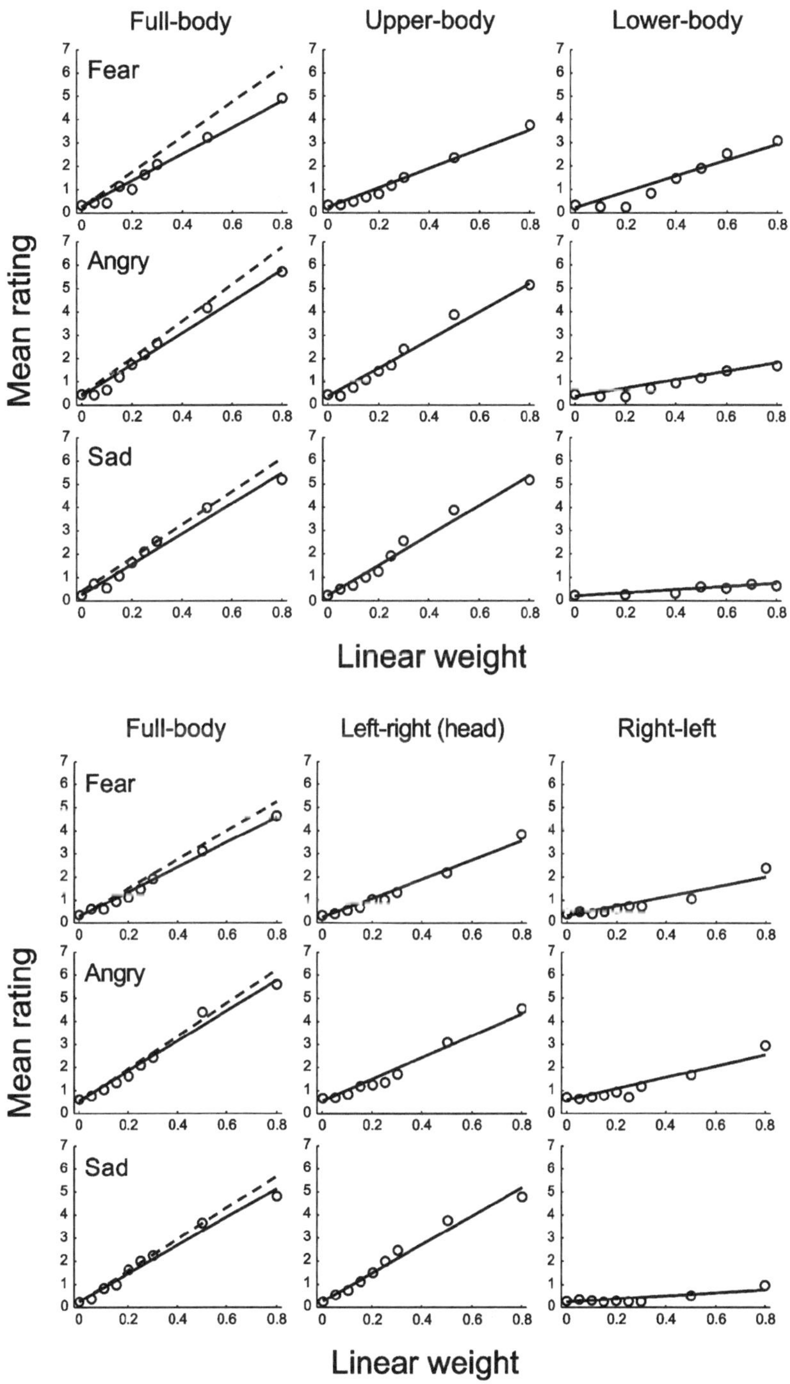
Full-body
Upper-body
Lower-body
Fear
Angry
Sad
Mean rating
Linear weight
Full-body
Left-right (head)
Right-left
Fear
Angry
Sad
Mean rating
Linear weight

Fig. 14.6

Table 14.3 Percentage of subjects with significant deviations (*F* test) between the ratings obtained for the full-body stimuli and the model prediction derived from the first- and second-component condition. In all cases of significant deviation the prediction by the ideal-observer model overestimated the results obtained with the full-body stimuli

Emotion	Upper–lower	Right–left
Angry	67.7%	38.5%
Fearful	67.7%	38.5%
Sad	18.2%	38.5%

was integrated across the spatial components. The predictions derived from the model (dashed lines) are close to the real data, but often slightly steeper. This indicates a close-to-optimal but slightly suboptimal integration of the information provided by the spatial components. Interestingly, the predictions for the Right–left stimuli are closer to the experimental data than for the Upper–lower components. This indicates that, opposed to the hypothesis that spatial components that match the ones extracted from motor behavior are more efficiently processed, we found a more efficient integration of the "holistic" components that included opposite arms and legs.

These results are confirmed by a statistical analysis of the goodness of fit of the predictions obtained from the first- and second-component conditions in comparison with the results for the full-body stimuli. Table 14.3 provides a summary of the significant F values from the model comparison. Significant F values indicate that the model prediction deviated significantly from a regression model estimated directly from the test data (see Appendix). The F values were in the range 0.02–66.5. The table shows that for more than half of the subjects the ideal-observer model significantly overestimated the emotional-expressiveness rating significantly for the Upper–lower components, while this happens only for about one third of the subjects for the Right–left components.

The slopes of the regression lines for the first- and second-component stimuli in the Upper–lower set provide information about how informative arms and legs are for the expression of different emotions. The slope for the first component, corresponding to the arms, was always significantly higher than that obtained with the second (all $t>6.53$, d.f.≥ 9, $p<0.001$), except for the expression of fear. This finding suggests a general importance of the movement of the upper half of the body for the expression of emotions. This is consistent with the idea that leg movements are relatively constrained in walking, making the upper body more important for the expression of emotional styles. However, leg movement seems to contribute significantly to the perception of anger and fear.

14.5 Discussion

The current study analyzed the relevance of spatio-temporal features in the production and perception of emotionally expressive body movements. On the one hand, we presented an algorithm, combining unsupervised and sparse feature learning, suitable for the automatic extraction of highly informative sets of emotion-specific spatio-temporal features from the joint-angle trajectories of human body movements. For emotional gait patterns, this method extracted features that were consistent with the literature reports of informative spatio-temporal features of emotional gait, obtained from perceptual ratings of human observers. This result implies that the perception of emotional body movements is sensitive to visual features that correspond to the dominant differences between the joint trajectories of emotional and neutral gaits. The algorithm was also applied to non-periodic body movements. We found emotion-specific changes in joint-angle amplitudes for emotionally expressive arm movements despite normalizing the movement time. Since it has been shown before (Pollick et al. 2001; Sawada et al. 2003) that average speed and movement time are efficient cues for influencing emotion ratings, our analysis reveals that, beyond these elementary cues, there are more subtle changes in the movement kinematics that contribute to the perception of emotional expression of non-periodic body movements.

The proposed algorithm can be applied for the extraction of spatio-temporal features that are characteristic for other motion styles that are not related to emotions. This makes it interesting for the extraction, for example, of features relevant for the perceived attractiveness or skill level of movements, and even for clinical applications for the detection of local spatio-temporal features that are characteristic of, for example, neurological movement deficits.

In addition, the proposed algorithm provides a highly compact generative model for the joint-angle trajectories of full-body movements, accurate enough to model subtle changes of motion styles. This property makes the algorithm suitable for the learning of structured models of stylized classes of movements for synthesis applications, for example, in computer graphics and robotics. An important step, which is presently addressed in ongoing work, is the mapping of the proposed generative model onto a real-time capable architecture that generates trajectories online, and that is suitable for a reactive modulation of timing and trajectory style in response to external events.

The second part of this study tried to investigate how the visual system integrates information about the emotional style of a movement over multiple spatio-temporal components. For this purpose, we devised a special motion-morphing technique that is suitable for modulating the information content about emotion separately in different spatio-temporal components of point-light stimuli. We specifically tested whether components in the visual stimulus that match the ones extracted from the motor behavior in the first part of our analysis are integrated with higher efficiency than components that are inconsistent with the structure of motor behavior. We found that a simple linear model, which predicts perceived emotional expressiveness as a linear combination of the information content in the spatial components, parameterized by the corresponding morphing weights, provides a reasonable fit to the data. This was particularly true if the emotional information of only one spatial component is varied. If information was changed simultaneously for two spatial components we found that perceived expressiveness was often slightly overestimated by the additive model. Interestingly, the simple additive model provided a better fit for the spatial components that were not congruent with the components extracted from motor behavior (Right–left) than for the components designed to be congruent with motor behavior (Upper–lower). This might be explained by the fact that the Right–left components are more spatially extended, potentially requiring a broader distribution of attention, while the Upper–lower components might sometimes result in focused attention on the upper body (arms), so that subjects missed the additional information provided by the cues in the lower body. It is well established that the perception of biological motion is strongly influenced by attention (Cavanagh et al. 2001; Thornton et al. 2002). However, further studies, potentially including the monitoring of attentional strategies with eye-movement recordings, are needed to clarify whether this hypothesis is true.

The detailed analysis of the contribution of the Upper–lower components to perceived expressiveness revealed that for all emotions the movement of the upper body is most critical for emotion recognition. This is consistent with the results of our feature analysis for the trajectories, and with previous studies in the literature (Wallbott 1998).

Interestingly, there was an indication of a difference between the movement of the left and right sides of the body, especially for arm movement (top left panel of Fig. 14.3). This asymmetry we observed might be partially caused by the fact that the left body side moves with higher energy and amplitude for emotional body movements than the right body side (Roether et al. 2009). The tendency toward higher expressiveness at a given morphing level for the component containing the left arm than for the component containing the right arm (cf. Fig. 14.6b) may be influenced by this fact, but this conclusion is confounded by the design of the components: the position of the head was only varied with the former of the components in the Left–right set.

The result that spatial features that match components extracted from the execution of emotional body expressions are integrated less efficiently than spatial features that are inconsistent with these components points against the hypothesis that recognition of emotional body expressions uses an internal representation reflecting the fine structure of motor behavior. In this case, one would expect a more efficient integration of the information from components that match the intrinsic structure of such potential internal models. However, it might be that recognition of emotional body movements uses a form of more abstract predictive internal simulation that does not reflect the fine structure of the movement trajectories.

An alternative quite simple account for our results is that the perception of emotional movements is based on visual learning. The visual system might learn informative visual features that are distinctive for different emotions, independent of the exact structure of the motor system. Such approaches have been very successful in computer vision. Since emotion-specific changes of the movement trajectories usually also induce changes of visual features, this explains the observed similarity between the components extracted from motor behavior and from visual judgments of human observers in the literature. The integration of different spatial features, however, might be governed by the general rules of feature integration in the visual system: feature efficiency might be determined, for example, by the overlap between stimulus components and the receptive fields of different levels of the visual pathway, and on the attentional state, rather than being critically dependent on the structure of motor programs of emotional movements.

14.6 Supplementary Materials (CD-ROM)

Movie 1 (MorphAngry): Morphing between neutral walking and angry walking. Bar represents linear weight with which emotional prototype contributes to movement pattern.
Movie 2 (UpperAngry): Angry walking restricted to upper half of body.
Movie 3 (LowerAngry): Angry walking restricted to lower half of body.

Acknowledgments We thank T. Flash for many interesting discussions, and for pointing our interest to synergies as classical concept of spatio-temporal components in motor control, and B. de Gelder and A. Berthoz for interesting comments. We are grateful to W. Ilg for help with the motion capturing. This research was supported by HFSP, EC FP6 project COBOL, and the Volkswagenstiftung. Further support by the Max Planck Institute for Biological Cybernetics and the Hermann und Lilly Schilling-Stiftung is gratefully acknowledged.

Appendix: Ideal-Observer Model

For the statistical model, we assumed that the perceived emotional expressiveness y is a linear function of the morph parameters of the individual spatial components m_1 and m_2. Assuming that the ratings obtained for a fixed value of the morph parameter m_i are normally distributed, the parameters of the model, w_0 are w_1 and w_2 can then be estimated by linear regression.

Data are given as triples of the two morph-levels $m_1^{\,l}$ and $m_2^{\,l}$ and the rating response y^l for each trial. All analyses are performed within subject only. Data for the first- and second-component conditions, that is, in which the emotion content was only varied in one of the spatial components, were used to determine the parameters of the model (Training data). The parameters were estimated by minimization of the quadratic error function:

$$R_{\mathrm{F}}(\mathbf{w}) = \sum_{l} (y^l - w_0 - w_1 m_1^{\,l} - w_2 m_2^{\,l})^2. \tag{14.12}$$

In the following, the parameters estimated from the Training data are denoted by $w_{i,Tr}$.

These parameters were used to predict the ratings for the full-body stimuli (i.e., varying both morph levels together: $m_1^{\,l} = m_2^{\,l}$). The residual of this prediction is given by the function

$$R_T = \sum_{l'} (y^{l'} - w_{0,\mathrm{Tr}} + w_{1,\mathrm{Tr}} m_1^{\,l'} + w_{2,\mathrm{Tr}} m_2^{\,l'})^2. \tag{14.13}$$

where $y^{l'}$ signifies the Test data. This residual is compared with the residual $R_{TF}(\hat{\mathbf{w}})$ that is obtained by minimizing (14.12) using the Test data instead of the Training data. The resulting estimated parameters are referred to as $\hat{\mathbf{w}}$ in the following.

To evaluate the goodness of fit between the Test data and the model's prediction from the Training Data we used a likelihood-ratio test. As shown below, the test compares the difference between the residuals of the prediction and the true fit of the Test data with the error variance of the fit of the Test data:

$$F = \frac{(R_T - R_{TF}(\hat{\mathbf{w}}))/3}{R_{TF}(\hat{\mathbf{w}})/(N_T - 3)} \tag{14.14}$$

where fitted parameter N_T denotes the number of data points in the Test data set. For the discussed assumptions, this quantity has an F-distribution with $(3, N_T - 3)$ degrees of freedom.

References

Alais D, Burr D (2004) The ventriloquist effect results from near-optimal bimodal integration. Curr Biol 14(3):257–262

Amaya K, Bruderlin A, Calvert T (1996) Emotion from motion. In: Proceedings of the conference on graphics interface '96, Canadian Information Processing Society, Toronto, Ontario, Canada

Andrew YN (2004) Feature selection, L1 vs. L2 regularization, and rotational invariance. In: Proceedings of the twenty-first international conference on machine learning, ACM Press, Banff, Alberta, Canada

Atkinson AP, Dittrich WH, Gemmell AJ, Young AW (2004) Emotion perception from dynamic and static body expressions in point-light and full-light displays. Perception 33(6):717–746

Atkinson AP, Tunstall ML, Dittrich WH (2007) Evidence for distinct contributions of form and motion information to the recognition of emotions from body gestures. Cognition 104(1):59–72

Bell AJ, Sejnowski TJ (1995) An information-maximization approach to blind separation and blind deconvolution. Neural Comput 7(6):1129–1159

Bernstein NA (1967) The coordination and regulation of movements. Pergamon Press, Oxford, New York

Boone RT, Cunningham JG (1998) Children's decoding of emotion in expressive body movement: the development of cue attunement. Dev Psychol 34(5):1007–1016

Brainard DH (1997) The psychophysics toolbox. Spat Vis 10(4):433–436

Brand M, Hertzmann A (2000) Style machines. In: Proceedings of the 27th annual conference on computer graphics and interactive techniques, ACM Press/Addison-Wesley Publishing Co

Bruderlin A, Williams L (1995) Motion signal processing. In: Proceedings of the 22nd annual conference on computer graphics and interactive techniques, ACM Press, pp 97–104

Cacioppo JT, Berntson GG, Larsen JT, Poehlmann KM, Ito TA (2000) The psychophysiology of emotion. In: Lewis R, Haviland-Jones JM (eds) The handbook of emotion. Guilford Press, New York, pp 173–191

Cavanagh P, Labianca AT, Thornton IM (2001) Attention-based visual routines: sprites. Cognition 80(1–2):47–60

Clarke TJ, Bradshaw MF, Field DT, Hampson SE, Rose D (2005) The perception of emotion from body movement in point-light displays of interpersonal dialogue. Perception 34(10):1171–1180

d'Avella A, Bizzi E (2005) Shared and specific muscle synergies in natural motor behaviors. Proc Natl Acad Sci USA 102(8):3076–3081
Darwin, Charles (1872) The expression of the emotions in man and animals, London: John Murray
de Gelder B (2006) Towards the neurobiology of emotional body language. Nat Rev Neurosci 7(3):242–249
de Gelder B, Hadjikhani N (2006) Non-conscious recognition of emotional body language. Neuroreport 17(6):583–586
de Meijer M (1989) The contribution of general features of body movement to the attribution of emotions. J Nonverbal Behav 13(4):247–268
de Meijer M (1991) The attribution of aggression and grief to body movements: The effect of sex-stereotypes. Eur J Soc Psychol 21(3):249–259
Ekman P (1965) Differential communication of affect by head and body cues. J Pers Soc Psychol 2(5):726–735
Ekman P (1992) Are there basic emotions? Psychol Rev 99(3):550–553
Ekman P, Friesen WV (1967) Head and body cues in the judgment of emotion: a reformulation. Percept Mot Skills 24(3):711–724
Ekman P, Friesen WV (1971) Constants across cultures in the face and emotion. J Pers Soc Psychol 17(2):124–129
Ekman P, Friesen WV (1972) Hand movements. J Commun 22(4):353–374
Elfenbein HA, Foo MD, White JB, Tan HH, Aik VC (2007) Reading your counterpart: the benefit of emotion recognition accuracy for effectiveness in negotiation. J Nonverbal Behav 31(4):205–223
Ernst MO, Banks MS (2002) Humans integrate visual and haptic information in a statistically optimal fashion. Nature 415(6870):429–433
Flash T, Hochner B (2005) Motor primitives in vertebrates and invertebrates. Curr Opin Neurobiol 15(6):660–666
Friesen WV, Ekman P, Wallbott H (1979) Measuring hand movements. J Nonverbal Behav 4(2):97–112
Gallese V (2006) Intentional attunement: a neurophysiological perspective on social cognition and its disruption in autism. Brain Res 1079(1):15–24
Giese MA, Lappe M (2002) Measurement of generalization fields for the recognition of biological motion. Vision Res 42(15):1847–1858
Giese MA, Poggio T (2000) Morphable models for the analysis and synthesis of complex motion patterns. Int J Comput Vis 38(1):59–73
Giese MA, Poggio T (2003) Neural mechanisms for the recognition of biological movements. Nat Rev Neurosci 4(3):179–192
Grezes J, Pichon S, de Gelder B (2007) Perceiving fear in dynamic body expressions. Neuroimage 35(2):959–967
Harel A, Ullman S, Epshtein B, Bentin S (2007) Mutual information of image fragments predicts categorization in humans: electrophysiological and behavioral evidence. Vision Res 47(15):2010–2020
Hietanen JK, Leppanen JM, Lehtonen U (2004) Perception of emotions in the hand movement quality of finnish sign language. J Nonverbal Behav 28(1):53–64
Hillis JM, Watt SJ, Landy MS, Banks MS (2004) Slant from texture and disparity cues: optimal cue combination. J Vis 4(12):967–992
Hojen-Sorensen PAdFR, Winther O, Hansen LK (2002) Mean-field approaches to independent component analysis. Neural Comput 14(4):889–918
Ivanenko YP, Poppele RE, Lacquaniti F (2004) Five basic muscle activation patterns account for muscle activity during human locomotion. J Physiol 556(Pt 1):267–282
Izard CE (1977) Human emotions. Plenum Press, New York
Jordan H, Fallah M, Stoner GR (2006) Adaptation of gender derived from biological motion. Nat Neurosci 9(6):738–739
Knill DC (2003) Mixture models and the probabilistic structure of depth cues. Vision Res 43(7):831–854

Knill DC (2007) Robust cue integration: a Bayesian model and evidence from cue-conflict studies with stereoscopic and figure cues to slant. J Vis 7(7):1–24

Landy MS, Kojima H (2001) Ideal cue combination for localizing texture-defined edges. J Opt Soc Am A Opt Image Sci Vis 18(9):2307–2320

Landy MS, Maloney LT, Johnston EB, Young M (1995) Measurement and modeling of depth cue combination: in defense of weak fusion. Vision Res 35(3):389–412

Lee DD, Seung HS (1999) Learning the parts of objects by non-negative matrix factorization. Nature 401(6755):788–791

Logothetis NK, Pauls J, Poggio T (1995) Shape representation in the inferior temporal cortex of monkeys. Curr Biol 5(5):552–563

Mezger J, Ilg W, Giese MA (2005) Trajectory synthesis by hierarchical spatio-temporal correspondence: comparison of different methods. In: ACM SIGGRAPH symposium on applied perception in graphics and visualization, A Coruna, Spain, pp 25–32

Montepare J, Koff E, Zaitchik D, Albert M (1999) The use of body movements and gestures as cues to emotions in younger and older adults. J Nonverbal Behav 23(2):133–152

Montepare JM, Goldstein SB, Clausen A (1987) The identification of emotions from gait information. J Nonverbal Behav 11(1):33–42

Omlor L, Giese MA (2007) Blind source separation for over-determined delayed mixtures. In: Schoülkopf, Bernhard; Platt, John; Hofmann, Thomas (ed) Advances in neural information processing systems, vol 19. MIT Press, Cambridge, MA, pp 1049–1056

Poggio T, Bizzi E (2004) Generalization in vision and motor control. Nature 431(7010):768–774

Pollick FE, Lestou V, Ryu J, Cho SB (2002) Estimating the efficiency of recognizing gender and affect from biological motion. Vision Res 42(20):2345–2355

Pollick FE, Paterson HM, Bruderlin A, Sanford AJ (2001) Perceiving affect from arm movement. Cognition 82(2):B51–B61

Roether CL, Omlor L, Christensen A, Giese MA (2009) Critical features for the perception of emotion from gait. Journal of Vision, 9(6):15, 1–32, http://Journalofvision.org/9/6/15/, doi:10.1167/9.6.15

Safonova A, Hodgins JK, Pollard, NS (2004) Synthesizing physically realistic human motion in low-dimensional, behavior-specific spaces. In: ACM SIGGRAPH 2004 Papers, ACM Press, Los Angeles, California

Santello M, Soechting JF (1997) Matching object size by controlling finger span and hand shape. Somatosens Mot Res 14(3):203–212

Sawada M, Suda K, Ishii M (2003) Expression of emotions in dance: relation between arm movement characteristics and emotion. Percept Mot Skills 97(3 Pt 1):697–708

Sogon S, Masutani M (1989) Identification of emotion from body movements: a cross-cultural study of Americans and Japanese. Psychol Rep 65(1):35–46

Thornton IM, Rensink RA, Shiffrar M (2002) Active versus passive processing of biological motion. Perception 31(7):837–853

Troje NF (2002) Decomposing biological motion: a framework for analysis and synthesis of human gait patterns. J Vis 2(5):371–387

Unuma M, Anjyo K, Takeuchi R (1995) Fourier principles for emotion-based human figure animation. In: Proceedings of the 22nd annual conference on Computer graphics and interactive techniques, ACM Press, New York, N.Y., USA, pp 91–96

Walk RD, Homan CP (1984) Emotion and dance in dynamic light displays. Bull Psychon Soc 22(5):437–440

Wallbott HG (1998) Bodily expression of emotion. Eur J Soc Psychol 28(6):879–896

Wallbott HG, Scherer KR (1986) Cues and channels in emotion recognition. J Pers Soc Psychol 51(4):690–699

Westermann R, Spies K, Stahl G, Hesse FW (1996) Relative effectiveness and validity of mood induction procedures: A meta-analysis. Eur J Soc Psychol 26(4):557–580

Wiley DJ, Hahn JK (1997) Interpolation synthesis of articulated figure motion. IEEE Comput Graph Appl 17(6):39–45

Wolpert DM, Doya K, Kawato M (2003) A unifying computational framework for motor control and social interaction. Philos Trans R Soc Lond B Biol Sci 358(1431):593–602

Yacoob Y, Black MJ (1999) Parameterized modeling and recognition of activities. Comput Vis Image Underst 73(2):232–247(216)

Index

A

Action-perception cycle
  coherent motion, 203
  corollary discharge information, 202–203
  efference copy information, 200
  orthogonal decomposition, 202
  outline chevron, 200, 201
  output stages, brain, 204
  perceptual coherence, chevron, 202
  retinal image smear, 203
  retinal velocity, chevron, 201–202
  spatial attention, 204
  spatial representation, 200
  stable world assumption, 202
  visual sensitivity suppression, 200
Ad-hoc retinotopic constraints, 74
Aperture problem
  biological solution dynamics, 166
  coarse-to-fine progression, 38
  definition, 284
  feature tracking signals, 284–285
  finite unitary contour, line-endings, 5
  1D homogeneous contour motion, 4
  mixing retinal and non-retinal cues, 175–177
  1D-to-2D motion dynamics, visual stimuli, 39
  motion integration, 29
  neural computation, 37
  neurophysiological support, 285
  predicted aperture problem solution, 285–287
  recurrent architecture, 174, 182
  retinal and non-retinal cues
    anticipatory eye movements, 175
    motion direction, 175–176
    predictive mechanisms, 175
    pursuit initiation, 175, 176
    retinal image velocity, 176
    steady-state eye velocity, 176–177
    target visual motion and trajectory prediction, 173, 176
  schematic representation, 163
  stimulus encoding, 6
  velocity–speed and direction recovery, 7
  visual system, 47
  V1-MT recurrent loop, 182
  vote weight, 6–7
Attention-deficit hyperactivity disorder (ADHD), 205

B

Barber poles, 167, 168
Bar texture stimuli analysis, 64
Bayesian models, 171, 172, 301–303
Bistable stimuli, 198
BOLD activity, 24
Brain reading approach, 74

C

CDS. *See* Component direction selectivity
Chopsticks illusion, 291
Component direction selectivity (CDS), 60, 61, 63
Contour motion perception
  adaptation paradigm, 11
  orthogonal perceived direction, 9–10
Contrast sensitivity reduction
  central origin interpretation, 224–226
  H-H paradigm, 223–225
  retinal origin interpretation
    contour masking, 226
    magnocellular pathway specificity, 228
    retinal light adaptation level, 229–230
    threshold values, 227
    time course, 228

Cortical space, neuronal population dynamics
flash-lag effect, 105–106
line-motion illusion
cortical VSD response, 109–110
facilitatory effect, 109
low- and high-amplitude activity, 110
non-moving stimuli, 108
non-retinotopic mechanism, 110
sequential suprathreshold activation, 109
spatio-temporal pattern, 110
long-range horizontal connections, 106–108
optical imaging, 106

D

Directional grouping, 299–300
Directionally selective neurons, 60
Direction discrimination task (DDT), 130
Dorsal motion and ventral form
global motion recovery, 22
invariant spatial structure, 21
lateral occipital complex (LOC), 22
proto shape, 24
pursuit dependency, motion coherence, 25
space of stimulus, definition, 15
spatial discontinuities processing issues, 21
spatial organization, 21, 22
Dorsolateral pontine nucleus (DLPN), 251
Drift-balanced motion (dbm), 118
Dynamic association field, 85, 89

E

Elementary motion detector (EMD), 117
Emotion recognition
communication, 2
current study analysis
joint-angle trajectories, 24
motion integration, 24
results, 24, 25
upper-lower components, 26
visual cognition, 25
emotional movements database
actors, 4
data processing, 5
recording procedure, 4–5
emotion perception, 3
eye-wink/thumbs-up sign., 3
ideal-observer model, 27–28
motor behavior features
application, 13–16
approximation, 11–12
blind source separation algorithm, 9–11
learning generative models, 6–9
spatio-temporal algorithm, 12–13
spatial components perception
component-based motion morphing, 17–18
cue-fusion model, 21–22
experimental design, 18–21
experimental results, 22–24
spatio-temporal features, 4
visual cognition, 3
Endstopped neurons, 48
Extra-retinal model
perceptual evidence
contrast sensitivity, 243
isoluminant stimuli, 242–243
luminance-modulated gratings, 242, 243
photoreceptors, 243–244
physiological evidence
active and passive condition, 244
MT/MST neurons, 246, 250
parietal cortex of alert, 244
response latencies, 247
spike arrival times, 245
Extrinsic terminators, 287–289

F

Feature tracking signals, 284–285
Figure-ground separation
motion induction, 295–296
perceptual grouping, 295
First-order motion energy
detectors, 143–144
1-D *mf* grating stimuli, 145–147
missing fundamental *(mf)* stimulus, 143
radial luminance modulation, 147
Formotion capture prediction
chopsticks illusion, 290, 291
cooperative–competitive model, 291
motion grouping, 292
motion perception, adaptation, 291
occluders, 291, 292
probabilistic decision making, 293
V2-to-MT motion selection, 290
3D FORMOTION model
form processing system
bipole cells, 294
cross-orientation competition, 294–295
distinct ON and OFF cell networks, 293
motion induction, 295–296
perceptual grouping, 295

functional projections and properties, 294
motion processing system
directional grouping, 299–300
LGN input, 296
long-range filter and formotion selection, 299
spatial and opponent direction competition, 298
transient cells, 296–298
ubiquitous circuit design, 300
primate visual system, 293
Fourier amplitude spectrum, 65
Fourier motion (fm), 117, 118, 121
Fourier transform, 65, 262
Frontal eye field (FEF), 178
Functional magnetic resonance imaging (fMRI), 123
Functional networks
aperture and correspondence problem, 5
aperture stimuli, 16
association field illustration, 8
chopstick illusion, 17–18
combination rule, motion integration, 14–15
contour motion perception, dynamics
adaptation paradigm, 11
orthogonal perceived direction, 9–10
depiction processes, 28
diamond aperture, 17
direction discrimination, 18
dorsal motion and ventral form
global motion recovery, 22
invariant spatial structure, 21
lateral occipital complex (LOC), 22
proto shape, 24
pursuit dependency, motion coherence, 25
space of stimulus, definition, 15
spatial discontinuities processing issues, 21
spatial organization, 21, 22
eccentric *vs*. foveal motion integration, 26–27
electrophysiological recordings, 15–16
electrophysiological techniques, 3
1D homogeneous contour, 4
integration, segmentation and selection
coherence thresholds, measurements, 13
drifting plaids, 12
global motion coherence, 13
perceptual process, assessment, 11–12
plaids intersection, 14
local-ambiguous-direction, 4
motion disparity effect, 18
perceptual hysteresis, 20
wave propagation, contour neurons
functional assembly, 9
gestalt criterion, good continuity, 8
perceptual association field, 7
speed ranging, 8

G
Gelstalt illusion visualization
beta phenomenon, 84
collinear *vs*. parallel sequence, 86
dynamic association field, 85–86
line-motion illusion, 86
perceptual grouping theory, 84

H
Hildreth's smoothing process, 9

I
Intersection of constraints (IOC), 14
Intra-saccadic motion
acceleration/deceleration profile, 218
blur perception, 215, 216
central origin interpretation, 224–226
extra-retinal suppression theory, 214
H-H paradigm, 223–225, 234
lateral geniculate nucleus (LGN), 254
middle temporal cortex (MT), 232–233
motion contrast-sensitivity reduction, 217, 218, 222
perceptual consequences, visual latency
chronostasis, 250
flashed stimuli, 248, 249
image motion, 248
temporal precision, 250
time compression, 248
post-saccadic enhancement
DLPN and VPFL, 251
motion perception, 252
MT/MST neurons, 251–253
ocular following, 251
retinal origin interpretation
contour masking, 226
magnocellular pathway specificity, 228
retinal light adaptation level, 229–230
threshold values, 227
time course, 228
saccadic suppression
acceleration and deceleration, 241
image motion, 240

Intra-saccadic motion (*cont.*)
perceptual evidence, 242–244
photoreceptors, 241
physiological evidence, 244–247
saccadic suppression of image displacement (SSID)
apparent motion, 230
forward and backward masking, 231
trans-saccadic integration issue, 232
spatial frequency content, 218
temporal filling-in, 215, 223
temporal masking
homogeneous and parsimonious process, 218
saccadic omission, 215
smear omission, 217
trailing effect, 220–221
vertical gratings, 222–223
trailing eye effect
direction-specific adaptation, 220
Nyquist frequency, 219–220
retinal temporal frequency, 220
spatial frequency grating, 219–220
trans-saccadic fusion issue, 213, 230–231
visual/retinal spatio-temporal processes, 214
Intrinsic terminators, 287–289

K
Kalman filters, 172, 174
Kanizsa square illusion, 198–199

L
Laminar cortical circuits, 287
Latency difference model, 102
Lateral geniculate nucleus (LGN), 74, 80
eye movements and saccadic modulation, 272–273
ON and OFF cell inputs, 296
short-latency subcortical neurons, 254
temporal filter, 273
Lateral occipital complex (LOC), 14, 22, 24, 29
LGN. *See* Lateral geniculate nucleus
Linear–nonlinear ("L–N") model, 58
Line-motion illusion
cortical VSD response, 109–110
facilitatory effect, 109
low- and high-amplitude activity, 110
non-moving stimuli, 108
non-retinotopic mechanism, 110
sequential suprathreshold activation, 109
spatio-temporal pattern, 110
Long-range filter, 299
Lorenceau–Alais displays, 292

M
Magno-cellular system, 218, 221, 232
Maximizing causal information
dynamic coding, 273
information/noise, 265
LGN temporal filter, 273
predicted optimal temporal filter, 273, 274
predicted spatiotemporal receptive field, 266, 267
regularities and statistical properties, 264
Shannon information theory, 266
Medial superior temporal (MST) areas, 41, 178
Middle temporal cortex (MT), 232–233
Motion-dependent action and perception
cortical visual system, 119
latency and stimulus dependency, 121
psychometric function, 120
psychophysical experiments, 120–121
pursuit initiation, 121–122
SPEM generation, 119–120
Motion detection, reflexive tracking
biphasic temporal impulse response function, 152–153
eye movements, 142–143
first-order motion energy detectors, 143–144
1-D *mf* grating stimuli, 145–147
missing fundamental *(mf)* stimulus, 143
radial luminance modulation, 147
neural mediation, 154
non-linear interactions
component motion, 150–152
opponent motion, 147–150
spatiotemporal characteristics, 154–155
Motion integration
Bayesian model, 171–172
inferential process, 172, 174
motion representation, 181
predictive mechanism, 175
recurrent networks, 174
relative dynamics, 169
segmentation
aperture problem and feature tracking signals, 284–285
form and motion streams, 289–290
formotion binding, laminar cortical circuits, 287
formotion capture prediction, 290–293

intrinsic and extrinsic terminators, 287–289
predicted aperture problem solution, 285–287
spatio-temporal dynamics, 165
temporal dynamics, 167, 172, 180
V1-MT loop, 180
Motion processing system
directional grouping, 299–300
LGN input, 296
long-range filter and formotion selection, 299
spatial competition and opponent direction competition, 298
transient cells, 296–298
ubiquitous circuit design, 300
Motor behavior features
application
extracted right-handed tennis swing, 15, 16
Fourier PCA, 14
trajectories algorithm *vs.* published data, 13, 14
blind source separation algorithm
Fourier transform, 10
Wigner-Ville spectrum (WVS), 9, 10
learning generative models
analyzed EMG data, 7
anechoic mixture, 9
extracted ICA components, 7
independent component analysis (ICA), 7
maximum cross-correlations, 8
PCA/factor analysis, 6
MT neuronal response
directionally selective neurons, 60
pattern and component correlation, scatter plots, 61–62
population average response, 63–64
Multiscale functional imaging, primary visual cortex (V1)
gelstalt illusion visualization
beta phenomenon, 84
collinear *vs.* parallel sequence, 86
dynamic association field, 85–86
line-motion illusion, 86
theory of perceptual grouping, 84
multi-scale imaging, 75–77
propagation visualization, orientation belief, 88–89
structure *vs.* function, topological paradox, 73–75
synaptic imaging
electrophysiological basis, 77–78
propagation waves reconstruction, echoes, 78–80
voltage-sensitive dye (VSD) techniques
brain imaging methods, 80
dense *vs.* sparse regime, 83
depolarization state, 81
non-linear response properties, cortical cells, 82
retro-propagating waves induction, 81
spatial *vs.* temporal properties, 83
visual stimulus, 81

N
Natural scenes, motion
contrast signal calculation, 277
eye movement classification, 278–279
eye movements and saccadic modulation, 272–273
Fourier transform, 262
human visual sensitivity, 264, 265
information theoretical approach, 261–262
maximizing causal information
dynamic coding, 273
information/noise, 265
LGN temporal filter, 273
predicted optimal temporal filter, 273, 274
regularities and statistical properties, 264
Shannon information theory, 266
spatiotemporal receptive field, 266, 267
neuronal processing, 264
non stationary correlations, contrast signal, 270–272
non-stationary velocity distribution, 269–270
power spectrum, 262–263
retinal image motion quantification, 277–278
spatiotemporal correlation calculation, 276–277
velocity distributions, 263
visual input statistics, retina
natural time-varying images, 269
smooth tracking, objects, 267
spatially decorrelated representation, 269
statistical properties, 266
Neural correlates
endstopped neurons, 48
MT neuron response, 46
selective integration process, 47
visual stimuli velocity, 45

Neural models, decision-making
3D FORMOTION model
form processing system, 293–296
functional projections and properties, 294
motion processing system, 296–300
primate visual system, 293
motion integration and segmentation
aperture problem and feature tracking signals, 284–285
form and motion streams, 289–290
formotion binding, laminar cortical circuits, 287
formotion capture prediction, 290–293
intrinsic and extrinsic terminators, 287–289
predicted aperture problem solution, 285–287
temporal dynamics
Bayesian models, 301–303
brain design, 305–306
motion capture, 301
trackable features *vs.* coherently moving dots, 305
two movement tasks, 303–305
Neuronal population dynamics
cortical space
line-motion illusion, 108–110
long-range horizontal connections, 106–108
optical imaging, 106
visual space
latency differences, 102–104
motion anticipation, 104–105
moving stimuli representation, PRF, 99–101
OLE-derived population representations, 99
population receptive field concept, 96–98
RF-derived population representations, 98
Neuronal processing, 264
Neuronal substrate
dbm stimulus, 124
medial superior temporal (MST) area, 123
middle temporal (MT) area, 123, 124
motion perception, 123
neuron responses, 125
object motion, 126
orientation detection, 123
SPEM execution, 124
tm and fm stimulus, 124, 126
Non-linear interactions, WTA
component motion
direction-selective neurons, 151
3*f* and 5*f* stimulus strips, 150
3rd and 7th harmonics, *mf* stimulus, 150
initial open-loop period, 151
ocular tracking mechanism, 152
spatial-frequency bandwidth, 152
opponent motion
3*f* and 5*f* 1-D vertical sine-wave gratings, 149
3rd *vs.* 5th harmonics, 147–149
random-dot stimuli, 150
Nyquist frequency, 219–220

O

Object motion tracking
extra-retinal cues
human subjects pursue, 177
MST and FEF neurons, 178
neuronal and behavioral responses, 178–179
perceptual target, 177
independent recurrent loops
cortico-cortical loops, 182
motion detection mechanism, 180, 182
motion representation, 181, 182
MT and MST neurons, 182–183
retinal and extra-retinal signals, 182
target trajectory, 180
initiation tracking and visual motion dynamics
Bayesian model, 171–172
biological solution dynamics, 166
initial pursuit direction, 165–166
motion computation, 166, 167
MT neurons, 169, 170
oculomotor loop, 169
receptive fields, 164
reflexive ocular, 168
retinal image motion, 164
spatio-temporal dynamics, 165
target velocity, 164
tracking direction error, 167–170
line-drawing stimuli, 162, 163
retinal and non-retinal cues
anticipatory eye movements, 175
motion direction, 175–176
predictive mechanisms, 175
pursuit initiation, 175, 176
retinal image velocity, 176
steady-state eye velocity, 176–177

target visual motion and trajectory prediction, 173, 176
smooth pursuit eye movements, 161–162
steady-state tracking, 162
temporal dynamics
1D and 2D likelihoods, 174
Maximum-A-Posteriori (MAP) computation, 172–174
motion cues, 172
oculomotor negative feedback loop, 174
A Posteriori distribution, 172
Prior distribution, 172
variance estimation, 174
Occluder, 284, 287–289, 291, 292, 295, 296, 298, 299
Ocular following response (OFR), 142
Optical imaging, 106
Optimal linear estimator (OLE), 96, 99–102
Orientation-tuned normalization mechanism, 59

P
Pattern direction selectivity (PDS), 60, 61, 63
Pattern motion computation, dynamics
bar and plaid stimuli relationship
grating and bar texture stimuli, 64–65
macaque area, 64
contrast effects, 67–68
filtered bar texture response, 66–67
MT neuronal response
directionally selective neurons, 60
pattern and component correlation, scatter plots, 61–62
population average response, 63–64
pattern motion detection models
cascade model, 59
decoding motion, complex patterns, 58, 59
linear-nonlinear (“L-N”) model, 58
primary visual cortex (V1), 56–57
PDS. *See* Pattern direction selectivity
Perceptual grouping, 295
Photoreceptors, 241
Plaid motions, 167
Point light walker (PLW), 131–133
Point of subjective stationarity (PSS), 122
Population code, 98
Population receptive field (PRF), 98–100, 102, 103, 106
Predicted optimal temporal filter, 273, 274
PRF. *See* Population receptive field

R
Radial flow vergence response (RFVR), 142
Random dot kinematograms (RDK), 12–13, 16, 21, 22
RCMF. *See* Retino-cortical magnification factor
RDK. *See* Random dot kinematograms
Receptive field (RF), 56
Recurrent networks, 174
Reichardt-detector. *See* Elementary motion detector
Retinal image motion quantification, 277–278
Retinal origin interpretation
contour masking, 226
magnocellular pathway specificity, 228
retinal light adaptation level, 229–230
threshold values, 227
time course, 228
Retinal temporal frequency, 220
Retino-cortical magnification factor (RCMF), 79–80
RF-derived interpolation procedure, 99

S
Saccadic suppression. *See* also Contrast sensitivity reduction
acceleration and deceleration, 241
image motion, 240
perceptual evidence
contrast sensitivity, 243
isoluminant stimuli, 242–243
luminance-modulated gratings, 242, 243
photoreceptors, 243–244
photoreceptors, 241
physiological evidence
active and passive condition, 244
MT/MST neurons, 246, 250
parietal cortex of alert, 244
response latencies, 247
spike arrival times, 245
static retinal stimulus, 233
visual perception, 214
Saccadic suppression of image displacement (SSID)
apparent motion, 230
forward and backward masking, 231
trans-saccadic integration issue, 232

Second-order motion stimulus
biological motion perception
human and monkey, 131–133
second order motion, 133
definition, 118
Second-order motion stimulus (*cont.*)
dynamic flicker, 128–130
motion-dependent action and perception
cortical visual system, 119
latency and stimulus dependency, 121
psychometric function, 120
psychophysical experiments, 120–121
pursuit initiation, 121–122
SPEM generation, 119–120
multimodal motion representation, 126–128
neuronal substrate
dbm stimulus, 124
medial superior temporal (MST) area, 123
middle temporal (MT) area, 123, 124
motion perception, 123
neuron responses, 125
object motion, 126
orientation detection, 123
SPEM execution, 124
tm and fm stimulus, 124, 126
random dot kinematograms (RDKs), 118
Selective integration process, 47
Sensorimotor processing, 200
Shannon information theory, 266
Short-latency visuomotor, 251
Short-range filter, 298
Smooth pursuit eye movements and visual perception
action-perception cycle, closing
coherent motion, 203
corollary discharge information, 202–203
efference copy information, 200
orthogonal decomposition, 202
outline chevron, 200, 201
output stages, brain, 204
perceptual coherence, chevron, 202
retinal image smear, 203
retinal velocity, chevron, 201–202
spatial attention, 204
spatial representation, 200
stable world assumption, 202
visual sensitivity suppression, 200
attention-deficit hyperactivity disorder (ADHD), 205
cognitively processed inputs, 204
motion-dependent action and perception, 119–120
neuronal substrate, 124
perceptual and cognitive influence
apparent motion, 195
diamond perception, 197
Kanizsa square illusion, 198–199
motion direction, reversal, 199
perceived object, 195
sparse retinal image representation, 195, 196
vertical apertures, diamond vertices, 196, 197
visual stimuli, grouping, 196
sensory processing, initiation
gap effect, 192
initial momentary perceived direction, 194
location cueing, 193
MT neurons, 193–194
perceptual stability, 195
psychophysical tasks, 194
retinal motion signal, 192
sensory motion perception, 194
sub-threshold electrical microstimulation, 193
vector averaging, 194
visual and non-visual cognitive events, 205
Space time histograms, 118
Spatial components perception
computational method
motion morphing, 17
point-light stimuli, 20
point light walkers, 19, 20
stimuli, 20
experimental design
ideal-observer models, 19, 20
strong right-left symmetry, 18
Spatial frequency spectrum, 5
Spatial lag method, 103, 104
Spatiotemporal correlation, 276–277
Spike density functions (SDFs), 244
Superior temporal sulcus (STS), 123, 128
Synaptic imaging
electrophysiological basis
integration and discharge field, 78
intracellular recordings, 77
propagation waves reconstruction, echoes
receptive and discharge field, 80
retino-cortical magnification factor (RCMF), 79
travelling wave hypothesis, 78

T
Target selection, 193
Temporal dynamics, motion integration
  aperture problem
    coarse-to-fine progression, 38
    1D-to-2D motion dynamics, visual stimuli, 39
    neural computation, 37
  computational model
    model response, 49–50
    normalization mechanisms, 48
    pattern cells, 50–51
  eye movements
    bar pursuit data, 44
    contour length, 45
    endstopping, 44
    ocular following (OF), 42
    psychophysical judgments, 41
    single horizontal grating, 43
    smooth pursuit, 42
    unikinetic plaid, 44
  motion perception, 57–58
  neural correlates
    endstopped neurons, 48
    MT neuron response, 46
    selective integration process, 47
    visual stimuli velocity, 45
  primary visual cortex, 56–57
  psychophysics
    component grating, 40
    direction-selective neurons, 39
    resultant vectors, 41
    stimulus parameters, 40
Temporal masking
  intra-saccadic motion perception
    homogeneous and parsimonious process, 218
    saccadic omission, 215
    smear omission, 217
    trailing effect, 220–221
    vertical gratings, 222–223
Ternus display, 8
Theta motion (tm), 118
Trailing eye effect
  direction-specific adaptation, 220
  Nyquist frequency, 219–220
  retinal temporal frequency, 220
  spatial frequency grating, 219–220
Transient cells, 296–298

V
Vector averaging, 167, 170, 176
Ventral intraparietal area (VIP), 128
Ventral paraflocculus (VPFL), 251
VICON 612 motion capture system, 5
Visual attention, 191, 203
Visual/retinal spatio-temporal processes, 214
Visual space, neuronal population dynamics
  latency differences
    peak latency calculation, 103
    population receptive field (PRF), 103
    spatial offset, 102
  motion anticipation, 104–105
  moving stimuli representation, PRF, 99–101
  OLE-derived population representations, 99
  population receptive field concept, 96–98
  RF-derived population representations, 98
Voltage-sensitive dye (VSD) techniques
  brain imaging methods, 80
  dense *vs.* sparse regime, 83
  depolarization state, 81
  imaging, 106
  line-motion illusion, 109–110
  non-linear response properties, cortical cells, 82
  retro-propagating waves induction, 81
  spatial *vs.* temporal properties, 83
  visual stimulus, 81

W
Waterfall illusion, 122
Wave propagation, contour neurons
  functional assembly, 9
  gestalt criterion, good continuity, 8
  perceptual association field, 7
  speed ranging, 8
Winner-Take-All (WTA)
  component motion
    direction-selective neurons, 151
    3*f* and 5*f* stimulus strips, 150
    3rd and 7th harmonics, *mf* stimulus, 150
    initial open-loop period, 151
    ocular tracking mechanism, 152
    spatial-frequency bandwidth, 152
  opponent motion
    3*f* and 5*f* 1-D vertical sine-wave gratings, 149
    3rd *vs.* 5th harmonics, 147–149
    random-dot stimuli, 150

Color Plates

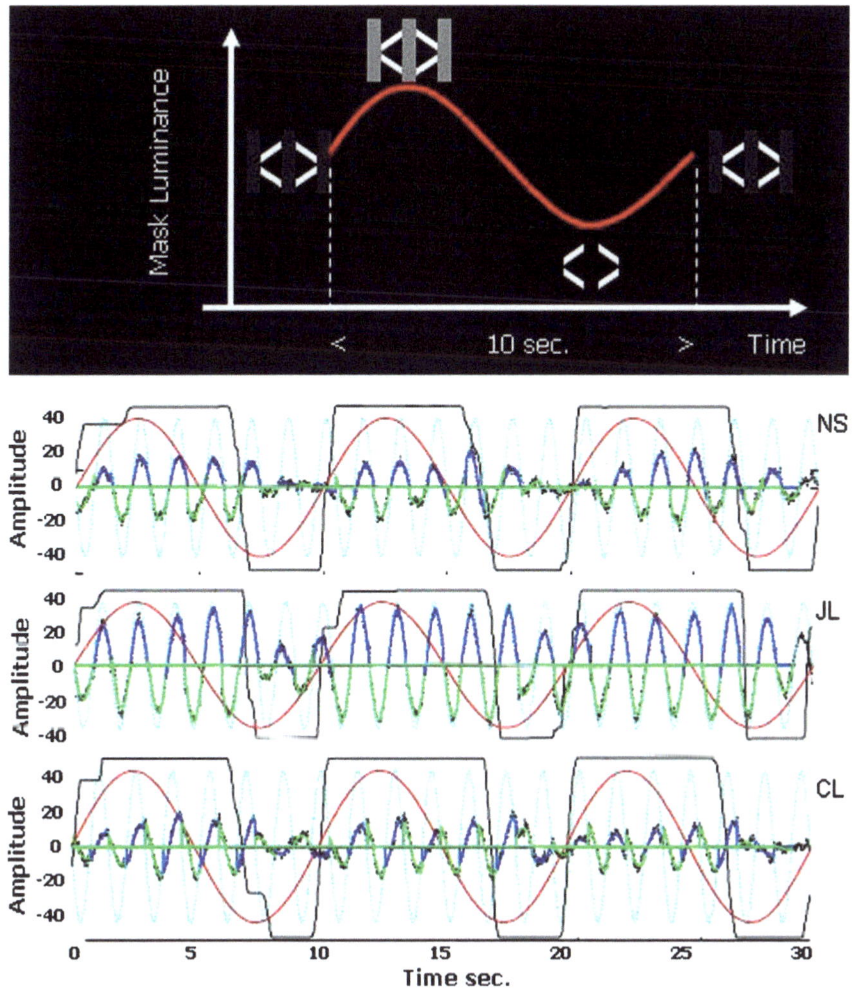

Fig. 1.9 *Top*. Illustration of the display used to study pursuit eye movements recorded during episodes of coherent and incoherent motion. Perceptual transitions were induced by smooth variation of mask luminance while the masked diamond rotated at 1 Hz. Observers were required to actively pursue the diamond's center while reporting their perceptual state with the pen of a tablet. *Bottom*: Results of three observers averaged across three periods of 30 s. *Green/blue* traces show the amplitude of horizontal pursuit eye movements as a function of time. Perceptual states are represented by the *black line*: *upward* for coherent states and *downward* for incoherent states. The *red line* represents mask luminance variations; the *dashed cyan line* shows horizontal stimulus motion. See text for details

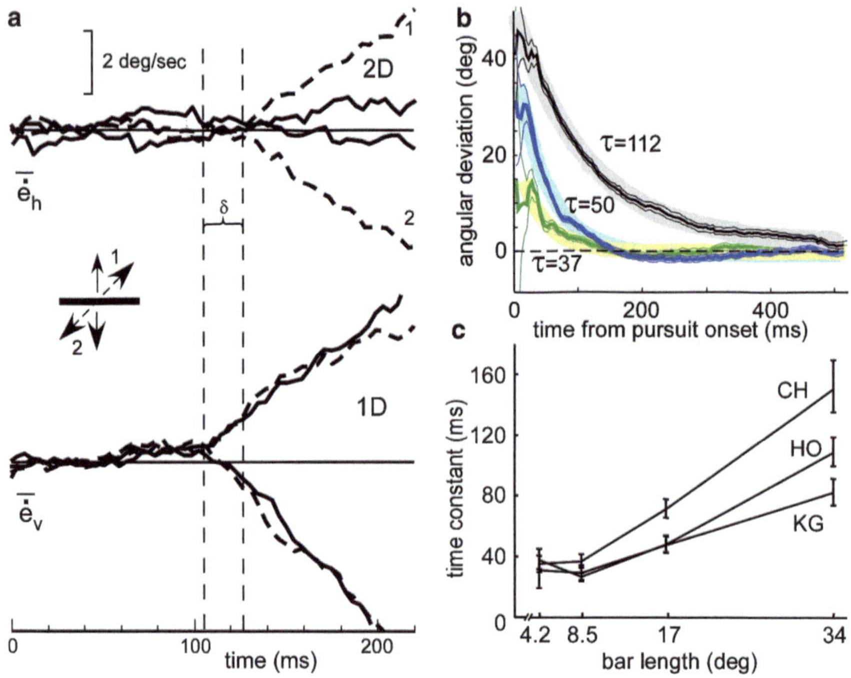

Fig. 2.3 Bar pursuit data from our own lab (Born et al. 2006) displayed so as to be directly comparable to the ocular following experiments of Masson and Castet (2002). (**a**) Monkeys actively pursued a horizontal bar that moved either vertically (solid lines), up (1) or down (2), or obliquely (*dashed lines*), up and to the right (1) or down and to the left (2). In both cases, the eye velocity response to the 1D component is reflected in the vertical component of the eye velocity ($\dot{e}_v$). For the obliquely moving bar, the responses to the 1D component seen in the vertical eye velocity traces are identical to those evoked by the vertically moving bar (*equated for speed*). The 2D responses, seen in the horizontal eye velocity traces ($\dot{e}_h$), show a longer latency (δ=20 ms). This data can be directly compared to that shown in Fig. 2.2b. (**b**) Time course of the angular deviation for tilted bars of different lengths (34, 17 and 4°) along with the best-fitting single exponential decay function for monkey HO. Values for τ are in meter seconds. (**c**) Time constant, τ, as a function of bar length for each of three different monkeys. Panels (**b**) and (**c**) were modified from Fig. 2.4 of Born et al. (2006)

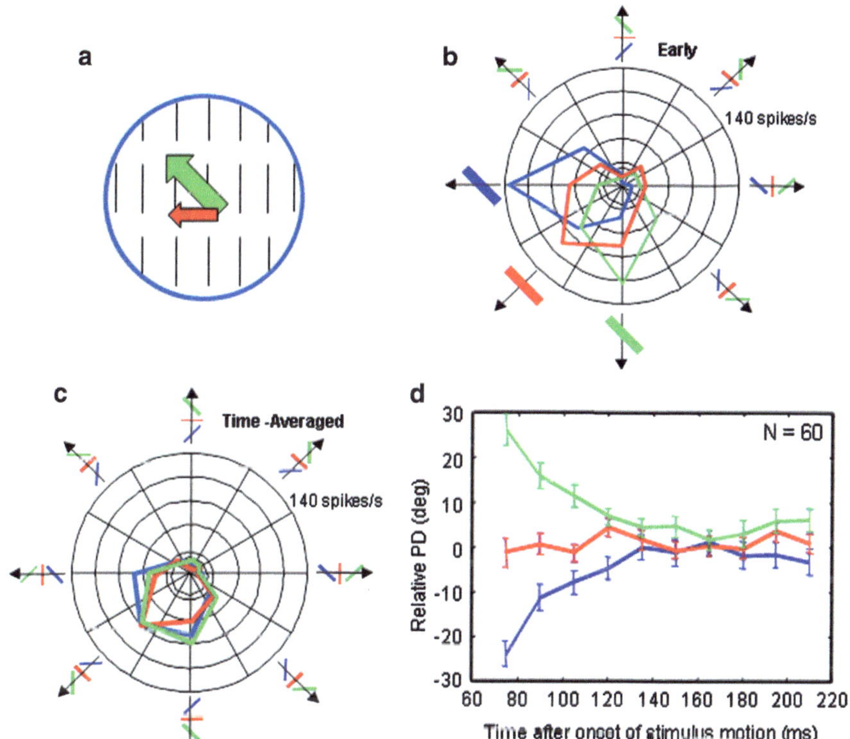

Fig. 2.4 Response of MT neurons to the bar field stimulus. (**a**) The bar field stimulus consisted of rows of bars with the same orientation, filling the receptive field (*blue circle*). When the bars moved obliquely with respect to their orientation (*green arrow*), the component of motion perpendicular to the bar motion (*red arrow*) differed from the true direction by 45°. (**b**) For a single MT neuron, the early part of the response depends on both the orientation and direction of the bar field. This neuron responds best whenever the bar has a left-oblique orientation and a leftward or downward motion component, indicating that it sees only the component of motion perpendicular to the bars. (**c**) The later part of the response depends only on the motion direction. (**d**) The transition from orientation-dependent responses to purely motion-dependent responses is evident in the population of 60 MT neurons. See also movies: *cu085c90.avi*, *cu085a45.avi*, *cu085b135.avi*, for a single-cell example of the temporal dynamics

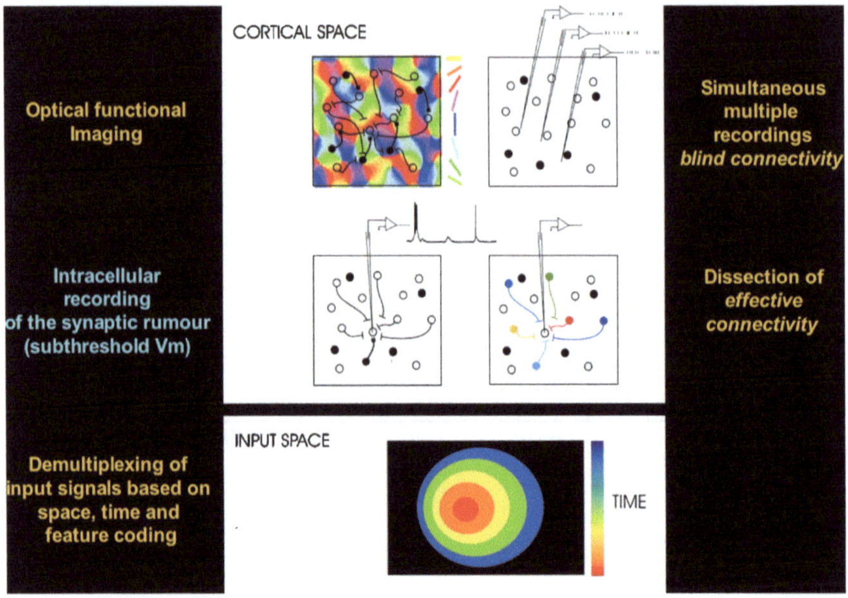

Fig. 4.1 Network dynamics imaging vs. synaptic functional imaging. Three methods for visualizing network dynamics are compared: (1) *top left*, optical functional imaging allows charting of cortical domains of iso-functional preference (color coded) on the basis of metabolic or hemodynamic signals; (2) *top right*, multiple simultaneous extracellular recordings are used to evaluate correlated activity patterns through the blind selection of potentially interconnected neurons; (3) *middle and bottom rows*, the reverse analysis of sub-threshold activity during long-duration intracellular recordings of the same single cell can be used to retrieve the effective network afferent to the recorded cell (see text)

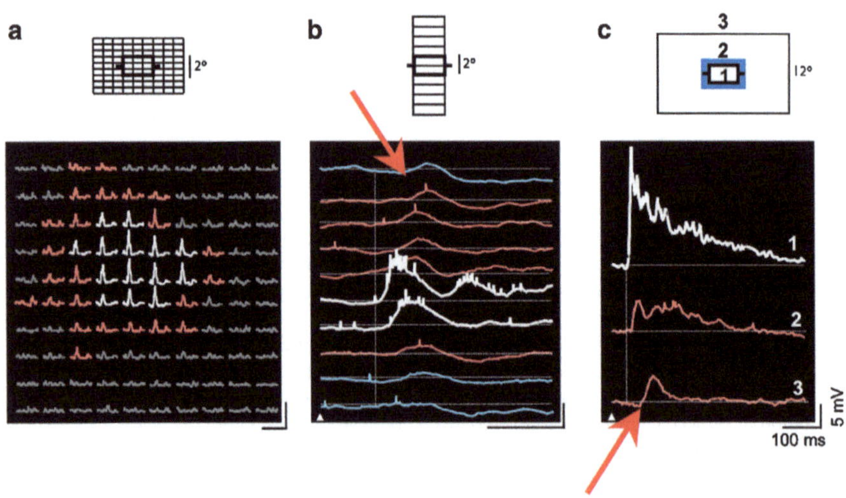

Fig. 4.2 Synaptic integration field and discharge field of visual cortical neurons. **a** to **c**: spatio-temporal maps of sub-threshold (*red*) and spiking responses (*white*) of V1 neurons. (**a**) sparse noise mapping; (**b**) 1-D bar mapping across the width axis of the receptive field; (**c**) co-centric gratings covering either the discharge field (1), the immediate (2) or the far (3) surround the periphery. Note the increase in the spatial extent of the sub-threshold RF with the level of temporal and spatial summation of the stimulus (which increases from **a** to **c**). Note also that the latency of the synaptic response (red arrows) increases as a function of eccentricity from the discharge field center. Adapted from [11], with permission

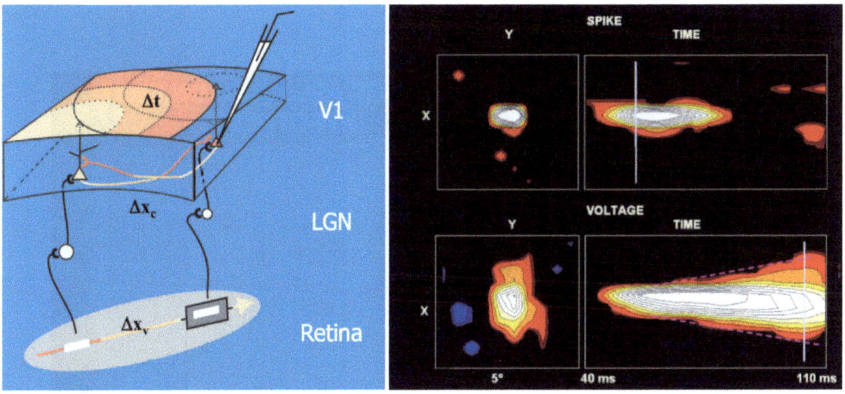

Fig. 4.3 Functional Synaptic Imaging. *Left*: Schematic representation of the hypothesis of reciprocal horizontal connections between two cortical cells (*left*, "sender"; *right* "receiver"). This schema allows to reconstruct a propagating wave (*circles*) from the intracellular measure of evoked latencies of synaptic responses in the "receiver"-cell (*right*, electrode). Δx_v is the eccentricity of the distal stimulus (*white rectangle, no outline*) from the central stimulus shown in the discharge field center (*grey rectangle, black outline*). Δt is the latency between the synaptic response onsets evoked through the two pathways. Δx_c is the intracortical distance between the cortical feed-forward impacts produced by the two stimuli, inferred from the know retinocortical magnification factor (see text for details). *Right*: Spatio-temporal (*left*: X-Y; *right*: X-t) maps of supra-threshold (spike, *upper panels*) and sub-threshold (voltage, *lower panels*) activations in the same V1 cell. The X-Y maps are presented for two specific delays corresponding respectively to the maximal extents of the discharge field (*upper*) and sub-threshold integration field (*lower*). The pink dotted lines show the average speed of the propagation of the reconstructed wave (0.2–0.4 m/s) [see Movie 1]

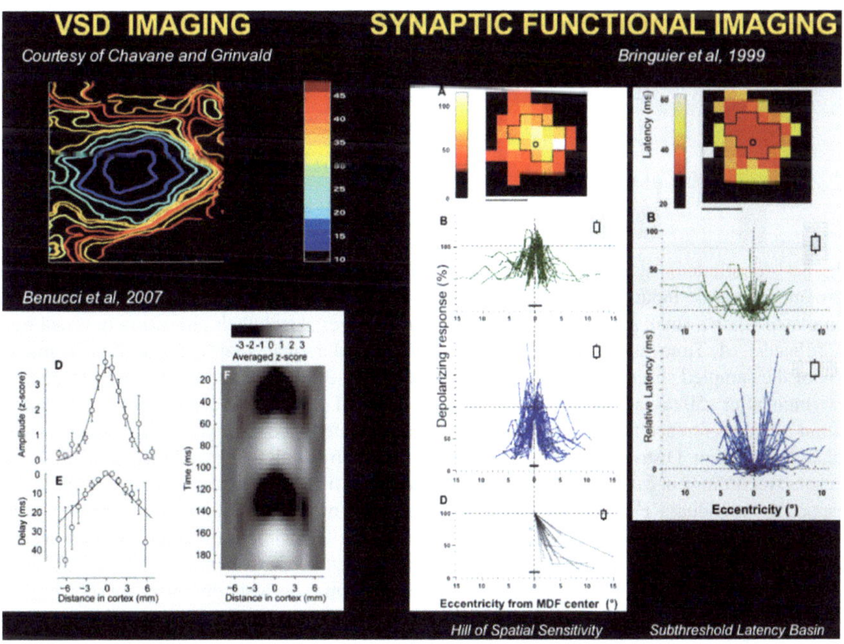

Fig. 4.4 Comparison of spatial and temporal properties of horizontal propagation, using network voltage sensitive dye imaging (*left*) and synaptic functional imaging (*right*). See text for details. *Left panel*, horizontal propagation wave monitored in the superficial layers of cat V1 with voltage sensitive dye imaging (*top left*, with courtesy of F. Chavane and A. Grinvald; *bottom left* [28]). *Right panel*, horizontal propagation inferred with intracellular synaptic functional imaging [10,11,47]. The same propagation speed (0.1–0.3 m/s) is measured by the two imaging techniques

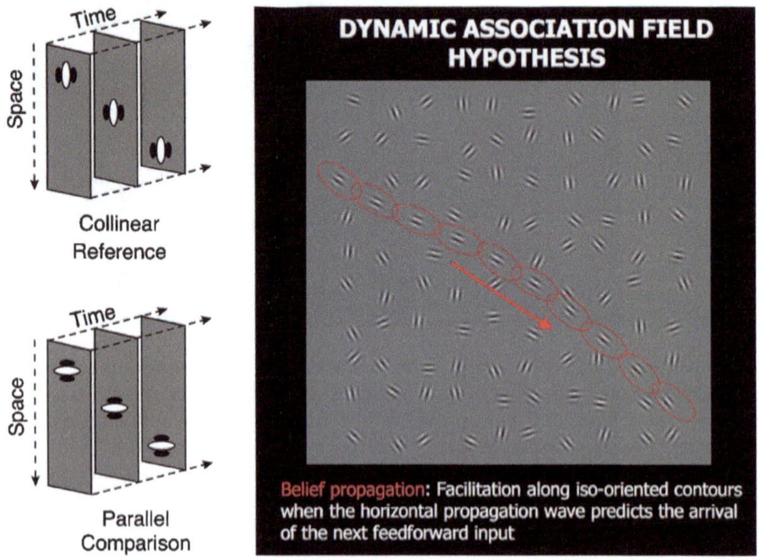

Fig. 4.5 Apparent motion and the hypothesis of the "dynamic association field". *Left*: two-forced choice apparent motion protocol. *Right*: the "dynamic association field" hypothesis; local oriented inputs (Gabor patches) induce a facilitation wave of activity traveling along horizontal connections intra V1. This wave tends to bind proximal receptive fields with co-linear preferred orientations, thus creating a contiguous path of temporal integration. The associative strength of the perceptual effect is maximal when the asynchronous feed-forward sequence produced by joint strokes of apparent motion (*arrow*) travels in phase with the visually evoked horizontal intracortical propagation

---

Fig. 5.1 (continued) symbolized spike traces) but are activated subthreshold due to long-range horizontal synaptic input. (**c**) Many neurons with densely overlapping receptive fields (RFs) are stimulated when a small square is presented (colored ellipses sketch their projection in visual field coordinates). (**d**) Time resolved population representation of a square that was flashed in the middle axis of the sampled space. Activity builds-up coherently ~30 ms after stimulus onset, reaches its maximum after ~50 ms, and decays homogenously centered on stimulus location. (**e**) Seven squares within a fixed reference frame (*white outline*) were presented independently of the neurons' individual RF location. Distributions of activity were obtained by pooling the spiking responses of 178 neurons (recordings in cat area 17). Each cell's normalized firing rate in response to the stimuli was mapped to individual RF-centers, resulting in distributions of activity. The responses were interpolated with a Gaussian, with Sigma matching the average RF widths. In summary, each neuron contributes to the entire population activity by its firing rate dependent on the location of its RF-center relative to the stimulus. Time averages of 30–80 ms after stimulus onset are shown (flash duration 25 ms). Note that the peaks of activity are closely around each of the stimulus positions. Thus the applied population approach enables sampling of visual space smaller than average RF sizes (size of squares was 0.4°). (**f**) Population representations derived by means of OLE for all seven elementary stimuli used (time average, 30–80 ms). The optimization procedure allowed the individual peaks to be accurately aligned with the position of each stimulus

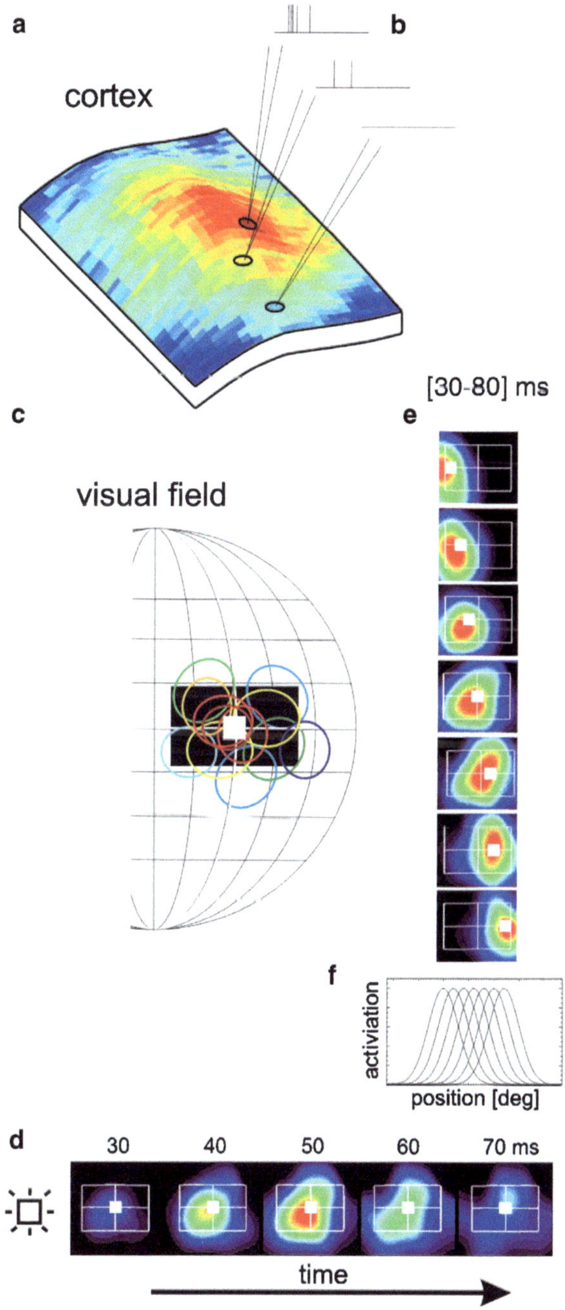

Fig. 5.1 Widespread cortical activity in response to small local stimuli. (**a**) A portion of visual cortex is sketched overlaid with an activity pattern that was evoked by a flashed square. Warm colors indicate high-amplitude activity revealed by voltage-sensitive dye imaging. (**b**) Extracellular measurements at three different recordings sites lead to different results: Whereas neurons that are located around the center of activation reach spiking thresholds, neurons further away do not (see

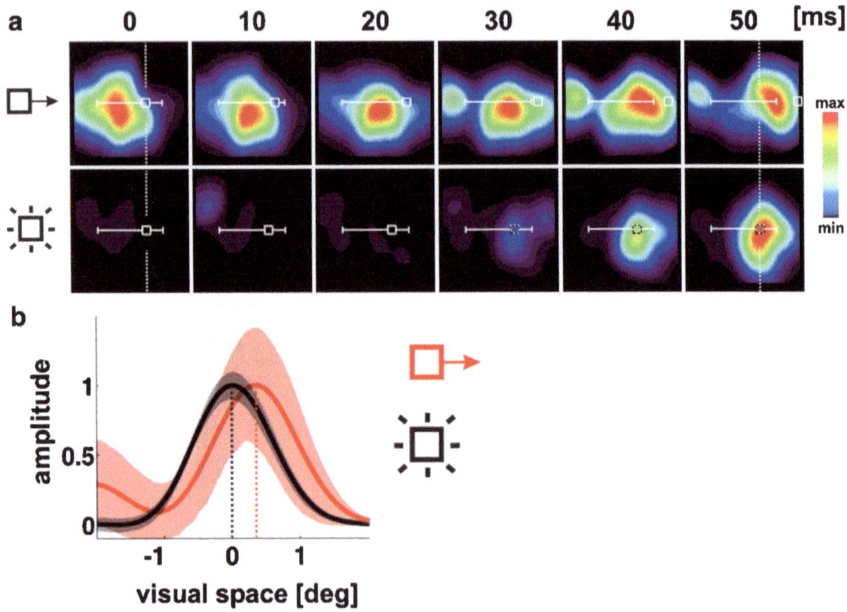

Fig. 5.3 The flash lags: impact of shorter latency for motion compared to a flash on the representation of position. (**a**) *Upper row*: moving square (38.4°s^{-1}). *Lower row*: flashed square (outlined in white; 25 ms on; shown stippled after stimulus off). Population representations are shown in time frames of 10 ms from *left* to *right*. The sampled space covered 2.8° of the central visual field (*horizontal white line*, cf. Fig. 5.1). *Color bar* indicates level of population activity; data were normalized separately for each stimulus condition. The moving stimulus started 3.2° outside the sampled visual space tracked by a peak of population activity that has been evoked some time before beyond the sampling region. At time zero stimulus position was identical for the moving and the flashed square (compare *vertical pointed lines*). During the next time steps activity in response to the flash emerged while the moving peak continued propagating. When activity for the flashed stimulus reached its maximum (50 ms), the moving peak already passed the mutual flash position due to its faster processing. (**b**) The spatial offset shown in (**a**) was confirmed applying the OLE approach in combination with a bootstrap analysis ($n = 1{,}000$). The time window shown depicts activity 50–60 ms after presentation of the flashed stimulus. Activity was separately normalized to the mean of each stimulus condition. *Gray shaded* areas show 99% confidence intervals for the flashed stimulus (*bright gray*; *black line* shows mean), respectively for the moving square (*bright red*; *red line* shows mean). A clear spatial offset ($p < 0.00001$) can be seen between the peaks of both conditions due to reduced time-to-peak latencies in response to motion (*vertical dotted line*).(Modified from Jancke et al. ©2004b, by permission from Wiley-Blackwell)

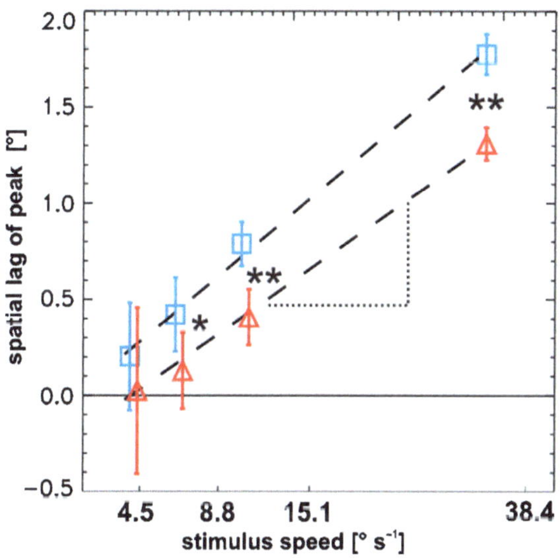

Fig. 5.4 Calculation of peak latencies for moving stimuli: speed-dependence of the spatial lag. The average spatial lag between the propagating peak of population activity and the actual position of the moving stimulus increased with stimulus speeds. Data points show mean spatial lags for each speed and direction (bootstrap, $n = 1{,}000$). *Blue squares* indicate centro-peripheral movement, red triangles indicate opposite motion direction. Latencies were calculated by linear regression. Latency for centro-peripheral motion was 42 ms, and 38 ms for peripheral-central direction (significance of direction difference: $** = p < 0.00001$; $* = p < 0.01$). (Modified from Jancke et al. ©2004b, by permission from Wiley-Blackwell)

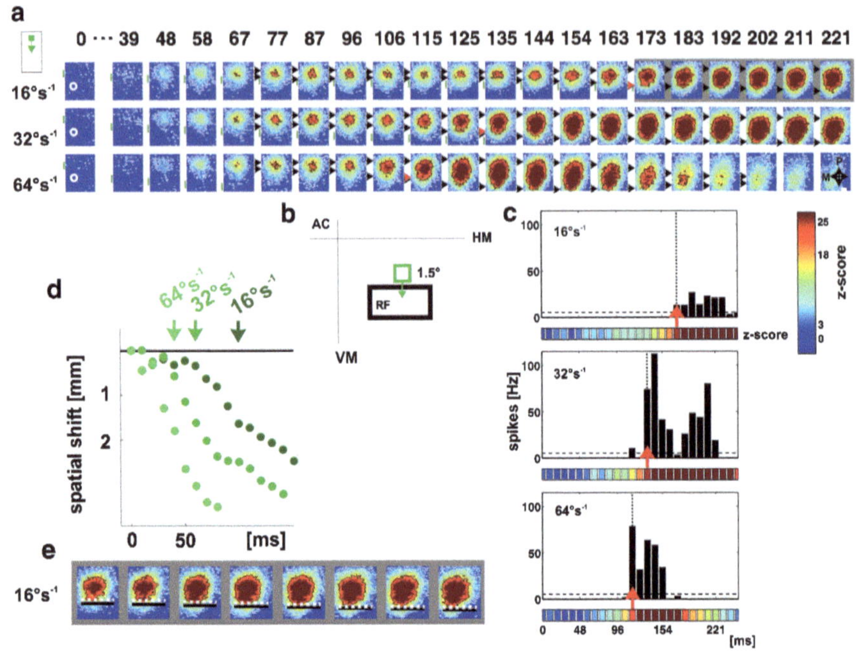

Fig. 5.5 Imaging cortical dynamics in response to moving squares. (**a**) Time courses of evoked cortical activity are shown in single frames (scale bar at *bottom right* 1 mm). Time after stimulus onset is indicated at the *top*. From *top to bottom row*: moving squares presented at different speeds, 16, 32, and 64°s^{-1}. *Green vertical lines* indicate the onset of stimuli and their estimated trajectory along posterior–anterior axis (M, medial; P, posterior). Extracellular recording (multi units) was performed simultaneously at the cortical location marked by *white circles* at frame zero (see (**b**) for receptive field (RF) of the recorded neurons. Spiking activity propagates with different speeds in anterior direction (*lower black arrows*). (**c**) To verify that the approaching wave front of VSD activity corresponds to spiking threshold, Post Stimulus Time Histograms (PSTHs) of spike events were derived. Horizontal stippled line depicts significance (*z*-score = 2, 30 repetitions). As the electrode was placed outside the retinotopic representation of the square's starting position, no spikes are evoked at the onset of motion, although the dye signal reports subthreshold spread of activity (*bottom row* shows actual z-score levels at the electrode position). Spikes are detected, however, as soon as activity reaches the spiking threshold at the position of the electrode (see *red arrows* in PSTH, and in **a**). With increasing stimulus speed, shifts in latency between the PSTHs can be seen because spiking level reached the electrode earlier for higher speeds. Thus squares moving at different speeds evoke consecutive spikes at a fixed electrode position. (**d**) Spatial shift of the wave front of spiking activity, slopes indicate speed of propagation across the cortex. Increasing hue of green depicts increasing stimulus speed. (**e**) Magnified section of imaging frames shown in (**a**), *upper row*. Note that the wave front of spiking activity (*white stippled lines*) closely matches the projected location of the stimulus (*black lines*) indicating "on-line" tracking of motion (see text). (Modified and reprinted by permission from Macmillan Publishers Ltd: Nature (Jancke et al. ©2004a)

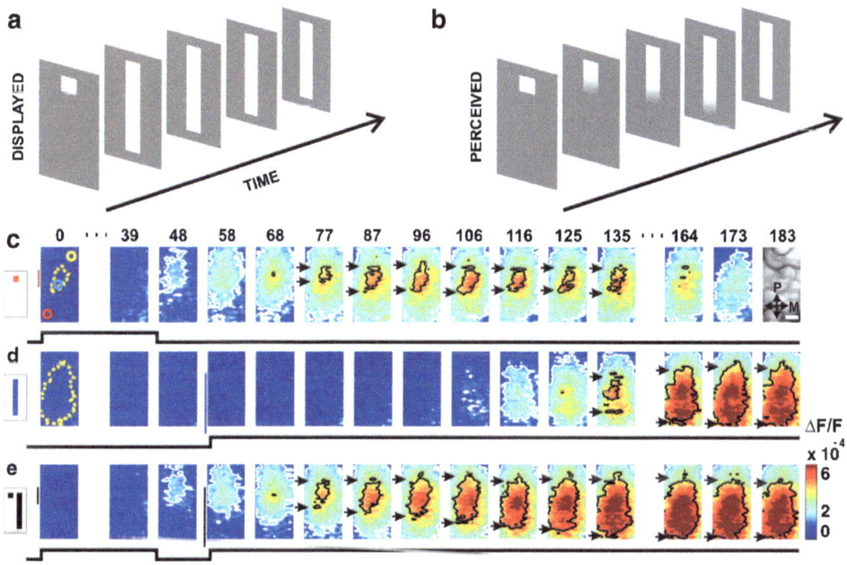

Fig. 5.6 Representations of stationary stimuli. (**a/b**) The line-motion paradigm. (**a**) Square ('cue') presented before a bar stimulus. (**b**) Subjects report illusory line-drawing (see supplementary movie 2). (**c–e**) Patterns of evoked cortical activity as a function of time. *Yellow dotted* contours approximate retinotopic representation of the stimuli; *coloured circles* indicate extracellular recording sites. *White contours* delimit low-amplitude activity (significance level; $p < 0.05$). The cortical area imaged is shown at upper right. Scale bar, 1 mm; P, posterior; M, medial. *Vertical lines* indicate estimated position of the stimuli (sizes: 1.5° square, 1.5° × 6° bar) along posterior-anterior axis. Time in milliseconds after stimulus onset is shown at the *top*. Stimulation period is depicted at the bottom of each row. Colour scale indicates averaged fractional changes in fluorescence intensity ($\Delta F/F$). Stimuli: (**c**) Flashed small square. (**d**) Flashed bar. (**e**) Line-motion paradigm. Average of 22 trials. (Modified and reprinted by permission from Macmillan Publishers Ltd: Nature (Jancke et al. ©2004a)

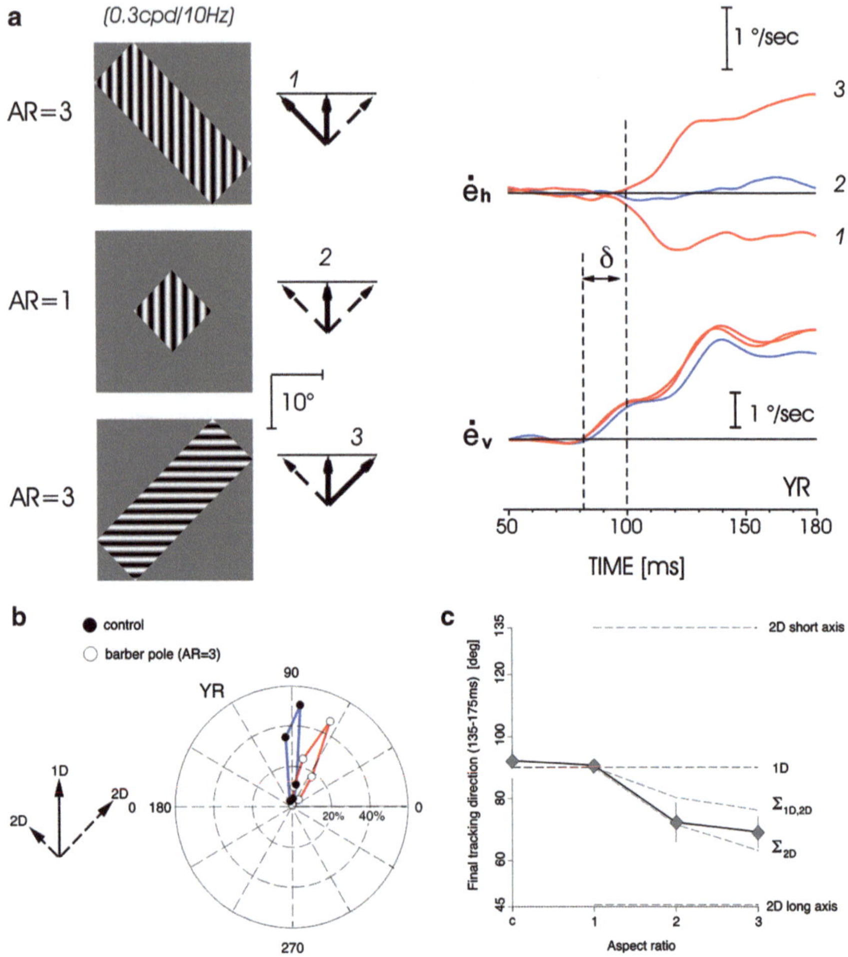

Fig. 8.3 Early dynamics of 2D motion integration. (**a**) A barber pole containing a horizontal grating moving upward is presented behind tilted apertures of aspect ratio (AR) either 1 or 3. With AR = 1, vector average of all 1D (grating) and 2D (line-endings) motion cues point towards the grating drifting direction (i.e. upward). With AR = 3, the vector average solution is biased towards the long axis of the aperture, so that global motion direction is diagonal along the upward-leftward axis (case 1) or the upward-rightward axis (case 2). Right-hand plot shows mean horizontal and vertical eye velocities for each case. Control condition (AR = 1) is shown in *blue*: reflexive tracking is initiated in the vertical direction at the usual latency of ~85 ms. No horizontal deflection is seen. The same vertical eye movements are initiated in cases 1 and 3. However, after a delay of 20 ms, eye velocity deviates towards either rightward (case 3) or leftward direction, indicating the influence of line-endings motion. (**b**) Direction distributions of late part of ocular following (i.e. ~150 ms after motion onset). Ones can see that an average 30° deviation is found with barber-pole of AR = 3, roughly consistent across all trials. (**c**) Mean tracking direction at the end of the open-loop period for a control condition (upright square aperture: 0) and different aspect ratios. *Dotted lines* show the predicted directions from the vector average of either 2D motions alone or 1D and 2D motions together. Modified from Masson et al. (2000).

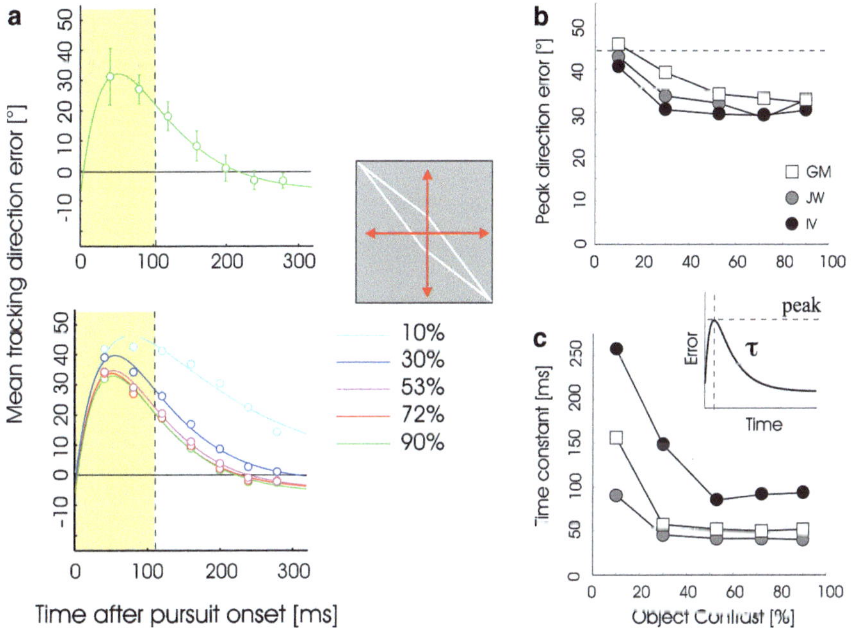

Fig. 8.4 Dependence of tracking direction errors on visual properties of moving object. (**a**) Temporal dynamics of the tracking direction error, obtained for rightward motion of a tilted line with different contrast. *Upper plot* shows the dynamics at full contrast (i.e. white bars against a black background). *Lower plots* illustrate the results with targets at contrast ranging from 10 to 90%. (**b**) Dependency of initial tracking direction error upon target contrast, for three subjects. (**c**) Relationship between time constant of the tracking direction error reduction and target contrast. Modified with permission from Wallace et al. (2005).

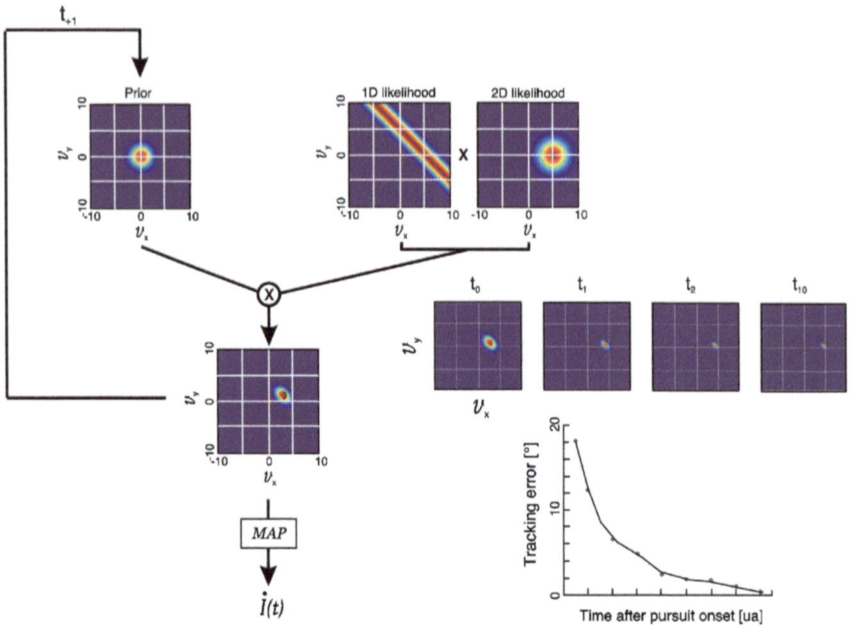

Fig. 8.5 A dynamical Bayesian model of motion integration. Ambiguous (i.e. edge) and non-ambiguous (i.e. features) motion information are represented as independent likelihood distributions in the velocity space. Following the earlier models by Weiss et al. (2001) and Barthélemy et al. (2008), these likelihoods are multiplied with a Prior distribution favoring slow speeds, to give the A Posteriori distribution of target velocity. The maximum of this distribution (MAP) is selected to give the best estimate of target velocity $\dot{I}(t)$ that can be used by the pursuit system. Such distributed A posteriori representation is then used as Prior information for the next iteration, assuming that likelihoods are left unchanged. Such recurrent processing is accumulating evidence about 2D target motion and therefore, the A posteriori distribution gradually evolves over time. Several time steps are illustrated (from t_0 to t_3), as well as the complete time course of tracking direction error. This model mimics two important aspect of the experimental data illustrated in Fig. 8.2: the initial tracking direction error and its time course (modified from Montagnini et al. 2007).

Fig. 8.7 (continued) Neuronal and behavioral responses to motion of real and illusory targets. When a real target (see Fig. 8.1d) is tracked, the observed pursuit-related activity can be sensory or motor in origin. It is impossible to reveal the nature of this activity. However, if the pursuit-related activity is also present when the monkey is pursuing the imaginary target (see Fig. 8.1d), the visual origin can be excluded. Eye (*black and red*) and target (*green*) position during pursuit of a real (**a**, **b**, **e**) and an imaginary (**b**, **d**, **f**) target are shown. Black represents the preferred direction of the recorded neuron, red gives the non-preferred direction. The neuronal activity is shown as raster display and spike density functions (width of Gaussian kernel 20 ms). Time zero is set to the onset of target motion. (**a**) and (**b**) details the response recorded from the middle temporal area (MT), (**c**) and (**d**) from the medial superior temporal area (MST), and E and F the frontal eye field (FEF). Note that the neuron recorded from MT did not respond during pursuit of the imaginary target. The activity of the MST and FEF neuron during pursuit of the real and imaginary target in preferred direction was not significantly different.

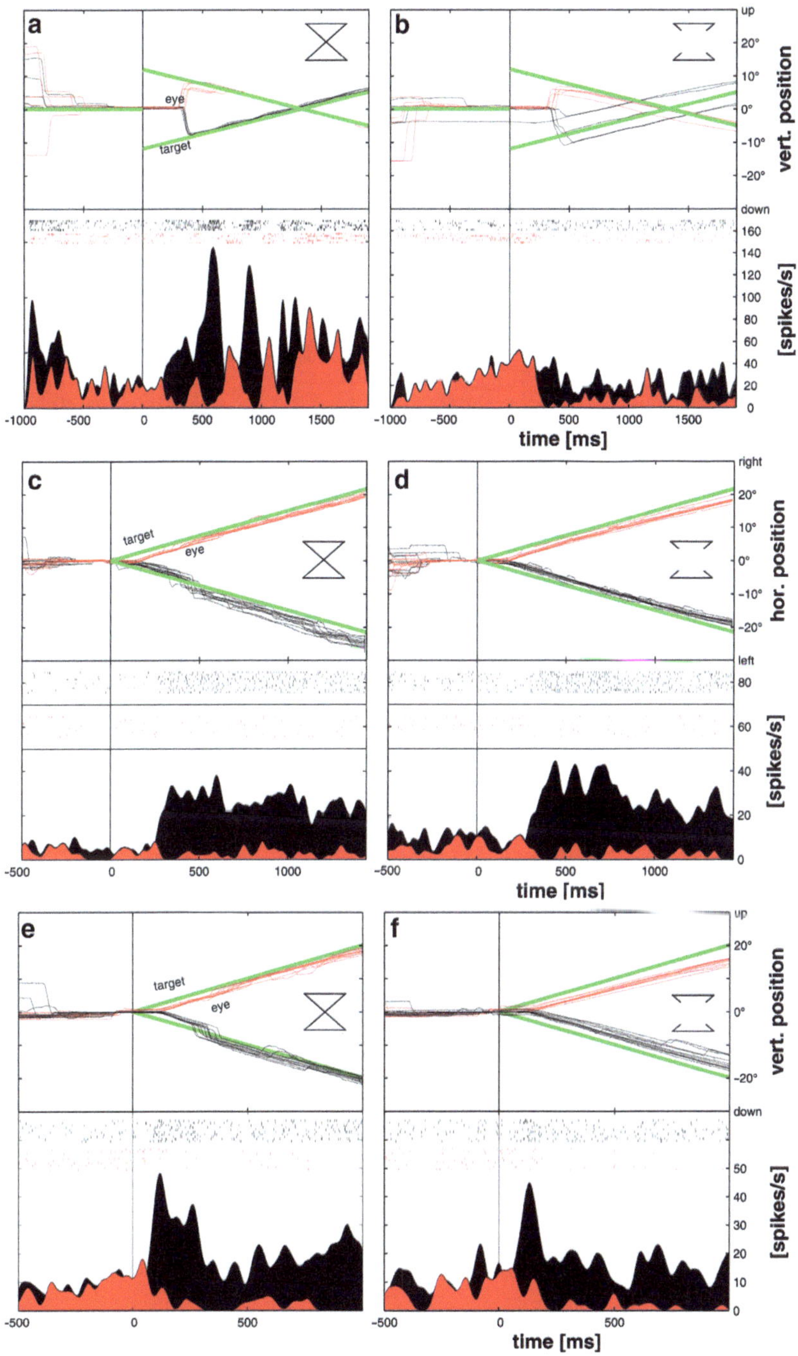

Fig. 8.7 (continued)

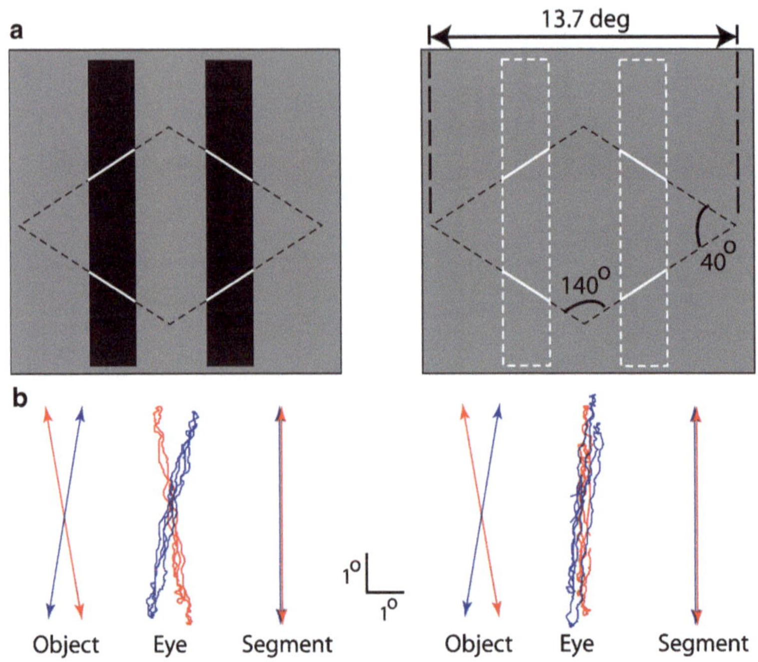

Fig. 9.3 Perceiving a rigid form allows smooth pursuit to track it correctly. (**a**) The scene on the left consists of a partially occluded outline of a diamond. The diamond translates in either a positive or a negative diagonal direction, such that its vertices are never visible though the two vertical apertures. Because of the apertures, each of the visible segments of the diamond moves in a purely vertical fashion (Lorenceau and Shiffrar 1992). The scene on the right is identical, but now the apertures reveal a background of identical luminance as the occluder. In this case, occlusion cues disappear, and it is difficult to perceive a rigid diamond (Lorenceau and Shiffrar 1992). See Supplementary Movies 3 and 4. (**b**) Eye movements track the motion of the diamond (*left*) when it is perceived correctly (**a**, *Left*) and the motion of the segments (*right*) when no percept of the diamond is possible (**a**, *right*). In both conditions, the physical motion of the visible line segments is identical, but the percept is different. Adapted from Krauzlis and Stone (1999)

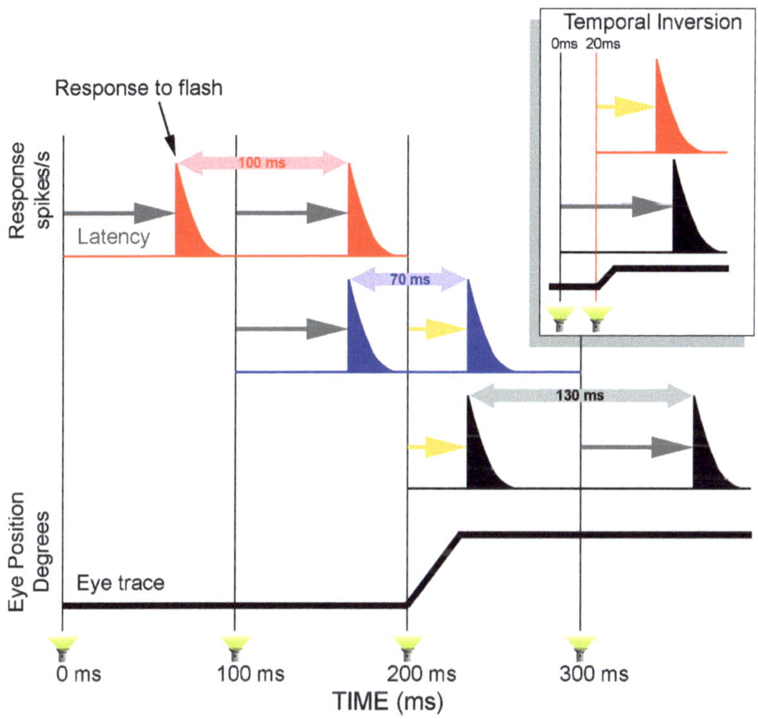

Fig. 11.4 Theoretical flashed stimuli are presented at 100 ms intervals (vertical black lines). A saccade begins at the onset of the third flash (deviation in eye trace). The cell responds to each flash with a burst of action potentials (filled response profiles). In the theoretical drawings, response amplitudes are not suppressed close to saccades, but it would be expected that responses before and during saccades would be smaller in reality. The latency to the response in the theoretical example is 65 ms (*horizontal gray arrows*) except for the response to the flash at saccade onset where the latency is 35 ms (*horizontal yellow arrows*). For the three flash intervals shown the perceived inter-flash intervals are 100 ms (control; red responses), 70 ms (peri-saccadic time compression; blue responses) and 130 ms (postsaccadic time expansion; black responses). The inset panel shows how time inversion could occur for two flashed stimuli with an inter-stimulus interval of 20 ms. In this case, the response to the second flash (red) arrives before the response to the first flash (black). Figure reproduced from original artwork with permission from Current Biology

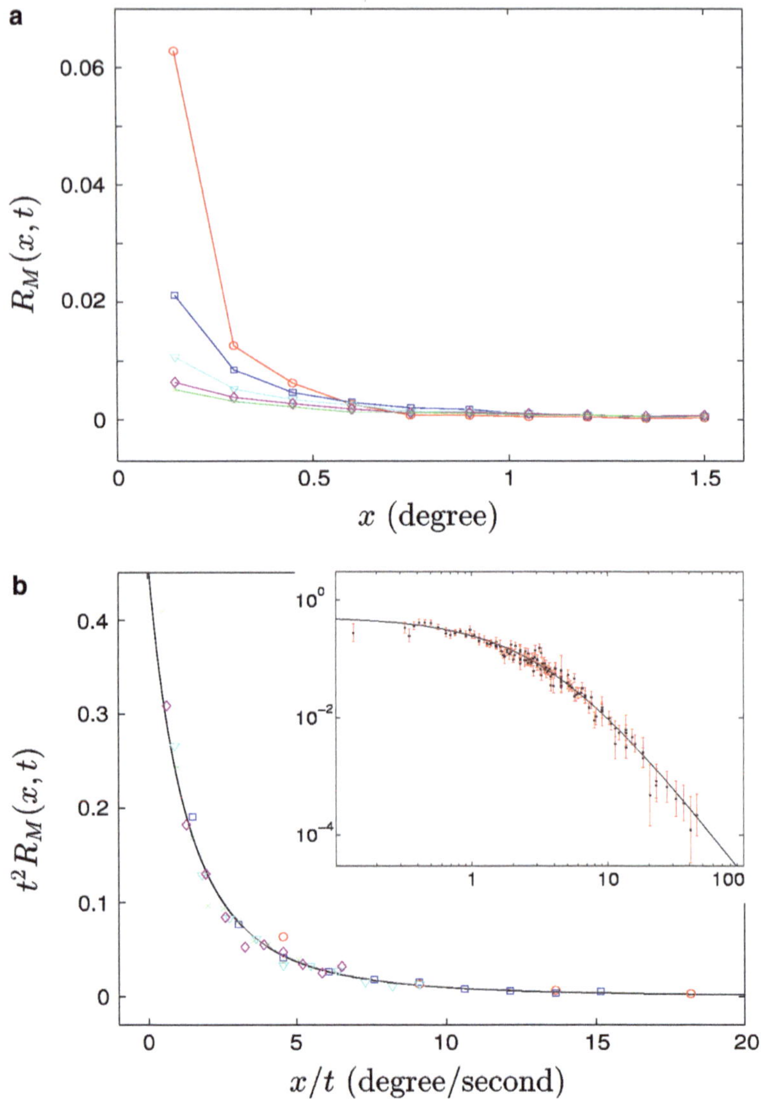

Fig. 12.4 Measured spatiotemporal correlation of the contrast signal $R_M(x,t)$. (**a**) the curves in red (*circle*), blue (*square*), cyan (*triangle*), purple (*diamond*), and green (*cross*), for $t=33$, 100,167, 233, and 300 ms, respectively. (**b**) all the data points in (**a**) re-plotted as a function of x/t after multiplication by t^2. Inset: t^2 $R_M(x,t)$ plotted in the log–log scale with standard error bars for all data points of $x<1.5°$ and $t<300$ ms. Also plotted in (**b**) and the inset is the curve $P(x/t)$, while $P(\upsilon)$ is the best fit curve for the velocity distribution

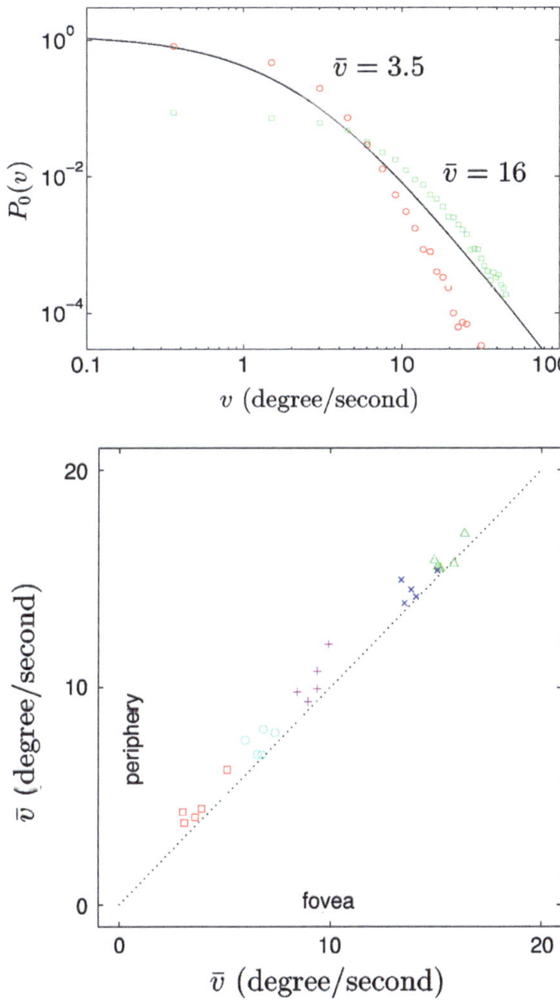

Fig. 12.5 *Top***:** the measured probability density distribution of the retinal image velocity $P_0(\upsilon)$ for two different scenes (*red circle*: watching birds on a river bank, green square: picking fruits in a forest). Also plotted is $P_0(\upsilon) = \upsilon_0/\pi\ (\upsilon+\upsilon_0{}^3$ curve with the best fit υ_0=2.3°/s for the velocity distribution averaged over five scenes. *Bottom*: the averages of the retinal image velocity υ of the fovea (*horizontal*) vs. the periphery (*vertical*). Different colors are for different scenes and different data points of the same color are for different observers. The average retinal image velocity at 3° from the fovea (the vertical axis) is clearly higher than on the fovea (the horizontal axis) for all scenes

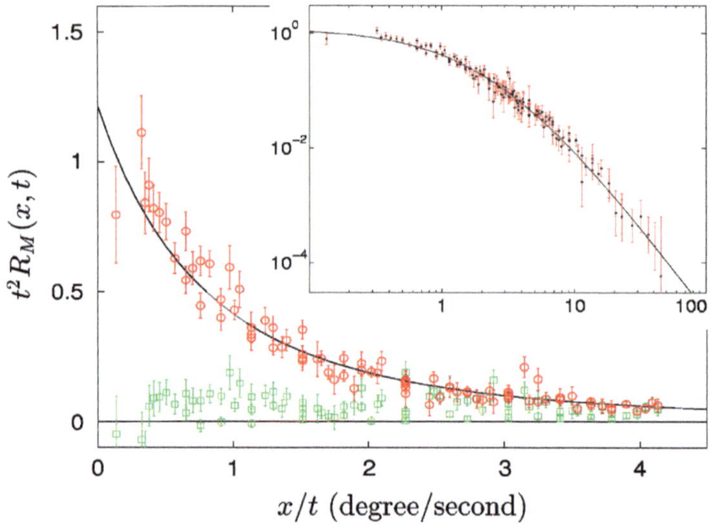

Fig. 12.6 Measured spatiotemporal correlation functions R_1 (with saccades) and R_0 (without saccades). The scaled correlation functions $t^2R_1(x,t)$ and $t^2R_0(x,t)$ in *green square* and *red circle*, respectively, shown for a small x/t range. Inset: $t^2R_0(x,t)$ shown in the log–log plot with standard error bars for all data points of $x < 1.5°$ and $t < 300$ ms. Also plotted in both the curve $P_0(x/t)$, while $P_0(v)$ is the best fit curve for the velocity distribution without any saccade in between

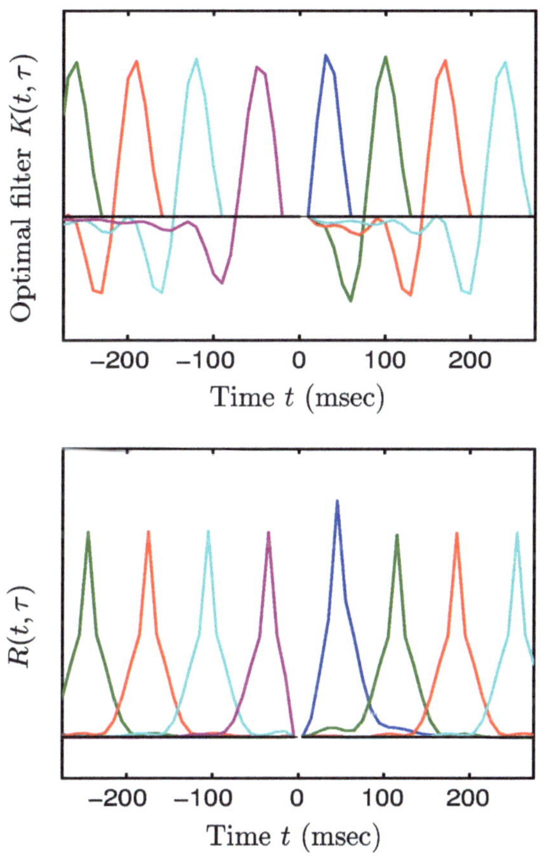

Fig. 12.7 The predicted optimal temporal filter $K(t,\tau)$ (*top*) and the corresponding output correlation $R(t,\tau)$ (*bottom*). The filter $K(t,\tau)$ at eight different τ are plotted, in which τ is the time relative to the saccade (τ=−245, −175, −105, −35; 35, 105, 175, 245 ms from *left* to *right*). It is clear that the optimal temporal filter of LGN depends on the time relative to saccades. In predicting the optimal filter on the left, the temporal correlation for the fourth scene in Fig. 12.3 (*right*) is used. The correlation $R(t,\tau)$ are cloe to zero except near $t=\tau$

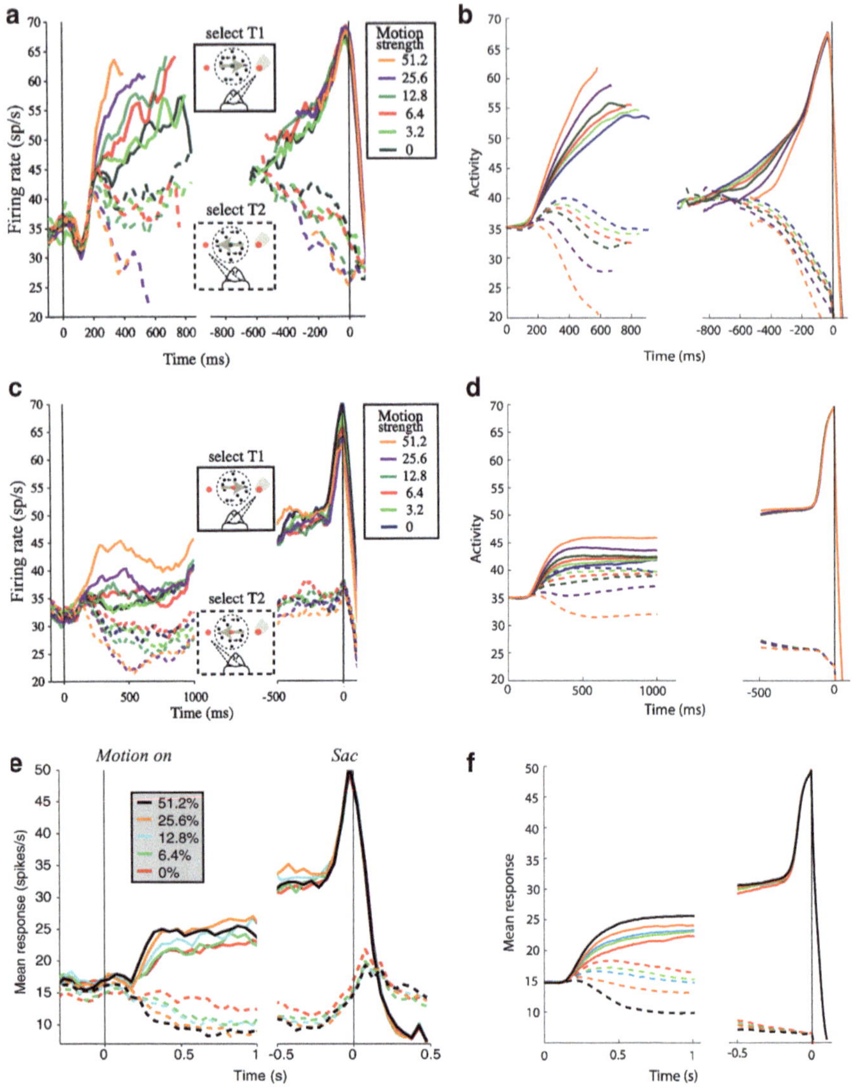

Fig. 13.9 Temporal dynamics of LIP neuronal responses during the fixed duration (FD) and reaction time (RT) tasks. (**a**) Average responses of a population of 54 LIP neurons among correct trials during the RT task (Roitman and Shadlen 2002). The *left part* of the plot is time-aligned to the motion onset, and includes activity only up to the median RT, and excludes any activity within 100 ms backward from saccade initiation (which corresponds to presaccadic enhancement). The *right part* of the plot is time-aligned to the saccade initiation, and excludes any activity within 200 ms forward from motion onset (which corresponds to initial transient dip and rise). (**b**) Model simulations replicate LIP cell recordings during the RT task. In both data and simulations for the RT task, the average responses were smoothed with a 60 ms running mean. (**c**) Average responses of a population of 38 LIP neurons among correct trials during the 2002 FD task (Roitman and Shadlen 2002), during both the motion viewing period (1 s) and a part (0.5 s) of the delay period before the saccade is made. (**d**) Model simulations mimic LIP cell recordings (See text for details)

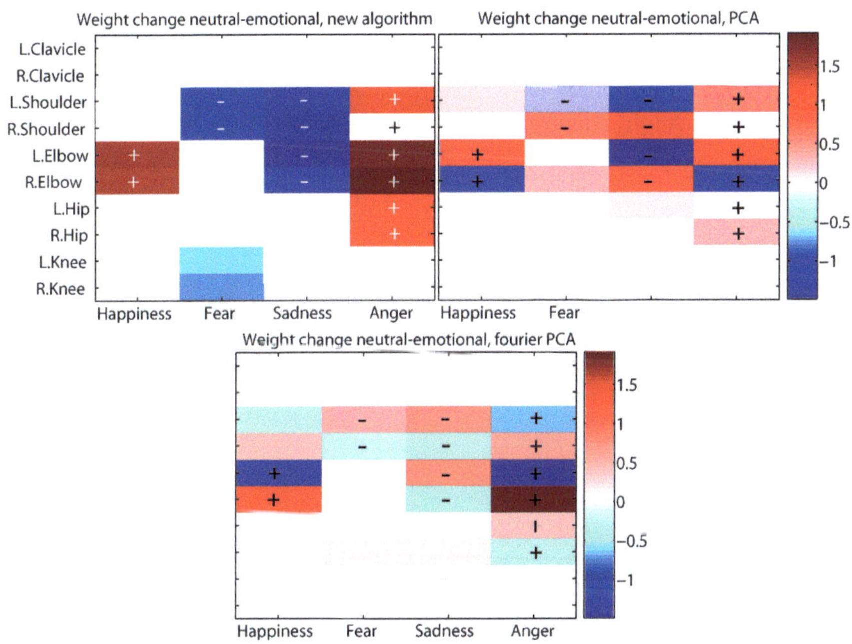

Fig. 14.3 Joint-specific changes of the contributions (weights) of the first source in emotional walking compared to neutral walking for three different generative models. Kinematic features that have been shown to be important for the perception of emotions from gait in psychophysical experiments are indicated by the plus and minus signs (details see text)

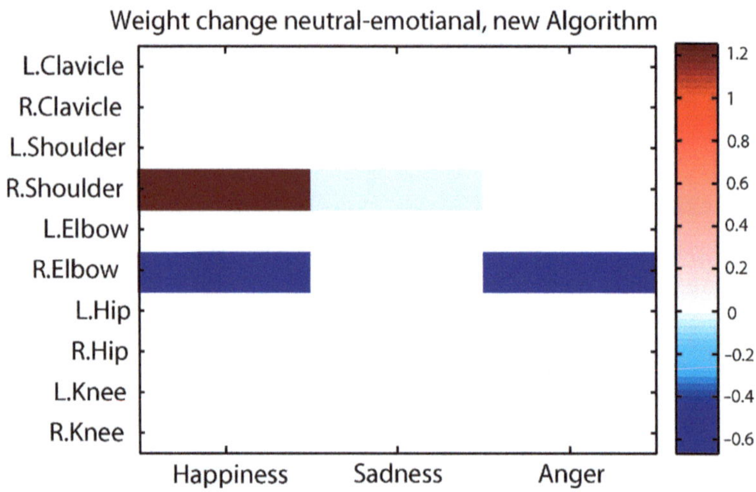

Fig. 14.4 Joint-specific change in the contribution (weight) of the first source in a right-handed tennis swing compared to the neutral movement

MIX
Papier aus verantwortungsvollen Quellen
Paper from responsible sources
FSC® C105338

If you have any concerns about our products,
you can contact us on
ProductSafety@springernature.com

In case Publisher is established outside the EU,
the EU authorized representative is:
Springer Nature Customer Service Center GmbH
Europaplatz 3, 69115 Heidelberg, Germany

Printed by Libri Plureos GmbH
in Hamburg, Germany